QUANTITATIVE METHODS IN GEOGRAPHY

HOUGHTON MIFFLIN COMPANY BOSTON
Atlanta Dallas Geneva, Ill. Hopewell, N.J. Palo Alto London

QUANTITATIVE METHODS in GEOGRAPHY

AN INTRODUCTION TO SPATIAL ANALYSIS

PETER J. TAYLOR *University of Newcastle upon Tyne, England*

TO CARL AND CLARE

Printed in the U.S.A.
Library of Congress Catalog Card Number: 76-13927
ISBN: 0-395-18699-4

CONTENTS

PREFACE

This book is intended for use in quantitative methods courses for geography students. As statistical and mathematical procedures become increasingly common in geographical study and research, a knowledge of quantitative techniques and their use in geography is important. The text teaches not only quantitative techniques but also when and why they can be used in geography.

Quantitative Methods in Geography: An Introduction to Spatial Analysis is first and foremost a geography book. It focuses on geography problems to which statistical and mathematical procedures can be applied. Most of these problems involve the analysis of maps, a traditional activity of geographers. In recent years, statistical and mathematical procedures have been newly applied in map analysis and contemporary map analysis has been rechristened "spatial analysis."

The topics developed in this book require only a basic knowledge of algebra and arithmetic. Some geometry and probability theory are introduced as they are needed. An introductory knowledge of geographical concepts is also assumed.

Because of the need to clarify the role of mathematics in geography before pursuing actual geographical research techniques, this book begins with a discussion of the nature of mathematics and its geographical applications (Chapter 1). Then we build upon that discussion by covering the major facets of contemporary map analysis in the remainder of the book. The overall organization and flow of ideas can be best illustrated by listing the topics covered:

The nature of mathematics—what it does and how it does it (Chapter 1)

How we can use mathematics in geographical research (Chapter 1)

The measurement of empirical (spatial) concepts—the link between mathematics and geography (Chapter 2)
Areal data—the raw material of geographical research (Chapter 2)
Probability theory—*the* language of modern geographical research (Chapter 3)
Probability inferences—techniques for geographical research decisions (Chapter 3)
Inferences about maps—analysis of single map patterns (Chapters 3 and 4)
Comparisons of pairs of maps (Chapter 5)
Functional relationships between several maps (Chapter 5)
Comparing large numbers of maps to derive common underlying patterns (Chapter 6)
Predicting future map patterns (Chapter 7)
Planning ideal map patterns (Chapter 7)

The techniques covered are statistical and mathematical but the problems remain essentially geographical: our map analysis progresses from consideration of single map patterns to pairs of maps, to several maps simultaneously, to considering very many maps simultaneously, and finally to consideration of maps that do not yet exist but may in the future. This geography book begins with a discussion of the nature of mathematics and ends, appropriately, on an applied note.

Each chapter concludes with "Work Tables" which show worked-out examples with every step in the calculations easily seen and followed. Simple exercises are also provided for the student; answers are given in the Appendix. In the later parts of the text where the techniques are less amenable to simple exercises, students are advised to work with packaged computer programs.

A brief section suggesting further readings follows each chapter. These references include some readings which parallel the coverage in the text and other readings which go beyond what is covered in the text. For instance, the student may wish to consult the references for further extension of some of the techniques through the use of calculus and matrix algebra. The Bibliography at the end of the book contains the full bibliographic citations for all references listed at the ends of chapters.

Through the years, many people have contributed in various ways to the creation of this text. I wish to thank those students in the United States and Great Britain who have attended my quantitative methods courses at the University of Iowa and the University of Newcastle upon Tyne. They made explicit suggestions in discussion and implicit suggestions through their attitudes and performance. I owe a large debt to numerous geographers with whom I have had little or no personal contact. Both William Bunge and Peter Haggett made a large impact on my thoughts with their books on theoretical geography and locational analysis. At a more technical level, the pioneering papers of Michael Dacey and Brian Berry must be singled out.

Mention should also be made of the immense intellectual pleasure I experienced while teaching at the University of Iowa. This was in part due to all the staff, although specific mention may be made of Gerry Rushton and Dave Reynolds, who, in their different ways, illustrated for me potentialities of geographical scholarship which I had begun to doubt could exist. In England my contacts with Graham Gudgin have been extremely beneficial as my ideas concerning aspects of this book, as well as many other matters, have been scrutinized by the most rigorously logical mind I have ever encountered.

I wish also to thank the following people who reviewed the manuscript during its development: Reginald G. Golledge, Brian J. L. Berry, Theodore K. Miller, Rowland Tinline, David R. Meyer, and Emilio Casetti.

Finally, I have dedicated this book to my two children, Carl and Clare, for the pleasures of family life which they control. However, I should not forget my wife, Enid, without whose help neither book nor children would have been possible.

Peter J. Taylor

THE ROLE OF MATHEMATICS IN GEOGRAPHY

Let's start with the simple proposition that we cannot begin to consider how mathematics can be used in geographical research until we have a basic understanding of what mathematics is. Such a statement seems logical and unexceptional; however, much recent debate on the role of mathematics in geography has been based on misunderstanding about the nature of mathematics. The purpose of this chapter is to spell out clearly the relationship between mathematics and geography in recent research.

The most fundamental question that arises in discussing the role of mathematics in geography is simply, "What is mathematics?" Many people may think this a strange question because "math" is generally thought to consist of numbers (arithmetic), letters without words (algebra), and neatly drawn diagrams (geometry). Although this briefly describes the content of a small part of mathematics, it says nothing about the *nature* of mathematics—what it is and how it works. It is only when we understand this that we can begin to consider the relationship between mathematics and geography and how maths can be properly applied in geographical research. Therefore, we shall consider the nature of mathematics in the first section of this chapter. We present the *axiomatic viewpoint* of the "new maths" that replaces the notion of mathematics being some body of absolute truths. The idea of mathematics as absolute truth goes back to classical Greece. Since the nineteenth century, statements such as "two and two equals four" have been considered true only in the context of their specific math or calculus—in this case, the real number system based on ten. An alternative number system based on different rules usually produces a different sum. There is clearly a fundamental difference between the modern view of mathematics and the more traditional view that has often figured in methodological debate in geography.

Once we have established what the nature of mathematics is, the next question that arises is, "How can we use mathematics in geographical research?" *Applied mathematics* is the subject of the second section in this chapter. One of the purposes of that section is to show that applied mathematics is relevant in a much broader field than merely that of the physicist. The most sophisticated applications of mathematics are found in physics, but these uses constitute just one part of the total field of applied mathematics. Such applications are better termed *physical mathematics* or even *classical applied mathematics* rather than simply applied mathematics. In this chapter we present mathematics as used generally in all science and then describe types of applied mathematics in geography, with specific emphasis on descriptive statistics. The remainder of this book is about applied mathematics in geography, or *spatial analysis*.

THE NATURE OF MATHEMATICS

Mathematics is, in many ways, a rather peculiar subject. The more we consider this statement the more it seems to be true. Mathematics appears within the boundaries of other seemingly distinct disciplines, and we are not surprised to hear of an engineer, for example, using mathematical equations. Because mathematics is used in a wide variety of contexts, attempts to define exactly what it consists of have proven difficult. Whereas members of other disciplines are concerned with particular sets of facts which can be observed, mathematicians seem not to deal with any observable facts in particular. We may thus distinguish between two basic types of science—*factual sciences* and *formal sciences*. Geography, in which there is particular concern with spatial distributions on the Earth's surface, and all other sciences of observable facts are factual sciences. Only a small number of sciences are formal, and mathematics is clearly one of them. Having made this distinction, we can attempt a definition of mathematics by considering major aspects of this discipline and improving successive definitions. In this way we should be able to conclude the first section of this chapter with a definition of mathematics that will serve for the remainder of the book.

What Is Mathematics?

Our initial definition will have to reflect our classification of mathematics as a formal science. What are mathematicians concerned with if not observable facts? A popular answer is that they are concerned with *symbols*, abstract objects rather than the more concrete subject matter of the factual sciences. Let's begin, therefore, with a preliminary definition of mathematics: *Mathematics is a discipline that is concerned with abstract objects or symbols.* To consider this definition is to consider the concept of abstraction.

Abstracting from Reality: Models

Abstraction can be defined as the process of forgetting unimportant details. If we accept this definition, then we see clearly that there are many cases in which we forget unimportant details but do not use mathematics. We spend our whole lives abstracting from reality. Every time we look at the world around us we perform some degree of selection of the events we perceive. Were we to attempt to collect information on every detail of every event we witness, then thinking—and therefore life—would be impossible. Thus, abstracting from reality is a vital part of our existence. The degree of abstraction in the observations we make in our everyday lives is normally smaller than that which scientists employ in studying their subject matter, however. In fact, scientists have formalized this procedure into a structured process that they term *modeling* or *model building*. A scientist's model is the result of varying degrees of abstraction from the real world; mathematical models constitute one end of a continuum whose range is from the least abstract to the totally abstract.

We may define a *model* as a carefully designed representation of reality. A model may be a copy of an object or of an event. The simplest way to build a model is to transform reality in terms of scale only. Makers of such models attempt to build a copy of a real object that essentially differs from it only in terms of size. An architect builds scale models of buildings by reducing the size factor while, in contrast, a physicist builds models of atoms by increasing the size factor. All such models are termed *iconic models*. They are the least abstract of models because the essential change involved is only one of scale. In geography, a relief model of a physical landscape would be an example of an iconic model. In Figure 1.1a we present such a model of an island dome.

Models become more abstract when we not only change the scale but also transform the properties of the real object or event. Such models are *analogue models*. Some geographers have experimented to find physical analogues that represent dynamic patterns in human geography. Thus, migration flows have been modeled as electric currents on a conducting sheet. In such a model it is understood that the "heat" at a certain time at a given location on the sheet is analogous to migration intensity at a given time and location in the real world. The historical westward movement in the United States has been modeled in just this way. The most important analogue model in geography is the map. Maps represent a particularly useful selection of reality. Not only is the scale altered but also the various elements are represented in a new form—a stream in the landscape becomes depicted as a blue line and a church is depicted as a simple cross.

(a)

(b)

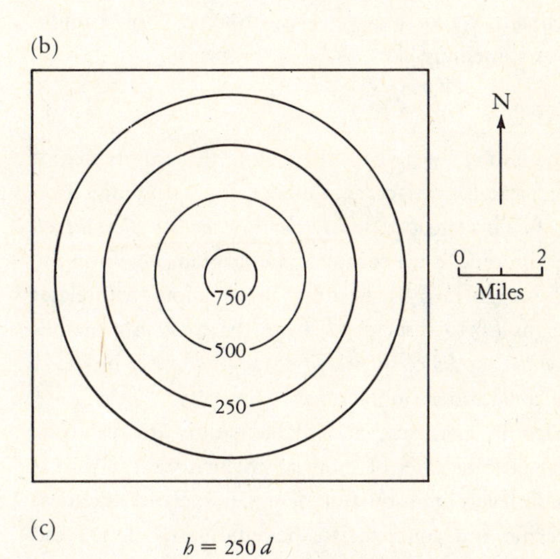

(c)

$$h = 250\,d$$

Figure 1.1 Alternative models of an island dome. (a) An iconic model: a scale relief model. (b) An analogue model: a contour map (contours in feet). (c) A symbolic model: the equation $h = 250d$, where h is the height or altitude (in feet) and d is the distance from the nearest coast (in miles). This equation simply says that altitude increases by 250 feet for every mile inland you move toward the center of the island. (This example is discussed in more detail in Chapter Five.)

The iconic relief model becomes a contour map on which the third dimension is represented by lines that join points of equal height. An analogue model representing the same area as the iconic relief model is shown in Figure 1.1b.

In the highest degrees of our increasing abstraction, the properties of the real world are represented by symbols. At this point we may seem at a great remove from the objects we are modeling but, curiously enough, it is remote models that scientists seem to prefer. A geographer who has tired of playing with his conducting-sheet migration analogue might well attempt to predict migrations by setting up an equation in which the size of a migration flow is represented simply by the symbol y and various factors thought to influence this flow (such as levels of income at the destination or of unemployment at the origin) are denoted $x_1, x_2 + \ldots + x_n$. Such a model might look something like

$$y = b_1x_1 + b_2x_2 + \ldots + b_nx_n$$

The b's are "weights" that denote the importance of the particular x factors in influencing y, the migration flow. This equation is a *symbolic* or *mathematical model.* In Figure 1.1c we complete the diagram begun in the other figures by presenting an equation that relates height (h) to distance (d) from the coast in a mathematical model of the island dome previously modeled in iconic and analogue forms.

Iconic, analogue, and mathematical are the three general categories of model by which we can distinguish levels of abstraction. Iconic models are the most specific and concrete; mathematical models are the most general and abstract. Geographers have employed most often the compromise analogue model, the map. Cartographic representation of features on the Earth's surface is certainly indispensable to most geography study. We must admit, however, that the three types of model can play complementary roles in geography. Iconic models are important for experimental work in physical geography, but as with maps their less abstract nature makes them particularly suited to illustrative purposes, particularly in teaching. We have already noticed that scientists generally favor the more abstract, mathematical models. This may seem to be a paradox if we consider factual scientists, because mathematical models are formal and at the greatest remove from reality. The paradox disappears when we consider exactly why geographers build models. Initially we build models of the real world in order to produce a clear picture of a particular feature that is more difficult to perceive when it is merely one small part of a complicated reality. This is the essence of abstraction. We can obtain a clear picture of the pattern of a stream network much more quickly by studying a map than by spending days tracing the streams in the landscape. Once we have a model, moreover, we may wish to manipulate it in order to find out more about the object that it represents. We do not have the resources to modify the actual attraction a certain city has for migrants, but we can manipulate a conducting-sheet migration analogue and record the effects of changing attractions on migration flows as they are modeled. The most important property of our threefold classification relates directly to this basic use of models. *The more abstract a model is, the easier it is to manipulate.* In fact, the main reason for changing the properties of an iconic model so that it becomes an analogue model is to substitute new properties that are easier to manipulate. A contour map is a much more flexible tool than a relief model. Furthermore, with the use of symbolic models, the whole process of manipulation takes on new meaning. When properties have been represented as abstract symbols, we can use the particularly rigorous methods of mathematical analysis.

Our initial definition of mathematics seems adequate to differentiate it from other, more concrete modeling, but it does not incorporate reference to the particular manipulative powers of mathematics. We will therefore revise our definition as follows: *Mathematics is a discipline that is concerned with deductive forms of argument that use abstract objects or symbols.* In order to clarify this definition, we must consider deductive inference.

Deductive and Inductive Inference

What do we mean by a deductive form of argument? We can best explain this by discussing the contrast between *deduction* and *induction.* In research there are basically two ways in which we can draw conclusions. One is to study some sample of cases, observe the behavior, and then infer that the same behavior is true of all cases, including those not observed. This is known as *inductive inference* and the result is an *empirical law.* A good example in geography is Ravenstein's first law of migration, stated in 1885. After studying the location of birthplaces of residents of various counties recorded in the 1881 British population census, Ravenstein made the inductive inference that "most migrants travel only short distances." Notice that this conclusion, made in the limited context of late-nineteenth-century British mobility, was stated as an empirical law to apply to migration patterns in general. As such, the law was subsequently incorporated into several aspects of theoretical geography. Such empirical laws are never certain, and it must be remembered that there may always be exceptions to the rule, even some that have yet to be observed.

The second way of drawing conclusions is to set out some statements that are accepted initially and then to infer other statements from them by using simple logic. This is *deductive inference* and it consists of reasoning such as:

Chicago is in Illinois;
Illinois is in the Midwest;
therefore, Chicago is in the Midwest.

Such conclusions are definite and certain *given the original assumptions.*

We can now see what is meant when we say that mathematicians use a deductive form of argument. Mathematical reasoning is like that in the Chicago example except that the subject matter is abstract rather than related directly to observable fact. A simple example would be:

$$A = B$$
$$B = C$$
$$\text{therefore, } A = C$$

It seems that we have exhausted the meaning of our previous definition of mathematics, although we might well feel that some essence of mathematics is still missing. Surely mathematics does not consist merely of small pieces of deductive argument such as the A, B, C example above; it consists, rather, of many such arguments put together into distinct areas of mathematics that we know as geometry, algebra, and so on. It would seem, therefore, that we need another revision of our definition of mathematics that incorporates this notion of groups of deductive arguments linked together into a logical whole. We define mathematics, now, as *a collection of artificial languages dealing with abstract objects and using deductive forms of argument.* We have here introduced the notion of mathematics as language. Clearly this requires some justification and explanation.

Mathematics as Language

To compare mathematics with languages, such as English or French, may at first seem strange. Closer inspection reveals many similarities. The natural languages we use in our everyday life consist of the myriad of statements we wish to make in our communications with our fellow human beings. Any two people from the same language group can converse with one another. Mathematics is also a form of communication. It consists of abstract statements and any two people acquainted with a particular part of mathematics can converse. Thus mathematics is like natural languages in that both have terms (words) linked together by statements (sentences) using relations (verbs). Mathematics has been more consciously created than natural languages, however, and so it may be categorized as *artificial language* or *calculus.*

Of course, mathematics differs from natural languages in several other important ways. Natural languages are more variable in nature. That is, they can involve both concrete and abstract terms, and they can employ deductive as well as any other form of argument. Natural languages are much richer in content: they have more terms and each term may have many shades of meaning. Consider the term "revolution," for example. Any dictionary presents us with several distinct definitions. This variable meaning of the same word allows for the following nonsense verbal deductions: Geography is the study of the Earth; the Earth experiences one revolution a day; therefore, it is not surprising that geography has been experiencing several revolutions in its research methods. Mathematical language has relatively few abstract terms and each has very clearly specified meaning, where it is thought necessary to define terms. Mathematical language is therefore more limited in content but it has the fundamental advantage of avoiding internal inconsistencies.

If mathematics is to be viewed as a language, what

sort of structure does it possess? This question is surprisingly easy to answer. Because all mathematical languages use deductive argument, their basic structures are similar and consist of four basic elements.

UNDEFINED TERMS In a deductive argument some terms remain undefined. Unless we are willing to accept circular reasoning, we must start somewhere with undefined terms. Thus we may define the term "line" in geometry as "that which has no breadth" but unless we go on to define the term "breadth" the statement is meaningless. In fact, it is sometimes simplest to use "line" as an undefined term.

DEFINED TERMS Other terms that are defined may use undefined terms. Thus, the term "triangle" might be defined by using the undefined term "line."

AXIOMS Axioms, also referred to as postulates or assumptions, are statements concerning relationships between terms. Their acceptability has nothing to do with absolute truth, self-evidence, intuition, observation, or common sense. They are simply initial postulates designated true for the purposes of the calculus. In theory, an axiom can contain any set of propositions while, in practice, axioms are very carefully chosen.

THEOREMS These are also statements concerning relationships between terms; they are proved on the basis of the axioms and definitions using deductive argument. Thus, they are no truer than the axioms themselves.

We have used examples from geometry in this discussion because of its general familiarity. Given our definition of mathematics, we can now see that the popular notion of geometry as a discipline merely concerned with neatly drawn diagrams is insufficient. The calculus commonly known as geometry is not about geometric figures; it is about relationships between a set of objects known by terms such as "point", "line," "area." The fact that these relationships can be depicted graphically has no influence on the logical structure of the language; geometric figures are therefore strictly unnecessary appendixes of deductive arguments. Graphic portrayal of ideas is helpful, however, even to the student able to cope with the most abstract ideas. Such a portrayal of relationships is known as a *concrete model* and is, obviously, often indispensable for teaching mathematics. The geometric figures that we are all familiar with are not themselves geometry; they are concrete models that serve as teaching or learning aids by visually representing the terms and propositions of geometry. Notice that such models are similar to analogue models in that they are less abstract than mathematical models *per se*. Concrete models differ from all other models that we have considered, however, because they are *not* representations of reality but are, rather, representations of mathematical relationships. As such they are a rather distinctive type of model.

Our latest definition of mathematics incorporates all the features of the discipline relevant to its consideration in relation to geography. This definition is so basic to our subsequent arguments that to repeat it is a virtue. *Mathematics is a collection of artificial languages (calculi) dealing with abstract objects (symbols) and using deductive forms of argument (deductive inference).*

Having talked so far rather generally about mathematics, let's now include a specific example of a mathematical language to help fix the ideas that have been introduced. We present a small abstract system of deductive argument that forms a simple miniature geometry.

A Miniature Geometry

We shall denote our miniature geometry as *S*. The structure of *S* is as follows.

UNDEFINED TERMS *point, line, on, in common*

AXIOMS

1. Every pair of lines in *S* has one point in common.
2. Every point in *S* is on two lines.
3. The total number of lines in *S* is four.

THEOREMS

1. There are exactly six points in *S*.

Proof:

(1) There are four lines (from Axiom 3).
(2) Therefore, there are six pairs of lines.
(3) Every pair of lines has one point in common (from Axiom 1).

(4) Therefore (from 2 and 3), there are six points.

2. There are three points on every line in *S*.

Proo :
(1) There are four lines (from Axiom 3).
(2) Therefore, every line is a member of three pairs of lines.
(3) Every pair of lines has one point in common (from Axiom 1) and every point is on two lines (from Axiom 2).
(4) Therefore (from 2 and 3), there are three points on each line.

This is a very simple example of a mathematical language with two relatively straightforward deductive arguments leading to theorems. It will help us understand this calculus if we construct a concrete model to represent the propositions stated and proved above. This model will no doubt help satisfy the more inductively oriented student that the axioms and theorems do in fact make sense—that is, that they are consistent.

We can set up a model simply by building a figure step by step from the axioms. First of all, we can interpret the undefined terms "point," "line," "on," and "in common" in what we may consider the traditional geometric way. Our next step is to identify individual points and lines. Let's start with axiom 3 and name the four lines A, B, C, and D. This gives us, from axiom 1, the following six numbered points that the lines have in common:

Pairs of lines	*Points*
AB	1
AC	2
AD	3
BC	4
BD	5
CD	6

We are now in a position to set up the model. We begin by drawing two lines, A and B, which meet at point 1 (Figure 1.2). If we add line C by joining A and B, respectively, at points 2 and 4, we shall nearly have completed the model. However, we need one more line, D, which if drawn to connect A to B by crossing C, allows us to identify points 3, 6, and 5. We now have a model that has four lines, six points, a point to every pair of lines, every point on two lines, every line having three points on it. The diagram is therefore a concrete representation of our previously constructed abstract calculus *S*.

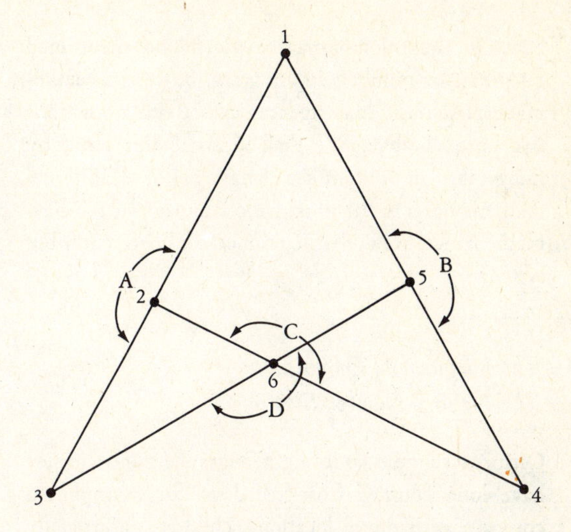

Figure 1.2 A concrete model of the miniature geometry, *S*.

This miniature geometry is the sort of thing we mean when we talk of a calculus or mathematical language. It has been presented here because of the simplicity with which it illustrates our previous, rather abstract discussion. The mathematical languages used in geographical research will be clearly larger and more complicated calculi. In the next two sections, we introduce further branches of mathematics that relate to topics of particular interest to geographers. We begin with the quantity and quality theme in mathematics before turning to the mathematical languages of spatial form (the geometries).

Scratch for now

Quantity and Quality in Number Systems

Misconceptions about the nature and use of mathematics have been particularly prevalent in the so-called quantity versus quality debate. This debate has revealed a common assumption that mathematics is synonymous with quantification and is therefore a discipline having to do solely with quantities expressed as numbers. Our definition of mathematics has not mentioned quantities or numbers in any way, however. Mathematics, like most other sciences, can be either quantitative or qualitative. We shall illustrate this point by using the most familiar language of traditional mathematics and the most basic language of the new mathematics, both of which can be considered number systems.

The Language of Quantification: The Natural Number System

Quantity suggests counting. Geographers and farmers have long counted frost-free days, for example, to compare agricultural locations; children traditionally learn to count by using their fingers as the concrete model. In both cases *cardinal numbers* are produced; this means simply that the numbers are the results of inductive observation of an object or event. This collection of numbers derived from the real world can be arranged in an abstract language known as the *natural number system* in which the numbers are represented as symbols. It is this language that is widely used in day-to-day existence in modern societies. Let's look briefly at its structure.

In his little book *New Mathematics*, Irving Adler presents the following structure for a natural number system excluding zero.

UNDEFINED TERMS

1. Members of the system of natural numbers are denoted x, y, and z.
2. A particular member is denoted 1.
3. Two operations (addition and multiplication) are denoted $+$ and $\cdot$.

DEFINED TERMS

1. The successor of $x = x + 1$.
2. The successor of $(x + y) = x +$ (the successor of y).
3. $x \cdot 1 = x$
4. $x \cdot$ (the successor of y) $= x \cdot y + x$

AXIOMS

1. The system contains a number called 1.
2. For every number of the system, there is just one other number called its successor.
3. Two distinct numbers do not have the same successor.
4. No number in the system has 1 as its successor.
5. If a collection of the numbers includes, first, 1 and, second, every successor of every number in the collection, then this collection contains the whole system.

THEOREMS

It is possible now to prove the basic rules of the natural number system.

1. *The commutative law of addition* states simply that it is possible to add two numbers in any order:

$$x + y = y + x$$

2. *The commutative law of multiplication* states that it is possible to multiply two numbers in any order:

$$x \cdot y = y \cdot x$$

3. *The associative law of addition* states that it is possible to add three numbers in any order:

$$x + (y + z) = (x + y) + z$$

4. *The associative law of multiplication* states that it is possible to multiply three numbers in any order:

$$x \cdot (y \cdot z) = (x \cdot y) \cdot z$$

5. *The distributive law of multiplication in respect of addition* states that the multiplier may be distributed among the elements in the expression it multiplies:

$$x \cdot (y + z) = (x \cdot y) + (x \cdot z)$$

This is the essence of the natural number system, the basic language of quantification. It should be noticed that the structure contains three undefined terms ("natural numbers," "addition," and "multiplication") and that we have simply introduced quantities into it as examples of *members* with which two *operations* may be used to produce a *number system.* Such a system can be identified simply from the theorems—the five basic laws that define a number system. With natural numbers and addition and multiplication, we can produce the quantitative natural number system.

With other elements and operations, we can produce other number systems. One such example is the qualitative language known as set theory.

Qualitative Mathematics: Set Theory

In *set theory*, the numbers of the natural number system are replaced by *sets*, which may simply be defined nonquantitatively as any collection of objects. For example, instead of counting the number of frost-free days, we might consider all frost-free days in a particular year as a set. Objects belonging to a set are known as the *elements* of a set. The sets are the members of the number system we are building. The operations that replace addition and multiplication are known as *union* and *intersection*, respectively. A union of two sets is formed by taking as its elements all elements in both sets. (This is the addition operation of the qualitative system.) The intersection of two sets (the multiplication operation) is formed by the elements that belong to *both* sets. If we have two sets A and B, then the two operations are written $A \cup B$ (union) and $A \cap B$ (intersection).

Whereas the natural number system uses dots, apples, or fingers as typical concrete models, set theory has specific models known as *Venn diagrams*. In these figures sets are represented simply by enclosed areas, usually circles. The operation union can be represented by the shaded area:

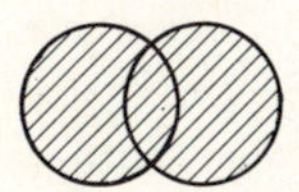

where the two circles depict any two sets. Similarly, the operation intersection is represented by a more restricted shaded area:

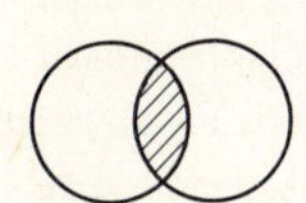

With sets, elements, unions, and intersections we can easily show that set theory obeys the five basic laws that define a number system.

1. *The commutative law with unions*:

$$A \cup B = B \cup A$$

Because $A \cup B$ includes all elements in both sets A and B and $B \cup A$ includes all elements in B and A, the two sets are obviously the same, and the commutative law is obeyed.

2. *The commutative law with intersections*:

$$A \cap B = B \cap A$$

$A \cap B$ produces the set with all elements common to both sets A and B, and $B \cap A$ produces all elements common to both B and A; these are obviously the same, and this law is obeyed.

3. *The associative law with unions*:

$$A \cup (B \cup C) = (A \cup B) \cup C$$

Both series of operations result in sets that include all elements in sets A, B, and C, and the law is obeyed.

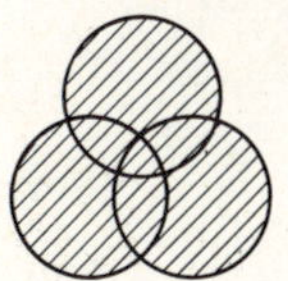

4. *The associative law with intersections*:

$$A \cap (B \cap C) = (A \cap B) \cap C$$

$A \cap (B \cap C)$ includes all elements in both B and C that are also in A; while $(A \cap B) \cap C$ includes all elements in both A and B that are also in C. The Venn diagram again illustrates that this is the same set in either case, and the law is obeyed.

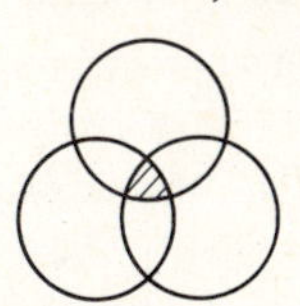

5. *The distributive law for intersections with respect to unions*:

$$A \cap (B \cup C) = (A \cap B) \cup (A \cap C)$$

$A \cap (B \cup C)$ produces the set with elements in B and C that are also in A; $(A \cap B) \cup (A \cap C)$ produces the set with elements in both A and B and also those elements in both A and C. The Venn diagram illustrates the equality of the two series of operations and that thus this final law is obeyed.

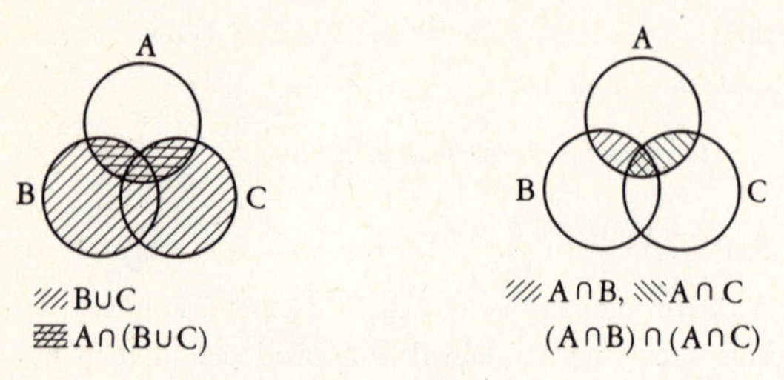

Thus we have a number system without numbers! Obviously we have come a long way from our original idea of numbers and counting to the idea of number systems as general abstract languages. The set theory calculus is a qualitative example of a number system that uses the same basic logical structure as the real number system. We shall use both quantitative and qualitative mathematics in the following chapters.

Geometries

Geometry literally means measurement of the Earth. It probably originated in ancient Egypt from the practical needs of surveying to help control the floods in the Nile delta. The operations and rules produced from this practical application were synthesized into an abstract system in about 300 B.C. by the Greek mathematician Euclid in twelve famous books known as the *Elements*. Euclid's *Elements* represented geometry until the nineteenth century, and it still constitutes geometry as it is taught in the majority of schools. About a century or so ago, however, some mathematicians showed that Euclid's geometry is not the only geometry. The set of *non-Euclidean geometries* that arose was one of the major breakthroughs of the new mathematics, because Euclid's geometry had seemed to fit the real world so exactly and was the most quoted example of "absolute truth." It was seen to follow naturally that, if Euclidean geometry is just one of several geometries, then mathematics is best interpreted as abstract language, the geometries being languages concerned with the structure of spaces.

The new spatial languages are not unrelated to Euclidean geometry and in 1871 the German mathematician Felix Klein produced his famous Erglanger Program, which related several of the geometries in a single scheme. He was able specifically to link Euclidean geometry with topology and, because these two geometries are the most widely used in spatial analysis, we introduce them here within this framework.

Transformations and Invariance

The Erglanger Program was concerned with the logical structure of the geometries and related them by using the concepts of transformation and invariance. The geometries differ in the transformation of a geometric figure, which is allowed so its new form does not affect the logic of the geometry under consideration (Figure 1.3). This property is known as *invariance.* In Euclidean geometry we can transform a square by rotating it through 30° and still recognize it as a square; we say that the two figures are *congruent.* We can also reduce the length of the sides of a figure in a transformation and still recognize it as a square; we say that the two squares are *similar* though not congruent. Thus Euclidean geometry allows two transformations—rotation and changes in size or scale. In fact, this defines two geometries—*equiarea geometry*, in which figures are invariant to rotation, and *similarity geometry*, in which figures are invariant to scale.

A basic difference between these two transformations is that the similarity criteria are less restrictive than the congruence criteria. Thus, all congruent figures are similar. We can continue with further transformations by introducing criteria even less restrictive than similarity. If we do not consider angles in geometric figures to be criteria in comparing shapes, we will not be able to distinguish between a square and a parallelogram of equal sides. This transformation leads us into *affine geometry*. If we further consider lengths of sides to be unimportant, we enter the realm of *projective geometry*, which does not differentiate between a square and any other quadrilateral. If we take this process one stage further, we reach the rubber-sheet geometry known as *topology*. Even the idea of numbers of sides is lost, and all we are left with is an inside and an outside identification of our original square, which is not

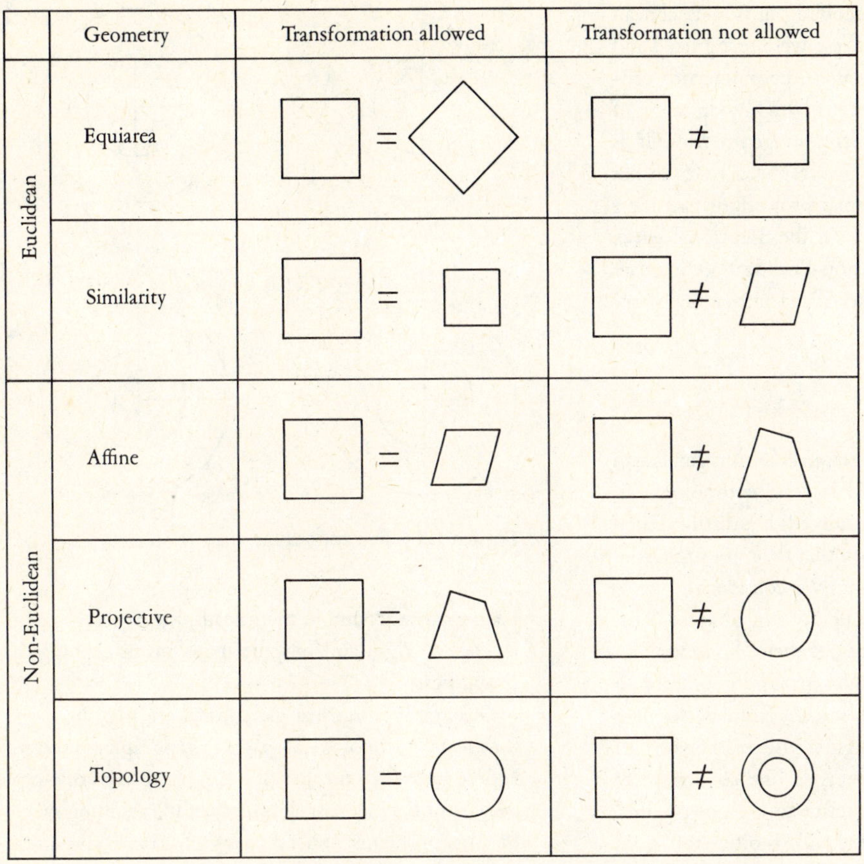

	Geometry	Transformation allowed	Transformation not allowed
Euclidean	Equiarea	=	≠
	Similarity	=	≠
Non-Euclidean	Affine	=	≠
	Projective	=	≠
	Topology	=	≠

Figure 1.3 Transformations and invariances of some geometries.

enough to distinguish it from a circle of infinite sides in Euclidean geometry. In fact, we can imagine the original square on a piece of rubber which may be stretched in any manner to transform the original square into any shape, and yet both figures will be identical if viewed within a topological framework. However, if the rubber sheet tears, producing a hole in the shape, then we no longer have a simple inside-outside pattern and the topological similarity with the original square will have broken down. The Erglanger Program is laid out diagrammatically in Figure 1.3.

What relevance have these geometries to geographers and their research needs? Spatial concepts that are invariant under a transformation can be studied within the context of that particular geometry. Thus, distance and area concepts depend on scale and therefore their analysis requires use of Euclidean equiarea geometry. The concept of direction, on the other hand, requires only Euclidean similarity geometry. Euclidean geometry has a multitude of minor applications relating to Euclidean concepts in geography. Normally distances as we refer to them will have been calculated with the Pythagorean theorem, for example.

Applications of non-Euclidean geometries involve problems in which distance, direction, and area are not to be preserved. This statement brings to mind the problem of representing a three-dimensional globe on a two-dimensional plane. Solutions to this problem have always had to sacrifice some of the properties on the globe's surface. This has meant that, in effect, cartographers have been dabbling in non-Euclidean geometries centuries before these were recognized and formalized by mathematicians in the nineteenth century. Map projections present us with a wide range of problems involving non-Euclidean geometries. It is perhaps ironic that modern geographical research has been little concerned with map projections and their associated geometries, at least in the traditional sense. This potential link between non-Euclidean geometries and geography has not materialized into a major research connection.

Topology

The interest of modern geographers in non-Euclidean geometry has been restricted largely to topology. In fact, a new type of map has emerged; it is distinguished from traditional projections in that the only properties it preserves are topological ones. We consider the use of such highly distorted representations in Chapter Four. Before we can begin to consider them, however, we must take a closer look at this strange geometry in which we cannot distinguish squares from circles. It is a geometry without the highly visible property of shape. This abstract space seems at first so devoid of properties as to be of no interest to the geographer. Euclidean space, on the other hand, is often too rigid for some problems in geography, and a topological approach gives us a more flexible space in which to work. Which kind of space is to be preferred obviously depends on the problem and its context. It should be noticed, for example, that the seeming "naturalness" of Euclidean space is not as universal as many people imagine. Some non-Western societies seem certainly to view space in non-Euclidean terms. In fact, Western concern for precise boundaries in the context of private property and political sovereignty may have led to a cultural predisposition to favor a Euclidean view of the world. It seems that other societies sometimes organize their space with much more flexible, topological properties. For instance, the Tiv of central Nigeria have a spatially fluid pattern of land ownership that's been described as topological in structure. It has been suggested that, even within Western society, objective maps may be Euclidean but the maps

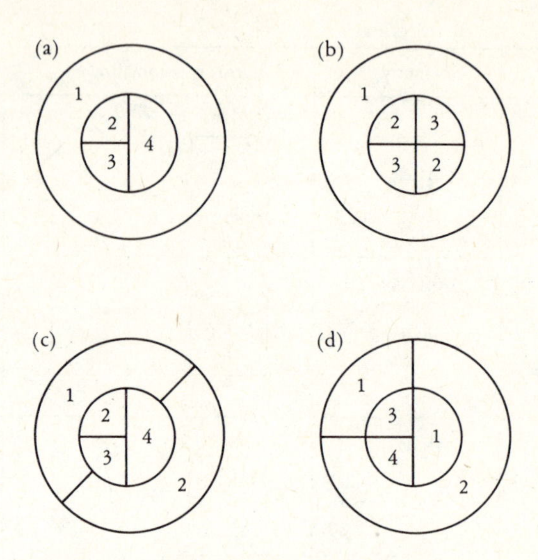

Figure 1.4 The four-color map problem.

people carry around in their heads, their mental maps, are not so rigid and may be made up of topological properties.

On largely empirical grounds, there may be scope for the use of topological spaces in geographic research. Moreover, it is quite often desirable to use topology because of the inherent simplicity of the resulting space. In some problems, concepts such as distance and direction are not relevant and need not be incorporated into the space being considered. This is the case with problems on the concept of contiguity—that is, the property of two regions touching one another by having a common boundary. To deal with a contiguity problem within a Euclidean rather than a topological context would be to complicate it, because a whole series of spatial properties would be introduced that have no relevance. Let us look at an example of a typical topological problem.

A brief time after the presentation of the Erglanger Program, the English mathematician Arthur Cayley published a short paper in the *Proceedings of the Royal Geographical Society*, describing the four-color map problem. This can be stated as follows. If a map is partitioned into areas, these can always be colored in such a way that, using four colors or less, contiguous areas do not have the same color. In Figure 1.4a, we have four areas colored 1, 2, 3, and 4. If we now add an area, to produce Figure 1.4b, we find that, rather than having to use an extra color, we need use only three for the five areas. In fact, no matter how hard we try, no matter how complicated we make our map, we will find that only four colors or less are required to complete it (Figures 1.4c and 1.4d).

This problem was presented by Cayley and demonstrated inductively as we have done. It is not a theorem of topology because it remains unproven from topology's axioms. It therefore remains within the realm of unsolved problems. It does give an idea of the essence of topology, however. We talk simply of a partition of a map rather than of any particular partition drawn accurately according to Euclidean principles. All properties of the map such as distances and directions are ignored in this context, except for the simple topological property of touching, which is, incidently essentially qualitative in nature. Thus, we have a qualitative geometry (topology) to complement our qualitative number system (set theory).

APPLIED MATHEMATICS

Our definition of mathematics in the preceding section has left us with something of a paradox. If indeed we regard mathematics as purely abstract, then how can it be of use in areas of knowledge, such as geography, about events in the real world? This problem does not seem to have restricted people from using mathematics in a wide variety of work. Theorems from geometry, for example, are used extensively in fields as far apart as cartography and physics. The "paradox" has been solved in practice, and it is the manner in which this has been accomplished that we turn to now.

Interpreted Languages

An abstract mathematical calculus that is not being used for real world problems is *uninterpreted.* It is part of *pure mathematics.* All such languages contain terms and relationship that can potentially be given real world interpretations, however. The statement 2 + 2 = 4 is pure mathematics, but we have the possibility of interpreting the numbers in terms of quantities of apples so that two apples + two apples = four apples is a statement in *applied mathematics,* at its most elementary level. Notice that when we make such a statement we are not moving from deductive to inductive inference; no experimentation is inferred and we are simply interpreting the terms in the abstract language as real world concepts. In this particular case, the number system has become an *interpreted language.* This is fundamentally no different from using any language, artificial or natural, to describe the real world. The procedure is essentially one of translation from real world to language terms and relations.

The Basic Research Model

This general procedure applies to applications of mathematics in many areas of empirical research covering both physical and social sciences. In geography, we can formulate the simple problem-solving model of applied mathematics shown in Figure 1.5. We begin with a geographical problem that we can translate into an abstract calculus; we next apply deductive logical arguments to the reformulated problem to get a mathematical conclusion; this can then be translated back into geographical terms to give us a solution to our original research problem. This procedure is the basic model for applying mathematics in any factual science. Notice that the translation arrows flow from geographic problem to mathematical language. Whereas pure mathematicians might search for concrete models to illustrate their calculi (that is, they move from abstract to nonabstract), applied mathematicians are concerned with finding mathematical languages for real world problems (they move from nonabstract to abstract). This difference of direction represents the basic distinction in the approaches of pure and applied mathematics.

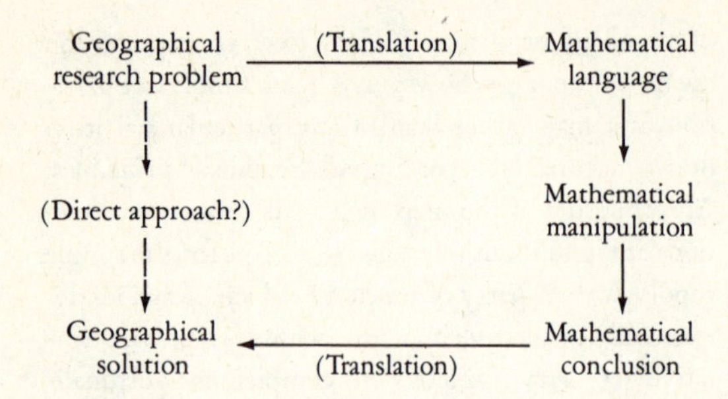

Figure 1.5 Applied mathematics in geographical research.

One might ask whether the research model represents a diversion along a rather long and unfamiliar road while a much more direct route seems open straight from geographical problem to a solution using, perhaps, cartographic descriptions, thus avoiding the mathematics. The answer is simply that some questions may be soluble to the satisfaction of a researcher using a nonmathematical approach but there are undoubtedly many cases in which a mathematical approach is essential. One simple nonmathematical approach to problems involves visual comparison of pairs of maps. Today many geographers compare large numbers of maps, perhaps more than one hundred, not just in pairs but simultaneously, to understand the underlying structures common to all the maps. This is patently impossible visually, but such comparisons are used with increasing frequency and the techniques are described in Chapter Six.

However valid this practical justification for the research detour may be, the ultimate advantage of the diversion lies in the logical clarity and deductive power of the abstract language being used. It is for this reason that the physical sciences, the most developed factual sciences, are the most mathematical. Results are obtained that are not always obvious given the axioms of the calculus, and therefore mathematics has a basic role in the extension of knowledge. This is much less true in other factual sciences, such as geography, in which mathematics has been used to less extent. Mathematical languages act ideally, in the words of the geographer David Harvey, as "a theoretical juice extractor." An interpretation of our simple miniature geometry will illustrate the derivation of nonobvious results by mathematical deductive inference.

Quantgeog Airlines

Quantgeog Airlines is a small fictional charter company, a subsidiary of a larger fictional consortium that specializes in organizing flights for academic personnel in major universities throughout the world. Its services are made economically viable through a grant from a fictional international educational organization whose sole aim is to promote academic discussion between leading researchers in various fields of science. Quantgeog Airlines transports geographers. When originally planning its flight schedule, the company contacted the geographical community to inform them that resources would restrict the proposed network so that each university served would have available only two separate flights, that there would be one connection between every pair of flights, and that there would be four flights in all. Given this limited information, the company was willing to hear from geographers proposals for the structure of the flight network.

This spatial problem can be tackled simply by translating the information into the miniature geometry. The points of the abstract calculus can be interpreted as universities served by the airline, the lines become the proposed flights, and the restrictions imposed by economic considerations can be translated directly into the three axioms of the calulus. We now have an interpreted language. In the next stage of the analysis we perform the mathematical manipulation—the deductive argument producing the two theorems, which in turn gives us a mathematical conclusion that tells us the network will have a structure basically the same as the concrete model in Figure 1.2. When these conclusions are translated back into a geographical context, we find that we can manage to produce a network that serves only six universities and that each flight will serve three universities only. These limitations to the flight plan are not at all obvious from the conditions laid down by the company. This is a "theoretical juice extractor" at work.

One of the geographical solutions that was proposed is illustrated in Figure 1.6. The correspondence be-

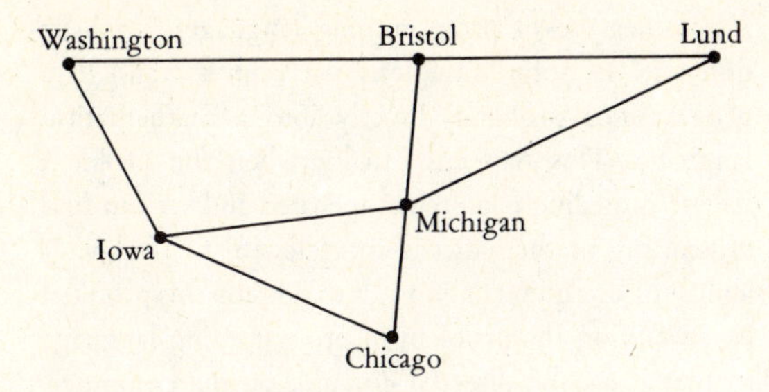

Figure 1.6 Quantgeog Airlines flight plan.

tween this flight network and the original concrete model is very direct. Lines and routes translate as follows:

Line	*Route*
A	Washington–Chicago–Iowa City
B	Washington–Bristol–Lund (Sweden)
C	Iowa City–Ann Arbor–Lund
D	Chicago–Ann Arbor–Bristol

Each university corresponds to one of the original six points in Figure 1.2. This is *one-to-one correspondence*, which means that there is an *isomorphism* between the network and the concrete model and, hence, the abstract geometric language. This example represents an ideal or perfect translation and we should aim to emulate this in a true research context.

This simple example of language interpretation leaves two important questions unanswered. First, what sorts of mathematical languages comprise the modern applied mathematics that geographers use? Second, what sort of geographical questions are asked that can be translated into this mathematics? We attempt to answer these two questions in the remainder of this section.

Fields of Applied Mathematics

From our basic research model it is clear that potentially all pure mathematics may become interpreted languages and, hence, qualify as applied mathematics. In practice, however, certain areas of mathematics have been widely interpreted and have become identified specifically as a field of applied mathematics. The outstanding example is physical mathematics or "classical applied mathematics," which has developed from applications of mathematics in the physical sciences. Fields of applied mathematics may evolve from applications in any factual science. In the twentieth century, we have seen the emergence of biometrics, econometrics, and psychometrics. Each of these deals with mathematical, and particularly statistical, applications in very specific areas, and they have rather dubious claim to separate identity as independent disciplines; they are best viewed as part of the more general field of statistics. In contrast, some fields of applied mathematics have evolved from quite broad areas of application and have grown and developed far enough to have a mathematical character of their own. It is in these broad fields of applied mathematics that we can identify recent geographical applications.

In 1968 the National Academy of Sciences in the United States published a report in which four major fields of applied mathematics are identified and discussed. Two of these fields, statistics and computer science, involve areas of study outside mathematics and may be described best as partly mathematical. The other two fields derive directly from applications in a broad subject area. These constitute classical applied mathematics, which is derived from physical applications, and operations research, a recent phenomenon developing from applications in business and organization studies. All four fields have been used in recent geographic research.

Classical applied mathematics, as we would expect, has been used most in various areas of physical geography. The mathematics of fluid dynamics makes its appearance in a wide range of geographical contexts, from meteorology and oceanography to hydrology. Classical applied mathematics has had impact on human geography also. Analytical geometry, first used in physics, has spread to all factual sciences in which functional relationships are of interest, including human geography. (We consider application of this abstract geometry in Chapter Five.) Furthermore, studies of flows of people and goods have long been modeled in analogy with Newtonian physics; the social gravity model has recently been developed to incorporate the mathematics of statistical mechanics. We describe elementary forms of this almost traditional application of classical applied mathematics in human geography in Chapter Seven.

Operations research has evolved from a wide range of economic, military, and engineering studies. It is essentially a set of practical problem-solving calculi and is therefore the tool of decision-makers who require optimum solutions to allocation problems. When these involve allocation in space, they come within the preserve of applied geography. In recent years many geographers have been turning to the techniques of operations research to answer questions such as where best to locate schools and hospitals. We consider location-allocation models in Chapter Seven along with the gravity model.

The rise of operations research has been made possible by the widespread availability of electronic computers. *Computer science* is much more than simply a field of applied mathematics. As the study of computers and computation, it includes the design of computer *hardware* and is therefore a specialized branch of electronic engineering. The *software*, the computation systems that convert the hardware into productive equipment, is equally inportant, for in this second role computer science includes the development of abstract languages. The computer programming languages such as Algol and Fortran have had two applications in geography.

Their most common use is to facilitate translation of existing mathematical languages into a form that a computer can understand. Thus, the mathematical procedures of operations research are programmed as one of the steps between the formalization of a problem and its solution. All the mathematics we deal with can be applied though computers and their languages. In modern geographical research, all but the simplest calculations are applied in this way. It is not our purpose to describe computer languages, however. The majority of the techniques we describe are standard procedures readily available in existing computer libraries. Use of the computer requires merely the ability to read the library instructions and accurately punch control and data cards. This ease of access clearly has advantages and disadvantages.

Another use of programming languages is a little different. In some situations we cannot translate a geographical problem directly into a mathematical language. This happens usually when the problem we are modeling is quite complicated and we can find no existing mathematics isomorphic to it. The availability of computers allows us to sidestep this problem by specifying the problem in programming language so that it can be directly *simulated* on the computer. The resulting program can be long and complicated and may include many separate formulas representing different details. In this way we can solve particular problems for which general solutions involving completely mathematical deductions are not possible. Such numerical solutions are not entirely new phenomena and are not restricted to the use of computers. Nonetheless the availability of computers has meant that the scope of such approaches has widened as vastly larger problems can be tackled. We illustrate an example of this approach in the next section.

Statistics is the final field of applied mathematics identified by the National Academy of Sciences. It is not usually thought of as being wholly mathematical. Although it is firmly based in the mathematical theory of probability, the major concern of modern statisticians has been to establish rules for drawing inductive inference in empirical research. As far as geography is concerned, however, and in terms of amount of use, statistics can be identified as the major area of applied mathematics. In fact, in the 1960s the phrase "statistical geography" was widely used to describe most modern research trends. The use of statistics in geographical research has ranged from simple descriptive statistics through sophisticated multivariate analyses. We consider simple means and standard deviations with their spatial equivalents in the next section; inferential statistics is introduced in Chapter Three; bivariate statistics is discussed in Chapter Five; large-scale multivariate statistics is represented by factor analytic techniques in Chapter Six. On top of this, the closely related topic of probability models is the subject matter of Chapter Four.

We can now begin to see why statistics is considered *the* applied mathematics of modern geographical research. The quantity of its use is such that we have already been able to identify levels of sophistication in the applications described above. It is useful to extend this notion to applied mathematics in general.

Levels of Mathematization

We have intimated that the application of mathematics to problems in modern geography has not always been highly sophisticated. The time has come to move from considering how geographers *may* use mathematics to considering how geographers *have* used mathematics. Much modern geography has been quantitative in nature, but we may doubt its mathematical credentials in terms of the use of abstract deductive reasoning. Russian geographers have given much thought to this argument in recent years in their debates over mathematical geography, and they have developed the notion of different levels of "mathematization."

The Russian population geographer Yuri Medvedkov has suggested that there are three levels of application of mathematics in geography relating to the differences in the complexity of the questions being asked. In Figure 1.5 we denoted geographical problem as our starting point, but we find it necessary now to distinguish between types of problem being solved. The three types of question Medvedkov has identified are: "How much?" "What are the relationships?" and "What mechanisms can be identified?" These questions may be viewed as increasing in complexity, the third involving the use of mathematics to model hypothesized processes. The first level of mathematical application is simply the introduction of measurements and indicators into geography. Much of recent so-called quantitative geography has been at this level. Typical examples are the quantitative derivation of shape measurements and the application of descriptive statistics to spatial distributions. We can easily argue that this is hardly mathematics at all but is best considered as measurement, which is the first step in translating a complicated geographical problem into a form amenable to more sophisticated mathematical analysis. We consider measurement from this viewpoint in Chapter Two.

The second level of mathematical application involves the identification and specification of relationships between measurements and quantitative indicators. This is essentially an inductive procedure in which the geographer searches for empirical regularities and patterns. The tools of this research are analytical geometry (graphs) plus the statistical techniques of correlation, regression, and inferential tests. (We have noted previously that all these topics are covered in Chapters Three and Five.) They represent an advance on simple measurement and pave the way for more sophisticated analysis.

The third level of application of mathematics is directly comparable to what is often viewed as the ideal role of mathematics. Research that employs abstract deductive reasoning to uncover mechanisms operating in a given situation is the fullest use of mathematics, and by means of it we produce a complete mathematical model of a geographical situation. The problems we deal with at this level are not merely one small part of a research design, as in the previous two levels, but the fundamental research problems themselves. Thus, the whole research is mathematized. Examples of such mathematical models are the application of linear programming to model trade flows and the use of probability process models to describe settlement patterns. Both involve derivation, by mathematical deduction, of theoretical patterns that may be compared with the actual patterns. They are described in Chapters Four and Seven.

These three levels of mathematization can also be thought of as stages in the development of modern geographical research. Statistical geography to the middle of the 1950s involved the use of much simple descriptive statistics; numerous correlation and regression studies have characterized studies since that time; probability models have been a feature of the more recent past. In many ways, one level can be said to lead inevitably to the next. Success in measurement stimulates questions about relationships, which in turn require more complete formulation to satisfy final research goals: to describe and understand how a given situation operates. The three-level scheme is only a classification, however, and like most classifications the boundary zones are blurred. Some of the approaches we describe below will be hard to categorize. Nonetheless the scheme is a useful way of considering the application of mathematics to geography because it extends our basic research model (Figure 1.5) so that we can relate it directly to recent research trends.

This section has been about interpreted languages in general. The most interpreted mathematical language is undoubtedly the natural number system. When applied to the description of numerical data, such application is generally termed descriptive statistics. We conclude this chapter with a more specific discussion concerning this most elementary form of statistical analysis. The purpose is first to give some concrete examples to illustrate and clarify our previous discussion of applied mathematics, and second, to introduce some basic concepts of statistics and spatial analysis, which we draw upon in subsequent chapters.

Descriptive Statistics

Descriptive statistics precedes the close links between modern statistics and probability theory. The rise of government machinery in Western countries in the eighteenth and nineteenth centuries was accompanied by the production of statistics concerning economic and social conditions. The outstanding example is, of course, the national population census, which originated in Europe and North America late in the eighteenth and early in the nineteenth centuries. The availability of such data led to public and academic interest in the formation of statistical societies in many countries. The nineteenth-century journals of these societies record the compilation, tabulation, and elementary numerical analysis of statistics. This is the origin of the descriptive statistics which still serves as a useful tool, at least in initial data analyses, in modern social science research. Let's consider its application in geography.

Simple Descriptive Statistics

The raw material of any statistical analysis is a set of data. We can define data generally as information about characteristics of objects. Thus, geographical data typically consist of information about characteristics of places or areas. Figure 1.7a shows geographical data in the form of percentages of land under wheat cultivation in nine counties. Often the first step in an analysis is to arrange data in the form of a frequency diagram, a simple graph on which the y axis (vertical) represents the frequency of occurrence of an event and the x axis (horizontal) represents the descriptions or measures of the event. In Figure 1.7b, the x axis is specified as percentage of land in wheat; the y axis is the number of counties. We can now plot the data from Figure 1.7a on the graph in Figure 1.7b in the usual manner. We use classes in intervals of 10 (0–9, 10–19, and so on), so that we find one county that has up to 9% of its land in wheat, 2 counties that have 10–19%, and so on. Two types of diagram can be drawn from this information—the histogram and the frequency polygon. In the histogram, the height of the class column represents the frequencies (Figure 1.7c). In the frequency polygon, the points are connected to form a curve (Figure 1.7c). Such diagrams can be constructed for most sets of data and are usually referred to as the statistical distribution of the variable.

The primary purpose of descriptive statistics is to describe quantitatively such sets of data with merely one or two numbers. The two most important statistical concepts that assist toward this end are *central tendency* and *dispersion*. The first leads to a measure of the location of the frequency distribution on the x axis, the second refers to the amount of variation about this location. We shall consider each concept in turn.

The most common measure of central tendency is the simple *average* or *arithmetic mean*. This is familiar to people in all walks of life because it is used in such disparate contexts as describing sports and weather conditions. To find an arithmetic mean, add all our figures and divide by the number of figures in the data set. In our example in Figure 1.7a, this is

$$\frac{25 + 32 + 45 + 22 + 28 + 38 + 5 + 12 + 18}{9} = 25$$

We say that the average percentage of land under wheat cultivation in the nine counties is 25%. This is our measure of central tendency. It tells us the central location of the frequency distribution on the x axis in Figures 1.7c and 1.7d.

We can use this well-known statistical concept to

(a)

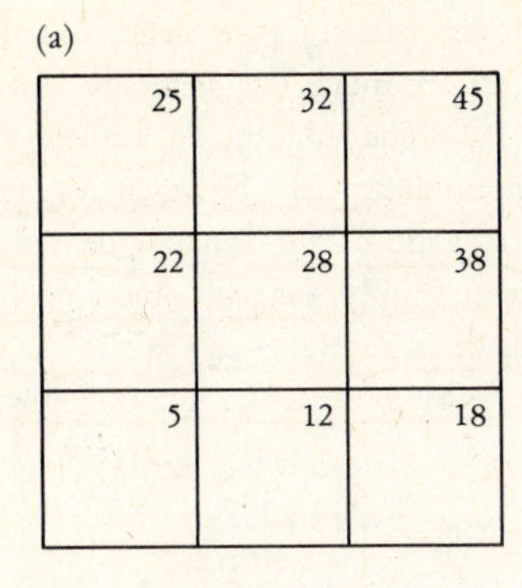

(b)

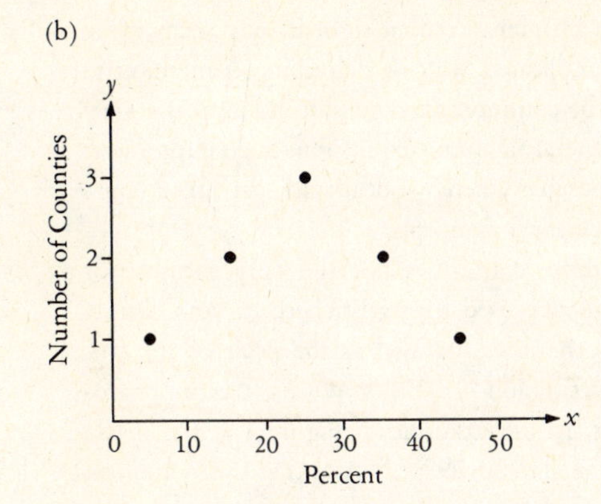

(c)

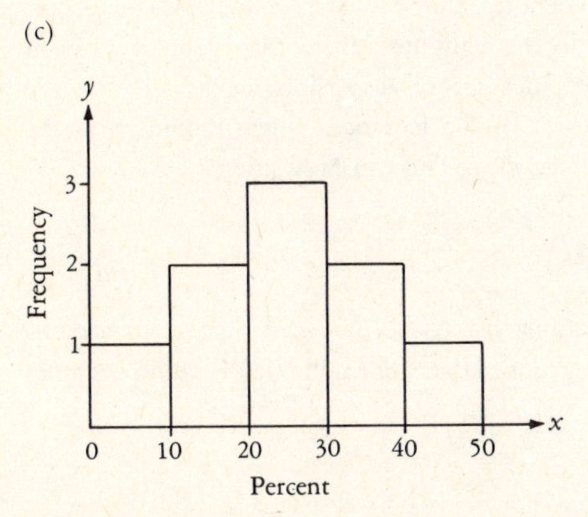

(d)

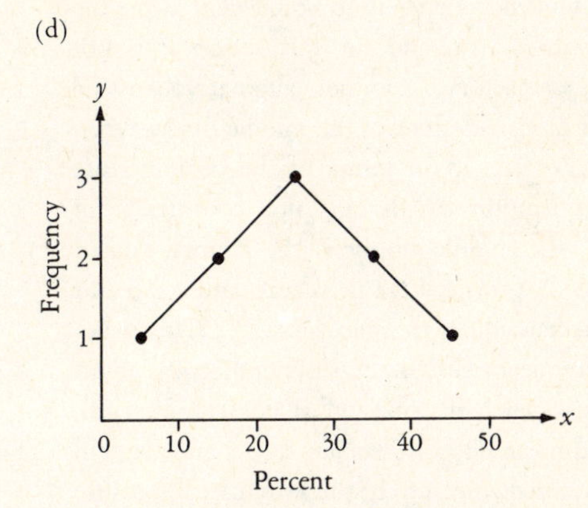

Figure 1.7 The statistical distribution of a variable. (Figures refer to the percent of farmland under wheat.)

introduce the way in which statisticians write down their formulas for defining such concepts. We begin by converting the specific arithmetic example into a more general algebraic form. We replace the numbers by x_1, x_2, up to x_9, and refer to the mean as $\bar{x}$ (which we read as "x bar"). Thus,

$$\bar{x} = \frac{x_1 + x_2 + x_3 + x_4 + x_5 + x_6 + x_7 + x_8 + x_9}{9}$$

or, as can be more generally stated with n objects,

$$\bar{x} = \frac{x_1 + x_2 + \ . \ . \ . + x_n}{n}$$

To simplify this equation, we employ a sigma sign, Σ, which is an instruction to add (or sum) as specified. The specifications are given around the sign, so that

$$\bar{x} = \frac{\sum_{i=1}^{i=n} x_i}{n}$$

reads "the mean equals the sum of all the x_i's from $i = 1$ to $i = n$, divided by n." In most common uses of this notation, we sum over all objects so that we write simply

$$\bar{x} = \frac{\Sigma x_i}{n}$$

with its being assumed that the specifications are from $i = 1$ to $i = n$. This is sometimes known as *sigma notation.*

We have traced the derivation of this notation carefully step by step because it is widely used for defining most of the statistics we shall come across in subsequent chapters. At first sight it may seem to be a quite complicated way of describing something as simple as the common average, but in fact it is a very succinct general algebraic expression. Such expressions are indispensable when we deal with yet more complicated statistical concepts.

The arithmetic mean is not the only measure of central tendency used by statisticians. A very simple measure is the *mode*, defined as the peak of the frequency distribution—in other words, the most frequent class. In the example in Figure 1.7, the mode occurs at the class 20–29%. Notice that this definition of central tendency depends to some degree on the class intervals chosen, and for this reason it is not altogether satisfactory. Another alternative to the mean is the *median*, defined as the middle object when all objects are ranked in terms of the characteristic under consideration. With our nine counties from Figure 1.7, the middle object is the county ranked fifth. It has 25% of its land in wheat, and so in this case the median equals the mean at 25%. This is because the frequency distribution is symmetrical about its center. When a distribution is not symmetrical, median and mean do not coincide. In general we will find the arithmetic mean to be the most useful measure of central tendency.

Let's assume now that the original county data from Figure 1.7a were not available and that all we had were the grouped data presented in Figure 1.7b. Can we still find the mean of the distribution? The answer is yes, although we have to modify our formula slightly to accommodate grouped data as follows:

$$\bar{x} = \frac{\Sigma x_i f_i}{\Sigma f_i}$$

x_i represents the midpoints of the classes and f_i is the frequencies in those classes. In Figure 1.7b, we take our class midpoints (the x_i's) as 5, 15, 25, 35, 45. The mean now becomes

$$\bar{x} = \frac{(1 \cdot 5) + (2 \cdot 15) + (3 \cdot 25) + (2 \cdot 35) + (1 \cdot 45)}{1 + 2 + 3 + 2 + 1}$$

$$= \frac{225}{9} = 25$$

Measures of the dispersion of a frequency distribution about the mean are much less generally known than the mean itself. The concept of *variance* has widespread use among statisticians, however. Variance is defined as the average of the squared deviations from the mean. Quite simply, for each object we find the deviation from the mean $(x_i - \bar{x})$, square the result, add, and divide by n. In sigma notation, this is written

$$\text{Var}(x) = \frac{\Sigma(x_i - \bar{x})^2}{n}$$

Var(x) denotes the variance. This concept occurs repeatedly throughout statistical theory, but if we want a simple measure of dispersion we usually use the square root of the variance, which is known as the *standard deviation*. This can be written

$$\sigma = \sqrt{\frac{\Sigma(x_i - \bar{x})^2}{n}}$$

The formula has to be modified slightly to accommodate grouped data. The standard deviation becomes

$$\sigma = \sqrt{\frac{\Sigma(x_i - \bar{x})f_i}{\Sigma f_i}}$$

(Both the ungrouped and grouped formulas for calculating the standard deviation are illustrated in Work Table 1.1)

The standard deviation is an absolute measure of dispersion in units of the characteristic under consideration. Thus, for the frequency distribution in Figure 1.7, $\sigma = 13.9$. This indicates simply the degree of spread about the mean. When we wish to compare dispersion levels for different groups of places, however, it is often found that the standard deviation is closely related to the mean value itself. This can best be illustrated with an example.

Suppose that we are comparing the frequency distribution of annual rainfall at two metereological stations. If one station has an average of 100 inches a year, then an absolute deviation of 20 inches (either 80 or 120 inches) may not be uncommon. If the other station has an average annual rainfall of only 25 inches, then an absolute deviation of 20 inches, representing as it does an 80% deviation, will probably be very rare.

Hence, the first station, with the larger mean value, is likely to have the larger standard deviation based on absolute deviations. In this situation, we might want to use a *relative* measure of dispersion. This is easily derived, simply by dividing the standard deviation by the mean. The result is known as the *coefficient of variation*:

$$V = \frac{\sigma}{\bar{x}}$$

Since both σ and $\bar{x}$ are in the same units, this ratio is said to be dimensionless—that is, it is independent of the units of the characteristic under consideration.

The work of economist J.G. Williamson illustrates use of the coefficient of variation. He has used the following approach to examine regional inequalities. If we have per capita income figures for a set of regions within a country, the standard deviation of these figures gives us a gross measure of regional inequalities in income levels. Overall mean values for income commonly vary appreciably between countries and over time, however, and for comparative purposes we use the coefficient of variation Williamson has calculated coefficients of variation for several countries over time. Some of his time-series results are shown in Table 1.1. The same area units are used for each calculation for each country, thus allowing time comparisons. However, comparisons are not possible between these countries because of the different sizes of areal unit used (it is obvious that the fewer number of units used, the less the variation that can be recorded). From these results, Williamson has been able to postulate a simple dynamic model of regional change in which it is shown that inequalities initially increase with economic development to a peak, after which they gradually decline as the economy matures. In Table 1.1, this whole sequence is shown for the United States and Brazil while only the latter trend is shown for the shorter time series for Norway and Sweden.

Williamson's study illustrates very clearly a useful application of simple descriptive statistics to areal data. Notice, however, that, in order to compute these statistics, he has taken the areal data out of their spatial context before performing his statistical analysis. By this we mean that the actual locations of the regions within the countries are not taken into account. This

Table 1.1 Coefficients of Variation for Regional Incomes over Time: United States, Brazil, Norway, and Sweden

United States		*Brazil*	
1840	0.231	1939	0.502
1880	0.321	1947	0.693
1900	0.299	1949	0.713
1920	0.291	1951	0.725
1930	0.338	1953	0.703
1940	0.263	1955	0.692
1950	0.193	1957	0.665
1960	0.176	1959	0.663
Norway		*Sweden*	
1939	0.424	1930	0.440
1947	0.253	1944	0.331
1957	0.233	1950	0.229
1960	0.186	1961	0.192

Source: J.G. Williamson. *Regional Inequality and the Process of National Development.* Chicago: The University of Chicago Press, 1965, Table 5. © 1965 by The University of Chicago.

is illustrated by Figure 1.8, in which three four-region countries have very different spatial patterns of income level. Country (a) has a northern problem, country (b) has a central problem, and country (c) has a southern problem. If we take these income levels out of their spatial context, we find that the data are identical so that all countries have *equal* levels of regional inequality. Such a conclusion is certainly likely to be important and relevant, but the geographer will typically remain somewhat concerned with the spatial pattern. As well

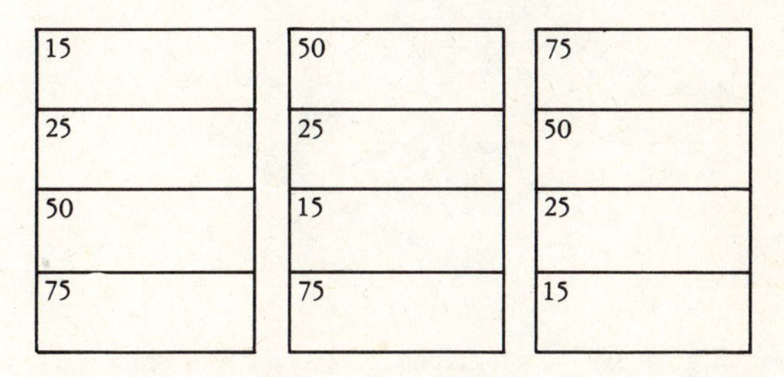

Figure 1.8 Regional problems in three countries. (Figures refer to regional income levels.)

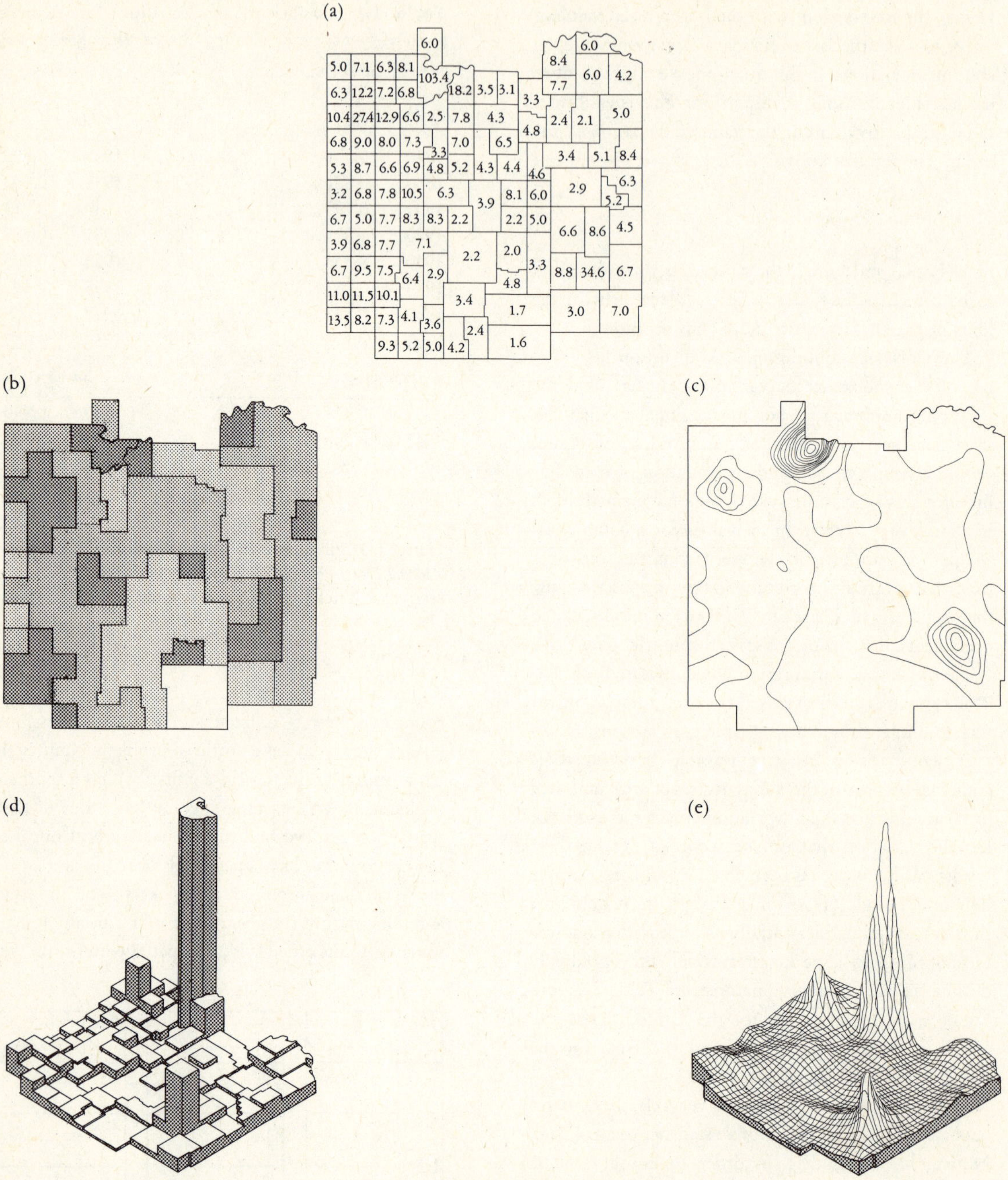

Figure 1.9 Cartographic portrayal of statistical data: Population densities for part of central Kansas. (a) Original data, (b) choropleth representation, (c) isoline representation, (d) stepped statistical surface, (e) smoothed statistical surface.
From G.F. Jenks. Generalization in statistical mapping. Reproduced by permission from the *Annals* of the Association of American Geographers, vol. 53, 1963.

as being interested in variations among places, he is concerned with *variations over space*. Techniques that treat of such variation have come to be known as *geostatistics* and are derived largely from a simple analogy between the statistician's frequency diagrams and the geographer's frequency map. The result is two-dimensional descriptive statistics.

Consider Figure 1.9a. It shows a set of geographical data over space. These are population densities by small civil divisions for eight central Kansas counties. We wish to describe this distribution without losing the locational information that the map contains. There are two ways in which cartographers have usually represented such data—as a choropleth map and as an isoline map (Figures 1.9b and 1.9c). These are the equivalent of the statistician's histogram and frequency polygon, respectively. This can be most easily seen by drawing a "third dimension" above the map surface. When this is done, the choropleth shadings become columns of varying heights and the isolines become an undulating surface. These are known as stepped and smooth statistical surfaces, respectively (Figures 1.9d and 1.9e), and are three-dimensional drawings of what cartographers represent every time they draw a choropleth or isoline map. The equivalence to histograms and frequency polygons is readily apparent.

We can conclude that statistical maps differ from frequency graphs on two counts. First, the maps show density rather than frequency. This simply reflects the need to allow for variations in areal unit size in presenting the data. If the areal units were the same size, we could use pure frequency maps and, conversely, if the class intervals on the graphs were of different sizes they would also have to be considered as densities rather than frequencies. Second, of course, our map data have a two-dimensional base upon which frequencies and densities are shown, whereas graphs have a one-dimensional base. We shall find that the importance of this change in dimension from a single x axis to a whole plane affects standard statistical concepts in different ways as we try to apply them in geostatistics.

In the change from a single x axis to a plane, the first requirement is to find an equivalent to the concept of deviation for the two-dimensional case. In standard statistics, deviation is simply the difference in magnitude between two observations. On a frequency graph such deviations are represented by distances between observations on the x axis. This gives us our clue to a meaning for deviation in geostatistics. It is quite simply that distances on the plane are treated in the same way as distances along the x axis, so that we have a spatial concept of deviation. This idea is the cornerstone of all geostatistics and is clearly illustrated for the two most commonly used descriptive geostatistics, the mean center and the standard distance.

The Mean Center

Let's begin by considering the statistical concept of central tendency. We can transfer immediately the concept of mode to the areal case. It is the highest frequency or density value in choropleth or stepped statistical surface that can be identified in Figure 1.9. The two other measures of central tendency differ in the ease with which they can be transferred to the areal case—the areal median is difficult to define while the mean is almost as easy to find as the mode. We consider the areal median below; let's now consider the areal mean or *mean center*.

The mean center is the location on a statistical map that can be used best to summarize the distribution. It can be found simply by computing the arithmetic means for the coordinates of the observations on the two axes. The point of these two means, $\bar{x}$ and $\bar{y}$, defines the mean center. If the data are not in point form but consist of frequencies as aggregate areal data, we have to compute the two arithmetic means by using the grouped data formula. In this case, in order to compute the mean center we must allocate coordinate values to the centers of the areas. If the areas are small, this may be done by visual estimation. The coordinate values are then used to compute $\bar{x}$ and $\bar{y}$ and hence define a mean center from grouped data. The definitions of mean centers for point and grouped data are illustrated in Work Table 1.2.

Mean centers have been found to be useful in two different contexts in geographical research. They can be used to trace the changing pattern of a single distribution over time. The first reported use of the mean center in this way is that of J.E. Hilgard in 1872 to summarize what he termed "the advance of the population in the United States." Mean centers have been reported

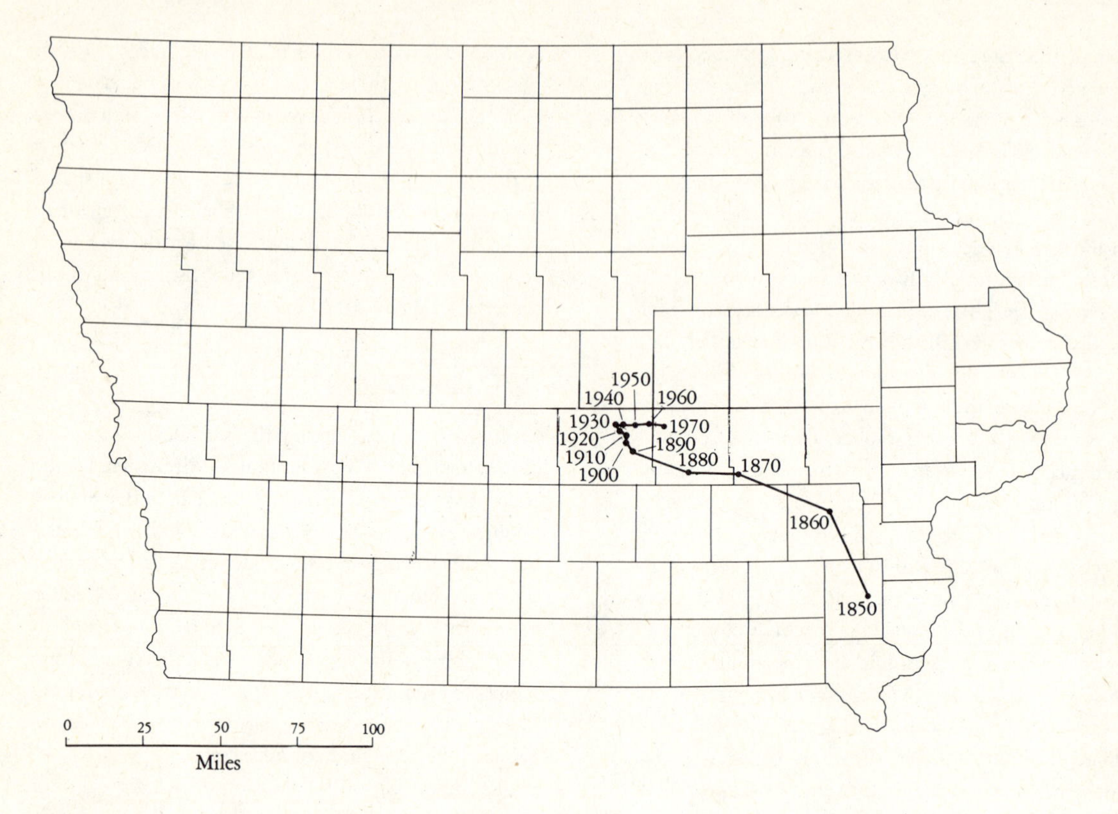

Figure 1.10 (a) Mean centers of Iowa population, 1850–1970.

for every census since 1870, and therefore the westward movement of the U. S. population can be traced through the movements of its mean center to the present day. Figure 1.10a shows this trend in Iowa for the years from 1850 to 1970. Starting in the southwest, the mean center moved northward and then westward in large steps to 1890. The large steps represent the fact that decennial changes in Iowa's population pattern were relatively major changes and reflected the expansion in settlement. The year 1890 marks the end of the frontier, according to historian Frederick Jackson Turner, and, sure enough, 1890 is a break point on the map between large, mainly westward movements of the mean center and small, largely northward movements of the center. Since World War II, the mean center has been tracing its steps eastward in response to urban growth along the Mississippi River. Thus, we can see that the mean center provides a method for succinctly summarizing a distribution in one location so that its aggregate movement over time can be easily discerned.

In statistics, of course, population refers to any collection of objects, not merely to human beings. The second context in which mean centers have been used in geographical research directly exploits this liberation from a strictly demographic interpretation of population. They are used to compare the distributions of several different phenomena in the same region. Thus, as well as tracing the evolution of Iowa's population to 1970, we can begin to look at its spatial structure in 1970. Three subpopulations defined by the 1970 census are shown as mean centers in Figure 1.10b. We find that the urban population differs in location from the total population to an extent greater than it does for either the very old (over 75) or the very young (under 5). This is reflected in the distances between the centers. The

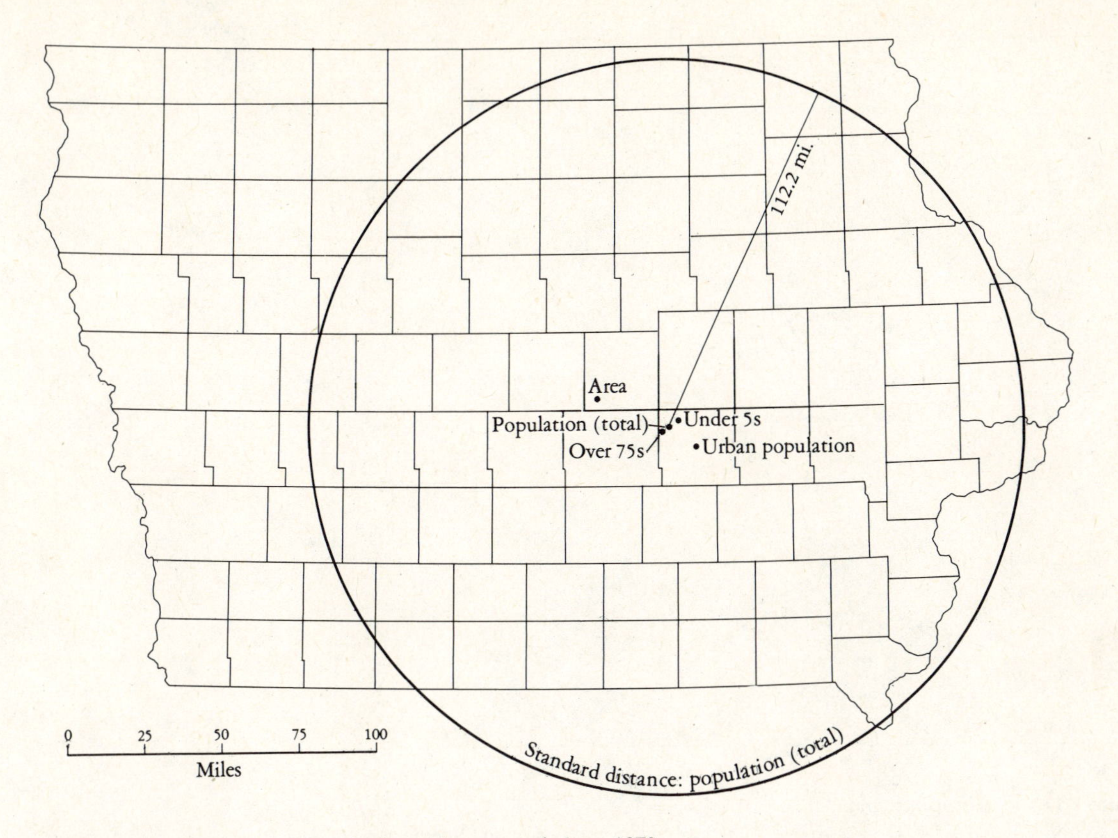

Figure 1.10 (b) Mean centers of selected Iowa populations, 1970.

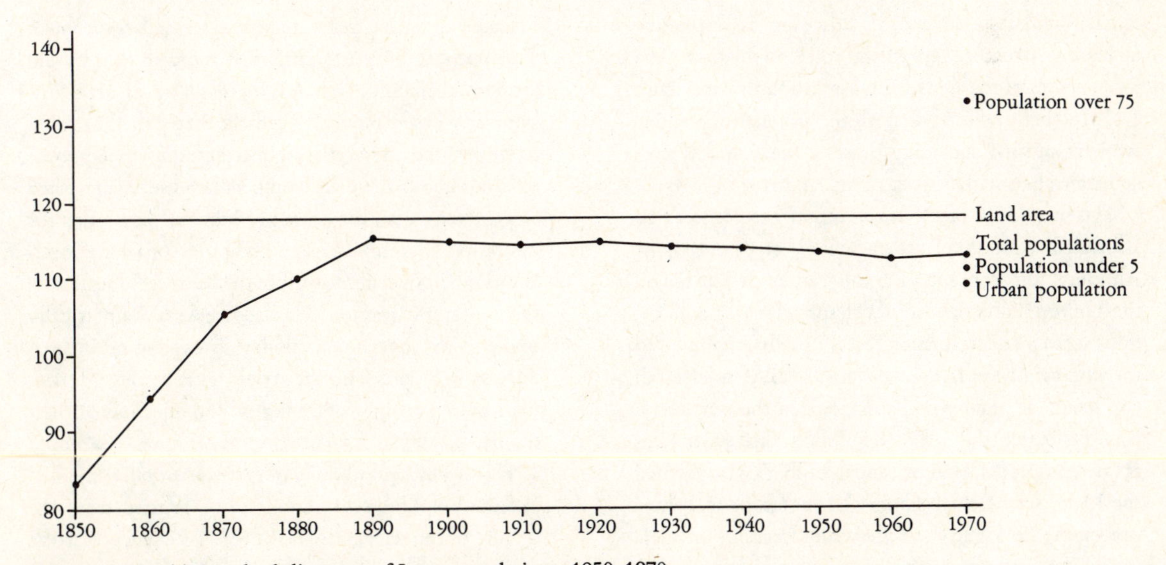

Figure 1.10 (c) Standard distances of Iowa populations, 1850–1970.

Figure 1.11 Ukraine centrogram, 1926.
From Sviatlovsky and Eells. The centrographic method and regional analysis. Reprinted from the *Geographical Review, vol.* 27, 1937, copyrighted by the American Geographical Society.

older population (over 75), however, is distinctive in having a center west of the total population. The mean center of the area of Iowa is also shown in Figure 1.10b, which indicates that all the population centers we have plotted, including those of the over 75s, have an eastern bias with respect to the land area of Iowa.

This second use of the mean center has come to be associated with the Russian school of *centrography*. Although the concept of mean center originated in the United States, its most widespread, early applications were made in Russia. This is usually attributed to the interest of the famous chemist D.I. Mendeleev in the center of gravity of Russia, late in the nineteenth century. After the 1917 Revolution, his work was followed up by a band of centrographers who formed the Mendeleev Centrographical Laboratory in Leningrad in 1925. Figure 1.11 shows an example of their work, with centers ranging from sugar to goats in the Ukraine in 1926. Such maps are termed *centrograms*. The purpose of such empirical research was to aid economic planning by developing laws of areal distribution based on mean centers. Early in the 1930s, however, the advice they gave the policymakers was to bring about their downfall. When they were asked to produce a plan for grain production, they called for limitation of commercial planting in Russia's traditional bread belt in order to ensure the "correct" location of the center of gravity. This advice was diametrically opposed to government policy to expand grain production in Siberia, and the group never recovered from the loss of prestige following on the rejection of their report.

This is a rather sad story of early quantitative applied geography. There are lessons to be learned from the demise of centrography, however. To expect to arrive at laws for planning purposes through the use of mean centers is obviously to oversimplify in the extreme. Any development of laws of areal distributions clearly requires more thought than contemplation of centrograms provides. In technical terms we can easily identify the basic flaw in centrography. We have already noted that, in simple descriptive statistics, measures of central tendency are only one way in which a distribution may be described. The same is true for geostatistics. In both cases, central tendency relates only to the location of the frequency graph or map. Other properties of these distributions require other measures. Let's consider how geostatistics treats of the property of dispersion or spread of distribution about the mean center. Such measures were independently introduced into the description of areal distributions late in the 1950s by David Neft and William Warntz in the United States and by Roberto Bachi in Israel. Their work constitutes the major step in transforming dishonored centrography into the more respectable geostatistics.

Standard Distance

Statistical concepts of dispersion are, on the whole, just as easy to transfer to the areal case as is the mean. Because distances represent deviations in the areal case, we can define a *standard distance* as the equivalent of the statistician's standard deviation. Distances from the mean center to every observation are found, squared, and divided by n, and the square root is taken. This can be written

$$SD = \sqrt{\frac{\Sigma r_{ic}^2}{n}}$$

SD is the standard distance and r_{ic} is the distance from the ith observation to the center. Alternatively, the standard distance can be calculated from the two variances derived by treating each axis of the map separately, as in finding the mean center. This is given by

$$SD = \sqrt{\sigma_y^2 + \sigma_x^2}$$

(Both formulas are employed in Work Table 1.3.) With aggregated data, our formulas must be modified, as was indicated earlier.

Finally, we can consider how the notion of relative dispersion can be incorporated into the areal case. This is not as straightforward as the procedure detailed above, because the mean center is a location on a plane rather than a magnitude and does not influence deviations in the same way that the arithmetic mean does. Several measures of relative dispersion nonetheless have been developed in geostatistics and we consider two of these in this discussion. We shall leave their description until we come to consider their application below.

Application of measures of spread are much less common than studies of central tendency in geostatistics. This is a pity, because the two statistical concepts obviously complement one another in the description of frequency distributions in two dimensions as well as one. We can illustrate this complementarity by returning to our Iowa example. Standard distances have been calculated for all the distributions for which we have plotted mean centers on Figures 1.10a and 1.10b. All these results could, of course, be drawn as circles about the centers, but the resulting map of interlocking circles is almost impossible to interpret. The standard distance for the 1970 population has been added to Figure 1.10b for illustrative purposes. We will discuss our results for the standard distances in terms of their graph in Figure 1.10c.

Figure 1.10c shows the trend in dispersion of the Iowa population from 1850 to 1970. The westward spread to 1890 has been shown by using the mean center (from Figure 1.10a) and the standard distances show the actual degree of dispersion about this changing center. As the center was moving westward, the population was becoming increasingly dispersed until this reached a peak in 1890. From this time new settlement gave way to population redistribution as the major form of change in the pattern of population, and this is reflected in a very gradual decline of the standard distances. Figure 1.10c also shows the standard distances associated with the mean centers in Figure 1.10b. The standard distance for the land area of the Iowa counties is of course constant for the period under consideration, and we see that the spread of the total population never reached the same level as the land area. This means simply that the settlement process approached, but did not reach, the situation of an even spread of population. For 1970, standard distances for the very young, very old, and urban populations are shown. The young and the urban populations are found to be more concentrated about their respective mean centers while the locations of population of those over 75 are highly dispersed. A standard distance above that of the land area indicates that this latter distribution is overweighted in the peripheral and western counties of Iowa at the expense of the central and eastern counties. We can see that, with just two statistics, the mean center and the standard distance, we can begin to build up a picture of the evolution and structure of a population distribution.

Although the complementary use of mean centers and standard distances is to be generally recommended, there are some situations for which standard distance is a far more useful measure than mean center. This is particularly the case in the analysis of commercial functions within cities. Some functions, such as insurance offices, are clustered in the central business district while others, such as grocery shops, are distributed throughout the city. The mean center does not necessarily distinguish between these two types of distribution. For example, for a centrally located business district in a roughly circular city, both types of function will have similarly located mean centers, but their differences in pattern will be illustrated clearly by their standard distances. Thus, in Table 1.2, standard distances are given for a set of commercial functions in Tel Aviv, Israel, in 1964. It is immediately apparent that these data order the functions in terms of their central concentration in the way we have argued above. Thus, cinemas have a standard distance very much like the population they serve, whereas insurance offices have a very low value.

Table 1.2 Standard Distances for Commercial Functions in Tel Aviv (1964)

Function	*Standard distance* (m)	*Function*	*Standard distance* (m)
Insurance	740	Laundry	2,020
Corporate office	1,150	Pharmacy	2,090
Hotel	1,300	Barber	2,230
Restaurant	1,970	Cinema	2,630
Population	2,870	Area	3,860

Source: A. Shachar, *Some Applications of Geo-Statistical Methods in Urban Research*. Philadelphia: Regional Science Association Paper XVIII, Table 1, p. 203, 1967.

Notice that we cannot use these findings directly to compare Tel Aviv's commercial patterns with those of other cities. Table 1.3 shows standard distances for several functions in Rome and Jerusalem as well as Tel Aviv. In all but one case, Rome's standard distance is largest and Jerusalem's is smallest. This merely reflects the relative size of the three cities. In order to compare these patterns, we clearly must use relative measures of dispersion. Because these commercial functions serve the population, we can define a measure of *relative dispersion* for distribution x (RD_x) as

Table 1.3 Absolute and Relative Measures of Dispersion for Commercial Functions: Jerusalem, Tel Aviv, Rome*

	Population		*Groceries*		*Apparel*		*Barbers*		*Finance and insurance*	
	SD	RD	SD	RD	SD	RD	SD	RD	SD	RD
Tel Aviv	2,870	1.00	2,440	0.85	1,300	0.45	2,230	0.78	780	0.27
Jerusalem	1,800	1.00	1,650	0.86	770	0.43	1,470	0.80		
Rome	4,180	1.00	3,850	0.92					1,520	0.36

*SD = standard distance (in meters). RD = relative dispersion = SD/SD for population.

Source: A. Shachar. *Some Applications of Geo-Statistical Methods in Urban Research*. Philadelphia: Regional Science Association Paper XVIII, Table 2, p. 203, 1967.

Table 1.4 Absolute and Relative Measures of Population Dispersion: Selected Countries

Country	*Standard distance* (km)	*Standard distance* (km)/r_A
Australia	615	0.63
United Kingdom	134	0.77
Brazil	697	0.68
Japan	256	1.20
United States	839	0.86
India	538	0.85
China	579	0.52

Source: Adapted from David S. Neft, *Statistical Analysis for Areal Distributions*. Philadelphia: Regional Science Association, 1966, Tables 4 and 16, pp. 66 and 106.

$$RD_x = \frac{SD_x}{SD_p}$$

SD_x is the standard distance of the distribution and SD_p is the standard distance of the population. When this measure is calculated for the three cities, we find remarkable uniformity among them, as shown in Table 1.3.

Other measures of relative dispersion are available. This method obviously cannot be used to compare standard distances for population because RD_x always has a value of one. For comparing population distributions, therefore, we must define a second measure of relative dispersion. The standard distance of a population reflects the size of the region or country for which it is computed. Hence, we find the large variations in standard distances for selected national populations in Table 1.4. We can overcome this problem of comparison simply by dividing these standard distances by a function of the areal extent over which they are measured. Thus, we define SD/r_A, in which r_A is the radius of the area of the country assumed to be circular. This results in the relative measures of the population dispersion given in Table 1.4, which are dimensionless. We can now compare these countries directly. Japan, it seems, has a very dispersed population about its mean center whereas China's population is relatively clustered about its mean center. We have here clearly a useful measure for comparing relative levels of dispersion about mean centers at different scales.

Let's conclude this discussion by considering some of the properties of mean centers and standard distances. All their major characteristics stem from the fact that they are macroconcepts. This means simply that they are computed from measures for all objects under consideration. They contrast with the concept of the mode in this respect. If we know that a location is at the mode of a distribution, we can expect high frequencies of population—the mode of the U.S. population is New York, for example. The same is not true of the mean or mean center—the mean center of the U.S. population in 1970 was in a field in southern Illinois. In fact, the mean center of a population need not even be within the same region as that population. The mean center of an evenly distributed population of palm trees on a circular atoll is in the lagoon. The mean center of the population of member nations of NATO occurs in the North Atlantic Ocean. Thus, we must treat the mean center as a macroconcept and not expect it to tell us about microproperties of particular areas. There are advantages that accrue from its being a macroconcept, however. By definition, it depends on all members of a population, and thus it is highly sensitive to changes in that population. It is this property of the mean center that we exploit when we use it to trace population changes over time. This is in stark contrast to the areal mode, which is highly insentive to population changes. The mode of the U.S. population has not moved from New York through all the westward migration since early in the nineteenth century.

These macroproperties of the mean center are shared by the standard distance. The fact that an area is crossed by a standard distance circle tells us nothing about the particular characteristics of that area. When we consider the sensitivity of this measure, we must remember that standard distance is computed from *squares* of distances about the mean. Its value is therefore greatly affected by extreme locations. An addition to the population one mile from a mean center will quite simply change the standard distance much less than a single addition one thousand miles away. In practice, this means that births in Oregon influence the

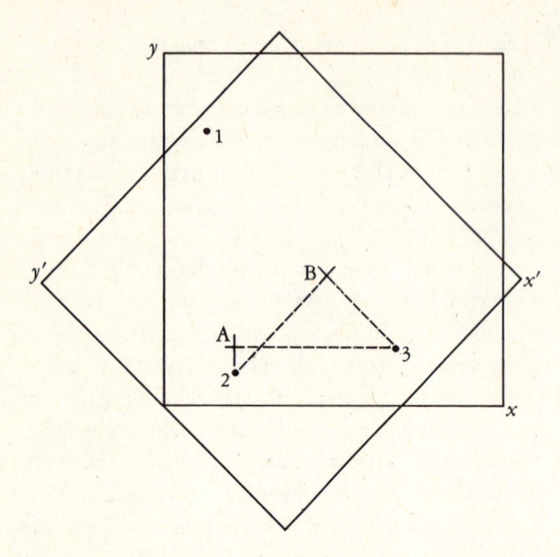

Figure 1.12 The areal median defined by ranking. Medians: y axis, 3; x axis, 2; median center, point A. Medians: y' axis, 2; x' axis, 3; median center, point B.

standard distance of the U.S. population far more than do births in Illinois. We shall see in the discussion that follows that this overemphasis of extreme locations is also a property of the mean center itself.

The Areal Median and the Generalized Weber Problem

Let's consider what happens when we attempt to apply the methods used above to define the median for an areal distribution. We can use the median value for each axis to define a center for the distribution, as is shown in Figure 1.12 (point A). But this does not define a unique areal median because when we rotate the coordinate axes we define a different center. With y and x axes, point A is defined as the *areal median* but with y' and x' axes, this changes to point B. This definition of a center is not invariant to the coordinate axes being used. As such, it is of very limited utility in spatial analysis.

Does this mean that there is no single equivalent to the median in the study of variations over space? Fortunately, the answer to this question is no. In order to define a unique areal median, however, we must consider further the properties of the statistician's median. Consider the case in which we have measures on three objects giving values of 0, 10, and 35. The mean is 15 and the median is 10. Now compute the sum of the absolute deviations about each measure of central tendency. For the mean, these are $15 + 5 + 20 = 40$, and for the median they are $10 + 0 + 25 = 35$. Notice that the sum of absolute deviations about the median is lower than about the mean. This is always the case for any set of data. In fact, the median has the lowest sum of absolute deviations for *any* value or location. In our three-measure example, the sum of absolute deviations about the value 9 is 36 and about 11 it is also 36. No matter how many other values we try, we shall never be able to get below the median's total deviation of 35. We can use this property of the median as an alternative definition to the ranking approach initially adopted. Our new definition of the median is the location that minimizes the sum of the absolute deviations. We shall use this definition to translate the statistical concept of the median into a spatial context. As an aside, we can notice here that, were we to repeat the exercise above with *squared* deviations, we would find that the mean has the smallest sum. This is a general property of the mean and applies also to the mean center: it is the location that minimizes the sum of the *squared* deviations.

Let's now relocate our three-point example by placing it on a two-dimensional plane. Except for the special case in which they still lie on a line in the plane, these points now define a triangle. If we weight these points as if they represented grouped data, we have constructed the classical location triangle of Alfred Weber's industrial location theory (Figure 1.13a). Weber's problem was to find the location for a factory that minimized transport costs between two sources of raw material and the market. If we make some simplifying assumptions, such as that distance is directly related to transport costs and the weights at the points represent the weight of either raw material or finished product, then it follows that to search for an optimum location for the factory is to search for the areal median.

How can we find a solution to Weber's problem? In two special cases, the solution can be found by inspec-

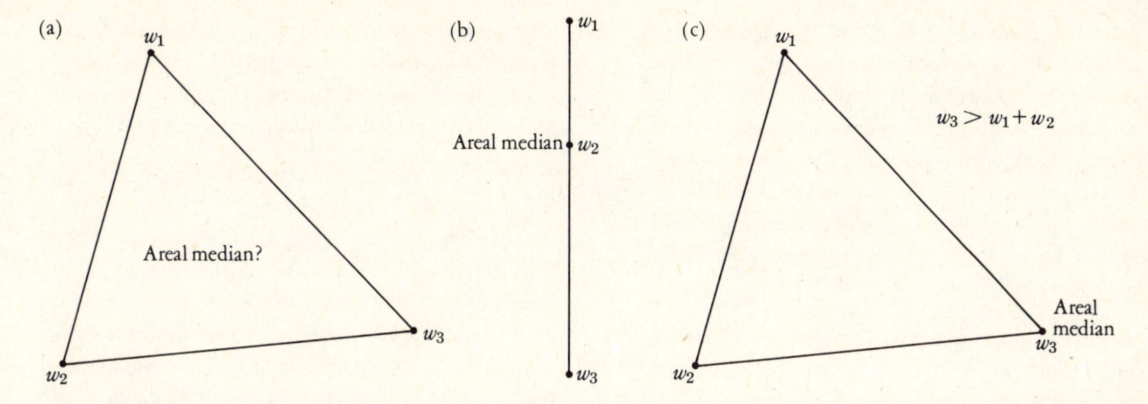

Figure 1.13 Weber's location triangle (the *w*'s refer to the weights).

tion of the diagram. If the points lie on a straight line in the plane, the median is simply the middle location as found by the usual ranking procedure (Figure 1.13b). Where a triangle is formed, there is no middle location in this sense, however. Where one weight is larger than the combined value of the two other weights, then the areal median is located at that point. In a sense, this is derived from the ranking procedure, because if a class contains over half a population it follows that it must contain the middle one. In the case of the Weber triangle, this is known as the *corner solution* (Figure 1.13c). Where these two special cases do not obtain, however, all we can say is that the solution lies within the triangle (Figure 1.13a). There remains no *mathematical* solution for this general case, but nonmathematical solutions have been devised.

There have been various original methods for finding solutions to the Weber problem. The earliest procedures involve analogue models of either the physical or the graphic variety. The most famous physical model is the *Varignon frame*, consisting of a system of strings, weights, and pulleys. The strings are joined at the center, passed through pulleys, and attached to weights. The strings represent the transport routes, the pulleys represent the sites of raw materials and markets, and the weights are analogous to the weights in the original problem. These weights seek their lowest level and consequently, pull the junction of the strings to the point of minimum transport costs. All this may be useful for demonstration purposes, but it is rather cumbersome for research. It is often replaced by graphic methods, notably the construction of the isodapanes, or lines connecting places of equal total transport cost. An approximate solution to the Weber problem can be found by identifying the lowest point on the surface of isodapanes. Both physical and graphic models can be extended beyond the simple three-point situation of Weber's triangle. Generalized Weber problems consist of several raw materials and several markets. Theoretically any number of locations can be considered simply by adding pulleys to the Varignon frame or by considering more points in the graphic method. In practice, however, large numbers of locations lead to highly complicated diagrams and tangled strings. In general we can identify a need for less cumbersome methods of solution.

We have already noted that a mathematical solution to nontrivial Weber problems does not exist. By this we mean that we cannot *deductively* derive a solution. We can attempt a numerical solution using inductive methods. This approach has been stimulated by the relatively recent, widespread availability of electronic computers. Earlier in this chapter we noted that the computer has made possible the development of procedures for finding solutions to problems that are not mathematically soluble. The Weber problem is a classic case of the application of this numerical approach. What follows illustrates both how to find an areal median and some basic elements of this type of numerical analysis.

The principle of this method is to devise a set of rules that provide an approximate solution. Such rules define a procedure known as an *algorithm*. The algorithm we deal with here is known as a *converging algorithm*, because it gets progressively closer and closer to the true answer. Such algorithms have a stopping rule to terminate the algorithm when an approximate solution reaches a specified level of precision. We shall describe briefly a very simple algorithm devised by David Seymour to solve the generalized Weber problem. The steps are as follows.

1. Regularly spaced trial points are laid down over the whole region.
2. Aggregate distances, weighted appropriately, are calculated for all raw material and market locations at each trial point.
3. The trial point with the lowest aggregate distance is identified and becomes the center of a new, smaller search region drawn around it.
4. A new set of trial points is laid down over the new search region. Because this is only part of the previous search region, it follows that the points are much closer together.
5. Steps 2 to 4 are repeated until a desired level of precision is reached. The final trial point with the lowest aggregate distance is the approximate solution to the generalized Weber problem.

Notice that no deductive argument is involved, although the result is a numerical solution. This algorithm normally requires a large number of calculations and, hence, is feasible only with a computer. In fact, the algorithm is quite long-winded and would therefore be classified by a computer scientist as inefficient. In order to devise a more efficient algorithm, we must know more about the areal median. This has been achieved, although it is beyond our discussion here. There is reference to such an algorithm in the further reading at the end of the chapter.

The Seymour algorithm has the following characteristics, typical of all converging algorithms:

1. They are repetitive; they iterate, using the same simple procedure many times.
2. They are inductive; their solutions are not derived deductively and, hence, are usually termed simply *numerical* (but not mathematical).
3. As a corollary of 2, they produce not exact solutions but only approximate solutions to some specified degree of precision; they are not mathematically elegant.
4. They enable the researcher to find acceptable solutions to mathematically intractable problems.

It is this latter characteristic that is the most important. Characteristics 2 and 3 would seem to be important only to mathematicians because, with the advent of high-speed computers, characteristic 1 has become a minor hindrance and acceptable solutions are available to all empirical researchers. There is one damper on this highly satisfactory situation. Because, the solutions are not derived deductively, it is not always possible to prove that we have in fact produced an approximation to the true solution. In simple problems such as the generalized Weber problem, this is not usually an important limitation. Thus, in this case, spatial analysts can put aside their analogue models and use their numerical algorithms confidently. In more complicated situations, this confidence must be relaxed somewhat. We consider this problem in Chapter Seven when we meet iterative converging algorithms again as we generalize the Weber problem further to consider defining more than one center at a time.

Further Reading

To suggest further reading for a topic such as mathematics could quite easily be an endless task. Clearly no professional mathematician would admit to having finished his reading on the subject. That we have been concerned with the nature of the discipline circumscribes us, however. The field remains massive nonetheless, and the books and articles referred to below reflect our particular route into the subject matter.

We are fortunate in the fact that the new math has stimulated many new elementary mathematics textbooks that deal with the methodological questions that have concerned us in this chapter. The introductory chapters of Adler (1959), Sawyer (1966), Shaaf (1969), and Jacobs (1970) are all rewarding. Adler (1959) is the source for the section on number systems, Sawyer (1966) is the source for the geometries, and Schaaf (1969) is the source of the miniature geometry. A recent "official" view of modern mathematics can be found in National Research Council (1968), which is the source for the four fields of applied mathematics identified above. The discussion of models and abstraction is based on a similar argument in Ackoff *et al.* (1962).

The debate concerning mathematics in geography can be followed in two sets of literature. English-speaking geographers have conducted a quantity versus quality debate that is well represented by Gregory (1971) and Spate (1960), respectively. Parallel to this has been a more fundamental debate on the role of mathematics in geography by Russian geographers; it is reported in English, mainly in *Soviet Geography*, in the following articles: Gurevich and Saushkin (1966), Medvedkov (1967), Smirnov (1968), Anuchin (1970), and Saushkin (1971). The whole topic of mathematics and geographical research is quite fully covered in Harvey (1969).

These references give the sources for the material in Chapter One. The starting point for extending these ideas is Harvey (1969). This remains the most complete and, in fact, almost unique discussion of mathematics in geography. Useful insights into the role of mathematics can sometimes be gained from looking at parallel developments in other disciplines. Thus, Arrow (1951) is a very early discussion of modern mathematics in social science, emphasizing the idea of mathematics as language. In sociology, in particular, Coleman (1964) follows up this theme; his final chapter, "Tactics and Strategies in the Use of Mathematics," is recommended particularly for consideration from a geographical viewpoint. Fararo (1969) has also presented a nontechnical statement on mathematics in sociology, emphasizing the abstract nature of mathematics. In another parallel field, Griffiths (1970) reviews developments in the use of mathematics in earth science.

For further, advanced consideration of the nature and role of mathematics we must turn to the mathematics literature itself. For people for whom the notion of a qualitative mathematics is particularly hard to swallow, the best reference is to Kemeny (1961), who discusses qualitative mathematics and gives examples of its application. More generally, two books can be particularly recommended. Courant and Robbins (1941) is a classic, early statement of the new mathematics and, although largely technical, it is written from the methodological stance used here. The second book is wholly methodological and philosophical and therefore nontechnical: Korner (1960) presents a clear, succinct statement on the nature of mathematics and its role in empirical science. His applied examples are drawn largely from physics, but the basic research model we have presented is most clearly developed in his book. In contrast, reference should be made to Etter (1963). This is a well-written antimathematical statement that dwells on the particular and discovers that mathematics is not like history. I mention it because it represents an alternative viewpoint and, as such, presents a challenge to refute. Discussion of the article in the light of this chapter is a useful exercise (whatever the student's age). This may be supplemented by considering the debate over the utility of framing regional division in set theoretic terms conducted by Golledge and Amadeo (1966) and Hamill (1966).

Finally, the example of applied mathematics developed in this chapter is covered by all elementary statistics textbooks in the nonareal case. Gregory (1968) is one example. The application of the coefficient of variation is by Williamson (1965). The basic reference for the simple spatial statistics is Neft (1966), which draws on the work of Warntz and Neft (1960) and Bachi (1963). Early Russian centrography is reported in Sviatlovsky and Eells (1937) and its demise is described by Poulsen (1959). A more general discussion of centrography can be found in Caprio (1970), where the circular standard distance is extended in terms of the standard ellipse. This has been considered further by Yuill (1971). The simple algorithm for finding the areal median is in Seymour (1968) and a more sophisticated algorithm is presented by Kuhn and Kuenne (1962). Examples of the use of geostatistics in research contexts can be found in Shachar (1967) and Hirst (1971b).

Work Table 1.1 Variance and Standard Deviation

Purpose

To measure the variability or spread of a statistical distribution about the mean

Data

From Figure 1.7

Definition

UNGROUPED DATA Variance = Var(x)

$$= \frac{\Sigma(x_i - \bar{x})^2}{n}$$

Standard deviation $= \sigma$

$$= \sqrt{\text{Var}(x)}$$

GROUPED DATA $\text{Variance} = \text{Var}(x) = \dfrac{\Sigma(x_i - \bar{x})^2 f_i}{\Sigma f_i}$

where f_i is the frequency and x_i is the midpoint of classes. Standard deviation is as above.

Arithmetic

UNGROUPED $(\bar{x} = 25)$

x	$x - \bar{x}$	$(x - \bar{x})^2$
25	0	0
32	+ 7	49
45	+20	400
22	− 3	9
28	+ 3	9
38	+13	169
5	−20	400
12	−13	169
18	− 7	49
		1,254

$$\text{Var}(x) = \frac{1254}{9} = 139.33$$

$$\sigma = \sqrt{139.33} = 11.8$$

GROUPED $(\bar{x} = 25)$

Class	*Midpoint* x	$x - \bar{x}$	$(x - \bar{x})^2$	f	$(x - \bar{x})^2 f$
0–9	5	− 20	400	1	400
10–19	15	− 10	100	2	200
20–29	25	0	0	3	0
30–39	35	+ 10	100	2	200
40–49	45	+ 20	400	1	400
				9	1,200

$$\text{Var}(x) = \frac{1,200}{9} = 133.33$$

$$\sigma = \sqrt{133.33} = 11.54$$

Comment

Notice that by grouping our data we have lost some of the variation as shown by lower variance and standard deviations. We have literally lost information by grouping and, hence, produce a different answer for the ungrouped calculations. This aggregation effect is obviously important to all other techniques based on variance concepts, and we treat this problem in detail in the context of correlation in Chapter Five.

Exercises

1. Recompute Var(x) and σ with the ungrouped data used above: (a) with two additional observations $x = 1$ and $x = 49$ and (b) with two additional observations $x = 24$ and $x = 26$.

What effect do these extra observations have on the variance and standard deviation in each case?

2. Repeat Exercises 1(a) and 1(b), using the grouped formula and allocating the extra observations to the original classes.

Work Table 1.2 The Mean Center

Purpose

To locate the center of an areal distribution

Data

UNGROUPED *Location coordinates*

x	y
1	2
2	3
3	2

GROUPED

Area	*Frequency*	*Location coordinates* x	y
A	10	1	4
B	100	2	4
C	10	4	2

Definition

In both cases the mean center is located at $\bar{x}$, $\bar{y}$.

Arithmetic

UNGROUPED

$$\bar{x} = \frac{1 + 2 + 3}{3} = 2;$$

$$\bar{y} = \frac{2 + 3 + 2}{3} = 2.33.$$

Mean center is located at $x = 2$, $y = 2.33$.

GROUPED

$$\bar{x} = \frac{(10 \cdot 1) + (100 \cdot 2) + (10 \cdot 4)}{10 + 100 + 10} = \frac{250}{120} = 2.08$$

$$\bar{y} = \frac{(10 \cdot 4) + (100 \cdot 4) + (10 \cdot 2)}{10 + 100 + 10} = \frac{460}{120} = 3.83$$

Mean center is located at $x = 2.08$, $y = 3.83$.

Exercises

1. Draw maps to represent both ungrouped and grouped data and locate the computed mean centers on these maps. N.B. The location coordinates for the grouped data refer only to the center of each areal unit, so that the shape of the units on your map may vary.
2. Add the point $x = 4$, $y = 4$ to your dot map from Exercise 1 and compute a new mean center. What effect does this additional point have on the location of the mean center on your map?
3. Recompute the mean center for the grouped data for the case in which area *B*'s frequency declines to 20. What effect does this reduction in frequency for one area have on the location of the mean center on your map?

Work Table 1.3 Standard Distance

Purpose

To measure the degree of spread about the mean center

Data

From Work Table 1.2

Definition

UNGROUPED DATA

$$\text{Standard distance} = \text{SD} = \sqrt{\frac{\Sigma(x - \bar{x})^2}{n} + \frac{\Sigma(y - \bar{y})^2}{n}}$$

GROUPED DATA

$$\text{SD} = \sqrt{\frac{\Sigma f(x - \bar{x})^2}{\Sigma f} + \frac{\Sigma f(y - \bar{y})^2}{\Sigma f}}$$

The f's refer to frequencies.

Arithmetic

UNGROUPED $(\bar{x} = 2, \bar{y} = 2.33)$

x	$x - \bar{x}$	$(x - \bar{x})^2$	y	$y - \bar{y}$	$(y - \bar{y})^2$
1	− 1	1	2	− 0.33	0.109
2	0	0	3	+ 0.67	0.449
3	+ 1	1	2	− 0.33	0.109
		2			0.667

$$\text{SD} = \sqrt{\frac{2}{3} + \frac{0.667}{3}} = \sqrt{0.889} = 0.943$$

GROUPED $(\bar{x} = 2.08, \bar{y} = 3.83)$

f	x	$x - \bar{x}$	$(x - \bar{x})^2$	$f(x - \bar{x})^2$	y	$y - \bar{y}$	$(y - \bar{y})^2$	$f(y - \bar{y})^2$
10	1	− 1.08	1.166	11.66	4	+ 0.17	0.029	0.29
100	2	− 0.08	0.006	0.6	4	+ 0.17	0.029	2.9
10	4	+ 1.92	3.686	36.86	2	− 1.83	3.349	33.49
120				49.12				36.68

$$\text{SD} = \sqrt{\frac{49.12}{120} + \frac{36.68}{120}} = \sqrt{0.715} = 0.846$$

Exercises

1. Draw the standard distances as circles about the first two mean centers you located on the maps for Work Table 1.2.

2. Compute the standard distance for ungrouped data with the point $x = 4$, $y = 4$ added as in Exercise 2, Work Table 1.2. Is the new four-point distribution spread out more or less than the original three-point distribution? Draw the circle about the mean center for the new distribution (from Work Table 1.2) on your map.

3. Compute the standard distance for grouped data with area *B*'s frequency at $f = 20$ as in Exercise 3, Work Table 1.2. Is this new frequency pattern areally concentrated more or less than before? Draw the circle about the mean center for the new distribution (from Work Table 1.2) on your map.

MEASUREMENT AND DATA FOR GEOGRAPHICAL RESEARCH

"When you can measure what you are speaking about, and express it in numbers, you know something about it; but when you cannot measure it, when you cannot express it in numbers, your knowledge is of a meager and unsatisfactory kind." So wrote Lord Kelvin a century ago, and his spirit is beginning to haunt geography in an overwhelming fashion. It is not that quantitative measurement is new to geography, since it shares many of its nineteenth-century antecedents with statistics; it is, rather, that geographers are becoming aware of the general methodology of measurement. It is our initial purpose in this chapter to present the basic elements of these ideas.

The role of measurement in any factual science is quite straightforward. *Measurement is the basic link between mathematics and empirical research.* In our applied mathematics research model in Chapter One (Figure 1.5), measurement constitutes the practical steps taken in the translation link from geographical problem to mathematical language. A little thought soon produces a predicament in terms of our normal understanding of the procedure of measuring. Measurement is usually closely associated with quantification. Is measurement not necessary, then, in qualitative mathematics? Our way out of the problem is to redefine our notion of measurement. In fact, this has already been achieved by psychologists in the separation of measurement from quantification in their theory of measurement. This theory allows modern geographers and other social scientists to accept Lord Kelvin's dictum within their own disciplines. Thus, like mathematics, measurement can be quantitative or qualitative. We pursue this matter in the first section of this chapter, with particular emphasis on the measurement of spatial concepts.

A collection of measurements constitutes the data upon which empirical scientists carry out their analyses. Areally based data are, in a very real sense, the raw material of the geographer as spatial analyst. In the second section of this chapter, we concentrate specifically on data sources and collection in geographical research in terms of both problems and techniques.

THE THEORY AND PRACTICE OF MEASUREMENT

It has sometimes been argued that the increase in quantification in geography has been at the expense of quality. What is usually meant by such statements is that some properties are intrinsically unmeasurable and cannot be incorporated into a quantitative geography, even when they are vital to the problem under consideration. Some properties of objects are quantitative; others are qualitative. We might argue with respect to a set of regions, for example, that whereas we can measure their respective areas we cannot measure their respective shapes. It is clear that one region can have more area than another region but it cannot have more shape. Quantity-quality debates in other sciences have always led to the conclusion that the ability to measure some property depends not on the property itself but on how we conceptualize it and what measurement procedure we use. Thus, it is said, everything is measurable, and it is only our ability to measure that varies. This type of argument can be sustained only if we adhere to a broad view of measurement. This is the position we take in this book.

Concept and Measurement

We define measurement as a system of rules that distinguishes one or more of the following relations between two objects A and B with respect to some property x:

Rule

1. $A(x) = B(x)$
2. $A(x) \neq B(x)$
3. $A(x) < B(x)$
4. $A(x) > B(x)$

This definition does not require that measurement involve numbers, and so it is not necessarily quantitative. Let's return to the areas and shapes of a set of regions. We have agreed that we cannot meet rules 3 and 4 for shape—region A cannot have more or less shape than region B—but we can identify whether their respective shapes are identical or different. This constitutes measurement under rules 1 or 2. Shape is not immeasurable, but it is measured at a very simple level. The areas of regions, however, can be related in terms of rules 3 and 4, so that we have more sophisticated measures for area. Thus we see that measurements may be qualitative or quantitative as reflected in terms of some sort of level of measurement.

Before we proceed to describe various levels of measurement, we must remind ourselves of the role of measurement. If measurement is the vital link in our research model, our first task is to consider exactly what it is that measurement is supposed to deal with when we translate data into mathematically amenable form. The measurement part of the translation involves interpretation of the empirical concepts contained in the statement of the geographical problem. Our discussion of geographical measurement will be approached by way of a consideration of empirical concepts.

Empirical Concepts

Empirical concepts are the terms that scientists use to describe the objects and events that they study in the real world. Any problem can be stated in terms of relations between empirical concepts. When scientists wish to make the problem unambiguous, they must carefully define these concepts. Such definition has rendered for discussion three basic types of concept—classificatory, ordered, and variable.

Classificatory concepts are the simplest because they involve little more than identifying objects in terms of class. Thus, land use categories in human geography are classificatory concepts. In rural land use mapping, for example, the empirical concept arable land use is employed simply to identify a field as arable and then allocate that field to the arable land use category on the map. Geographers use classificatory concepts in their systematic studies, ranging from weather types in climatology to functional classifications of towns in urban geography. In spatial analysis, we can identify shape as a classificatory concept. We commonly allocate shapes to classes based on similarities to some ideal shape. In this way, the shapes of settlements have been classified as circular, linear, or starlike.

Sometimes concepts involve more than simple categories. Several classes are often related to one another in a definite order. Thus, as well as carrying out a land use survey, an agricultural geographer might find it necessary to make land quality assessments in a given study area so that soil samples may be classified in a range of types from fertile to infertile. Such classification involves ranking classes in terms of specified criteria and the classes constitute *ordered concepts*. Ordered concepts are often found in research problems involving hierarchies. For instance, an urban geographer might be concerned with the central place hierarchy for allocating towns not by general function such as resort or seaport but by classes ranging downward from metropolitan center to market town or village. Notice that these concepts are usually less general than simple classificatory concepts and that they result often from the development of more specific concepts. Thus, whereas spatial analysts might deal generally with shape classes such as circular or starlike, alternatively they can concentrate on one aspect of shape, such as the degree to which shapes are circular. In such case, *all* shapes would be allocated to ordered classes from negligible circularity or compactness to very high compactness.

Variable concepts are those to which variable quantities can be assigned. We can recognize area as such a concept when we say, for instance, that region *A* has 217.3 square miles. Similarly, although we cannot conceive of more or less shape, the idea of one region being more compact than another is quite plausible. In the example, we have made our concept of shape more specific so that we may move from a general classificatory concept of shape to a variable concept of compactness. We consider below ways of assigning numbers to regions to measure their compactness. We can notice here that variable concepts usually relate to the most specific terms in a research problem. Thus, urban hierarchy may be compared to a concept of population size as recorded by a census in which population totals represent a variable concept. In rural studies we can compare the very specific variable concept of pH levels of soil acidity to the ordered concept of soil quality and to the classificatory concept of land use. Thus, we can identify three basic types of concept likely to occur in geographical problems. To translate them into a mathematical language, we must measure them. This can be achieved by using the developments of measurement theory pioneered largely by psychologists in the last fifty years.

Scales of Measurement

We have used the terms quantitative and qualitative following our definition of measurement. In fact, measurement theory involves a scheme of four levels or scales of measurement that relate very closely to the types of empirical concepts identified above.

The most elementary form of measurement involves simply denoting differences and similarities between objects or events. Such an operation belongs to the *nominal scale* of measurement; in common language it would probably be termed identification rather than measurement. It is measurement according to our definition above, however, because we can distinguish one or more of the relations designated in the first two rules of the definition. By denoting differences and similarities, we can convert classificatory concepts into items on an empirical nominal scale of measurement that can range from simple presence-absence to more complex classifications. An example of presence-absence measurement is the recording of whether or not settlements possess certain facilities such as post offices and hotels. A more complex classification is allocation among land use category concepts of a set of fields within a parish to produce a nominal measurement of land use within the parish.

This form of measurement is essentially qualitative in the operation of identifying objects and allocating them to classes, but once the operation has been set in motion classes build in content, and numbers of objects in each class can be counted to produce quantitative frequencies. (We illustrate in Chapter Three how these quantitative frequencies can be used in statistical analysis.) Nominal scales are the lowest level of measurement in the sense that a nominal measurement gives us least information about an object. All we can say is that it either does or does not belong in a particular class or group.

Ordinal scales of measurement give more information about an object with respect to some property. If we are allocating objects to classes, then the classes themselves are ranked in a specified manner. Ordinal measurement clearly relates directly to ordered concepts. Thus, we may allocate central places, river valleys, or occupation groups to various systems of ranked classes within a hierarchy. In these cases we would be using *weak ordering*. The measurement is weak in the sense that, when objects belong to different classes, we can specify all four of the relationships in our definition of measurement; nevertheless, when objects belong to the same class, we can specify only the first two relationships. When two river valleys are at different levels in the valley system, we can order them in terms of importance, but when two valleys are at the same level in the hierarchy we can specify only that they are equal. This is clearly a higher level of measurement than nominal scales, but we can aim for much stronger ordinal measurement.

If we are able to rank every individual in a sequence, then we have *complete ordering*. For example, we can determine the relative hardness of rocks by actually observing which rock scratches which other rock. Such experimentation with a collection of rocks can produce a complete ordering in terms of the empirical concept of hardness. Another example can be found in migration and perception studies, when questionnaire respondents are asked to rank a set of locations in terms of how much they would like to live in them. Such residential preference scales usually form a complete ordering of locations. Such measurement gives us clearly the maximum amount of information we can expect from any ordinal scale.

The two highest levels of measurement correspond to what we may call measurement in the restricted sense of common language. Such measurement allows direct translation of empirical variable concepts into quantitative number systems. The first is known as the *interval scale*. Its basic characteristic is that objects are ranked not only in terms of some property but also the differences or intervals between objects in terms of that property are known. Three towns with average summer temperatures of 94°F, 93°F, and 50°F can be ordered and the intervals between them known. Interval scales thus give us more information than ordinal scales do. In this particular case, we can see that the first and second towns are similar, differing only by 1°F, but the third town is much colder (by 44°F and 43°F respectively) than the two other towns.

Interval scale is distinguished from *ratio scale* in that it possesses an arbitrary rather than a natural origin or zero. In the summer temperature example, the zero quite arbitrarily being used is in the Fahrenheit scale (it is the freezing point of brine). We might have used another popular scale, the centigrade, establishing another, equally arbitrary zero (the freezing point of water). Rainfall, on the other hand, has a natural zero—no rainfall is the starting point or origin of this variable concept—so that it can be measured on a ratio scale. This means simply that, as the name of the scale suggests, we can use the numbers in statements involving ratios between two measurements. If in a given year one rainfall station receives 50 inches of rain and another receives 100 inches, we can calculate the ratio $100/50 = 2$ and conclude that the second station receives twice as much rain as the first. We cannot make such statements with simple interval measurement. 100°C is *not* twice as hot as 50°C. This can be demonstrated clearly by converting these centigrade degrees of heat into an alternative, Fahrenheit scale (212°F and 127°F), and the conversion obviously does *not* result in a ratio of 2.

This distinction between ratio and interval scales is

Table 2.1 Scales of Measurement

Level	Basic empirical operations	Descriptive statistics
Nominal	Determination of equality	Mode
Ordinal	Determination of greater or less	Median
Interval	Determination of equality or differences of interval	Mean, standard deviation
Ratio	Determination of equality or differences of ratio	Coefficient of variation

Source: S.S. Stevens, On the theory of scales of measurement. *Science* 103:677–680, Table 1, 7 June 1946. Copyright 1946 by the American Association for the Advancement of Science.

obviously very important for our translations into mathematical languages, not least because the abstract symbols resulting from the two scales appear to be isomorphic. In fact, we can notice that the distinctions between all four types of scale are important in terms of techniques available for using the information they provide. This is illustrated clearly by the simple descriptive statistics considered in Chapter One. Table 2.1 summarizes the distinctions between the scales and specifies what descriptive statistics are appropriate at each scale. The rows are cumulative downward: equalities can be determined and modes can be found for all levels of measurement.

Measurement Practice

Our measurement theory has given us scales of measurement, but we must still consider putting this theory into practice in empirical research. Measurement practice has normally been concerned simply with technical problems relating to specific measurement instruments in different disciplines. Let's make some general comments on measurement practice, illustrating them in terms of area and shape measures, before considering the measurement of more fundamental spatial concepts in the next section. It can be shown quite easily that the way in which we measure a concept can have important implications for how we treat the results. This is most explicitly apparent in the distinction between direct and indirect measurement.

Direct and Indirect Measurment

Direct measurement involves a straightforward operation for measuring a concept involving the measurement of no other concept. We might place a ruler alongside a straight line and pronounce on its length. Indirect measurement, on the other hand, involves measuring other concepts in order to derive a measure for the particular concept under consideration. Population density is a good example. We obtain measures for two other concepts, population and area, and their ratio gives us a measurement of population density. Notice that we are using the natural number system and its operation of division to produce our measure. This corresponds to the most elementary use of mathematics as identified in Chapter One. The question "How much?" seems at first an easy one, but practice in measurement soon reveals otherwise. The mathematics involved can range from simple ratios to quite sophisticated mathematics. In Chapter Six we describe factor analytic techniques, sophisticated measurement or scaling procedures. Here we restrict ourselves to simpler examples and implications of various measurement practices.

It must be emphasized that the distinction between direct and indirect measurement involves the measurement procedures and not the concepts themselves. This distinction is unlike the differences in scales of measurement. In fact, the *same* concept can be measured at the same level of measurement *both* directly and indirectly. Measurement of area neatly illustrates this point.

Area can be measured directly if we place a fine grid over a specified region and count the number of cells of the grid that fall within the region. Border cells with more than half their total area in the specified region are usually counted; those with less than half in are not. The final total of cells counted as being within the region represents a measure of the concept of area. Other direct measures of area involve light-ray interception of mechanical scanning of a map. For instance, J.T. Coppock and J. Johnson (1962) have pointed out that if the area to be measured is shaded on a translucent material, the amount of light that is intercepted by the area can be found by using a sensitive photoelectric cell; this amount of light represents a direct measure of the shaded area. Both these approaches produce straightforward and direct measures.

Let's consider the measurement of area in geometry. To find the area of any regular geometric figure, we simply use some given formula. The area of any rectangle, for example, is the product of two adjacent sides, and the area of a circle is πr^2. In these cases, the concept of area is clearly being measured indirectly with measures of distance. This is also the case when we measure areas of irregular shapes with a planimeter. By tracing the perimeter of an area with the wheel of the planimeter, we are actually measuring distance as well as continuously relating the recorded distances to the direction in which the wheel is traveling at any particular point. We then have a measure of area based on the distance and directions the perimeter wheel has traveled. This, too, is indirect measurement.

Let's consider measurement practice with respect to shape concepts. In nominal measurement we must specify rules explicitly for allocating objects to classes. The nature of the rules will determine clearly whether the measurement is direct or indirect. The rules are often so simple that classification is by inspection. For instance, we might classify the shapes of administrative areas into the following classes:

1. Separated: there are detached portions.
2. Punctured: there is one portion and it encloses other areas.
3. Concave: there is one portion with no enclosed areas but indentations.
4. Convex: there is one portion, and there are no enclosed areas and no indentations.

This classification is both comprehensive and nonoverlapping so that all administrative areas can be allocated by inspection to one, and only one, class.

More sophisticated measures of shape require indirect measurement. Consider the concept of compactness that we discussed previously. Several ways have been proposed for measuring compactness and all involve relationships between other spatial concepts. The simplest measures relate properties of the area under consideration to properties of a circle of equal area. Thus, we can compute

$$e = \frac{d}{l}$$

l is the long axis of the shape (the longest line between two points on the shape's perimeter) and d is the diameter of a circle (its long axis) of equal area. (d is thus defined as $2\sqrt{a/\pi}$ where a is the area of the shape.) Notice that a circle scores one and all other shapes score less than one. The more circlelike or compact a shape is, the smaller will be its long axis in relation to the diameter. Thus, the ratio approaches unity. The limiting case in terms of lack of compactness is simply a line, which has no area. In this case the diameter of the circle of equal area is zero, so that e reduces to zero. Hence, the measure ranges from one for a circle (maximum compactness) to zero for a line (minimum or no compactness). This is a simple but quite effective indirect measure of compactness on a ratio scale. The use of this index is illustrated in Figure 2.1 to specify changes in compactness resulting from the political reorganization of part of central Europe after the first and second world wars. The trend toward increasing compactness for Czechoslovakia and Poland is visibly apparent and our measures have quantified these changes. On the other hand, the changing boundary of Germany visually hides the similarities in compactness, which is also shown by the values of e.

The measurement of shape and area illustrate that there is a practical distinction between direct and indirect measurement which is separate and independent from the theoretical scales of measurement identified earlier. It seems reasonable to ask whether this

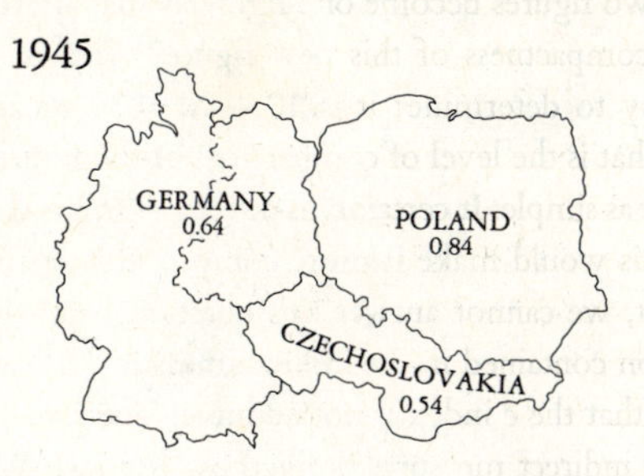

Figure 2.1 Boundary changes in part of central Europe. (Figures refer to compactness as measured by *e*.) The figure for Germany in 1920 is found by excluding East Prussia, and the figure for Germany in 1945 includes both east and west sectors.

distinction is important at all when we measure concepts in empirical research. There are two reasons for answering this query in the affirmative. The first is that indirect measures may compound errors existing in the direct measures they use. (We deal with errors in measurement in some detail below.) The second is that indirect measures have properties relating to the mathematical procedures used in their derivation. We can illustrate these differences by considering how variations in the use of the natural number system affect the resulting measures.

All four operations of the natural number system—addition, subtraction, multiplication, and division—may be used to derive indirect measures. For many ill-defined concepts in social science, it is not always clear which operation is appropriate. We might assume, for example, that the political power of a state is related to its energy consumption, population, and land area. How do we put them together? Should we simply add them up or should we multiply them? The answer has important implications because multiplication favors states (such as the United States) with reasonably high values for all three direct measures while addition favors states (such as China) with only one or two very high direct measures. Without further theoretical specification of how and why political power is related to population, area, and energy consumption, our measures must be interpreted very carefully. Thus, although the final value is based on amalgamation of natural numbers, we probably would not want to place too much weight on exact numerical results. On the face of it, we seem to have achieved the highest level of measurement, the ratio scale, because we have a natural zero: having no population, no area, and no energy consumption certainly means having no political power! Our knowledge of measurement practice tells us something quite different, however. We have seemingly exact results, but we might well decide to treat the scale as an ordinal one. Were we to do so, we could say that our procedure ranks countries correctly, but intervals between the scores would not be meaningful. Given the crudity of the procedure, we might even go as far as to reduce the results to a simple ordered scale that weakly classifies countries by ranked groups such as very powerful, powerful, and so on.

Even when the results from an indirect measurement are acceptable as measures of a clearly specified empirical concept, we must never forget the procedure used to produce the numbers so derived. We can illustrate this point by using shape and area concepts again. Suppose that we have two rectangles. Their respective areas are found by multiplying their sides; let's say that they are of 10 and 20 square inches. Their compactness is measured by using the *e* ratio, which produces results of 0.5 and 0.7. Now let's place the two rectangles together so that they share one side. If we erase this side, the two figures become one figure. What are the area and compactness of this new figure? Area is of course easy to determine; it is $10 + 20 = 30$ square inches. What is the level of compactness of the figure? This is not as simple. It certainly is not $0.5 + 0.7 = 1.2$ because this would make it more compact than a circle. In fact, we cannot answer this question from the information contained in the indirect measure. All we can say is that the e index is not additive. The same is true of all indirect measures derived by division. We cannot add population densities together, for example.

These examples may seem rather simple, but a little thought soon shows that they are of general relevance. The use of percentages is widespread in all types of geography; percentage is a standardized way of presenting simple ratios. Percentages are not usually additive. Furthermore, these examples illustrate the general rule of measurement practice that the method of computation of an indirect measure must be considered in all applications. This is true equally for simple ratios and the more sophisticated measurement procedures we deal with in Chapter Six.

Before we leave the topic of derived measures, we must consider their units of measurement. With direct measures the units are specified beforehand as standard units, such as inches. Units are not always easy to specify after an indirect measure, however. For a measure derived by division, we define new units in terms of the original units. Velocity, for example, is measured as distance (miles) divided by time (hours), and we talk of miles per hour as units of velocity. Similarly, population density is given in persons per unit area. When the direct measures of an object or a concept are in the same units, the resulting derived measure is a pure ratio. It has no units and is dimensionless, as is the coefficient of variation in Chapter One. This is the case with e, our measure of compactness. Furthermore, all proportions (including percentages) are dimensionless because they relate to totals in the same units as the measure under consideration.

Measurement Error: Accuracy and Precision

Quantification is often thought to be a good thing because of a supposed exactness it brings to research. This suggests the unlikely situation of little or no error in measurement. It is more correct to say that the new methodology of geographical research does not necessarily avoid error but does give us more opportunity to identify and rectify any errors that do occur.

Errors that occur in measuring concepts can be of two distinctive types—*compensating* or random and *noncompensating* or systematic. Compensating errors are the easiest to deal with and cause least problems. A good example of the use of the compensating property of some errors is the method of measuring area by counting grid squares in an overlay. It is assumed that, by counting squares that are more than half covered and ignoring those that are less than half covered, the accepted errors that arise cancel one another out so that the final result is a reasonable estimate of the true area. On the other hand, if the procedure consisted of counting only squares that completely cover the area, then errors of omission would not be compensated and the true area would be clearly underestimated.

These ideas are epitomized in the two terms "precise" and "accurate." These are interchangeable in normal usage, but their use in technical measurement requires careful definition. Accuracy is related to the degree of bias in measurement; precision is related to the range of results produced by measurement. If we wish to measure room temperature, for example, we might ask a group of students to read a thermometer. The results will probably differ from student to student. If the actual temperature is 20°C, then most students will read the thermometer as recording a temperature of between 19.5°C and 20.5°C. One or two students might read the temperature as lying beyond these wide limits. If a more finely calibrated thermometer is now brought into use, student readings may vary only between 19.9°C and 20.1°C. We say that the second set of readings is more *precise* than the first set. If there are no systematic biases in these readings (that is, there are as many readings above 20°C as there are below it), then *both* sets of readings are *accurate*. If, however, one of the thermometers is incorrectly calibrated so that the majority of students are recording temperatures between 20°C and 20.1°C, then the measurements are considered inaccurate because of the bias, although the results remain reasonably precise.

If we return to the example of area, we can see that

compensating errors produce accurate results that have no biases but may be quite imprecise. Different placements of the grid, for instance, are likely to give slightly different results. The degree of precision depends on the grid size, which directly controls the sizes of the compensating errors. Noncompensating errors, on the other hand, produce biased results and are thus inaccurate. Their precision continues to depend on the grid size, however. With a small grid size and a procedure for counting only fully covered squares, we would produce results that are precisely wrong!

Imprecision may be troublesome, but normally it can be taken into account and easily overcome. Often, we need only repeat the operation several times and take the average. Inaccuracy is a far greater problem. The researcher who is unaware of it may produce results that are downright misleading. If biases are known to exist, normally they cannot be adjusted easily. We might know that our thermometer always measures at 0.5°C higher than the true temperature, and we need only deduct the systematic error. But such specific knowledge about bias in a measurement procedure is unlikely, and customarily we have to revise our procedure. We must recalibrate the thermometer or start counting some of the grid squares that partially cover an area being measured. A good practical example is the history of measuring rainfall, which consists largely of redesigning gauges to counteract splash.

Problems of error are generally more acute in indirect measurement. This is particularly true of measures derived from multiplication and division. A derived measure is normally less precise than the direct measures upon which it is based. Consider two direct measures with true values of 10 but with an observed range between 9 and 11. The range of results is 20% of the true value in each case, and we can take this as the degree of precision. If we derive a new measure by multiplication, the new range is from 81 $(9 \cdot 9)$ to 121 $(11 \cdot 11)$ around the true value of 100. The degree of precision is thus 40%. A ratio measure derived from these figures produces a range from 0.82 (9/11) to 1.11 (11/9) around the true value of one. The degree of precision is thus 29%. In both cases, the original precision is compounded. Furthermore, systematic biases in direct measures are carried over into any indirect measurement in which they are involved. Consideration of accuracy and precision is therefore doubly important in indirect measurement.

This aspect of measurement practice relates to all scales of measurement, although our examples have had to do with interval and ratio levels. Imprecise ranking refers to an ordering system that involves large random errors; inaccurate ranking occurs when systematic biases enter the procedure. The use of income data to specify social status is usually understood as producing systematic biases if, for example, it consistently ranks some manual occupations (production line work) above nonmanual occupations of higher status but lower pay (priests). The result would be a precise but inaccurate ordinal scale for measuring social status. Similarly, for the nominal scale, rural land use mapping from a hilltop is likely to produce an accurate but imprecise classification, whereas classification by a conscientious student who confuses wheat with barley will give systematic biases and result in a precise but inaccurate classification.

Fundamental Spatial Concepts

Thus far we have been concerned largely with two spatial concepts—area and shape. We used them to illustrate aspects of measurement theory and practice. When we consider spatial concepts for their own sake, we soon realize that there are more fundamental concepts to be identified. Early in the 1960s, John Nystuen considered this problem explicitly when he attempted to define a small, basic set of concepts necessary to build any spatial pattern. He termed these building blocks "fundamental spatial concepts." We need not enter into discussion of his work except to report his conclusions. He identified three fundamental concepts: distance, direction, and a notion of relative position. We consider the measurement of each concept in turn.

Distances and directions are involved in any spatial analysis that uses Euclidean geometry. Other concepts such as area and shape are derived from them. Although they are familiar concepts to geographers and laymen alike, they do not fit easily into general measurement theory and practice as presented above. In fact each concept requires careful individual consideration if it is to be related to existing measurement

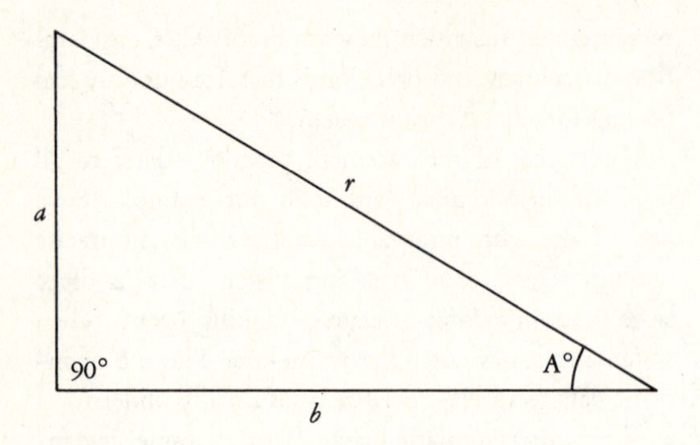

Figure 2.2 Geometric relationships and measures of distance.

theory; for each concept we must isolate its peculiarities important in the translation from spatial concept to mathematical language.

Distance

Distance seems at first to be a very simple concept to measure. Certainly no problems are incurred in its consideration in the context of the regular figures of Euclidean geometry. This ease of measurement is transferred into a geographical context when we deal with concepts such as migration distance, measured as the distance along a straight line joining origin to destination. The geometric solution to this measurement problem involves relations between sides and angles within right-angled triangles. This approach derives from the practical surveying needs originated before the formulation of geometry by Euclid. Thus, for any right-angled triangle, if the lengths of two sides are known then the length of the third side can be found. This, of course, paraphrases the famous Pythagorean theorem:

$$r^2 = a^2 + b^2$$

in which r is the length of the side opposite the right angle (the hypotenuse) and a and b are the lengths of the two other sides (Figure 2.2). In surveying, it is often easier to find the length of one side and one of the angles other than the right angle. The lengths of the two other sides can then be found by the use of elementary trigonometry, the study of the relations between sides and angles of right-angled triangles. It is based on the similar geometric property that ratios between sides remain constant for any pair of similar triangles. These constant ratios have been tabulated for different right-angled triangles and can be found published with tables of logarithms. Constant ratios of sides are termed sines, cosines, and tangents. In Figure 2.2 the following relations obtain for an angle of $A°$:

$$\sin A° = \frac{a}{r}$$

$$\cos A° = \frac{b}{r}$$

$$\tan A° = \frac{a}{b}$$

If we know the value of $A°$ and the length of one side, we can use these trigonomic relationships to specify the triangle completely. The measurement of straight distance in a plane has been studied comprehensively by geometricians, and all salient problems have been solved. Thus trigonometric relations suffice for most surveying and geographical purposes.

We can generalize our observation of the ease of computation of straight distances to include all distances between points on a curve that can be described by mathematical formula. This includes regular, smooth curved lines in a plane as well as straight lines. Such measures are clearly at a ratio level, because they possess the natural origin of zero distance. Not all distances have such straightforword ratio scale measures, however. If a curve is highly irregular, as in the case of many physical and human lines on the Earth's surface, then the measurement of its length is not nearly as simple. Questions such as "What is the length of Norway's coastline?" bring to light all sorts of problems of definition and measurement. Apparent discrepancies begin to appear. Spain describes its border with Portugal as being 987 km long, but Portugal views the same border as 1214 km. Holland has a 380 km boundary with Belgium whereas Belgium has a 449 km boundary with Holland! Does this mean that one or both of the governments cannot measure distance? This seems highly unlikely, and we are left with a conclusion that at first may seem even more unlikely—namely, that neither country of either pair is wrong. We can, in fact, justify this conclusion in terms of the peculiar properties of irregular curves.

When an irregular curve or *empirical distance* is measured by a number of different people, they tend to get slightly different results because of varying degrees of precision. This feature is common to all measurement, as we have suggested previously. We may be able to observe a range of precision from such measurements, but identification of accuracy and bias are more difficult to interpret in this context. This is because there seems to be no true value about which we might expect the measures to cluster in either an accurate or a biased manner. In fact, we find that the more precise we make our measurement, the longer the distance becomes. This unique measurement phenomenon is known as the *Steinhaus paradox*, after the mathematician who first pointed it out. If we make a very rough estimate of Norway's coastline, using a piece of string, we will calculate a distance far shorter than if we steadfastly trace around every little inlet. It is quite clear that the length of the empirical distance depends on the scale of operation or *degree of resolution* of the measuring procedure. The more indentations that are ignored, the shorter the reported line will be until, at the minimum level of resolution at which we can still measure the line, the distance is the shortest (straight-line) distance between the two end points of the line. The search for accuracy in such measures becomes meaningless whereas precision enjoys a peculiar, but important, role in defining the degree of resolution.

The Steinhaus paradox has some rather frightening implications for spatial analysis. If any measurement is to be useful in a scientific context, it must be intersubjective. This means that different people measuring the same property of any object should produce comparable values. When this is not the case, we do not satisfy our definition of measurement (we cannot even be sure of equalities and inequalities) and analysis becomes meaningless. The evidence of the political boundaries suggests that empirical distances may not in fact result in intersubjective measures. There is a theoretically simple way out of this dilemma, however. We have concluded that the problem of empirical distance is related directly to degree of resolution; if we specify the scale, then we might expect that different people measuring the same distance *at the same scale* should produce the same value. By holding the degree of resolution constant, we are able thus to produce intersubjective measures of empirical distance.

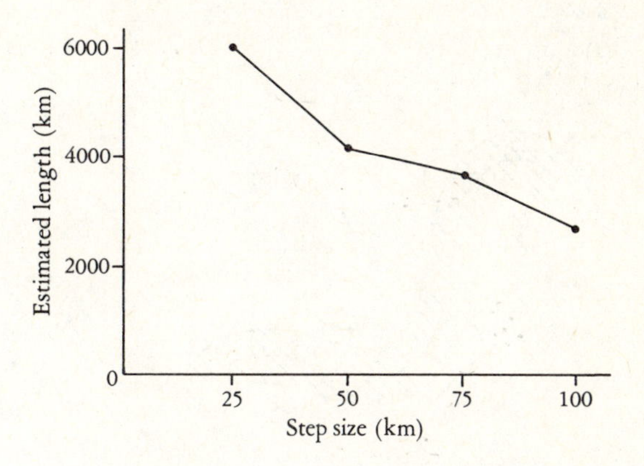

Figure 2.3 The increasing length of Norway's coastline.

Two ways of measuring empirical distance at specified scales have been proposed. The simplest method is to set a pair of dividers at a certain distance and to step along the line so that the measured distance is made up of the number of steps multiplied by the step distance. In this procedure, we have converted our irregular curve into a line made up basically of a finite number of straight lines whose distance measurements are known. The degree of resolution is obviously the length of the lines, the steps, because it is this distance that controls which details of the empirical curve are lost (that is, stepped over in the measurement). The increase in the length of the Norwegian coastline with decreasing step size is illustrated in Figure 2.3.

An alternative approach to measuring empirical distance is to declare the degree of resolution in terms of a circle of radius ε being rolled along the line, its center following the path of the line. Such an operation

(a)

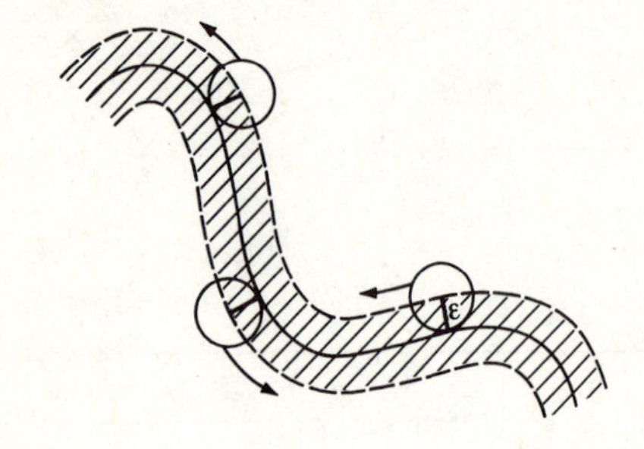

(b)

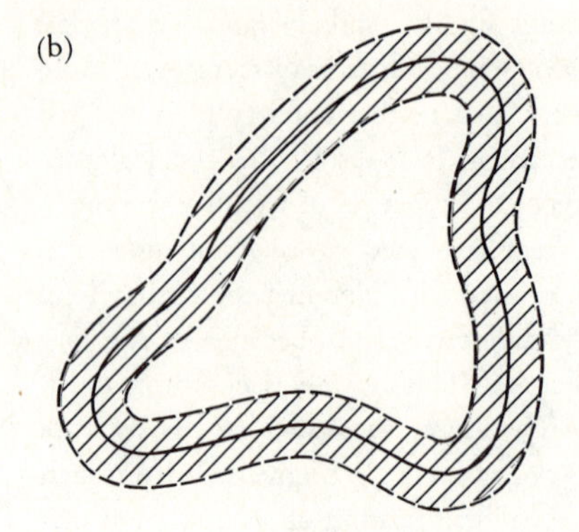

(c)

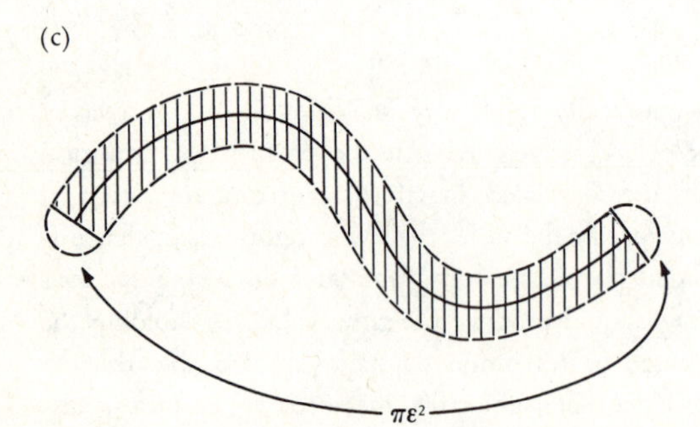

(d)

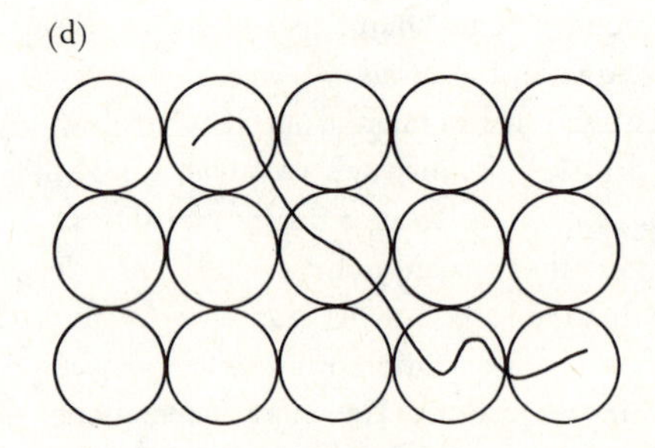

Figure 2.4 Epsilon measurement of distance.

defines an area that can be measured and used to estimate the length of the line (Figure 2.4a). This is done by using the relation between width (w), length (l), and area (a) in a regular rectangle:

$$a = wl$$

$$l = \frac{a}{w}$$

If we write the epsilon length as $l_\varepsilon(x)$ and the area as $a_\varepsilon(x)$, the width being known (2ε), then

$$l_\varepsilon(x) = \frac{a_\varepsilon(x)}{2\varepsilon}$$

This formula is for a closed curve for which there are no ends, such as on a lake shore or an island perimeter (Figure 2.4b). In the case of an open curve such as on a coastline or common political boundary, the two semicircles around each end must be subtracted to estimate the length (Figure 2.4c). Thus,

$$l_\varepsilon(x) = \left(\frac{a_\varepsilon(x)}{2\varepsilon}\right) - \pi\,\varepsilon^2$$

This measure is known as the *Perkal epsilon length*, named after the Polish mathematician who devised it. The scale influence on epsilon lengths is simple to appreciate. The smaller that the circle is, the more indentations there will be that the circle rolls into, and therefore the measurement is more precise and the line becomes longer.

One problem that arises in using this method is that

Table 2.2 Estimates of the Spanish–Portuguese Boundary Using Perkal's Epsilon Method

Radius of circles (km)			
$\varepsilon=1$ (6.4)	$\varepsilon=2$ (12.8)	$\varepsilon=6$ (38.4)	$\varepsilon=8$ (51.2)
955	919	894	821
922	876	841	868
1008	1029	964	867
1001	946	835	841
983	946	833	829
1103	927	887	918
1114	998	989	854
1078	927	845	982
1078	1030	848	841
1014	964	823	829
1099	997	989	854

it relies on the measurement of another empirical concept—area ($a_\varepsilon(x)$)—which is itself subject to problems of degree of resolution. Indeed, many approaches to measuring area rely in part on measuring distance (the perimeter) to estimate area. This can be avoided by employing the epsilon scale. By constructing a square lattice of circles of radius ε and placing it over the empirical line, the number of circles cut by the line can be counted. In the placement example in Figure 2.4d, four circles are cut by the line. This procedure is repeated 2ε times, and the total number of circles that contain a line are counted. Notice that ε must be an integer (whole number) in order to specify the number of placements. This can be achieved by using only selected ε values of whole numbers of kilometers or miles or other units in distance or by using special distance units (say, 6.4 kilometer units). In either case,

$$l_\varepsilon(x) = n - \frac{\pi}{2}\varepsilon$$

in which n is the total number of circles containing a line. Thus, we see that area has been omitted from the calculation and, therefore, the problem of measuring area is avoided. The lattice can be placed in an infinite number of orientations, however, and the 2ε placements actually carried out are not the only set of placements possible. This method tends to produce similar values, but a range of estimates is to be expected. Even though a single level of resolution is used, there will consequently still be some degree of imprecision. In Table 2.2, for example, the Spanish–Portuguese border has been measured eleven times for four sizes of epsilon. The distance units were 6.4 kilometer segments, so that 2, 4, 12, and 16 placements were carried out. Distance estimates therefore increase as ε declines, as expected, although there is also some degree of overlap. The low level of precision exhibited in the table is the result of sampling error, which will be discussed in Chapter Three. A full example is given in Work Table 2.1.

The most interesting feature of Table 2.2 is that, if we consider any one level of resolution, we might find the level of imprecision rather high but at least we have a measure of empirical distance that does in fact behave in a manner similar to other measures. At each scale, the notion of accuracy can be considered, because the measures seem to be fairly evenly spread about central values. The most detailed level of resolution in Table 2.2, for example, gives measures clustering around a value of approximately 1010 km. This level of resolution and the three other scales seem to offer no evidence of bias, so that we may suggest that the measures are accurate but relatively imprecise.

Thus we begin to see empirical distances as similar to other measurements after all. We are now in a position to return to the original, official distances given for the Spanish–Portuguese boundary and interpret them not as incompatible but as related to different levels of resolution. The Portuguese measurement of 1214 km seems to be at a level of resolution much greater than we have attempted in our experiments; the Spanish estimate seems to suggest a degree of resolution equivalent to our ε between 6.4 and 12.8 km. By giving this explanation to seemingly contradictory boundary measurements, we have clearly a solution to an apparent measurement discrepancy.

Direction

Distance exhibits some rather peculiar properties when it is measured. Its Euclidean partner, direction, also has some unusual characteristics that are no less disconcerting.

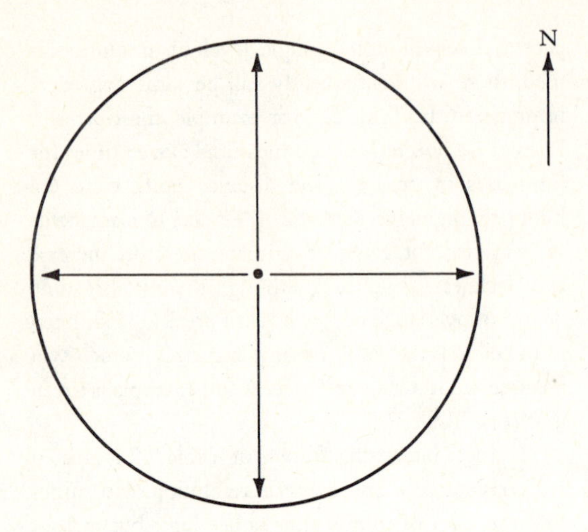

Figure 2.5 Directions of migration flows.

Directions, or bearings, are measured as angles and seem to be directly measurable at a ratio scale. If we perceive angles in terms of the size of the gap between two intersecting straight lines, for example, the idea of "no angle" as no gap between the lines constitutes a natural zero. An angle at 45° is visually and clearly recognizable as half a right angle of 90°. Direction implies something more than an angle of Euclidean geometry, however. All bearings must relate to some frame of reference. A migration flow of 45° tells us nothing unless we know from where the bearing is being taken. All directions are thus relative measures. Because the starting point for measuring bearings as angles is usually the north direction, it may seem that the relativity of direction is unimportant. However, it is possible to have different frames of reference for different research contexts. In intraurban migration, a standard north starting point may mask the flow patterns being sought. Consider the following four flows in a hypothetical city (Figure 2.5). Our measure will show *no* directional biases in the pattern because there are equal numbers of migrants in all four directions. We can observe clearly a directional bias in the pattern however, and we should call this "movement outward from the center." This phrase indicates that our use of the term "direction" in this context refers to angles measured from a frame of reference based on the city center and using a separate line reference (or radial) for each measure. When we employ this frame of reference, we identify a directional bias in which all four flows are outward and no flows are centerward.

Unfortunately, measurement of the concept of direction cannot be seen as simply a relative ratio scale. Like distances, directions have a property that seems unique in measurement theory. When we rotate an intersecting line about the point at which it crosses a frame of reference line, the gap between the two becomes larger, so that the angle increases. The size of the gap reaches a maximum at 180°, however. Using our usual frame of reference, we consider south to be opposite to north, after which the gap decreases again. The bearings 1°, 180°, and 359°, although seemingly real numbers, cannot be translated into the natural number system for mathematical manipulation. 1° and 359° are clearly similar directions and both differ greatly from 180°. The number sequence is not linear but has a unique circular pattern to it. Therefore, in all translations of direction measures into mathematical language, this feature must be taken into account. Ways of overcoming this problem have been devised using vector arithmetic. We illustrate this approach by computing the average of a set of directions.

Consider Figure 2.6a, showing bearings of 60° and 330°. If we treat these as real numbers, we can compute their average as $(60 + 330)/2 = 195°$, which is a southerly direction. This is clearly the opposite of the correct answer, and it results from the circular nature of direction measures. We can overcome this problem by treating each direction as a *vector*, which is defined by a direction and a length or magnitude. We assume unit vectors (magnitude of 1) to simplify discussion. The addition of two vectors can be portrayed graphically as in Figure 2.6b. The *resultant* of this addition is another vector whose bearing (the angle $\theta°$) is clearly the average of the first two vectors. This gives us a graphic solution for finding the average direction. We simply measure, producing an answer of 15°. This procedure is simple when we deal with two directions, but it becomes cumbersome when we have many directions. A direct mathematical solution can be found by using simple trigonometry.

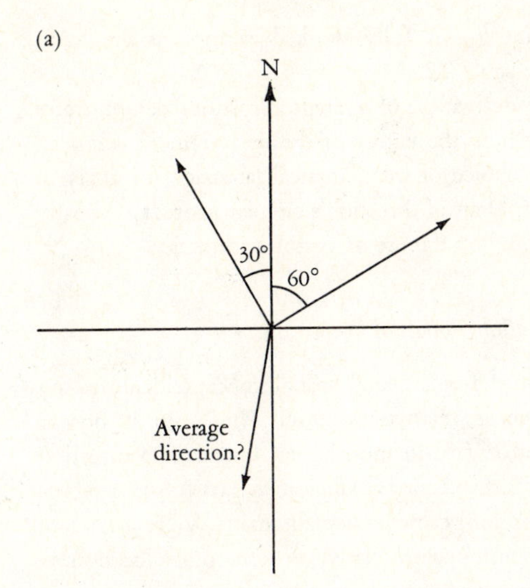

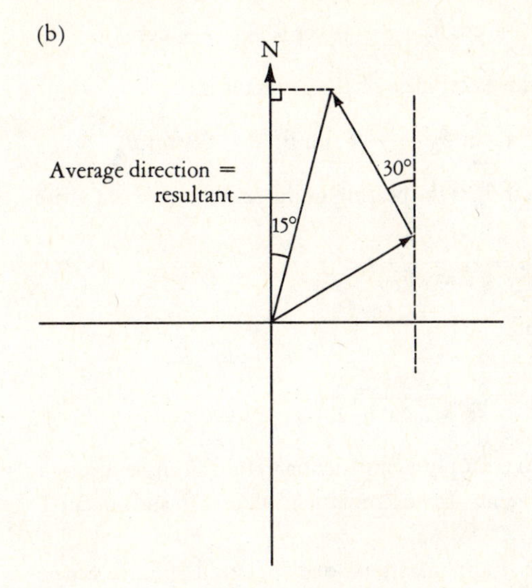

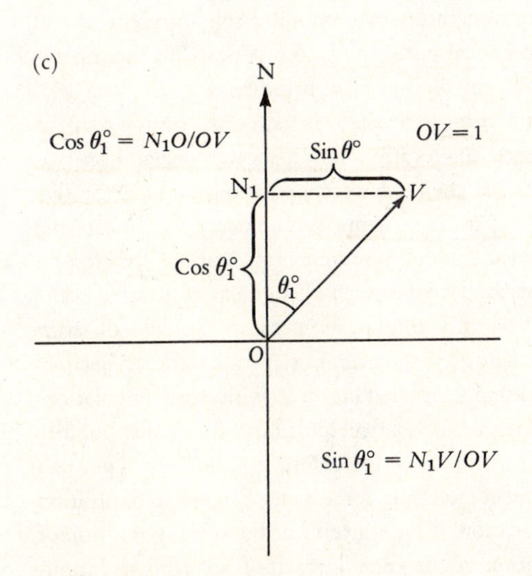

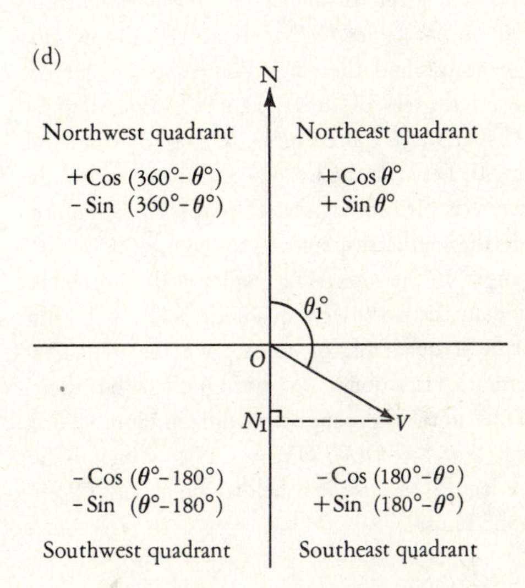

Figure 2.6 Directions as vectors.

Let's view Figure 2.6b in another way. If we construct a right-angled triangle for which the resultant is the hypotenuse, then the tangent of the bearing of the resultant is given by the ratio of the opposite side over the adjacent side. The ratio for the problem at hand can be defined as the distance eastward that the resultant reaches over the distance northward that it reaches. Thus the resultant bearing can be derived by algebraically finding ways of specifying northerliness and easterliness of vectors. Consider the single vector in Figure 2.6c. Its bearing is θ_1°, and its distance is 1 unit. Its easterliness is thus defined by N_1V, which equals $\sin \theta_1^\circ$, because OV = 1. Similarly, its northerliness is defined by N_1O, which equals $\cos \theta_1^\circ$ because OV = 1. Thus we have specified both distances in terms of the original bearing θ_1°. Suppose we have several such bearings $\theta_1^\circ, \theta_2^\circ, \cdots, \theta_i^\circ, \cdots, \theta_n^\circ$. The northerliness of the resultant is

$$\cos \theta_1^\circ + \cos \theta_2^\circ + \cdots + \cos \theta_i^\circ + \cdots + \cos \theta_n^\circ$$

and the easterliness of the resultant is

$$\sin \theta_1^\circ + \sin \theta_2^\circ + \cdots + \sin \theta_i^\circ + \cdots + \sin \theta_n^\circ$$

Thus, if θ_R° is the bearing of the resultant, we can write

$$\tan \theta_R^\circ = \frac{\Sigma \sin \theta_i^\circ}{\Sigma \cos \theta_i^\circ}$$

so that

$$\theta_R^\circ = \arctan\left(\frac{\Sigma \sin \theta_i^\circ}{\Sigma \cos \theta_i^\circ}\right)$$

Arctan is simply an instruction to find an angle through its tangent—for example, tan 45° = 1, and arctan 1 = 45°.

We must mention one practical point. Sines, cosines, and tangents of angles are given only from 0° to 90° in the tables. What, therefore, do we do when we must find the sine, cosine, or tangent of bearings of degrees of more than 90°, as in Figure 2.6d? The answer is quite simple. We use the angle of less than 90° between the bearing and the north-south reference axis. The bearing defined by θ_1° in Figure 2.6d is in the southeast quadrant, so we use $180^\circ - \theta^\circ$ as our angle. In the southwest quadrant, $\theta^\circ - 180^\circ$ is used, and in the northwest quadrant $360^\circ - \theta^\circ$. (In the northeast quadrant, of course, we use θ°.) The trigonometric functions are distinguished by their signs. Thus, in the right angled triangle in Figure 2.6d, $\cos (180 - \theta_1^\circ) = ON_1/OV = ON_1$, which is a negative length, because it is below the origin, O, on the graph. Thus,

$$\cos \theta_1^\circ = -ON_1$$

From the same triangle, N_1V is positive (on the right-hand side of the origin), so that

$$\sin \theta_1^\circ = +N_1V$$

Similar arguments can be made for right-angled triangles in each quadrant. The results of these arguments are displayed in Figure 2.6d. These signs are obviously to be taken into account when we employ the formula above for finding the bearing of the resultant θ_R°. A fully worked example is shown in Work Table 2.2.

The derivation of average directions demonstrates clearly how the nature of the measurement is crucial to the choice of mathematical language for analysis. The problem of direction's circular property is neatly overcome by the use of vector arithmetic.

Population Potential

The third fundamental spatial concept identified by Nystuen is relative position. This term is not as straightforward in meaning as distance and direction are. In fact, it is not at all clear that its converse (absolute position) can ever actually exist. We can measure no location except relative to some other location or locations. Specific notions of relative position have played nonetheless an important role in recent studies in theoretical geography. The outstanding example is the concept of *population potential.*

Population potential is based on the notion that the influence one location has on another depends on two factors: (1) their relative populations and (2) the distance separating them. By influence we mean the general social and economic impact of one location on another, reflected in such phenomena as flows of commodities and migration streams. Studies of such movement have shown consistently that impact declines with distance and that locations with larger populations have greater impact than locations with smaller populations. Such findings are plausible intuitively. They can be incorporated into a measure of the relative position of a location, if we conceive of the relative position of a location as the total impact of all other locations on that location. This is basically the idea behind the measurement of population potential. Thus, we define population potential (v_i) at location i as

$$v_i = \sum_{j=1}^{j=n} \frac{p_j}{d_{ij}}$$

in which there are n locations influencing location i whose populations are given by p_j and whose distances from i are given by d_{ij}. We take each location in turn (from $j=1$ to $j=n$) and compute the ratios p_j/d_{ij}, which

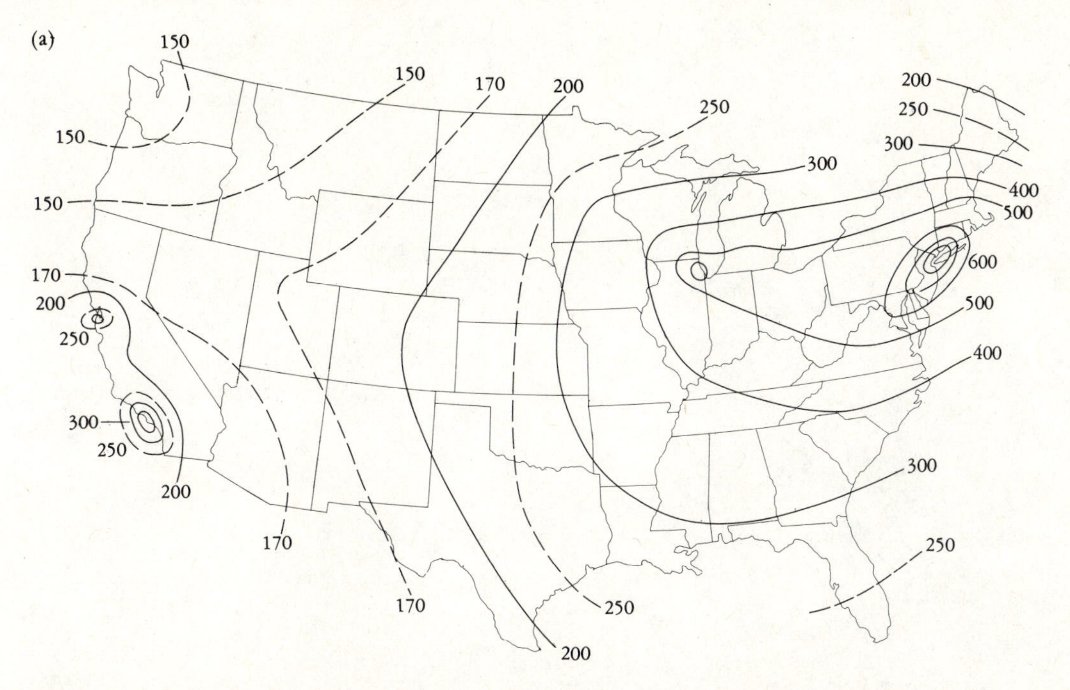

Figure 2.7 Population potential surfaces: (a) United States, 1960 and (b) United Kingdom, 1961.
From David S. Neft. *Statistical Analysis for Areal Distributions*, Monograph Series No. 2. Philadelphia: Regional Science Research Institute, 1966.

are then summed to give the potential at *i*. This result is units of persons per mile. (See the example in Work Table 2.3.) If the population potential is worked out for the set of all *n* locations, we can produce an isoline map of the pattern of population potential, referred to sometimes as a surface of potential. The population potential surfaces of the United States and the United Kingdom are illustrated in Figure 2.7. Notice the way in which both maps neatly identify the economic core areas of each country, usually termed the manufacturing belt and the axial belt, respectively. It should be noticed further that population potential tells us nothing about the particular characteristics of specific locations. It is a macro spatial concept, just as mean center and standard distance are. Thus, a derelict farm in upstate New York has a higher population potential value than the city of Los Angeles. This may seem strange until we remember that we are concerned with position relative to the *whole* of the United States population. In this sense, the derelict farm is clearly closer than Los Angeles to more people in the United States.

Let's now consider some applications of the population potential concept. This measure was popularized by Chauncy Harris in the mid-1950s in his attempts to specify the spatial pattern of the American domestic market. His work illustrates the essential flexibility of the measure. It is flexible in two ways. First, we can vary the populations we deal with, as shown in the descriptive statistics discussed in Chapter One. Second, we can replace the distances by more directly meaningful influences on flows and impacts. In economic studies, distance is usually replaced by transport costs. Harris employed both of these modifications of the potential concept. The overall U.S. market pattern is described by means of a potential formula in which the population is specified as total retail sales and the distances are replaced by estimated land transport costs.

(b)

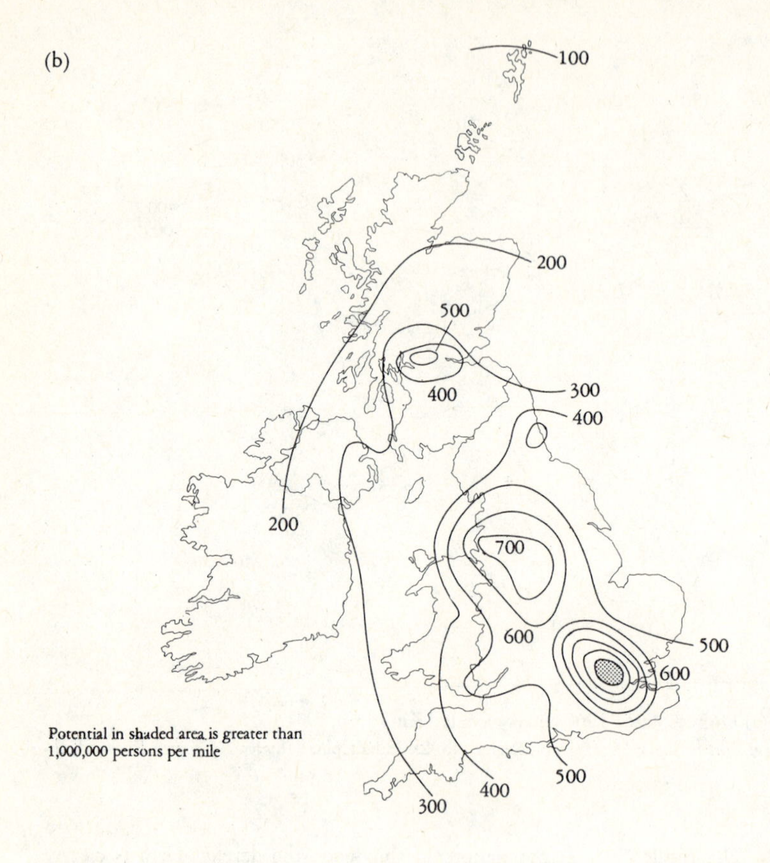

Figure 2.7 *(cont'd.)*

The cost estimates were derived empirically for U.S. transportation conditions early in the 1950s. Thus, for trips of under 300 miles, truck movement is assumed to range from $6 for local trips to $18 per ton for trips of 300 miles. Cost figures for longer distances were similarly estimated for railway shipments. The resulting map of market potential is shown in Figure 2.8a. Values of potential have been converted into percentages of the peak value over New York. This method allows for comparison of patterns of potential defined by different criteria.

Figure 2.8a shows the potential for a measure of the total U.S. market. We can repeat this exercise for specific sectors of the market. Thus, Harris produced maps of manufacturing potential (Figure 2.8b) and farm potential (Figure 2.8c), replacing retail sales by manufacturing employment totals and numbers of farm tractors, respectively. These two new populations produce new patterns of more specialized markets in the United States, the former leading to more emphasis on the manufacturing belt, the latter highlighting the Midwestern market in the agricultural sector.

Further flexibility in the distance factor is illustrated in Figure 2.8d. Here Harris allowed for sea transport between the East and West coasts. He added $18 for terminal costs but then used the very low cost rate of $0.025 per ton mile. Costs were calculated between locations for land and land-sea transport, and the cheapest (that is, "shortest") was used to compute the transport costs. The resulting map for the total market shows a quite distinctive pattern with the West Coast no longer as peripheral and advantages within the manufacturing belt shifting eastward toward the coast.

These examples illustrate the flexibility of this ap-

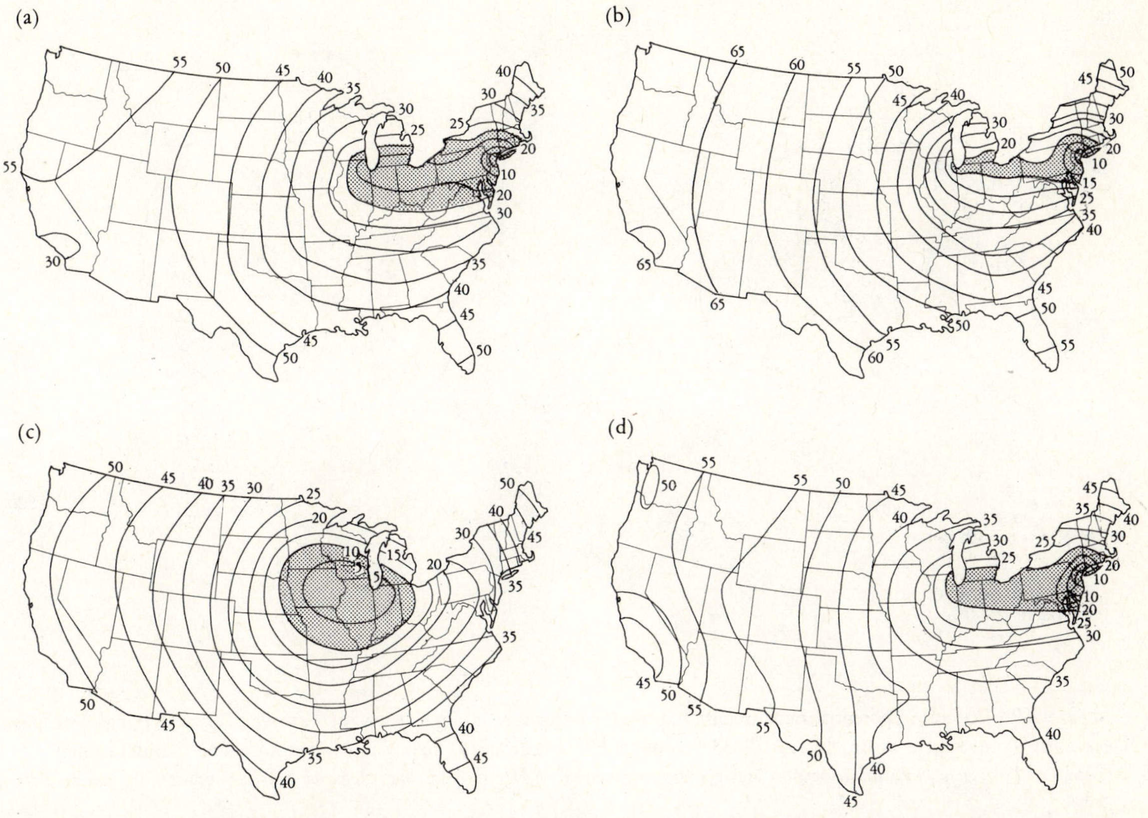

Figure 2.8 Market potentials for the United States. a. Market potential: retail sales (percent below New York City by land transport). b. Manufacturing potential: employment (percent below New York City by land transport). c. Farm potential: tractors (percent below Peoria, Illinois). d. Market potential including sea transport (percent below New York City by combined land and sea transport).

From C.D. Harris, The market as a factor in the localization of industry in the U.S. Reproduced by permission from the *Annals* of the Association of American Geographers, vol. 44, 1954.

proach to measuring relative position. If we avoid strict applications of human population and distance, we can provide ourselves with measures of relative position designed specifically for any problem at hand. The examples above are, however, a little disappointing; they are illustrative rather than convincing demonstrations of the quantitative approach to relative position. We conclude this section with an example of research concerned directly with problems of changing relative position.

One criticism that can be made of Harris's work is that he was concerned solely with the national market whereas we are all aware that much American production finds its way to foreign markets. In fact, Harris concluded by considering how much the patterns on his maps would be changed if Canada were included in the analysis. This sort of question is much more intriguing if we cross the Atlantic and consider western Europe, where there has been conscious effort to increase the size of national markets by the institution of the European Economic Union or Common Market. This union is reflected not only in a larger total home market but also the cutting of tariffs between member countries produces a change in the pattern of relative location within the area. The economic union is highly favorable to some areas and less favorable to

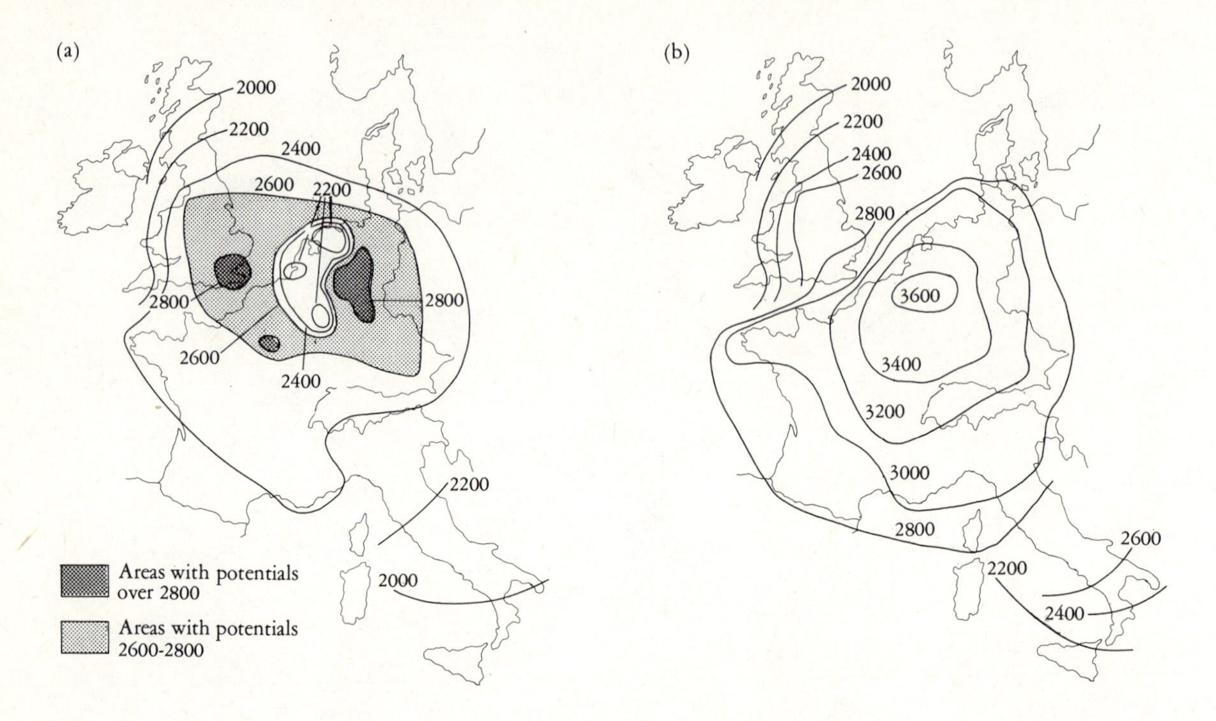

Figure 2.9 Economic potential experiments for western Europe. a. Pre-Treaty of Rome. b. The original European economic community. c. An enlarged European economic community, including Norway. d. The channel tunnel.
From C. Clark, *et al.*, Industrial location and economic potential in Western Europe. *Regional Studies* (London, Pergamon Press), 1969.

others. These differences in relative gains from the Common Market are obviously of considerable interest to prospective members of the union. Thus the debates on the entry of Britain to the Common Market in the 1960s included predictions of Britain's becoming an isolated island on the fringe of Europe in stark contrast to its earlier location at the center of a worldwide empire. This is an interesting problem in changing relative position.

Economist Colin Clark and his associates have attempted to predict changing patterns in relative position in Europe by using economic potential concepts. Clark defines economic potential specifically with gross regional income as the population and total transfer costs as the distance. The transfer costs are of particular interest, because they allow for experimentation. The costs include three elements: handling and local distribution, transport costs, and tariffs. Manipulation of tariffs allows for assessment of the influence of economic union on relative positions. Manipulation of transport cost allows for assessment of the effect of transport innovations. Clark and his associates carried out five experiments, four of which are reported here. Transfer costs were estimated empirically in a way similar to Harris's, and all figures relate to 1962 conditions. From western European countries, 103 regions were used as the areal units for computing potentials, although all parts of the world contribute to these potentials in the calculations.

The first experiment assumes that there is no European Economic Union. Thus, transfer costs between all regions in Europe incur the tariff costs for crossing any national boundary. The result is a patchwork pattern of economic potential (Figure 2.9a) with three

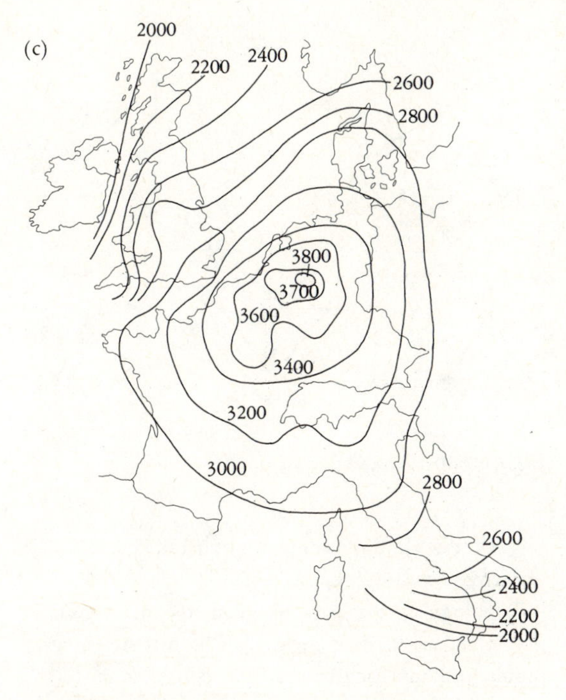

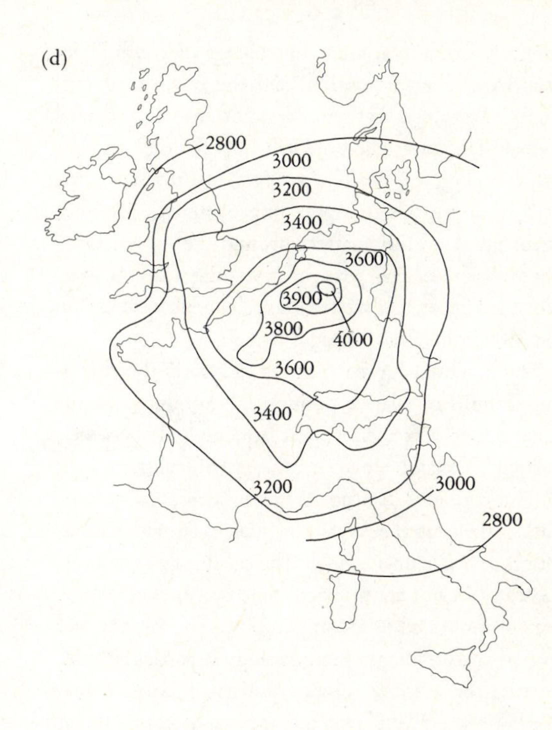

Figure 2.9 *(cont'd.)*

separate peaks in Germany (the Ruhr), Britain (London), and France (Paris). The smaller countries Belgium, Holland, and Luxembourg are represented by hollows in the surface because of their small home markets. This is our base map for comparison of changes in relative position. Notice in particular that the 2,800 isoline defines the three peaks of potential.

The second experiment assumes the situation that prevailed in the 1960s, with France, Germany, Italy, Belgium, Holland, and Luxembourg all in the economic union and Britain outside it. In the context of our economic potential measure, this means that transfer costs for regions within the Common Market do not incur tariff additions. Because the distance element in the potential is lessened for Common Market areas, we should expect higher potentials here. This is confirmed in Figure 2.9b. This map has a much smoother pattern and the whole of the Common Market, with the exception of middle and southern Italy, is above the 2800 isoline. The position of Britain has hardly changed but rather than being a peak of potential the London area now becomes equated with central Italy in terms of economic potential. The major change is in the small countries of Belgium, Holland, and Luxembourg as they now form the peak of potential in the lower Rhine Valley. These two maps describe emphatically the change in relative position in terms of economic location that has been brought about by the Common Market.

The third experiment sets up an enlarged European Economic Community that includes Ireland, Denmark, Norway, and Britain. With the exception of Norway, this situation has existed since 1973. In calculating potentials, no tariff costs between regions of countries are included in the enlarged union. The result is not startling (Figure 2.9c); although Britain's potential rises, the formerly perceived concentric pattern about the lower Rhine is confirmed. In fact, the addition of Britain ensures the centrality of the lower Rhine area by adding a northeast element to the Common Market's configuration. Thus, the peak of potential rises to 3,800.

Finally, Clark carried out two experiments to assess the effects of improved transport. One changed the transport costs to account for savings that would accrue from containerization, and the other used estimates of savings that would result from a Channel tunnel. The economic potential map for the tunnel is shown in Figure 2.9d. The potential of Britain is raised somewhat by this transport improvement, but once again the concentric pattern around the lower Rhine is emphasized. The peak of potential rises to 4000 while a large area of western Europe is enclosed within the 3600 isoline that excludes Britain.

The potential approach shows us clearly the changing patterns of relative position in varying economic situations. The predictions that Britain would become a peripheral island are to some extent substantiated; Britain certainly does not seem to be able to become part of the central peak of potential, even with the addition of a Channel tunnel. The substantive results of the study are not our principal concern, however. What we are interested in are application of technique and overall methodology. Methodology is particularly interesting, because it involves a strategy of experimentation: we set up different situations and measure their effects on patterns of relative position. This represents probably the most sophisticated approach to the study of relative position.

Topological Concepts

We have treated Nystuen's fundamental concepts as measurement exercises in Euclidean geometry. Distance, direction, and relative position, as we have defined them, clearly can have no place in any topological approach to measurement. Geographers have found use for topological spatial concepts in one particular research context, however. Consider the railway network in Figure 2.10a. How far is it from station A to station B? In Euclidean geometry, this can be specified as an exact distance. To the train traveler, the obvious answer may be that there is a direct rail link with no stops between A and B. This answer refers simply to connections between points, not to their distance or direction. It can be formulated in a topological framework. We consider Nystuen's spatial concepts, especially relative position, within

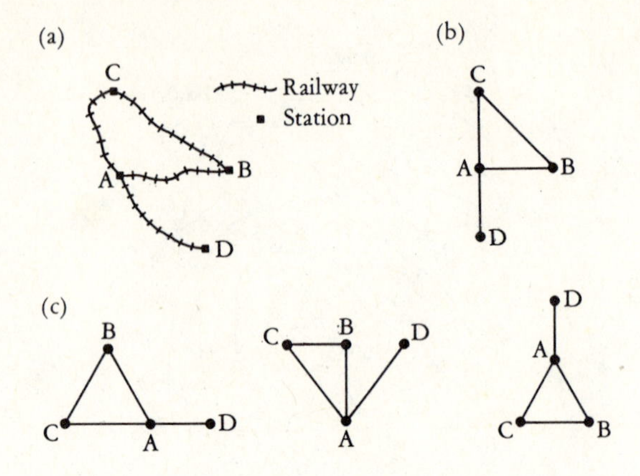

Figure 2.10 Abstraction to graphs.

this framework in the context of what has come to be known as *network analysis.*

Topology is the part of geometry that reduces everything to its bare essentials. The first step in any topological approach to networks is to abstract them to a simple pattern of points for intersections or stations and lines for routes or connections. The railway network in Figure 2.10a can be viewed as the diagram shown in Figure 2.10b. Here all that is maintained is the relationship between lines and points; it shows which point is connected to which other point. This diagram is known as a topological graph. The topology that deals with such line and point patterns is usually referred to as *graph theory.*

We must distinguish now between *planar* and *nonplanar* graphs. The graph in Figure 2.10b is planar—it can be depicted on a plane. Stream networks, railway networks, road networks, and many line patterns studied by geographers can be modeled as planar graphs. Consider an airline network, on the other hand. Airline routes commonly cross one another without being connected because they are separated in the third dimension. When they are depicted on a plane, their line junctions do not necessarily represent points of contact. Such graphs are nonplanar. It follows that a planar graph must have a point recorded at every line junction and that a nonplanar graph can have junctions with no point on the plane diagram. Planar

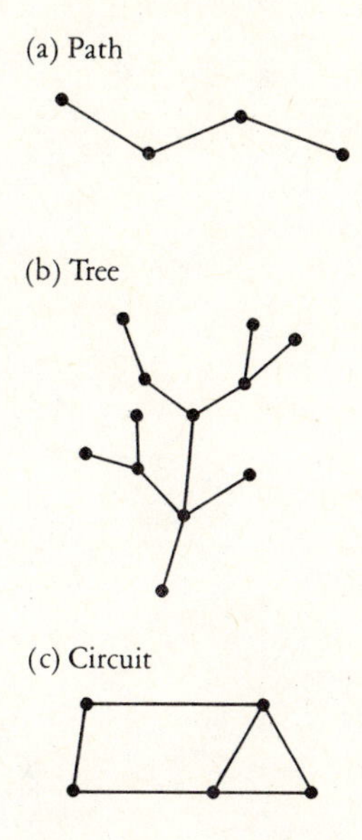

Figure 2.11 Basic types of graphs.

graphs are more common in geographical research than nonplanar graphs. In what follows, we discuss only planar examples and equations, although the equivalent equations for nonplanar graphs are given in the summary table at the end of the discussion (Table 2.6).

In graph theory, distance, direction, and all other Euclidean concepts are ignored. All the graphs in Figure 2.10c are equivalent topologically to Figure 2.10b, despite their obvious Euclidean differences. In graph theory, only three types of basic structure are identified: paths, trees, and circuits (Figure 2.11). A path is a series of one or more lines connecting points such that each vertex is connected to only one other point—that is, there are no branches (Figure 2.11a). A branching network or tree has no closed set of lines—that is, you cannot get back to where you started without retracing your steps (Figure 2.11b). A circuit graph has one or more closed loops (Figure 2.11c). The example in Figure 2.10 can be identified as a simple circuit graph. In Figure 2.12, the development of the Nigerian railway network is shown as it evolved from a simple path to a tree and then to a circuit network.

How can abstract graphs contribute to the measurement of spatial concepts? Let's consider the general spatial notion of nearness. On a Euclidean plane, the nearness of two points is defined by straight-line distance, the shortest route between them. On a topological graph, nearness is also defined by the shortest route but here it is measured by numbers of links between two points. Thus, in the simple graph in Figure 2.10b, point C is nearer to A than it is to D because it is one link from A but two links from D. We can find the shortest path between two points on any graph to define their nearness in this topological sense. Given shortest paths, two measures of the relative position of points can be defined. The first is the most elementary and is known as the *associated number* of a point. This is simply the longest "shortest path" from a point. The associated number for B in Figure 2.10b is two—the shortest paths to C and D. Similarly, D and C have associated numbers of two. However, A has an associated number one because it is connected directly to the three other points. Thus, the more accessible that a point is, the lower is its associated number. This simple measure is far from satisfactory, even in the example. From simple inspection of the graph, we might suggest that C and B are more accessible than D because they both have two links with other points whereas D is linked only to A. In fact, by concentrating on one of the shortest paths, the associated number does not take into account all the relevant information contained in the graph. This criticism can be overcome easily by computing the index of accessibility (A_i) given by

$$A_i = \sum_{j=1}^{j=n} d_{ij}$$

in which d_{ij} is the shortest path from point i to point j. In our example, we find that $A_A = 3$, $A_B = 4$, $A_C = 4$, and $A_D = 5$, which corresponds to our intuitive notions of accessibility on the graph—A is most accessible, D is least accessible, and C and B are equally accessible between A and D.

When we deal with graphs larger than our simple four-point example, A_i cannot be obtained by simple visual inspection. With very large graphs, computer

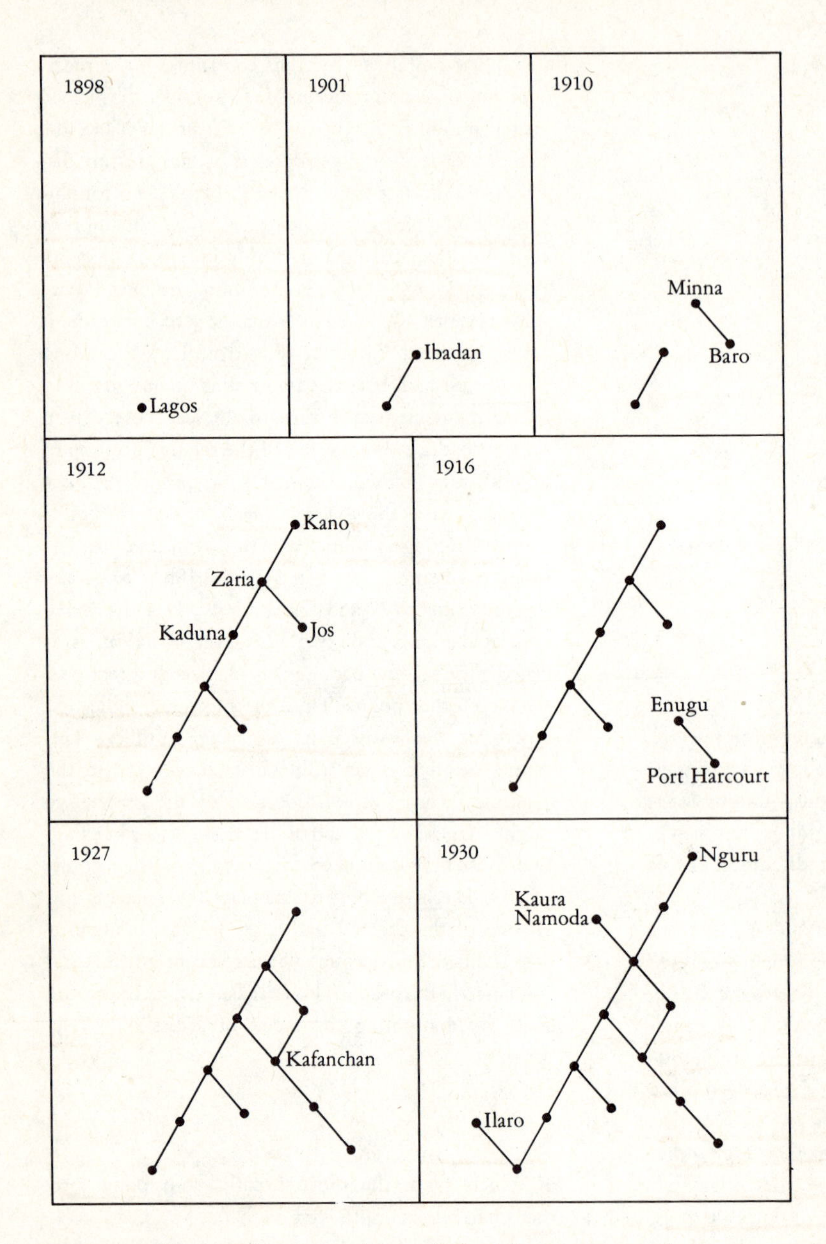

Figure 2.12 The topological evolution of Nigeria's railway network.

assistance is required first to find the shortest paths and then to compute accessibilities. With medium-sized graphs, such as for the Nigerian railway network, accessibility can be computed from a shortest-path matrix, as in Table 2.3. The sums of the columns are the respective A_i values. These can be located on the graph and the pattern of accessibility can be drawn. This is illustrated for the Nigerian network in Figure

Table 2.3 Shortest-Path Matrix for the Nigerian Railway Network (1930)

	1	*2*	*3*	*4*	*5*	*6*	*7*	*8*	*9*	*10*	*11*	*12*	*13*	*14*
1 Lagos	0	1	2	3	3	4	5	5	5	6	4	1	5	6
2 Ibadan	1	0	1	2	2	3	4	4	4	5	3	2	4	5
3 Minna	2	1	0	1	1	2	3	3	3	4	2	3	3	4
4 Baro	3	2	1	0	2	3	4	4	4	5	3	4	4	5
5 Kaduna	3	2	1	2	0	1	2	2	2	3	1	4	2	3
6 Zaria	4	3	2	3	1	0	1	1	3	4	2	5	1	2
7 Kano	5	4	3	4	2	1	0	2	4	5	3	6	2	1
8 Jos	5	4	3	4	2	1	2	0	2	3	1	6	2	3
9 Enugu	5	4	3	4	2	3	4	2	0	1	1	6	4	5
10 Port Harcourt	6	5	4	5	3	4	5	3	1	0	2	7	5	6
11 Kafanchan	4	3	2	3	1	2	3	1	1	2	0	5	3	4
12 Ilaro	1	2	3	4	4	5	6	6	6	7	5	0	6	7
13 Kaure Namoda	5	4	3	4	2	1	2	2	4	5	3	6	0	3
14 Nguru	6	5	4	5	3	2	1	3	5	6	4	7	3	0
Total $(= A_i)$	50	40	32	44	28	32	42	38	44	56	34	62	44	54

2.13, which shows a simple concentric pattern of declining network accessibility about Kaduna, the most accessible place on the network.

The index of accessibility allows us to describe the relative positions of individual points and the pattern of accessibility within the network. When we wish to know the overall accessibility of a network so that it can be compared with other networks, the concept we measure is *connectivity*. We ask: "How well-connected are the points in the network?" There are several ways of answering this question, all of which use the three properties of the network: e = the number of lines or edges, v = the number of points or vertices, and p = the number of separate subgraphs. We illustrate these measures in Figure 2.14 by considering the whole range of four-point examples which are shown.

The simplest way of measuring connectivity is to compute the *cyclomatic number* (μ), defined by

$$\mu = e - v + p$$

This tells us how many basic circuits there are in the network and, for simple networks, it can be found by inspection. Thus, paths and trees score 0 and the value increases as the number of circuits increases. With our four-point example, μ rises to 3, the maximum number of circuits possible with four points in a planar graph. There is an obvious deficiency in this measure—it confuses the concept of connectivity with size. The larger the network is, the higher its cyclomatic number can become. We may argue that connectivity is intrinsically related to network size, but it is generally more useful to have a *relative* measure of connectivity that makes allowance for size rather than an absolute measure such as the cyclomatic number. We consider the cyclomatic number to be a measure of absolute connectivity, and we need an alternative measure of relative connectivity. This is given by the alpha index (α), which is the ratio of the number of circuits (μ) to the maximum number of circuits that could occur given the number of points. The alpha index ranges from 0 to 1, where the maximum number of circuits obtains. The equation for alpha is

$$\alpha = \frac{\mu}{2v - 5}$$

For the four-point example, we have a maximum of $(2 \cdot 4 - 5) = 3$ circuits. Our examples in Figure 2.14 illustrate α from 0 to 1.

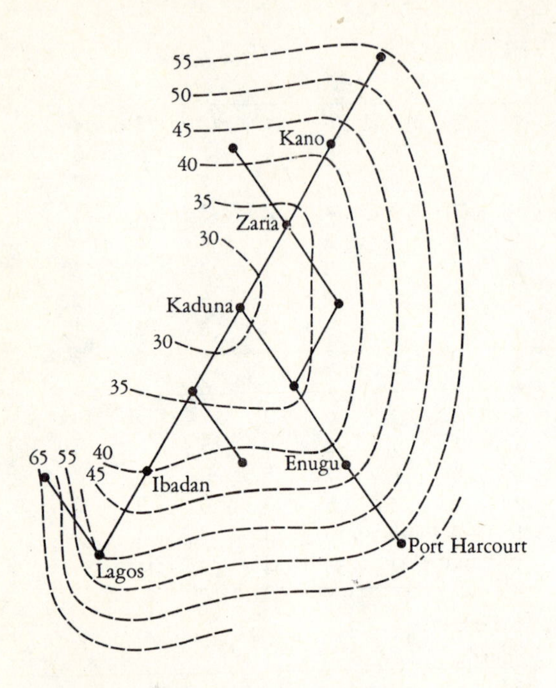

Figure 2.13 The pattern of accessibility on the Nigerian railway network. Isolines refer to points on the network in terms of values of indices of accessibility.

Although alpha is a relative measure of connectivity, it incorporates one limitation from its association with the cyclomatic number. All paths and trees automatically score 0 on its scale because they have no circuits by definition. We would normally consider a tree to be more connected than a path, which in turn is more connected than a pattern of no links between points, as in the first of our four-point examples. There are situations in which we wish to differentiate between these types of graph, and for this μ and α are unsuitable. The simplest way of overcoming the problem is to compute the beta index (β) as the ratio between the number of lines and number of points. Beta is defined as

$$\beta = \frac{e}{v}$$

in units of points per line. In this case the nonconnected graph continues to score 0 but the paths and trees are ranged between 0 and 1. All circuit networks have a score of 1 or more. Thus, beta differentiates between precircuit situations as well as circuit graphs. The only problem with this index is that its maximum value varies with the number of points. In a four-point example, the maximum is 1.5, but with larger graphs β can reach higher values. This problem is overcome by using a strategy similar to that for the alpha measure. We define a gamma index (γ) as the ratio of the number of lines to the maximum number of lines possible given the number of points. The equation for gamma is

$$\gamma = \frac{e}{3(v - 2)}$$

in which $v > 2$. Since e cannot be greater than its maximum, this index, like alpha, ranges from 0 to 1.

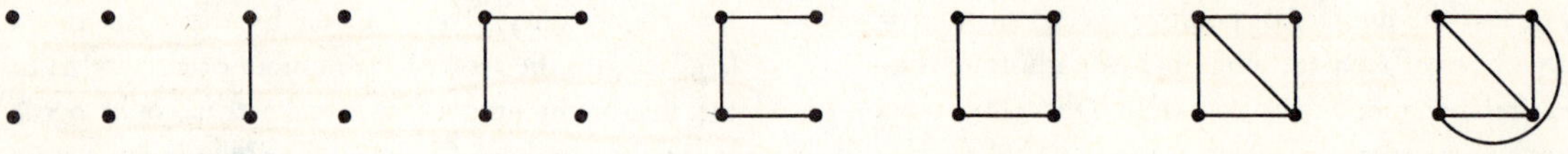

$v = 4$	$v = 4$	$v = 4$	$v = 4$	$v = 4$	$v = 4$	$v = 4$
$e = 0$	$e = 1$	$e = 2$	$e = 3$	$e = 4$	$e = 5$	$e = 6$
$p = 4$	$p = 3$	$p = 2$	$p = 1$	$p = 1$	$p = 1$	$p = 1$
$\mu = 0$	$\mu = 0$	$\mu = 0$	$\mu = 0$	$\mu = 1$	$\mu = 2$	$\mu = 3$
$\alpha = 0$	$\alpha = 0$	$\alpha = 0$	$\alpha = 0$	$\alpha = 0.33$	$\alpha = 0.67$	$\alpha = 1.00$
$\beta = 0$	$\beta = 0.25$	$\beta = 0.50$	$\beta = 0.75$	$\beta = 1.00$	$\beta = 1.25$	$\beta = 1.5$
$\gamma = 0$	$\gamma = 0.17$	$\gamma = 0.33$	$\gamma = 0.50$	$\gamma = 0.67$	$\gamma = 0.83$	$\gamma = 1.00$

Figure 2.14 Four-point graphs.

Table 2.4 Parameters and Measures of the Nigerian Railway Network (1898–1930)

	1898	*1901*	*1910*	*1912*	*1916*	*1972*	*1930*
e	0	1	2	7	8	11	14
v	1	2	4	8	10	11	14
p	1	1	2	1	2	1	1
μ	0	0	0	0	0	1	1
α	0	0	0	0	0	0.59	0.43
β	0	0.50	0.50	0.88	0.80	1	1
γ	—	—	0.33	0.39	0.33	0.41	0.39
γ'	0	0.03	0.06	0.19	0.22	0.31	0.39

γ' is based on the final set of points—that is, $v = 14$ for all calculations.

Unlike alpha, however, it enables us to differentiate between paths and trees, because it is based on lines rather than circuits. In many ways, gamma is the most useful of our measures of connectivity.

The discussion indicates that connectivity is an "over-identified" concept. We have one concept but several ways of measuring it. This can be a dangerous or, at least, a misleading luxury. It means that we have to refine our notion of connectivity to incorporate several different concepts. Notice that it is measurement practice that has forced the redefinition, not the geography that was originally the source of the empirical concept. We can best describe these concepts as follows:

cyclomatic number	absolute circuit connectivity
alpha index	relative circuit connectivity
beta index	simple linkage connectivity
gamma index	relative linkage connectivity

In any research context, we must consider exactly what it is we wish to measure before we choose a concept. We consider an example with this in mind.

We have already introduced Nigeria's changing railway as an example of network evolution. Our purpose now is to monitor the changing connectivity in the network for the years from 1898 to 1930. The changing parameters (e, v, p) and measures $(\mu, \alpha, \beta, \gamma)$ for the graphs are shown in Table 2.4. This table shows that circuit connectivity is a particularly inappropriate concept because a circuit is merely the end product of an evolution through path and tree stages. Hence, μ and α record no changes in connectivity until 1927. The beta index, on the other hand, closely monitors the changes from path through tree to circuit. We see a major change from tree to circuit in 1927, reflected in a sharp rise in β, and similarly the earlier change from path to tree in 1912. Notice also the fall in β as the network changed from a single tree to a tree and a path in 1916. This index records no change in connectivity when the network changes from one path to two paths in 1910.

Let's consider the gamma index for this example. If we use e and v frequencies from Table 2.4, we find that we cannot produce values for 1898 and 1901, when there were only one and two points, respectively. A set of fluctuating results can be determined for 1910 onward, however. These particularly emphasize the ups and downs of our connectivity measures for β and γ values, despite the fact that there are no rail closures. We might reasonably expect the measures consistently to increase in value as new lines are built. The builders of the railway were certainly not adding lines intentionally to reduce the network's connectivity. The reason that connectivity measures sometimes decline is that they are relative to the number of points that exist at that period in time. Gradually increasing connectivity is produced if we consider *all* points that were finally brought into the network for each date. We set $v = 14$ for all descriptions, as we have done in recalculating gamma as γ' in the bottom row of the table. These results measure the increasing linkage

Table 2.5 International Comparisons of Railway Network Connectivity Using the Beta Index

Multiple circuits		*Single circuits*		*Trees*	
Poland	1.43	Iraq	1.00	Malaya	0.95
France	1.41	Nigeria	1.00	Ghana	0.92
Hungary	1.39	Sudan	1.00	Bolivia	0.91
Czechoslovakia	1.37	Thailand	1.00	Iran	0.90
Chile	1.35			Ceylon	0.87
Sweden	1.29			Angola	0.66
Cuba	1.21				
Yugoslavia	1.17				
Rumania	1.16				
Bulgaria	1.11				
Mexico	1.06				
Tunisia	1.05				
Algeria	1.03				
Turkey	1.03				
Finland	1.02				

Source: K. J. Kansky. *Structure of Transportation Networks.* Geography Research Paper No. 84. Chicago, 1963, Table 8, p. 56.

Table 2.6 Measures of Connectivity for Planar and Nonplanar Graphs

Symbol	*Measure*	*Planar equation*	*Nonplanar Equation*	*Description*
For individual points				
AN	Associated number	$AN_i = \max d_{ij}$	$AN_i = \max d_{ij}$	Longest "shortest path" from point i
A	Accessibility index	$A_i = \Sigma d_{ij}$	$A_i = \Sigma d_{ij}$	Sum of shortest paths from point i
For complete network				
μ	Cyclomatic number	$\mu = e - v + p$	$\mu = e - v + p$	Number of basic circuits in the graph
α	Alpha index	$\alpha = \frac{\mu}{2v - 5}$	$\alpha = \frac{\mu}{(v(v-1)/2) - (v-1)}$	Number of observed circuits as a proportion of the maximum number of circuits, range 0–1
β	Beta index	$\beta = \frac{e}{v}$	$\beta = \frac{e}{v}$	Ratio of lines to points, range varying with v
γ	Gamma index	$\gamma = \frac{e}{3(v - 2)}$	$\gamma = \frac{e}{v(v - 1)/2}$	Number of observed edges as a proportion of maximum number of edges, range 0–1

d_{ij} is the shortest path between points i and j; e is the number of lines or edges; v is the number of points or vertices; p is the number of subgraphs.

among all points at every date, and consequently we are able to see a consistently increasing trend. It is clear that the number of points we are interested in comprise the total number finally brought into the network. Thus, the values shown in the bottom row of Table 2.4 are based on finally connected points (that is, 14) rather than the number connected at any one time. Such a measure is useful because all values are relative to the final network rather than to their own particular set of connected places.

In comparative studies of networks in different countries, there is no way of making our measures relative to some final set of points. In studies of this type, we normally use the simple β ratio. Such comparison is shown in Table 2.5, in which the β values for the railway networks of 25 selected countries are shown. Nigeria is among the four countries that have one basic circuit and hence have beta values of unity. Most of the countries have multiple circuit networks with beta values ranging up to 1.43 in Poland. There are also several countries whose railway networks have no circuits. Beta values for these range down to Angola's 0.66, indicating several tree networks but no simple path railway networks ($\beta < 0.50$). It is apparently not accidental that the highest value found is for a European country and the lowest for an African country: it seems highly plausible that railway network connectivity should be related to level of economic development. (In Chapter Five we show how these beta values can be used to test this hypothesis.) All the measures of relative position presented above are laid out along with the equivalent set of equations for nonplanar graphs in Table 2.6.

DATA SOURCES AND DATA COLLECTION

The measurements of spatial concepts discussed above have all proceeded on the assumption that we have some published map to use as a basic source of data to translate into measurement. We consider here data sources and data collection in geography. Geographers use more than simply spatial concepts, of course, and they are interested in data sources other than published maps. We can distinguish two fundamentally different types of data source used by geographers—field observations and archival material. Some geographers, particularly physical geographers, have advocated for a considerable period of time fieldwork as the basic data source. *Field observations* have been and remain important in geographical research. The nature of field observation has undergone considerable change, however. Sir Dudley Stamp's mammoth pre-World War II Land Utilization Survey of Great Britain recorded the land use of *every* field in the country; since that time, total enumeration of information relevant to a research topic has given way to sampling approaches.

Sampling is not new to geography, of course; it includes the traditional case study method in which a few typical examples of quite numerous phenomena are chosen for study. We can term this approach purposive or hunch sampling. A study of agricultural patterns in the American Great Plains may be carried out by choosing a few typical farms that are assumed to illustrate reasonably the farming variations in that area. A problem arises when the typicality of the examples is challenged. The degree to which a case study is typical of the region it represents is a function of the ability and experience of the researcher. Without recourse to further data collection involving other approaches, there is no way of satisfactorily checking how far conclusions drawn from a case study can be extended to a wider context. This problem can be overcome by *probability sampling*, which may be defined as a sampling procedure in which individuals or sampling units are selected in such manner that each individual has a known chance of appearing in the sample. Given this property, we can relate conclusions based on sample data to the situation existing in the total population of individuals. (This approach is a prerequisite for much subsequent analysis that involves probability theory and inferential statistics as we describe them in Chapter Three.) Probability sampling complements several recent trends in geographical research and it is increasingly applied. But probability sampling and field observations as a whole still lag far behind *archival sources* in terms of actual use in research.

An archive is a data source provided directly for the researcher, usually by some government institution. In archival use, collection and sometimes measurement

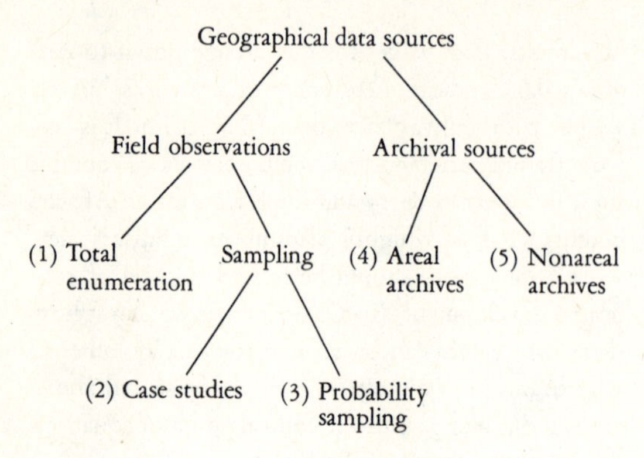

Figure 2.15 A typology of data sources.

are rolled into one; everything except final analysis is done for the researcher. The archival sources used traditionally by geographers are areal maps, aerial photography, and remote sensing data from Earth satellites. Nonareal sources are also widely used when they can be transferred easily to an areal base map. Thus, figures from national population censuses and agricultural statistics commonly provide data for geographical research.

This discussion of data sources is based largely on the thoughts of the British geographer Peter Haggett. In Figure 2.15, we present a modified version of his typology of data sources. The two categories field observation and archival source have been broken down into more specific categories based on the different criteria discussed above. These criteria are interchangeable. Field observation may be collected in an areal or a nonareal framework, and archival sources, consisting usually of total enumeration, may sometimes be based on samples. This is the case with part of the British population census, for example. Areal sources are often the basis for sampling operations; the initial data source may involve a total enumeration, but the data collected for analysis may be only a sample. (Toward the end of this section of the chapter, we illustrate sampling from a map.) With these reservations concerning the typology in Figure 2.15, it remains a useful scheme for considering data sources, because it identifies the five most common data sources in geographical research. These are total enumeration in the field, case study, probability sampling, areal archives, and nonareal archives.

Total enumeration in the field leads to few problems other than time and cost feasibility. In studies in which the number of individuals is not large—studying the central places within a region, for example—this approach is common. (We consider it further not in this chapter but in later chapters.) Detailed case studies selected by hunch have no place in modern spatial analysis and we do not consider them further. Probability sampling, with its direct links to probability theory and inferential statistics, is of great relevance to the content of many subsequent chapters, where we take it up in some detail. Before discussing modern sampling procedures, however, we consider some of the advantages and disadvantages of archival sources for spatial analysis.

Archival Sources

Because research workers do not actually measure the data they collect from archival sources, it follows that they have little or no control over the specific content of their source or over how the concepts they use have been measured and collected. The two basic problems in using archival sources of data are that we cannot control the precision and accuracy of the measures we use and that the archival source may not give us data of the right type in the right form for our analysis. We take each problem in turn.

Problems of Precision and Accuracy

Archival sources are found typically in official government publications that have an air of authority about their content, which leads to an assumption of inherent correctness. Governmental archive sources do not always have a high degree of precision or accuracy, in fact. Economist O. Morgenstern has made a scathing attack on the presentation of American economic statistics in which apparently precise figures, presented in some cases to one or more decimal places, in fact exhibit quite alarmingly high levels of imprecision when looked at critically. The level of precision is not usually reported with statistics, and consequently this data

(a)

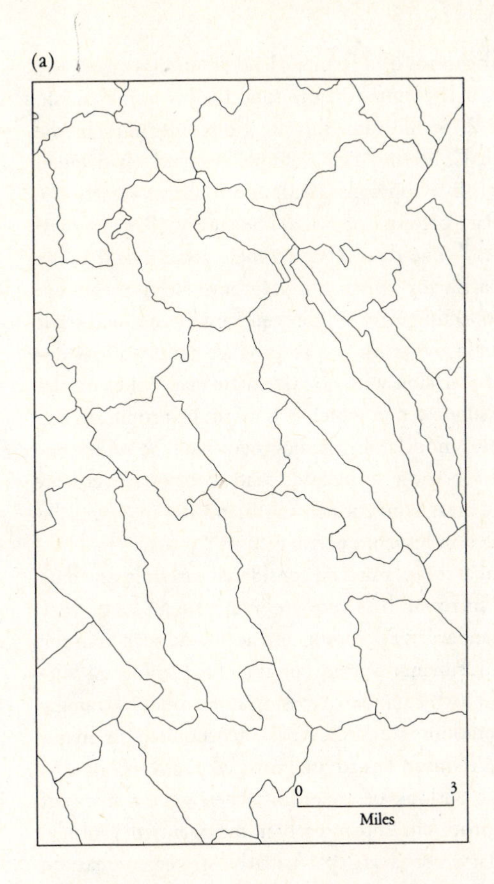

(b)

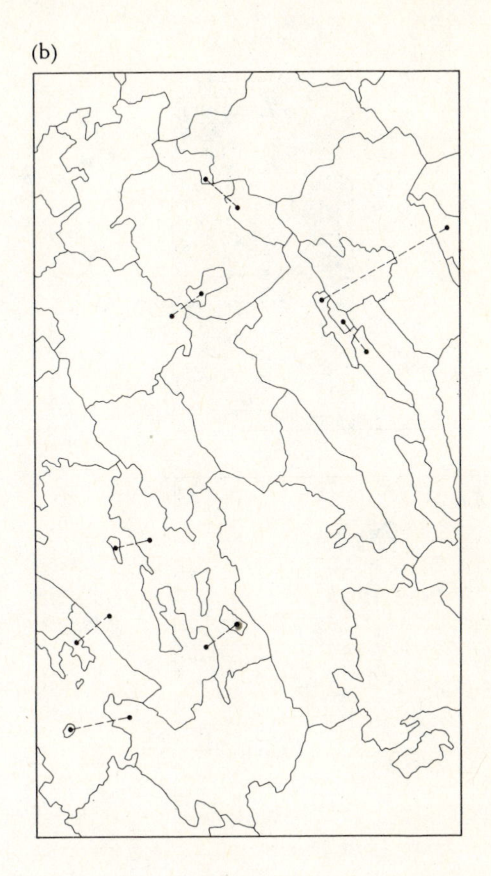

Figure 2.16 (a) Civil and (b) "agricultural" parishes in part of the English Chilterns, 1941.
From J.T. Coppock, The parish as a geographical statistical unit. *Tijdschrift voor Economische en Sociale Geografie*, 1960, vol. 51, p. 318.

source is far less useful, and even more dangerous than others, for economist and economic geographer alike. Official statistics may also have systematic biases in them. A case in point is the Nigerian censuses of 1962 and 1963, for which it had been previously decided that, after national independence, such features as federal money and parliamentary seats would be allocated to regions on a per-capita basis. The result of the census in both years was a phenomenally large statistical growth rate of population since 1952. The 1962 census was canceled after prolonged controversy that consisted of charges and countercharges of inflating population totals by the various states. A new count was ordered for 1963, but the new results were not accepted by the Eastern Nigerian government, which took the federal government to court over its handling of the census. The case was lost by Eastern Nigeria, but the suspicion of inflation of population totals remains, with Nigeria's population reportedly having risen from 30 million in 1952 to nearly 56 million in 1963; this will remain the official population of Nigeria until the next census.

Systematic bias may be far more subtle than is indicated in this example. British annual agricultural returns are presented by civil parishes, but they are collected by *farm* and, because farm boundaries do not necessarily coincide with parish boundaries, the published figures do not usually relate exactly to the parish to which they are allocated. The parish totals are found simply by adding together acreages of various crops as reported by farmers resident in each parish. Even if

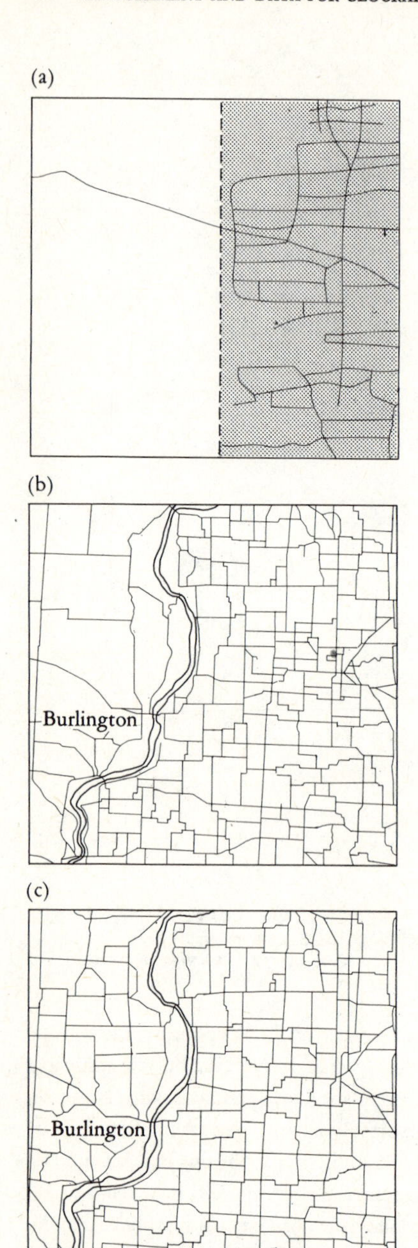

Figure 2.17 Political boundaries and road networks: (a) the Ontario-Ottawa boundary, (b) the Iowa-Illinois boundary, (c) the Iowa-Illinois boundary.

(a) From R.I. Wolfe, Transportation and politics. Reproduced by permission from the *Annals* of the Association of American Geographers, vol. 52, 1962. (b) and (c) From R.J. Chorley and P. Haggett. *Models in Geography*. London: Methuen & Co., 1969.

some, or most, of a farmer's land lies in more than one parish, it is counted in the total for his home parish. Figure 2.16 illustrates this areal disconformity in the English Chilterns. The example is based on detailed research by Coppock (1960) and it shows clearly the irregular pattern of what he terms "agricultural parishes." The data relate to these zones but it is the civil parish for which the data have been presented. This collecting error might seem largely to cancel itself out parish by parish and so produce simply a low degree of precision with no systematic bias, but examples can be suggested in which bias might be produced. For example, moorland parishes may well be underrepresented, because farmsteads tend to be on lower land in hilly areas; thus, much moorland use may well be presented in lowland parish returns.

Areal archives must be considered carefully as a data source in research. There are many examples of very high standards of precision and accuracy, such as in the British Ordnance Survey and the U.S. Geological Survey, but lower standards exist in some others. Strategic considerations are sometimes incorporated in maps. H.R. Wilkinson has advised us to take into account the patriotic outlook of an author when we use maps of ethnic groups in Europe early in the twentieth century. Boundary zones are particularly strategic areas on maps. Geographers often notice major changes in human spatial patterns as important political boundaries are traversed. The Ontario–Quebec boundary in Canada shows contrasting road patterns on adjacent sides of the border, as we see in Figure 2.17a, and these are clearly the result of boundary location. The Iowa–Illinois boundary seems to have had a similar effect, as depicted in Figure 2.17b, but the effects shown here may seem a little surprising because the lower road density suggests that there is less agricultural activity on the Iowa side of the border. Figure 2.17c justifies our surprise. The road network in Figure 2.17b is from an official Illinois map, but Figure 2.17c is from the map of an oil company that is much less concerned with the boundary than is the state government of Illinois. Thus we see that boundary influence can be purely artificial and result from arbitrary decisions on the part of map publishers. The example in Figure 2.17 is clearly extreme but it is equally certain that in any attempted analyses of road networks from map sources we must be sure of data accuracy before we attempt sophisticated analyses.

The Basic Problem of Purpose

Although problems of precision and accuracy are important when we consider archival sources, their most fundamental disadvantage is simply that they involve measurement and collection for purposes other than any particular researcher's specific needs. Research workers cannot choose the variables they will consider but can deal only with those that are available. Some of the large-scale multivariate analyses that use techniques we describe in Chapter Six seem to include, with no prior selection, almost all variables available in a census volume. Why a particular set of variables, selected presumably for administrative needs, should produce findings relevant to geographical theory is not usually explained. Sometimes the purpose of a data source may be even the antithesis of what the researcher would like. British trade directories, for example, give names and addresses of individual inhabitants of specified towns over a series of consecutive years. This data source is ideal for tracing intraurban mobility in a historical context, but the purpose of the directory is as an information source for a middle-class market of professional people and tradesmen. Most if not all lawyers and grocers are listed in the directory, but we cannot say the same for blacksmiths and other manual laborers. There will be a definite middle-class bias in any migration pattern we obtain from this source, and we must admit that any conclusions we may draw from subsequent analysis refer to a segment but not the total of the town population.

The problem of purpose is shown most clearly in the problems that arise in spatial analysis using the irregular base areas for which archival data are given. These irregular areas of various sizes are usually administrative districts relating to the administrative purpose of the data collection. A single variable is measured at a spatial scale slightly different in every separate collection unit. This has many serious ramifications but none more important than in the case of derived measures using area in their calculation. This can be illustrated with the concept of population density. If we accept the fact that population density varies over the Earth's surface, then it follows that the smaller the area of a collection unit is, the more likely it is to be either wholly full of population or completely empty. The range of densities in a set of measures depends on the size of the collection units. At the international level, small states invariably have denser populations than large states. Similarly, if we measure the population density in populous urban areas, this leads to declining density as we increase the areal definition beyond the central city. In Table 2.7, the population densities of Chicago and Detroit for 1950 are listed at three scales: city, urbanized area, and standard metropolitan area. Density clearly declines as the size of the collection area increases. Questions such as "What is the population density of Chicago?" have to be carefully specified. To compare densities, we must draw different conclusions at different scales, because at the largest scale Detroit has a denser population than Chicago but at the two smaller scales Chicago is more dense than Detroit. The concept of population density is difficult operationally, especially when the problem of variations in collection units arises.

Table 2.7 Population Densities of Chicago and Detroit (1950)

Scale	*Chicago*	*Detroit*
City	17,450	13,249
Urbanized area	7,713	6,734
Standard metropolitan area	1,519	1,535

Source: O.D. Duncan, R.P. Cuzzort, and B. Duncan. *Statistical Geography*. New York: Macmillan, 1961. Copyright © 1961 by Macmillan Publishing Co., Inc.

Another problem arises with archival descriptions of spatial interaction data. In almost all cases, movements are recorded only when they cross some administrative boundary. Thus, most short-distance migrations and commuter trips are not recorded in archival sources. Furthermore, the patterns that are recorded may give misleading impressions because of variations in administrative area properties. In Table 2.8, we list the recorded percentage of numbers of commuters in Belgium and Holland for 1947. Remember that a commuter is someone who resides in one district and works in another. From the first column, it seems

Table 2.8 Commuting Distances in Belgium and Holland (1947)

	Percentage commuters	*Average administrative area (hectares)*
Belgium	40.0	1,880
Holland	15.2	6,670

Source: Data from M.D.I. Chisholm, The geography of commuting. *Annals* of the Association of American Geographers, vol. 54, 1960.

clear that more people commute to work in Belgium than do in Holland. Consideration of the operational definition of the interaction leads us to a very different conclusion. The pattern of journey to work might be identical in both countries, but if one set of data were produced using larger administrative districts than in the other, we would expect very different results. It is obvious that the larger a district is in terms of area, the fewer will be the movements across the boundary, and recorded commuting numbers will be less even though actual travel patterns may be similar. In the second column in Table 2.8, we list the average size of the administrative areas used in compiling the data for Belgium and Holland in 1947, and here we find major differences. It seems quite reasonable to attribute Belgium's higher interaction rate to smaller administrative areas than to any difference in patterns of actual journey. The important point, however, is simply that given the archival data in this form, we do not know whether Belgium has a higher rate of journey to work than Holland.

Torsten Hagerstrand has contrasted the irregularity in collection areas for archival data to the treatment of time series data. Whereas spatial series are typically defined in terms of administrative areas, temporal data are arranged in terms of natural time periods such as monthly statistics or annual reports, thus enabling the sort of straightforward comparison over time that we have found difficult over area. Time equivalents of spatial administrative units would seem to be periods in office of administrative committees and other administratively arbitrary time segments. Time series statistics given in such form would prevent many of the simple trend comparisons that we are so familiar with from the news media. Fortunately, we do not have to contend with variable and irregular time periods, but this is not true with areal data. One possible solution to this problem is to convert a set of data from irregular to regular base areas. Let's examine this procedure.

Given a set of population totals recorded on an irregular pattern of base areas, the first step is to place a grid overlay on the base map. The regular grid may be a lattice of squares or hexagons. Taking each regular cell in turn, we allocate population to it in terms of the proportion of area in each original district that falls within the cell. Thus, in Figure 2.18, the population of the central cell consists of 35 percent of the population of base area A, 40 percent of the population of base area B, and 25 percent of the population of base area C. The three percentages represent the proportion of area of each base unit lying within the grid square under consideration. If the populations of A, B, and C were 1,000, 200, and 600, respectively, then the population total of the grid cell would be estimated as

$$\left(1{,}000 \cdot \frac{35}{100}\right) + \left(200 \cdot \frac{40}{100}\right) + \left(600 \cdot \frac{25}{100}\right)$$

which comes to $350 + 80 + 150 = 580$.

This population total of 580 for the central grid square differs from the population total of 1000 for base area A in a fundamental manner. Whereas 1000 represents a direct measure of population involving no other empirical concepts, the calculated value 580 is an example of a derived measure of population that uses measures of another concept (area) in its calculation. We have noted previously that derived measures are usually subject to greater levels of imprecision, and this is a case in point. The measurement approach here involves assumptions about the relationship between population and area that are usually violated in reality. In order to produce the true population total for our grid square, it is necessary that the population in each original base unit be spread evenly. This is the only assumption we can make really, given the absence of population data for the internal distribution of base areas, but it leads inevitably to imprecision in the resulting population measures. For example, 750 of area A's population

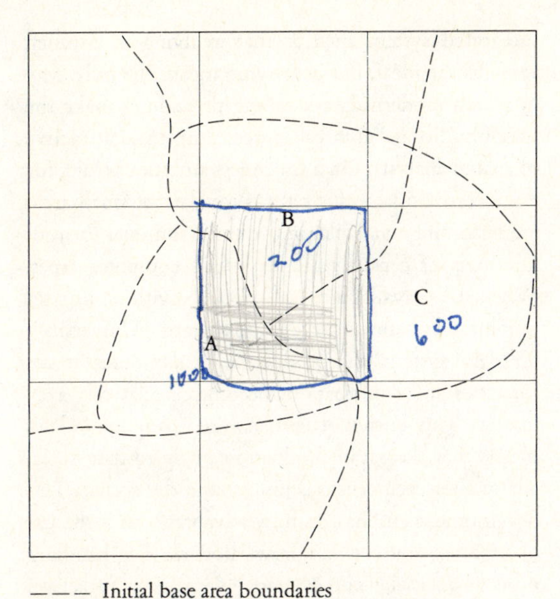

——— Initial base area boundaries

Figure 2.18 Transferring data to a regular grid.

in Figure 2.18 may be in a small town located within our grid cell although we allocate only 350 of area A's population to our derived total. There seems to be no systematic bias in this procedure, because some grid cells gain population and others lose, depending on the arbitary location of grid cell boundaries. Imprecision will always be present, however. It can be overcome to a certain extent by increasing the size of grid cell in relation to the original base units. It is clear that error arises where the grid cell boundaries partition original base areas. As the grid cell gets larger, fewer and fewer base units are partitioned, more and more lie wholly within a grid cell. Because these base units require no area measures to apportion their population, they involve no derived errors leading to imprecision. Thus, the larger the grid cells are relative to the original base areas, the greater is the degree of precision. This precision is bought at a price, however, because the data are aggregated into far fewer observations. Research workers have to choose between the levels of precision they think are adequate and the number of observations that remain for analysis.

Our solution to the problem of irregular data collection areas with archival sources is not wholly satisfactory. Solution of this problem involves not just tinkering with the end product but managing the initial data collection and measurement. In some countries, geographers have become involved at this stage of data production, and with the help of electronic computer equipment they have helped evolve large-scale geographical data banks.

Geographical Data Banks

We have concentrated largely on the difficulties of using archival sources. One overriding advantage they have is the basic fact that we can use data that are already available. This convenience obviously accounts for the general popularity of archival sources, despite the problems we have discussed. It is therefore of some importance that the convenience of archival data is on the threshold of major extension by computer-based information banks.

A *data bank* is simply a method of storing information in such a way that it is easily available for manipulation. In recent years, the widespread availability of computer technology has enabled data banks to become highly efficient systems of storing information, making data immediately available with a minimum of manual effort. Many separate data sources can be combined into a single system, thus facilitating easy comparison. An interesting aspect of this type of bank from the gegoraphic viewpoint is that often the only element common to various sets of data is location. Thus, to bring social, economic, political, and physical data into a single system, we can use location as the integrating property. In several computer data banks, location is therefore the key to the combined storage of information, and we may term such systems location-based or *geographical data banks*.

Sweden has been a pioneer of location-based data banks; experiments there go back to 1954. This country has a long tradition of innovation in archival sources. The first official national Swedish census was in 1749, and from it data were available at the individual scale. Unfortunately, this advantage of easily available data on individuals has been fully utilized only rarely in Swedish geographical research, because of the enormous task of abstracting this data from official returns. Thus in the 1950s Hagerstrand's research on patterns of individual migration flow over time had to be limited to a few selected parishes because there was simply too much data. Since then Hagerstrand has helped overcome this practical problem by making data at the individual level the basic element in building up a new computerized system. Using new maps based on aerial surveys, each plot of land and each dwelling is given x, y locational coordinates, thus identifying them with unique locational references. Furthermore, this absolute locational property can be extended, because the coordinate system enables the relative location of every element to be known. That is to say, the location of every element relative to every other element can be found by using simple geometry. The data can be aggregated into any specified pattern. If we desire regular areas for analysis, we can aggregate to grid squares without the assumptions inherent in the method of transfer from irregular areas as described above. In this case, because all cells have the same area, a simple map of population totals is equivalent to a derived density map. Cross-tabulation of data by computer before output means that information such as the areal distribution of people over age 65 and below a specified income level can be obtained with a minimum of effort. This is just one of an almost infinite number of possible patterns that can be produced. Data arranged in this way are readily available for comparison using mathematical manipulations, making data analysis much simpler than it is usually in archival sources. We need not aggregate to regular areas but we can still extract data for administrative units; this is, of course, an important need, because administrative units remain a basic decision-making unit in many areas of social and economic planning. Geographical data banks based on data of individual level are highly convenient and may be made to be extremely flexible.

The Swedish experience with data banks has occurred in a country with a long tradition of individual level data, and it is admittedly exceptional. Experiments with geographical data banks are currently going on in many other parts of the world. A fully integrated system such as that evolving in Sweden remains the ideal, but it is by no means the only way in which modern data-handling procedures make the use of traditional archival sources simple and attractive to spatial analysts. On a much less ambitious scale, for instance, British census returns for data of small areas are becoming available, not in the traditional form of hundreds of census volumes but as computer tapes. When this availability is coupled with computer graphics systems, such as Harvard University's SYMAP system, the production of computer atlases for cities and conurbations becomes a relatively easy matter. This current trend toward computer atlases means that the rapid production of up-to-date social atlases after each census is just around the corner. This development emphasizes how advantageous is the use of archival sources and how modern data-handling procedures make convenience even more attractive. Thus, although traditional problems of archival data remain, archival sources will undoubtedly maintain their importance in geographical research.

Probability Sampling

All field observations, including sampling, have the advantage of being specifically designed for the research in hand. With them the basic disadvantage of archival sources is avoided. Social geographers can ask questions on topics not available in the census—about perception, for example—at the individual scale and aggregate to suit their own purposes. Following the example of the vast majority of archival sources, however, geographers have often favored total enumeration for data collection. Why, in fact, should geographers attempt anything less than total enumeration?

Why Sample?

There are six basic reasons why geographers may choose to sample. The first three are of a rather practical nature, but in terms of research feasibility, they are no less important than other reasons.

1. By sampling, we can cut the costs of data collection. Research workers clearly do not have the financial

resources of traditional total enumerators such as the national census organizations.

2. Sampling requires less manpower. This is clearly related to the first point, but even if a research worker is fortunate enough to have research assistants, it would seem that there should be jobs more intellectually rewarding than data collection.

3. Sampling enables us to collect data more quickly than total enumeration does. This may be important in surveys of personal attitudes that are subject to change. Political pollsters are particularly aware of this factor.

These three practical reasons for sampling correspond to what Peter Haggett (1965) terms geography's "coverage problem." He has suggested that if it is the purpose of the geographer to provide an accurate description of the Earth's surface then, in 1960 at least, each of the world's 3000 professional geographers would have had to cover some 5000 square miles. One solution is to have more professional geographers; the other is to sample the variable character of the Earth's surface instead of attempting to describe every part. The first solution requires some elaborate self-justification for regional geography; the second solution is feasible given the present situation and humor of geographers. The final three reasons for sampling are less practical and concentrate on some of the more fundamental advantages of sampling.

4. Samples sometimes enable us to obtain information that it would not be possible to collect in total. Soil samples, for example, may be drawn and the collected soil may be broken down chemically so that it is no longer a soil material. Carrying out this experiment by total enumeration is clearly impossible, and farmers would certainly argue that it would be an undesirable goal.

5. By concentrating on fewer individuals, sampling allows more comprehensive data to be collected, given the same resources. Even national censuses turn to sampling for additional information on separate census forms. The British census, for instance, now conducts a 10% sample to obtain more detailed data than is obtained in general, total enumeration schedules.

6. A sample may be even more precise and accurate than a total enumeration. This is because, with fewer individuals to be concerned with, the collection can be controlled more tightly. In an underdeveloped country with a poor map base on which to plan a census, for example, a total enumeration might lead to imprecision in that many people are omitted and others are counted twice. At worst, this situation could lead to serious biases because the remoter part of a population under-represented in census results typically has characteristics that do not match the rest of the population—they will probably be poorer, for example. Thus, estimates of income by a national census that omits these people will produce an overestimate of average incomes. Very detailed coverage of a limited number of sample areas, on the other hand, may produce unbiased estimates of the properties of the total population within a known degree of precision.

Sampling, which may at first appear illogical, can be justified on several grounds over a wide variety of situations. In the remainder of the chapter, we deal with data collection by sampling.

Sampling Designs

Once we have decided to collect data by sampling, we do not go out immediately into the field and sample. Instead we think about the problem in terms of probability sampling procedures. We can identify four basic steps in the planning of a sampling operation.

1. First of all, we must define the population of objects that are to be sampled. This may sound easy, but in practice many difficulties can arise. A climatological study of a region might have as its ultimate population all locations in the region but, data may in fact be available only for the limited number of locations in which meteorological stations are maintained. If we draw a sample of meteorological stations, we are not sampling our *target population* (all locations in the region); we have instead a *sampled population*. Notice that our target population is one step removed from the sample, and problems of imprecision or even biases may enter, even at this first stage in the operation. The precision of isolines on climatological maps clearly varies with the density of meteorological stations. In much spatial analysis, we can fortunately define our

population simply in terms of all locations represented on a given map, which then becomes the sampled population. If our archival data source is reasonably accurate, we have no problems in linking our sample to the target population.

2. Once we have defined the population, we must set up a *sampling frame*. This is some specified ordering of the objects to be sampled. It may be simply a numbered list of all households in a district. If we are sampling from a map, the coordinate system of the map will usually serve as the sampling frame.

3. With a sampling frame set up, we are in a position to begin sampling. We have to specify the *information* we wish to collect. This simply relates to the research problem's empirical concepts and the measurement techniques we use to measure them. A pilot sample is often taken at this point to test the measurement procedure. At this stage we must also decide on the degree of precision we require from our results. Our sample results will be only estimates of the population's characteristics, and we have to decide how precise they ought to be. This is partly a question of sample size, because larger samples produce more precision. Precision is also a function of our choice in step 4.

4. Finally, we have to choose our *sample design*. There is a wide variety of designs to choose from, although most are complicated variations of four basic types: random, stratified, systematic, and cluster or nested sampling. We consider each approach in turn, paying particular attention to the properties of the data that each produces.

Table 2.9 is a list of six items that make up a small population. We use this population to illustrate how we draw different types of sample and their resulting properties. To make the hypothetical example more realistic, we consider the quantities to be the number of central place functions available at each of six small settlements. The list numbers are, of course, our sampling frame. Notice that we would not normally sample in such simple situations, because we have all the information at hand. It is useful as an example, however, because we have the total population as a standard against which to compare our sample results. We shall sample to find the mean value to compare against the known true population mean of 9.167.

Table 2.9 Hypothetical Sampled Population (Mean = 9.167)

Settlement	*Quantity of central place functions*
1	13
2	8
3	2
4	3
5	9
6	20
	Σ 55

The most common form of sampling is probably the simple *random* approach. Random numbers are arbitrarily selected from prepared lists. In the sampling frame, the object whose number coincides with the chosen random number is selected for the sample. (A table of random numbers can be found in the appendix as Table A.1.) Let's employ a random sample of three to estimate the mean of the population in Table 2.9. If we start in the top lefthand side of Table A.1, we find the number 6. This coincides with object 6 in our sampling frame, which has a quantity of 20 in Table 2.9. Continuing down this column, we come to number 8, which we reject because it is not represented in our frame. The next number is 4, so the object listed as 4 goes into our sample with its quantity of 3. The next number is 9, which we reject before selecting object 1, with quantity 13. Thus our random sample is 6, 4, and 1 with quantities 20, 3, and 13, giving a mean value of exactly 12. This compares with the true mean of 9.167. We have overestimated slightly but the sample result seems reasonably precise.

The basic property of such random sampling is *independence* of selection. The fact that one object has been selected has no influence on the selection of any other object. In our example above, 6 was selected first, and any of the five others could have been selected next. This property is a result of the use of random numbers, which are themselves defined in terms of independence. Any digit may occur at any place in a random number table irrespective of the digits in adjacent rows and columns. This means that we can start selecting numbers from any part of the table, and use columns, rows, or even diagonals. This independence property has important implications when we translate our geographical problem into abstract calculi, because certain statistical procedures require independence of selection as a condition of use.

This independence of selection has a disadvantage, however. Every possible set of objects may constitute the sample, and therefore there is the possibility of sometimes producing samples overweighted in terms of extreme quantities. In our example, the extreme sets are (1, 5, 6) and (2, 3, 4), which give means of 14 and 4.33 respectively. These are clearly very poor estimates of the true population mean, but each is nonetheless as likely to occur as our original sample with its estimate of 12. Random samples tend to be imprecise compared with other, more controlled sample designs. Notice that these extreme samples do not reflect biases, however. Both high and low estimates are equally likely, and the result is imprecision, not inaccuracy.

Stratified sampling is a way of overcoming the imprecision of simple random sampling. In this procedure we divide the sampling frame into *strata*, which then serve as subpopulations from which samples are drawn. If the strata are selected to produce reasonably homogeneous groups of objects, then by sampling each stratum the whole range of the total population is evenly represented. This procedure assumes some prior knowledge of the population but, if we have this knowledge, we also have the basic advantage of being able to produce more precise results than in simple random sampling. In our example, we might divide the population into three homogeneous strata (1, 6), (2, 5), and (3, 4). We now sample within each sample, and by picking one object from each we produce a total sample of three again. We can use random number tables for this purpose. In our previous use of Table A.1, more random numbers were considered than we actually used because of the rejections. With only two objects in each stratum, this problem would be an even greater nuisance. There is no need to use random numbers strictly in this way, patiently waiting for a 1 or a 6 to appear before sampling in the first stratum. In this example, it is much quicker to select the first object in a stratum if the random number is odd and the second object if the random number is even. In this way, we shall have no rejects and, because the random numbers are independent, it follows that their oddness and evenness also has this property. If we start in the top righthand corner of our random number table, we have the sequence odd, even, odd, going across the row, so that we select objects 1, 5, and 3 from our three strata. This time we produce an estimated mean of 8. The fact that this estimate is more precise than the simple random sample illustrates the property that stratified samples are usually more precise. This follows from the homogeneous definition of strata and the way in which homogeneity prevents extreme objects from entering the sample. In the present example, the lowest possible estimated mean is 7.67 from sample (1, 2, 3) and the highest possible estimated mean is 10.67 from sample (6, 5, 4). Both are reasonably precise compared to the extremes possible with a random sample. Objects 1 and 6, with quantities of 13 and 20, quite simply cannot occur in the sample with this design.

Operationally the simplest sampling procedure to use is the *systematic* sample. We select an initial object from the frame and then choose all subsequent members of the sample in regular sequence starting with the initial object. The initial selection may be random: starting in the top righthand corner of our table would give us object 1. If we require a sample of three, we select every other object in sequence, continuing through to 6. Objects 3 and 5 are chosen thus to join 1 in the sample. Once again, an estimated mean of 8 results. Such sampling is usually very simple and inexpensive. It does not produce independent selections, however, because once the initial object has been chosen all other selections follow. This may limit the usefulness of the resulting data. Furthermore, we cannot always assume accuracy with such sampling. We take up this point when we consider the areal case below.

Cluster sampling is termed nested sampling in the areal case. Just as in stratified sampling, we initially divide up the sampling frame, but the resulting groups are based on heterogeneity. That is, the total population is divided into groups so that each covers the whole range of properties exhibited by the total population itself. In our current example, we might divide the population into two groups of three objects—(1, 4, 5) and (2, 3, 6)—so that each cluster has one high, one medium, and one low quantity object. We now select one of these clusters as our sample. If the first is chosen, we produce an estimate mean of 8.33; if the second is chosen, our result is 10. The cluster can actually be chosen by using random numbers as before. This cluster has total enumeration of the objects in the cluster. We illustrate below a two-stage cluster design in which objects within selected clusters are themselves sampled. All types of cluster sampling are usually inexpensive and quick to carry out. If the groups are locationally contiguous, we may have to visit only a small part of each region to carry out a sample survey. Precision depends entirely on how well the groups reflect the total population in properties. If this is done poorly, very rare objects may occur all in the same group in a single sample and thus produce results as imprecise as is possible with random sampling.

We have referred several times to the areal case in sampling. This is because in spatial analysis we often sample from a map source. In the final section of this chapter, we illustrate the four sampling designs and related approaches, using examples of areal sampling from base maps.

Point Sampling from Maps

In areal sampling, either in the field or from maps, we have a choice of sample unit in terms of dimensions. We can sample small areas on the map and measure properties within these units. This is sometimes termed *quadrat sampling* and we deal with this approach in some detail in our discussion of point pattern analysis in Chapter Four. Alternatively, we can use lines as sampling units. Such areal *traverses* are common in field surveys and can be employed also when sampling from a map. These two approaches use sampling units that are two-dimensional (quadrats) and one-dimensional (traverses), but we can go one stage further and use zero-dimensional units—that is, points. This is the most common sampling unit in map analysis.

We introduce point sampling as a means of measuring areas in quite complicated areal patterns, as typified by any land use category on a detailed land use map. Total enumeration is represented by careful measurement of the areas with planimeters. This can be very slow and time consuming. By sampling, we can estimate quickly, and with varying efficiency, areas within even the most complicated areal patterns. The degree of efficiency depends on the point sampling design.

All the sample designs we have considered can be transferred quite easily to a two-dimensional areal situation, as has been illustrated in the work of Brian Berry. The basic difference is that we need *two* numbers to define a point rather than a single number as for the usual sampling frame. If our population is a map, then our sampling frame is the coordinate system of the map to some specified degree of accuracy. To sample a point on the map, we must have two numbers to give the x and y coordinates that define any point. To produce a random point sample on a map, we draw two random numbers from the table and let the first represent the x coordinate and the second the y coordinate, to locate our first random point at x_1, y_1. The point located at x_2, y_2 is selected in the same manner, and the procedure continues until the required number of locations has been sampled. A random sampling pattern is shown in Figure 2.19a.

The three other basic sampling designs can be similarly transferred. For stratified point sampling, it is common to use grid squares defined by the map coordinates as strata. If this is done, the random point sampling procedure may be repeated within each grid cell to produce an areally stratified random sample, as shown in Figure 2.19b. An areal systematic sample is shown in Figure 2.19c; the points are regularly spaced in terms of both x and y coordinates. The two-stage cluster sample shown in Figure 2.19d uses grid cells, as in the stratified sample, but then selects a sample of cells that are then point sampled quite densely. This represents the two-stage cluster sample design in the areal case.

One further point sampling design is illustrated in Figure 2.19f, a design made up of a combination of the simpler designs and developed specially for point

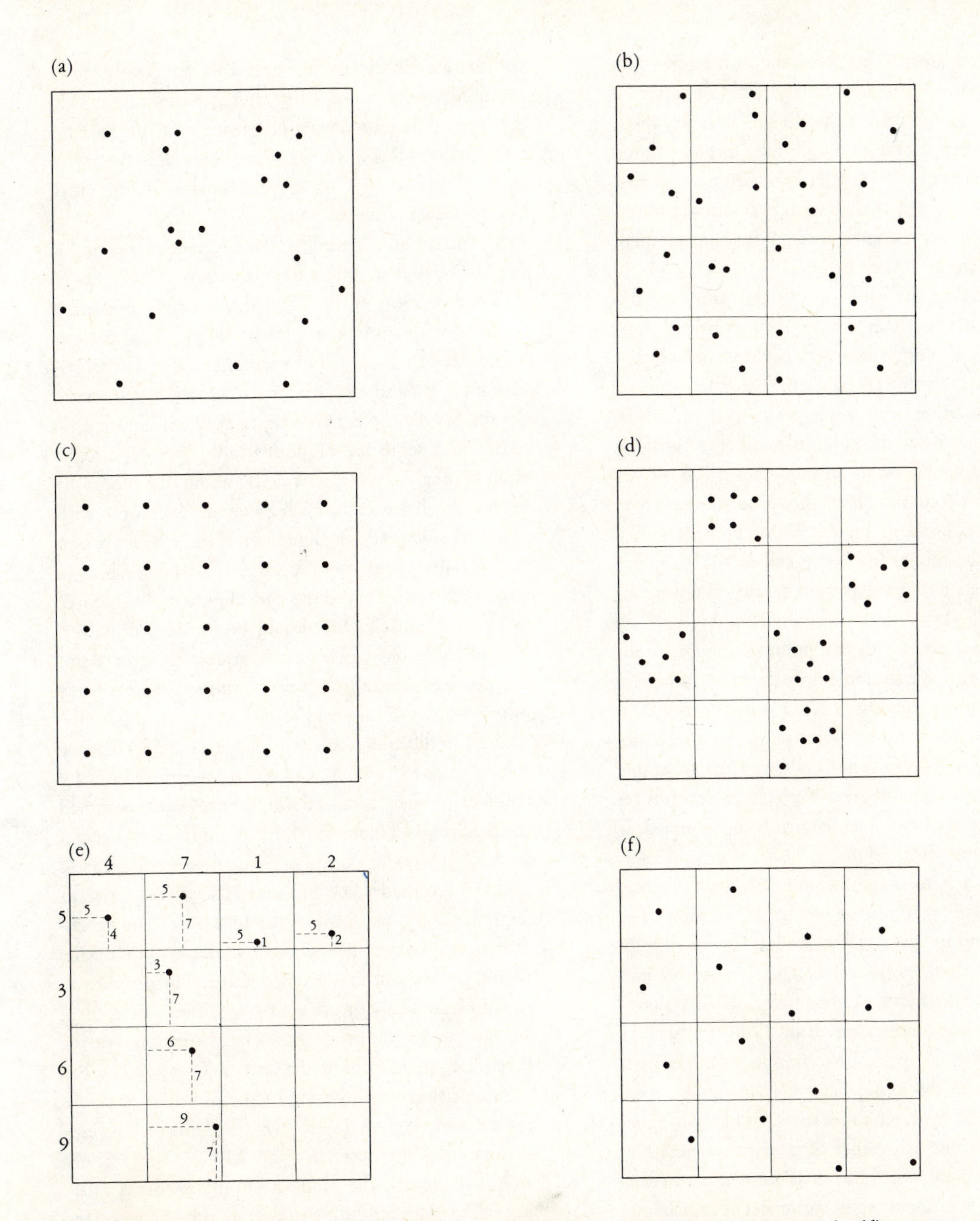

Figure 2.19 Point-sampling designs: (a) simple random, (b) areally stratified, (c) systematic, (d) two-stage areal cluster, (e) locating points for design (see part f), and (f) stratified, systematic, unaligned.

sampling over areas. It possesses the rather long title *stratified systematic unaligned sample*, which is actually an accurate description of the design. It is stratified, because the first step is to divide the sampling frame into grid cells; it is systematic, because once the first set of points has been located all other points automatically follow; and yet, unlike other systematic samples, it produces an unaligned pattern.

The procedure for carrying out this sampling design is fortunately much simpler than its title. A random number is selected for every cell row and another is selected for every cell column. The row number defines the location of a sample point on the x axis of all cells in that row; the column number defines the location of a sample point on the y axis of all cells in that particular column. This is shown for the first row and second column in Figure 2.19e. Thus, x and y coordinates are given for every cell in the sampling frame, and an unaligned pattern results, as in Figure 2.19f.

We find ourselves with five point sampling designs to choose from. We should emphasize that they share their basic properties with their simpler nonareal counterparts. The random point sampling design has the property of independence as the systematic sample design does not. The degree of precision depends on the distribution of the areal pattern being sampled. If the underlying distribution has no pattern, then it would seem not to matter which design we choose, because all will give similar levels of precision. The areal distributions studied by geographers are rarely devoid of pattern, however. In fact, it is sometimes said that "the first law of geography" is that things located near one another are more likely to be alike than are objects far apart. Two contiguous fields often have the same soil type, land use, or slope aspect whereas a third field some miles away is likely to differ from them in one or another of these properties. This property of areal distributions is known as *autocorrelation*, and we discuss it in more detail in Chapter Three. Its presence in most geographic problems that deal with areal data means that grid cells make reasonably good strata in terms of homogeneity, so that stratified samples typically lead to more precise estimates than do simple random samples. Autocorrelation also means that grid cells make very poor groupings of locations for cluster sampling, which depend on heterogeneity of group content for the best results. Thus it would seem likely that the two-stage cluster design using grid squares in the areal case produces rather imprecise estimates.

Systematic areal sampling gives even coverage of the population but it suffers from the basic disadvantage that one or more of the aligned point sequences may coincide with some linear feature in the areal pattern and thus we way greatly overestimate its relative importance. A land use survey based on a systematic design, for example, may lead to very inaccurate estimates if a sequence of points falls along a straight stretch of a major highway or along the floors of several parallel valleys. Such biases are overcome with stratified systematic unaligned samples, which contain the advantages of stratification, producing an even coverage, while avoiding the alignment of simple systematic sampling. It should be noticed, however, that observations based on this sample design are not independent, because they do not possess the property of independence for every selection.

We can illustrate the variations in precision that we have predicted for our various sample procedures by considering the results of a simple experiment designed to compare data from the different sample approaches. In this experiment, sampling is used to estimate areas within a specified shape on a map .The shape represents approximately one third of the map area. The procedure is simply to notice whether a sample point occurs within or without the specified shape. The proportion of sample points lying within the specified area is, then, an estimate of the proportion of the total area of the map that occurs within the specified shape. The experiment consists of drawing ten samples of 36 points for each of the five designs described above. The estimates of area are shown in Table 2.10

As we expect, the random sample produces fairly imprecise estimates, although they do not seem too different from the stratified sample estimates. The greatest degree of imprecision results from the cluster sampling design. The area being estimated consists of a single tract of contiguous land, so that autocorrelation exists in the pattern under investigation. In this situa-

Table 2.10 Frequencies of Sample Estimates of Map Areas

Sample type	*Percentages of total map area* 10–15.9	*16–21.9*	*22–27.9*	*28–33.9*	*34–39.9*	*40–45.9*	*46–51.9*	*52–57.9*
Random	0	1	2	4	1	1	0	1
Stratified	0	0	1	2	4	1	1	1
Systematic	0	0	0	3	5	2	0	0
Stratified, systematic, unaligned	0	0	0	5	5	0	0	0
Two-stage cluster	3	0	1	1	1	1	3	0

tion, a cluster of points obviously may fall totally within or without the area to be measured, and a wide range of area estimates is to be expected. Table 2.10 confirms this. The systematic samples are reasonably precise, but they produce two estimates of more than 40 percent. In this experiment, the stratified systematic unaligned sample produces the best results, with ten consistently precise estimates.

We find that our sampling designs provide estimates of varying degrees of precision. Although our experiment is based on ten samples of each design, the results are such that we can suggest that systematic and stratified systematic unaligned samples are to be preferred. In fact, the latter design has been used for updating land use data in the Chicago standard metropolitan statistical area, using a set of air photographs as the primary data source; limited availability of funds and time led to the adoption of this sampling procedure rather than the usual total enumeration of land use. It is clear that if the Northeastern Illinois Planning Commission must sample when collecting data, the individual geographical researcher, with far fewer resources, will find this approach amenable to the collection of areal data for many geographical problems.

Further Reading

The measurement theory we have presented is drawn largely from the pioneer work of Stevens (1946). This work has been reported widely, and a particularly good exposition of it can be found in Siegel (1956). The various empirical concepts identified in this chapter are based on Hempel (1952), and the idea of measurement as the link between these concepts and mathematics is most explicitly developed in Ellis (1966). Ellis is basic for measurement in general, although his work may be supplemented by Ackoff *et al.* (1962) and Wilson (1951). More specific treatments of measurement in fields related to geography can be found in Selltiz *et al.* (1962) for sociology, Hodge (1963) for town planning, and Krumbein (1958) for geology. Much of this work has been brought together for a geographical audience by Harvey (1969). Discussion of distance, direction, and relative position as "fundamental spatial concepts" is given by Nystuen (1963), developed subsequently by Papageorgiou (1969).

Supplementary references for extending the discussion on the specific problems of measuring spatial concepts are as follows. (1) For area, Coppock and Johnson (1962) describe new methods of measurement. (2) Bunge (1966) presents a sophisticated nominal measure of shape, and Boyce and Clark (1964), Lee and Sallee (1970), and Taylor (1971) suggest alternative approaches to measuring compactness. (3) For distance, Nystuen (1967) introduces the Steinhaus paradox into geography and illustrates the epsilon approach.

(4) Vector measurements for direction have been the concern largely of geologists such as Pincus (1956), although the recent interest of statisticians is reflected in Mardia (1972). (5) There is a very large literature on population potential as part of a social physics. The flavor of this school can be tasted in Stewart and Warntz (1958); a good example of this work is represented by Warntz (1967). The more flexible approach to potential measures was pioneered by Harris (1954) and followed up by Clark *et al.* (1969) in the European experiments described in this chapter. A similar experimental strategy of computing potential for alternative situations was employed earlier by Neft (1961) in a completely different political context for Southeast Asia. A critical assessment of the potential concept is presented by Houston (1969). (6) The first major study to employ numerous graph theoretical measures of connectivity was Kansky (1963). An interesting application in historical geography is presented by Pitts (1965). Subsequent developments in this field are described by Chorley and Haggett (1969).

Much of the discussion on data in geography covered in this chapter is based on Haggett (1965). Particular problems of nonareal archival sources are drawn from Morgenstern's (1965) discussion of economic statistics, Coppock's (1960) study of agricultural returns in Britain, Chisholm's (1960) comment on commuting patterns in Holland and Belgium, and the discussion of Duncan *et al.* (1961) on population densities. The problems of the Nigerian census are described in Udo (1968), and the utility of the British census and the U. S. census is discussed by Robertson (1969) and Fay and Klove (1970), respectively. The transformation of data to a regular pattern of base areas is employed by Robinson, Lindberg, and Brinkman (1961). Areal archival sources are critically examined by Board (1966) and specific political biases in Europe are described by Wilkinson (1951) and Sinnhuber (1964).

The advantages of archival sources when linked with computerized data handling are described and illustrated by Hagerstrand (1967), with particular reference to the Swedish experience. A much more local and less ambitious data bank for an English county is described by Jay (1966). Lee (1971) presents a general discussion of geographical data banks in an applied context. The production of computer atlases by linking archival sources with computer graphics packages is illustrated admirably by Rosing and Wood (1971). The general implications of data handling by computer for geographical research are discussed by Haggett (1969) and Gould (1970).

The initial presentation of sampling procedures in this chapter is drawn largely from Slonim (1960), and the discussion of areal sampling is based on Berry and Baker (1968). Examples of actual use of sampling in geographical research can be found in the work of Wood (1955) and Blaut (1959). More advanced discussion of areal sampling is given by Holmes (1967 and 1970) and similarly advanced discussion of sampling strategies in general are presented by Cochran (1953) and Yates (1953).

Work Table 2.1 Epsilon Measurement of Distance

Purpose

To measure the length of an irregular curve (empirical distance) to a specified degree of resolution

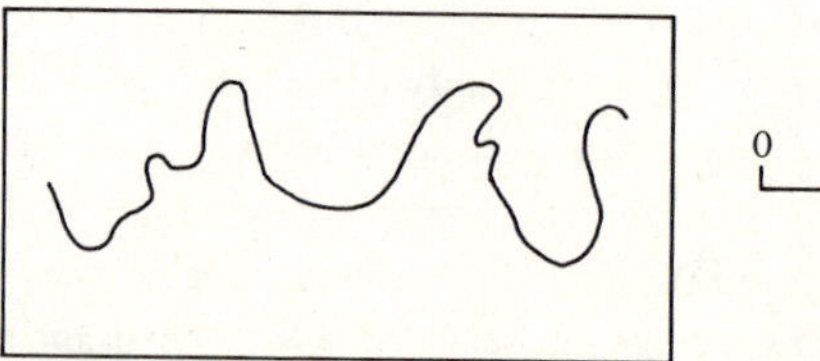

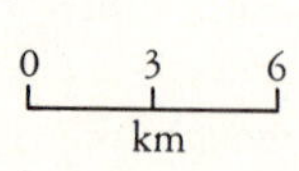

Data

Method

1. Define the degree of resolution as $\varepsilon = 1\frac{1}{2}$ kilometers.
2. Construct a square lattice of circles of radius ε.

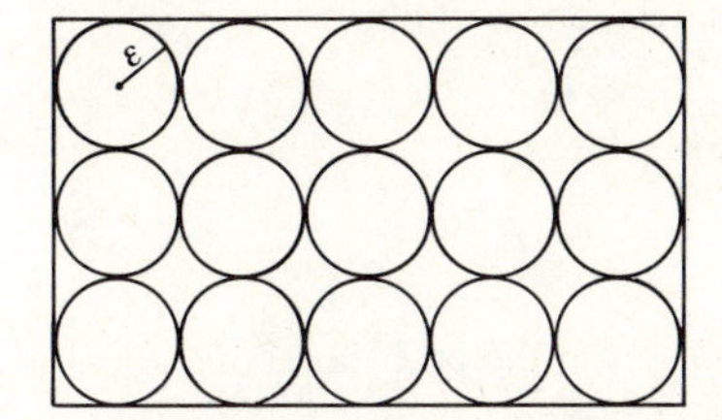

$\varepsilon = 1\frac{1}{2}$ km

3. Transfer the lattice to tracing paper and carry out 2ε arbitrary placements of the lattice over the data. In each case, count the number of circles cut by the irregular curve. In one example of such placements, the line cuts 9, 12, and 10 circles.
4. Compute $l_\varepsilon(x)$ from

$$l_\varepsilon(x) = n - \frac{\pi}{2}\varepsilon$$

in which n is the total number of circles cut. In this example, $n = 31$, so that

$$l_\varepsilon(x) = 31 - \frac{3.142 \cdot 1.5}{2} = 31 - 2.357 = 28.643\text{km}$$

Exercise

With this method, different results will be produced with different placements of the lattice of circles. An exercise with an exact answer to check against cannot be devised. An approximate check is possible, however, if the Spanish–Portuguese boundary is measured by using this method. Carefully trace this boundary from a relatively large-scale map of the Iberian peninsula, construct a lattice of epsilon radius circles (using ε = 6.5 km, 12.8 km, 38.4 km, and 51.2 km), and replicate the exercises reported in Table 2.2. Do your results show a consistently increasing length as ε becomes smaller? For each level of resolution, does your result fit in *among* the results presented in Table 2.2?

Work Table 2.2 Vector Measurement of Direction

Purpose

To find the average direction of a set of bearings from 0°–360°

Data

Bearings 60°, 125°, 195°, and 340°

Definition

The average direction is given by the resultant

$$\theta_R^\circ = \arctan \frac{\Sigma \sin \theta_i^\circ}{\Sigma \cos \theta_i^\circ}$$

Arithmetic

Bearing	*Cosine*		*Sine*	
60°	+Cos 60° =	+0.5000	+Sin 60° =	+0.8660
125°	−Cos 55° =	−0.5736	+Sin 55° =	+0.8192
195°	−Cos 15° =	−0.9659	−Sin 15° =	−0.2588
340°	+Cos 20° =	+0.9397	−Sin 20° =	−0.3420
		−0.0998		+1.0844

$$\theta_R^\circ = \arctan \frac{+1.0844}{-0.0998} = \arctan 10.8657$$

Because tangents are defined as the ratio or the opposite over the adjacent in a right-angled triangle, the following relationships hold.

1. If the ratio is $\frac{+\sin}{+\cos}$,

$$\theta_R^\circ = \arctan \frac{\sin}{\cos}$$

in a northeastward direction.

2. If the ratio is $\frac{+\sin}{-\cos}$,

$$\theta_R^\circ = 180^\circ - \arctan \frac{\sin}{\cos}$$

in a southeastward direction.

3. If the ratio is $\frac{-\sin}{-\cos}$,

$$\theta_R^\circ = 180^\circ + \arctan \frac{\sin}{\cos}$$

in a southwestward direction.

4. If the ratio is $\frac{-\sin}{+\cos}$,

$$\theta_R^\circ = 360^\circ - \arctan \frac{\sin}{\cos}$$

in a northwestward direction.

Thus, the signs attached to the sums of cosines and sines indicate in which of the four quadrants the resultant lies. Negative signs are then ignored in computing the ratios. In our case, relationship 2 obtains, so that

$$\theta_R^\circ = 180^\circ - \arctan 10.8657 = 180^\circ - 84^\circ\ 45' = 95^\circ\ 15'$$

Exercises

1. Derive the resultant in the problem above graphically by adding unit vectors.
2. Two further bearings of 185° and 190° are added to our initial set of bearings. Derive the revised resultant for all six bearings, using both vector arithmetic and the graphic method. What effect have these two similar bearings had on the average direction?

Work Table 2.3 Population Potential

Purpose

To measure the aggregate or gross position of a location relative to a set of other locations.

Definition

Population potential at i is $v_i = \Sigma(p_j/d_{ij})$, where p_j is the population of unit j, and d_{ij} is the distance from i to j.

Data

	A	B	C	D
POPULATION	20	10	5	5

DISTANCE $d_{AB} = d_{AC} = d_{CD} = d_{BD} = 1$

$d_{AD} = d_{BC} = 1.4142$

$d_{AA} = d_{BB} = d_{CC} = d_{DD} = 0.5$

A	B
C	D

Arithmetic

$$v_A = \frac{20}{0.5} + \frac{10}{1} + \frac{5}{1} + \frac{5}{1.414} = 58.536$$

$$v_B = \frac{20}{1} + \frac{10}{0.5} + \frac{5}{1.414} + \frac{5}{1} = 48.536$$

$$v_C = \frac{20}{1} + \frac{10}{1.414} + \frac{5}{0.5} + \frac{5}{1} = 42.072$$

$$v_D = \frac{20}{1.414} + \frac{10}{1} + \frac{5}{1} + \frac{5}{0.5} = 39.144$$

Exercises

1. Draw a map of the study region, plot the four potential values, and sketch in an approximate popula-

tion potential surface. (Use isolines for 55, 50, 45, and 40.)

2. Compute revised potentials for the four areal units for the situation in which area A's population grows to 100. Draw the new potential surface. What effects has this relatively large population growth had on the magnitudes and pattern of population potential?

PROBABILITY THEORY AND GEOGRAPHICAL RESEARCH INFERENCES

In carrying out empirical research, the geographer like any other scientist is confronted with a set of evidence to compare with some hypothesis that has been stated previously. The hypothesis may have been derived from geographic theory or it may be simply an intuitive notion. The empirical evidence consists of measures of the concepts that occur in the hypothesis and for which data have been collected. The question that must be faced is, "Does this evidence support the hypothesis?" How to answer is the subject of this chapter. Because the tool used increasingly to help us make this decision is probability theory, we are concerned with research inferences in a probabilistic framework.

Probability theory has its modern origins in seventeenth-century Europe in the correspondence between mathematicians Pierre de Fermat and Blaise Pascal concerning fair odds in games of chance. Such considerations were of direct relevance to their aristocratic patrons. From such trivial origins, a whole new way of thinking opened for scientific research. The concepts of probability were soon applied in social science with the development of life expectancy tables for insurance companies. Of much more general importance was the bringing together of various independent findings into a coherent probability theory by another French mathematician, the Marquis de Laplace (1749–1827). As an integral part of the new mathematics, probability theory had now begun to be used as a language within science. More recently, mathematician Lancelot Hogben (1957) identified four basically separate applications of probability concepts, which he termed the calculi of aggregates, errors, exploration, and judgment. We briefly consider each in turn.

The calculus of aggregates concerns the use of probability theory to make statements about probabilities of events occurring in situations involving very large populations. The two important applications are in the statistical mechanics of physics and in Mendelian genetics.

The calculus of errors is concerned with the distributions of measurement in the manner we have already introduced in our discussions of precision. From observations in astronomy, certain empirical laws of error distribution have been developed by using probability concepts. The major contribution is the development by Karl Friedrich Gauss of the normal probability distribution which, as we shall see, has been applied subsequently in many other contexts.

The calculus of exploration is concerned with the estimation of the properties of populations from samples. We have already illustrated this on a very elementary level in the sampling experiments in Chapter Two. We return to this problem in a more rigorous manner as we link it to our earlier discussion of descriptive statistics. The father of descriptive statistics is Sir Francis Galton; he reversed the traditional concern for eliminating variability in measurements in the calculus of errors by studying the variability itself. This was the key step in the development of all modern statistical analysis.

The calculus of judgment, or language of inductive inference, consists of the use of probability concepts to help in making decisions in empirical research. The problem of deciding whether data support or reject a hypothesis can be approached with probability theory. There has developed a wide range of statistical tests that comprise the part of statistics known as *inferential statistics.*

Of these four calculi, it is only the first that corresponds to our definition of complete mathematization of research. The calculus of aggregates involves processes directly modeled by probability distributions so that the whole research is mathematized. The three other probability calculi are more inductive than deductive in their application. In fact, the calculus of errors and the calculus of exploration remain concerned with simple "How much?" questions while the calculus of judgments is asking "What are the relationships?" They define research procedures rather than mathematical models, as we have defined them. The procedures remain nonetheless rigorously rooted in probability theory as an abstract deductive language. We deal with applications of probability theory to model processes in geographical research in Chapter Four. Here we present the probability background necessary for that discussion. We present the probability calculus as a branch of pure mathematics and then we show how it can be interpreted in an applied probability theory. Such interpretation has developed into the methodology of inferential statistics. The application of statistical tests has been at the forefront of the recent trend toward more rigorous approaches to geographical research. Therefore, we consider this topic in the second part of this chapter.

PROBABILITY CALCULUS

Probability theory was not fully axiomatized—that is, formalized into an abstract structure such as those described in Chapter One—until the first half of the twentieth century. Here we present a selected part of this impressive mathematical structure under the title pure probability theory. Our initial discussion is largely in terms of the set theory we presented in Chapter One, but we begin by introducing some further mathematics that will be useful later.

Some Preliminary Numerical Relationships

In most discussions of probability theory, it is necessary to introduce the basic properties of permutations and combinations and the binomial theorem. These are not in themselves part of probability theory because they relate generally to what mathematicians term "theories of counting." Such theories deal with numerical assessments of frequencies of outcome of events; seen in this light, they can be related directly to the probability experiments we deal with.

Permutations and Combinations

A *permutation* is a set of r objects in a given order. Consider objects A, B, C, and D. One permutation that could occur is (A, B, C, D); another is (B, D, C, A). In each case, we have permutations of 4 from a total of 4 objects. The number in a permutation does not have to equal the total number of objects, however. Thus, permutations of 3 from our 4 objects might be (A, B, C) or (C, D, A). These are referred to as permutations of 4 objects taken 3 at a time. In general, we talk of permutations of n objects taken r at a time.

Having defined a permutation, the next question is: How many different permutations of n objects are possible taken r at a time? Let's consider an example in which we can list all the permutations—specifically, 3 objects taken 2 at a time. The permutations are:

(A, B), (A, C), (B, A), (B, C), (C, A), (C, B)

Thus, with 3 objects taken 2 at a time, there are 6 permutations possible. We can write this as

$$P(3, 2) = 6$$

$P(3, 2)$ is read "the number of permutations of 3 objects taken 2 at a time."

We soon find that, as n increases, $P(n, r)$ becomes very large, and the method of listing all possible permutations becomes very cumbersome and time-consuming. It is clear that we need a mathematical solution to this problem of finding numbers of permutations. Such a solution is given by

$$P(n, r) = \frac{n!}{(n - r)!}$$

The exclamation mark signifies a *factorial*, so that $n!$ is read "n factorial." A factorial is simply the product of all numbers between 1 and the number under consideration. Thus,

$$4! = 4 \cdot 3 \cdot 2 \cdot 1 = 24$$

and, generally,

$$n! = n \cdot (n - 1) \cdot (n - 2) \cdots 2 \cdot 1$$

This rule covers all factorials except 0!. This is defined as 1 rather than 0, for reasons that need not concern us here. We use $0! = 1$ in what follows.

We are now in a position to return to our example, $P(3, 2)$. From our rule, this is

$$P(3,2) = \frac{3!}{(3 - 2)!} = \frac{3 \cdot 2 \cdot 1}{1} = 6$$

which, of course, is the answer we obtained by listing all permutations. We can now go beyond simple examples to cases in which listing is virtually impossible. If we consider all the letters in the alphabet, for example, we can ask how many different three-letter "words" can be made up. We have $n = 26$ and $r = 3$, so

$$P(n, r) = \frac{26!}{(26 - 3)!} = \frac{26!}{23!} = 26 \cdot 25 \cdot 24 = 15{,}600$$

which we do not attempt to list here!

A *combination* of n objects taken r at a time is *any* subset of r objects. Combinations differ from permutations in that order is not taken into account. (A, B, C) and (C, B, A) are two different permutations of 3, but they are classed as the *same* combination of 3 letters (A, B, C). It is clear that for any n objects there will be fewer combinations than there are permutations. For 3 objects taken 2 at a time, there are only the 3 combinations (A, B), (A, C), and (B, C), compared with the 6 permutations listed previously. The number of combinations of n objects taken r at a time is given by

$$C(n, r) = \frac{n!}{r!(n - r)!}$$

so that with $n = 3$, and $r = 2$, we have

$$C(3, 2) = \frac{3!}{2!(3-2)!} = \frac{3\cdot2\cdot1}{2\cdot1\cdot1} = 3$$

as we have demonstrated.

The Binomial Theorem

Combinations and permutations can vary in the number of objects they include, from 0 to n. Hence, there are $n + 1$ values of r possible. The $n + 1$ numbers of combination for each r are known as *binomial coefficients.* In this context, we find the expression $\binom{n}{r}$. However, $\binom{n}{r}$ differs from $C(n, r)$ only in terms of the notation employed; mathematically they are identical. Notice that $\binom{n}{r}$ is not the ratio n/r but, from the argument above,

$$\binom{n}{r} = \frac{n!}{r!(n-r)!}$$

When $n = 4$, the 5 binomial coefficients are

$$r = 0; \quad \binom{n}{r} = \frac{4!}{0!(4-0)!} = 1$$

For $r = 0$, there is 1 combination with no objects in it.

$$r = 1; \quad \binom{n}{r} = \frac{4!}{1!(4-1)!} = 4$$

For $r = 1$, there are 4 combinations with 1 object in them.

$$r = 2; \quad \binom{n}{r} = \frac{4!}{2!(4-2)!} = 6$$

For $r = 2$, there are 6 combinations with 2 objects in them.

$$r = 3; \quad \binom{n}{r} = \frac{4!}{3!(4-3)!} = 4$$

For $r = 3$, there are 4 combinations with 3 objects in them.

$$r = 4; \quad \binom{n}{r} = \frac{4!}{4!(4-4)!} = 1$$

For $r = 4$, there is 1 combination with 4 objects in it. They are called binomial coefficients because they are the coefficients that result from the expansion of binomial expressions of the form $(a + b)^n$. These coefficients therefore relate to the expansion of $(a + b)^4$ in the following manner:

$$(a + b)^4 = a^4 + 4a^3b + 6a^2b^2 + 4ab^3 + b^4$$

Thus the coefficients (or numbers preceding each product) are 1, 4, 6, 4, 1, which are identical to the coefficients derived above. This relationship is known as the *binomial theorem.*

We do not usually solve for $\binom{n}{r}$ in expanding binomial terms, because we can refer directly to what is known as *Pascal's triangle.* This is a triangular array of numbers produced by adding adjacent numbers to generate a new number in the row below them. To begin the process, we allocate 1 to the first and 1 to the last number in every row, with a 1 also for the apex of the triangle. The 1s in the second row add to 2, which is allocated to the middle of the third row. This 2 in turn adds to 3 with each of its neighboring 1s, to produce two 3s in the third row. The procedure can be continued indefinitely. The first 6 rows are:

1	$(a + b)^0$
1 1	$(a + b)^1$
1 2 1	$(a + b)^2$
1 3 3 1	$(a + b)^3$
1 4 6 4 1	$(a + b)^4$
1 5 10 10 5 1	$(a + b)^5$

The binomial expressions are shown on the right. We can recognize the binomial coefficients from our previous example $(a + b)^4$ in row 5. We shall see that this triangle, as well as having obvious algebraic utility, has very direct applications in probability theory.

Pure Probability Theory: Selected Aspects

We build our language of probability upon our previous knowledge of set theory with two additions. We are concerned with sets, unions, and intersections as defined in Chapter One and also with the concepts "contained in" and "complement." The set theory definition of these concepts is congruent with their usage in common language. A set is contained in another set when all the members of the first set are also members of the second set (written $A \subset B$). The complement of a set comprises all the members of the universal set (that contains all members) that do not belong to a particular set (written A^c). Given this knowledge of set theory, we can describe our language of probability under the usual headings—terms, axioms, and theorems. As we do so, we illustrate with two concrete models. The models we have chosen are the games of chance from which the theory originated. We consider a deck of playing cards and dice in imaginary sampling experiments.

Language of Probability

UNDEFINED TERM

Probability—it is convenient to leave this basic term undefined at this stage because it is interpreted in different ways in different schools of thought. In the context of our concrete models, we can make the following interpretation of probability. If we ask what is the probability that a 3 will occur when we throw a dice, we might argue as follows. A 3 is one of 6 possible outcomes when we throw a dice, and so the probability of its occurring is 1 in 6 or $\frac{1}{6}$. Similarly, the probability of drawing the ace of clubs from a deck of 52 cards is $\frac{1}{52}$.

DEFINED TERMS

Sample space (S) is the set of all outcomes of some given experiment. An example of a sample space is all the outcomes that may occur when we pick a card from our deck or throw our dice. Both these operations represent simple experiments, and all outcomes therefore constitute sample spaces.

Sample point is a particular outcome, a member of the sample space. Two examples from our concrete models might be the drawing of the 2 of hearts or the throwing of a 6 on the dice.

Event (A) is a set of outcomes, a subset of the sample points ($A \subset S$). Such an event might be, for example, any club occurring when we choose a card or any even number thrown on the dice.

Probability function is the probability that event A will occur. It is written $p(A)$. The probability that a club will be drawn from a deck of cards is 13 in 52, and it equals a 25% chance. Do not confuse this notation for probability function with the permutation notation above.

Conditional probability is the probability of one event A given another event E. It is written $p(A|E)$. Given that our dice has produced an even number, we might ask: What is the probability of that number being 4 or greater than 4? Or, given that we have drawn a picture card from our deck, what is the probability that this card will be a queen?

Finite equiprobable space is a sample space in which all outcomes have equal probability of occurring. Both of our concrete models involve finite equiprobable spaces, if we assume that cards and dice are fair (not stacked or loaded).

Independence specifies the situation in which the probability that event A will occur is not influenced by whether event B has occurred. This may be written $p(A) = p(A|B)$. The probability that the card we draw will be any specific number is independent of suit, because all suits have the same set of numbers. Thus, the probability that a 9 will occur is $\frac{4}{52}$, which reduces to $\frac{1}{13}$. If we know that the card is a club, the probability that it will be a 9 is $\frac{1}{13}$, which is the same.

Mutually exclusive events are two or more events whose sets do not intersect. The occurrence of a club and of a heart are mutually exclusive events in our card-drawing experiments. Similarly, even numbers and odd numbers constitute mutually exclusive events when we throw a dice.

AXIOMS

1. For every event A, $0 \leqslant p(A) \leqslant 1$.

This simply specifies the range of values we are to allocate to probabilities. Henceforth we talk not in terms of percentage chance but of probabilities ranging from 0 to 1. The probability that a club will occur becomes 0.25, the probability that an even number will be thrown becomes 0.5.

2. $p(S) = 1$

The total probability in a sample space is one; that is, the sum of probabilities of all the sample points is

one. All sample points in the card experiment sum to 52/52 = 1, and in the dice experiment they sum to 6/6 = 1.

3. If A and B are mutually exclusive events, then $p(A \cup B) = p(A) + p(B)$.

The probability of getting a 1 *or* a 2 in one throw of the dice is $\frac{1}{6} + \frac{1}{6}$, which equals $\frac{1}{3}$. The probability of getting a heart or a club when we pick a card from a deck is $\frac{1}{4} + \frac{1}{4}$ which equals $\frac{1}{2}$ This axiom is commonly known as the *addition law of probability*.

4. $p(A|E) = \dfrac{p(A \cap E)}{p(E)}$

This axiom simply tells us how to calculate conditional probabilities. We use our dice example from the definition for conditional probability to illustrate this axiom. The probability that a number 4 or greater will be thrown if we know that the outcome is an even number is the probability of the intersection between the two events divided by the probability that an even number will occur. The event "4 or greater" has two sample points (4 and 6) in common with the event of an even number, so that the probability that the intersection of these two events will occur is $\frac{1}{3}$. The probability that an even number will occur is $\frac{1}{2}$. Therefore, the probability of a 4 or greater, given the even property of the outcome, is $\frac{1}{3} \div \frac{1}{2}$, which equals $\frac{2}{3}$.

THEOREMS

1. $p(A^c) = 1 - p(A)$

Proof: By definition, $A \subset S$.
Therefore, $A \cup A^c = S$.
By Axiom 3, $p(A \cup A^c) = p(A) + p(A^c)$.
Therefore, $p(S) = p(A) + p(A^c)$.
By Axiom 2, $p(S) = 1$.
Therefore, $p(A) + p(A^c) = 1$.
Therefore, $p(A^c) = 1 - p(A)$.

This theorem has been included as a simple example of deductive logic in probability language. It says simply that if the probability of drawing a diamond card is $\frac{1}{4}$ then the probability of not drawing a diamond card is $1 - \frac{1}{4}$, which equals $\frac{3}{4}$.

2. For any two events A and E, $p(E \cap A) = p(E) \cdot p(A|E)$.

Proof: By Axiom 4, $p(A|E) = \dfrac{p(A \cap E)}{p(E)}$.
From set theory, $A \cap E = E \cap A$.
Substituting and multiplying through by $p(E)$, $p(E \cap A) = p(E) \cdot p(A|E)$.

This is known sometimes as the *multiplication theorem of conditional probability* and simply relates to the probability that both events will occur in the same experiment. The probability that an even number will occur *and* a number 4 or greater will occur is $\frac{1}{2} \cdot \frac{2}{3}$ (from Axiom 4), which equals $\frac{1}{3}$ (that is, 4 or 6).

3. For two independent events A and B, $p(A \cap B) = p(A) \cdot p(B)$.

Proof: From Theorem 2, $p(A \cap B) = p(A) \cdot p(B|A)$.
By definition for independence,
$p(B) = p(B|A)$.
Therefore, $p(A \cap B) = p(A) \cdot p(B)$.

This is termed sometimes the *multiplication theorem for independent events*, but it is commonly called simply the *multiplication law of probability*. It relates to the probability that two specified independent events will occur in two experiments. Thus, the probability of throwing a 1 and then a 2 is $\frac{1}{6} \cdot \frac{1}{6}$, which equals only $\frac{1}{36}$.

This description constitutes our language of probability interpreted by using familiar games of chance as concrete models. In most applications of probability theory, we find at the very least the defined term "probability function" lurking in the background. Before we consider applications, we expand our discussion of this term into the realm of the probability distribution.

We have defined a probability function as simply the probability that an event will occur in a specified experiment. If we array all possible events in order, we produce a *probability distribution*. Thus, in our six-faced dice example, we produce the following:

Event = number on the dice	1	2	3	4	5	6
Probability	$\frac{1}{6}$	$\frac{1}{6}$	$\frac{1}{6}$	$\frac{1}{6}$	$\frac{1}{6}$	$\frac{1}{6}$

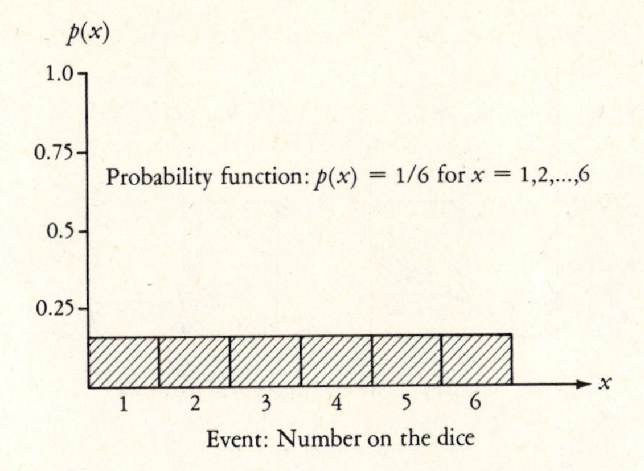

Figure 3.1 A truncated rectangular distribution.

This function can be written algebraically:

$$p(x) = \tfrac{1}{6}, \quad \text{for } x = 1, 2, \cdots, 6$$

in which $p(x)$ is the probability and x is the resulting event. The distribution is called a *truncated rectangular distribution*, because if we plot it on graph paper we produce a horizontal line from $x = 1$ to $x = 6$ (Figure 3.1).

The Binomial Distribution

Rectangular probability distributions are the simplest that occur in probability theory. Other, less simple distributions are theoretically more interesting. Consider the experiment of tossing a coin and recording the number of times that it falls heads. We assume that the coin is unbiased, so that the probability that heads will occur is 0.5. If we toss the coin twice, there are four possible outcomes: HH, HT, TH, and TT. These are all the permutations of two events taken two at a time. We are interested not in the order—whether heads comes before or after tails—but simply in the number of times heads occurs. Thus, we are interested in three combinations only: HH = 2 heads, HT (or TH) = 1 heads, and TT = 0 heads. These events are not equiprobable, however, and they do not define a rectangular distribution. We can demonstrate this easily by using the addition law and the multiplication law for independent events. The probability that heads will occur twice in two tosses is the product of two individual probabilities, which gives $(0.5)^2 = 0.25$. Similarly, the probability of obtaining no heads is $(0.5)^2 = 0.25$. The probability of obtaining one head is 0.25 for the permutation HT and 0.25 for the permutation TH so that, by the addition law, the probability is $0.25 + 0.25 = 0.5$. This can be summarized:

Event = number of heads	0	1	2
Probability	0.25	0.5	0.25

It does not define a rectangular distribution. In fact, the probability function is given by the expression

$$p(x) = \binom{2}{x} \cdot 0.5^x \cdot 0.52^{2-x}$$

which defines a *binomial distribution* in which $p(x)$ is the probability again and x is the number of heads. Thus when $x = 0$,

$$p(0) = \binom{2}{0} 0.5^0 \cdot 0.5^2 = \frac{2!}{0!(2-0)!} \cdot 1 \cdot 0.25 = 0.25$$

as we have previously shown. Similarly, when $x = 1$, then $p(1) = 0.5$, and when $x = 2$, then $p(2) = 0.25$.

This distribution is not called a binomial distribution simply because it includes binomial coefficients. The binomial theorem is concerned with the expansion of the binomial expression $(a + b)^n$, as we have already seen. We can interpret our tossing experiment above in this same way, however. If we let p equal the probability that the single event (heads) will occur and q equal the probability of its not occurring, we have $p = q = 0.5$. We can now state our previous argument, by using the multiplication law, as:

$$p^2 + pq + qp + q^2 = 1$$

which simplifies to

$$p^2 + 2pq + q^2 = 1$$

which is, of course, the expansion of
$(p + q)^2 = 1$
We have a probabilistic version of a binomial expression, set to unity because the total probability is 1.

Thus far we have considered only the case in which

$n = 2$. In this case, there are two tosses of the coin. The binomial expansion allows us to generalize this to cover any number of trials. We therefore define any binomial probability distribution as

$$p(x) = \binom{n}{x} p^x q^n$$

for n trials. When $p = q = 0.5$, this distribution gives us a mathematical model for our coin-tossing experiment. With Pascal's triangle, we do not have to compute the binomial coefficients. For 4 tosses, the probability of getting heads 0 to 4 times is as follows. From the fifth row of Pascal's triangle, we have coefficients of 1, 4, 6, 4, 1, so that the binomial expansion is:

$$p^4 + 4p^3q + 6p^2q^2 + 4pq^3 + q^4 = 1$$

and, because $p = q = 0.5$, we have

$$0.0625 + (4 \cdot 0.125 \cdot 0.5) + (6 \cdot 0.25 \cdot 0.25) + (4 \cdot 0.5 \cdot 0.125) + 0.0625 = 1$$

which simplifies to

$$0.0625 + 0.25 + 0.375 + 0.25 + 0.0625 = 1$$

These numbers are the probabilities of obtaining 0, 1, 2, 3, and 4 heads, respectively and therefore they define the binomial distribution for $n = 4$ with $p = q = 0.5$. It is clear that with 4 tosses the chances of getting no heads is very small ($p = 0.0625$). This binomial distribution is illustrated in Figure 3.2a.

It is not necessary that $p = q = 0.5$. Binomial distributions can be derived from any value of p between 0 and 1 with $q = 1 - p$. In this situation we may wish to continue with our coin-tossing model with a biased coin, although it is more usual to think in terms of the probability theorist's urn model. This is a hypothetical container from which white and black balls are drawn. The experiments consist of drawing a ball and recording whether it is black or white. If there are equal numbers of black and white balls, and we are concerned with the probabilities of drawing various numbers of black balls, then $p = q = 0.5$; here p is the probability of drawing a black ball. With four withdrawals from the urn, the result is identical to that in our coin-tossing example, and the probability of getting no black balls, like the probability of getting no heads, is

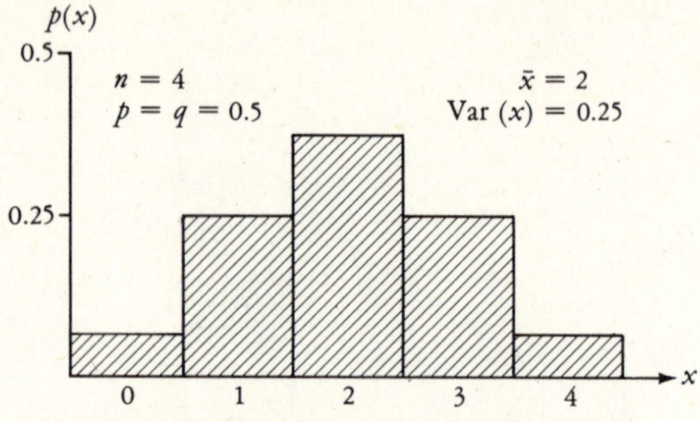

(a) Event: Number of "heads" to occur in n throws

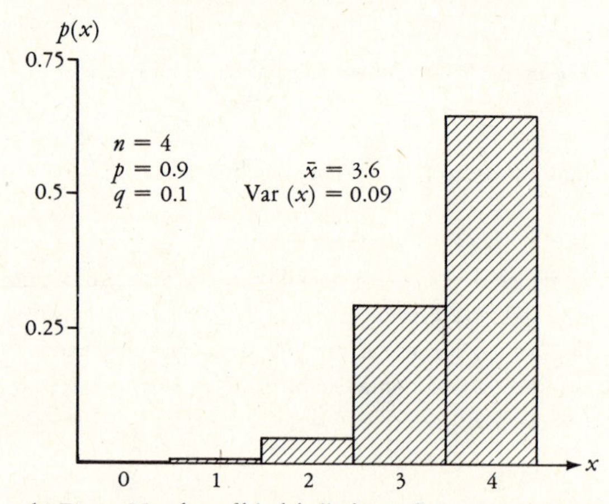

(b) Event: Number of black balls drawn from urn in n draws

Figure 3.2 Binomial distributions.

0.0625. We do not have to hypothesize equal numbers of black and white balls, however. With 9 black balls to every 1 white ball in the urn, $p = 0.9$ and $q = 0.1$. Thus, with four withdrawals from the urn, the probabilities of obtaining from 0 to 4 black balls are given by:

Event = number of black balls	0	1	2	3	4
Probability	$q^4 =$ 0.0001	$4q^3p =$ 0.0036	$6q^2p^2 =$ 0.0486	$4qp^3 =$ 0.2916	$p^4 =$ 0.6561

which clearly reflects the rarity of drawing white balls. In fact, 4 white balls appearing in 4 withdrawals (that is, no black balls) has a probability of only 0.0001, only 1 time in every 10,000 experiments!

We can see that binomial distributions are not always symmetrical (Figure 3.2b). In fact, symmetry is associated with the special case $p = q = 0.5$ and very large n. The shape of the binomial distribution depends on p and n. Hence, we can define the mean and variance of any binomial distribution in terms of these two parameters:

$$\bar{x} = np \qquad \text{and} \qquad \text{Var}(x) = pq$$

and as always, the standard deviation is the square root of the variance. These statistical concepts are interpreted in the same way as when we calculate empirically for frequency distributions. Thus, for our coin-tossing experiment, the mean of the resulting distribution is $4 \cdot 0.5 = 2$ and the variance is $0.5 \cdot 0.5 = 0.25$ (Figure 3.2a). In the urn model experiments, on the other hand, the mean and variance differ because of the different probabilities used. The mean is $4 \cdot 0.9 = 3.6$ and the variance is 0.09 (Figure 3.2b).

The Normal Distribution

Both the rectangular and binomial distributions are *discrete* probability distributions. The probabilities are given only for integer numbers 0, 1, 2, 3, and so on, and not for real numbers. We cannot draw 3.5721 black balls from an urn, no matter how many withdrawals we make. A second family of probability distributions are *continuous* in that probabilities are given for all points along the x axis. By far the most widely used of this type of distribution is the normal distribution; it has been said that the normal distribution is to statistics what the straight line is to geometry.

We mentioned briefly the normal distribution at the beginning of this chapter in the context of the calculus of errors. In fact, the calculus of errors is the original source of this distribution. A wide variety of measurement procedures have been found to have a distribution that is bell-shaped about the true value. This finding is associated with French mathematician Karl Friedrich Gauss, and the normal distribution is referred to sometimes as the Gaussian distribution. Its major application today is in inferential statistics, and we must consider this distribution in some detail here.

The normal distribution is given by the probability function

$$p(x) = \frac{1}{\sigma\sqrt{2\pi}}\, e^{\left[\frac{-(x-\bar{x})^2}{2\sigma^2}\right]}$$

in which π and e are constants (3.142 and 2.718, respectively). This formula looks quite complicated but on closer inspection we find that probabilities $p(x)$ depend on only two parameters, the mean $(\bar{x})$ and the standard deviation (σ). In fact, people using the normal distribution very rarely compute values of $p(x)$ from this formula, because they are able to look up required values in tables of the normal distribution (see, for example, Appendix Table A.2a). Suppose we want to construct a normal distribution in which $\bar{x} = 100$ and $\sigma = 10$. The first step is to standardize our measurement scale. This means that we merely convert our measures into units of their own standard deviation about their mean. In this case, we convert $x = 100$ to $z = 0$, because $\bar{x} = 100$. Now, $x = 105$ is half a standard deviation above the mean, and so this becomes $z = +0.5$. On the other hand, $x = 80$ is two standard deviations below the mean, so that it is converted to $z = -2$. All x values are converted to z values in this way. Table A.2(a) gives us probabilities for different locations on this z axis. Thus, at $z = 0$ ($x = \bar{x} = 100$), $p(z) = 0.3989$; at $z = \pm 1$ ($x = 95$, $x = 105$), $p(z) = 0.2420$, and so on. These are plotted in Figure 3.3a to give us our normal distribution on the z and x scales.

We are now in a position to illustrate graphically what we mean when we say that the normal distribution depends on the mean and standard deviation. In Figures 3.3b and 3.3c, we use the procedure above to draw two further normal distributions, defined by $\bar{x} = 100$, $\sigma = 2$, and $\bar{x} = 110$, $\sigma = 10$. The mean clearly controls the location of the distribution on the x axis while the standard deviation controls the spread. This is consistent, of course, with our definition of mean and standard deviation in Chapter One in the context of descriptive statistics. In terms of the x axis, there is not one but there are many standard deviations, whereas on the standardized z scale there is only the

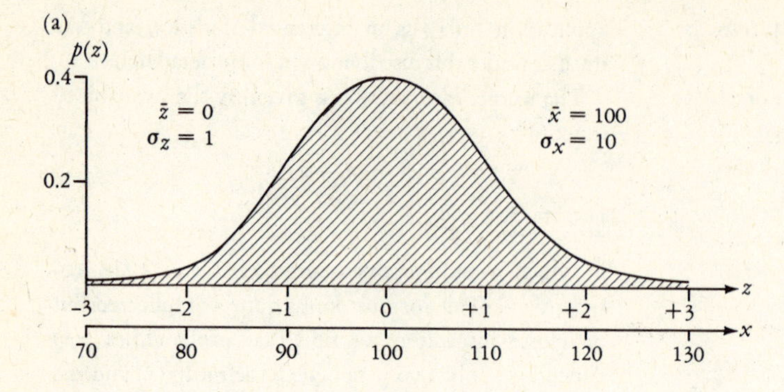

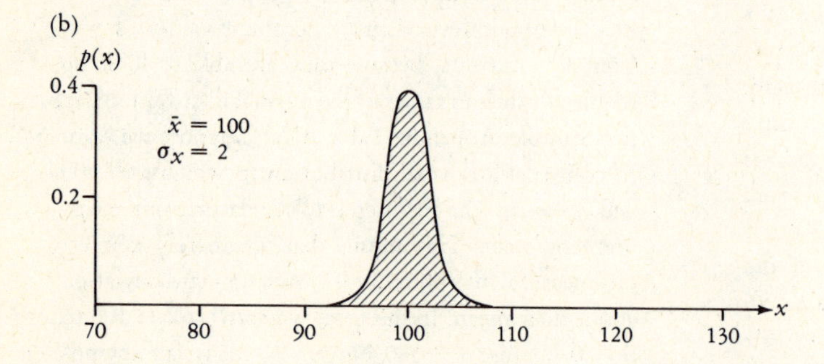

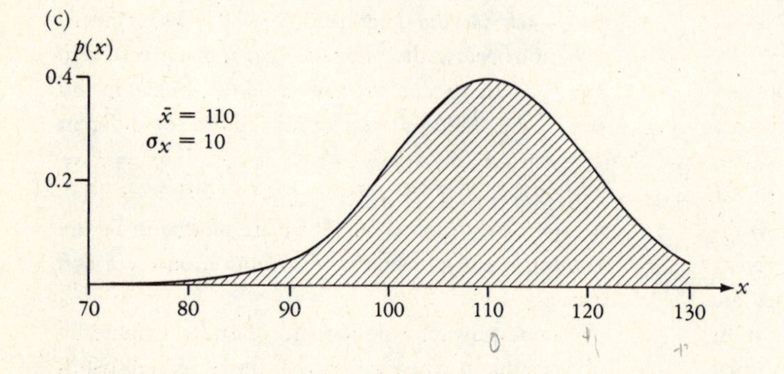

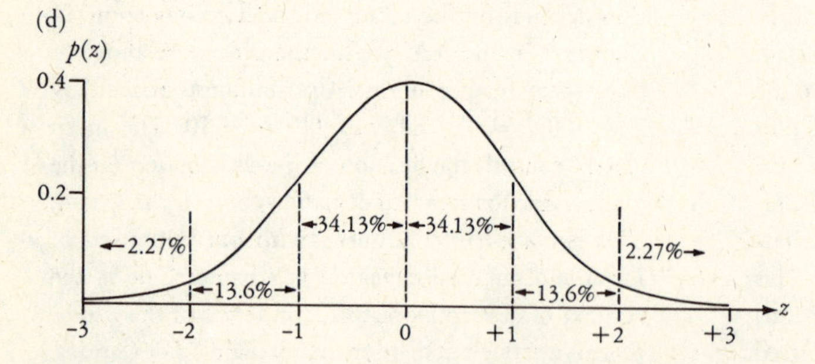

Figure 3.3 Normal distributions. Percentages given in part d refer to proportions of events occurring in various parts of the distribution.

one distribution given in the standard tables, shown in Figure 3.3a. The ease of translation between x and z scales makes application of the distribution particularly straightforward, as we have seen. Some properties of this distribution on the z scale are shown in Figure 3.3*d*. These properties are of particular importance in its application in inferential statistics.

In Chapter One we introduced the empirical frequency distribution of descriptive statistics. In the discussion here we have introduced theoretical distributions—theoretical in that they have been derived independently of empirical data. These theoretical distributions can be defined further as probability distributions, because they distribute the total probability of a sample space over different sample outcomes. Many such distributions have been developed by mathematicians, and we introduce some examples as process laws in Chapter Four. The truncated rectangular, binomial, and normal distributions are sufficient for our discussion of inferential statistics. Our next task, therefore, is to find the link between empirical distributions and probability distributions, between hypothetical sampling experiments and sampling in the real world and between probabilities in theory and probabilities in practice.

Applied Probability Theory: Probabilistic Inferences

By probabilistic inferences we simply mean making inferences concerning probabilities in the real world. The statement "it will probably rain today" is a probabilistic inference. Such statements clearly predate mathematical probability theory; they are probably as old as language itself. This phrase, too, is another probabilistic inference. We use the adverb "probably" to cover our ignorance of the history of the concept of probability in linguistics. Probabilistic inferences need not be based on ignorance or lack of information, however. If we have information concerning a subject, we can make more refined probabilistic inferences. Our statement "it will probably rain today" may become "there is a 50% chance of rain this afternoon" in the words of the meteorologist with weather forecasting equipment. The most refined probabilistic inferences are those that bring probability theory into contact with the real world. The full development of this type of inference is very much a twentieth-century development. It clearly relies on finding the link between probability theory and the real world that we referred to above. The first vital step to this end is to define empirically the concept of probability. The second step is to define probabilistic inferences rigorously in the context of our language of probability. We take each step in turn.

The Relative Frequency Interpretation of Probability

In natural language, the word "probability" is associated with notions of chance. We use this association in our empirical interpretation. A cotton farmer or an agricultural geographer might ask the question, "What is the chance of a certain small area in Tennessee experiencing more than two hundred frost-free days in the coming year?" The farmer will answer this question by using his past experience in Tennessee; the geographer might turn to the records of the nearest meteorological station. Suppose we find that, with records existing for the last 80 years, there have been 40 years having fewer than 200 frost-free days and 40 years having more than 200 frost-free days. Assuming no major climatic changes in the very recent past, we would think it reasonable to extrapolate from these records and suggest a 50 percent chance or 0.5 probability of more than 200 frost-free days in the coming year. How would we reach this rather precise figure? We divided the number of times that the event we are concerned with occurred (the 40 years with more than 200 frost-free days) by the total number of years; we divided 40 by 80. In a more northern meteorological station that has recorded that only 20 of the last 80 years have experienced 200 frost-free days, we would infer that the probability that 200 frost-free days will occur is 20/80, which gives us a 0.25 probability.

This interpretation of probability is usually referred to as the *relative frequency* approach. It may involve either theoretical or empirical definitions. The frost-free days example is purely empirical because it is based on events that have happened; we calculated probabilities in our games of chance models in the same way but we based them on events that had not actually occurred although we imagined that they might occur. In both cases, probability is defined as

$$p(\mathrm{A}) = \frac{\text{frequency (A)}}{\text{frequency } (T)}$$

in which A is the event under consideration and T is the total number of events. This definition allows us to translate events in the real world as studied by geographers into mathematical abstract probability theory.

We have presented many new ideas and concepts in just a few pages in this chapter. Even the brightest student will find quick absorption of all the ideas somewhat difficult. Now that we have an empirical definition of probability, we can take a break from our discussion and consider the position we have reached. In the following examples, all the probabilities are based on evidence of relative frequencies. Each question relates to one definition, axiom, or theorem from our language of probability. The purpose of the questions is for the student to become familiar with our new language in the empirical contexts faced by geographical researchers. Students should consider each question in turn and attempt an answer that specifies the part of the language of probability they are drawing on.

1. If the probability that a rainfall station will record fewer than 20 inches of precipitation is $p(\mathrm{A}) = 0.43$, what is the probability that the station will record 20 inches or more?
2. If the probability that a consumer will visit a specified shopping center ($p(\mathrm{A})$) is assumed to be independent of the age of the consumer and we know that $p(\mathrm{A}|\mathrm{B}) = 0.6$ (in which event B is a consumer whose age is between 20 and 30 years), then what is the probability that any consumer will patronize the shopping center?
3. The probability that a migrant will move a distance of up to 1 mile was found to be $p(\mathrm{A}) = 0.2$. In the same study, the probability that a migrant will move between 1 and 2 miles was found to be $p(\mathrm{B}) = 0.3$. What is the probability that a migrant will move any distance of up to 2 miles?
4. In a detailed study of an individual farm, it was found that 20% of the farmland had clay soil. Furthermore, 80% of this clay soil was covered by permanent grass. If we were to point sample randomly on this farm, the probability that a point would fall on clay soil is $p(\mathrm{B}) = 0.2$, and the conditional probability that the point would fall on permanent grass, given that it had fallen on clay soil, is $p(\mathrm{A}|\mathrm{B}) = 0.8$. What is the probability that a random point would fall on both clay and permanent grass?
5. In a study of urban areas having populations of over 100,000, it was found that a greater than average death rate is unrelated to population size. If the probability that a town that has a greater than average death rate is $p(\mathrm{A}) = 0.5$, and if the probability that an urban area that has a population in excess of 500,000 is $p(\mathrm{B}) = 0.1$, then what is the probability that any town we might choose to study will have a population in excess of 500,000 and a greater than average death rate?

Before you read the answers below, check to see that you have employed the correct part of our probability language. The correct parts of the calculus are: (1) Theorem 1, (2) definition for independence, (3) Axiom 1, (4) Theorem 2, and (5) Theorem 3. If you have used these parts of the calculus correctly, you should have obtained the following answers:

1. $p(\mathrm{A}^c) = 1 - 0.43 = 0.57$
2. $p(\mathrm{A}) = p(\mathrm{A}|\mathrm{B}) = 0.6$
3. $p(\mathrm{A} \cup \mathrm{B}) = p(\mathrm{A}) + p(\mathrm{B}) = 0.2 + 0.3 = 0.5$
4. $p(\mathrm{B} \cap \mathrm{A}) = p(\mathrm{B}) \cdot p(\mathrm{A}|\mathrm{B}) = 0.2 \cdot 0.8 = 0.16$
5. $p(\mathrm{A} \cap \mathrm{B}) = p(\mathrm{A}) \cdot p(\mathrm{B}) = 0.5 \cdot 0.1 = 0.05$

If you did not obtain these answers, you should read the previous section again. If you did obtain these answers, you are in a position to begin consideration of more elaborate uses of probability in geographical contexts.

Deterministic and Probabilistic Inferences

In Chapter One we identified two types of inference—deductive and inductive. Deduction consists of drawing conclusions from stated assumptions using logical reasoning. Deductive inferences are certain and definite, if we accept the assumptions, and therefore they are known sometimes as *deterministic inferences.* All mathematical conclusions, including those of probability theory, are of this type, and so are many empirical arguments. In fact, late in the nineteenth and early in the twentieth centuries, geographers commonly employed deterministic arguments in their search for environmental causes for human behavior. "Climatic determinism," as this was called, has been rejected in geography, largely because the assumptions of the arguments were greatly oversimplified. Determinism has long been out of favor in geography, but not all empirical deterministic statements are invalid. We illustrate this point with an example from the work of two philosophers of science, R. Cohen and Ernest Nagel.

We can construct a very simple deterministic argument to show that there are at least two people in New York with exactly the same number of hairs on their heads. It can be shown that the maximum density of hairs on any scalp will not be as great as 5000 per square centimeter. Furthermore, it can be accepted reasonably that no person will have a scalp with an area larger than 1000 square centimeters. Therefore, it follows that the maximum number of hairs anybody can have on his or her head is $5000 \cdot 1000$, which gives a total of 5 million. There are more than 5 million people living in New York, however. Therefore, if we find one inhabitant of New York to correspond with every possible number of hairs from 0 to 5 million, there will still be people left over who must therefore have the same number of hairs on their heads as some of the first 5 million. Thus, we can say that there are at least two people living in New York who have exactly the same number of hairs on their heads.

This conclusion is certain; it is deterministic. Not all conclusions can be so conclusive. We suggested in Chapter One that inductive inferences are usually open to doubt in some way. There is usually some evidence that can be brought forward that contradicts the conclusion to some degree. In fact, we can think of all inferences in terms of probabilities of their validity. Deterministic inferences are seen consequently to be a special case in which the probability is unity. Below this there exists a range of probabilities upon which we can place any inductive inferences we make. In geography, the reaction against climatic determinism took the form of "possibilism," which allowed for a wide range of possible outcomes in the relations between people and environment. It has been suggested by Fred Lukerman (1965) that this, largely French, school of geography was influenced by the long tradition of probability theory in French mathematics, dating back to Fermat, Pascal, and Laplace. In 1953, O. Spate coined the term "probabilism" as a sort of compromise between possibilism and determinism. None of these schools of geography actually used the calculus of probability, even though they might knowingly or unknowingly have used the concepts. Let's illustrate how we can make inferences that are less than deterministic using our simple language of probability.

In June 1964, a woman was robbed in San Pedro, California, and a blonde woman with her hair in a ponytail was seen to leave the scene of the crime and get into a yellow car driven by a bearded, black American. Some time later, Malcolm and Janet Collins were picked up because they fitted this description down to the last detail. They were not personally identified, however, and there was therefore only circumstantial evidence to link them with the crime. The prosecution used probability theory to show how strong the circumstantial evidence actually was. A series of probabilities were allocated to the various elements of the description. These ranged from $\frac{1}{4}$ for the probability of Janet Collins having blonde hair to 1/1,000 for the probability that a white girl would be accompanied by a black man. Considered separately, some of these probabilities represented quite common events (1 in 4 women were assumed blonde), but when they are combined together to make up the complete set of circumstantial evidence they become much more impressive. By using the multiplication law, the prosecution was able to show that there was only one chance in 12 million that Malcolm and Janet Collins were not in fact the couple seen leaving the crime. Malcolm Collins received a sentence of imprisonment for from one year to life; Janet Collins was sentenced to not less than one year in prison. The case made legal history because it was the first time that probability theory had been employed to specify the broad notion "beyond reasonable doubt."

There are no doubt several morals to this story; we concentrate on the methodological implications. Notice that it was not proven conclusively that Malcolm and Janet Collins were at the scene of the crime. One chance in 12 million might be beyond reasonable doubt, but it is not a deterministic inference. Probability theory was used precisely to specify the degree of doubt and, the degree having been specified, a decision of guilty was made. This is exactly how probability theory is used in making decisions in research. It enables us to specify just how certain we can be that a conclusion we have made is correct. In most cases in geographical research, we are not at all near the degree of certainty of the legal example. By using the language of probability, however, our research inferences can be taken beyond the philosophical opinions of individual researchers—possibilists, probabilists, or determinists—and can be given more general objective criteria. With a methodology based on probability theory, any single researcher can deal with the whole range of probabilities from zero (possibilists) to unity (determinists), depending on the empirical data collected to support the hypotheses.

Simple Probabilistic Inferences in Geographical Research

Let's show how we can apply the principles described above in geographical research contexts. We translate a geographical problem into the language of probability theory, deduce a probability conclusion, and then draw probabilistic inferences with regard to the original geographical problem. Let's see how this procedure works in practice.

Geographer Barry Garner (1967) has suggested a model of the internal structure of major shopping centers consisting of five concentric zones about the point of maximum land value (Figure 3.4). Each concentric zone is associated with some single or mixed hierarchical level of shops. Zone I includes the highest level of shopping outlets, such as large department stores, and is therefore termed the regional zone. This model was developed from detailed work on Chicago's retail structure, and we might test the model in another city to check on its more general validity. We use probability theory to help us carry out this research project.

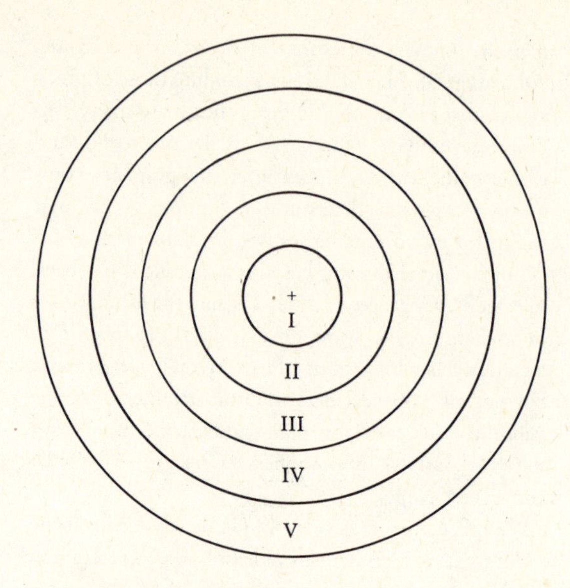

+	Zones: Peak land value
I	Regional level
II	Mixed regional and community level
III	Community level
IV	Mixed community and neighborhood level
V	Neighborhood level

Figure 3.4 A concentric ring model of major shopping centers.

A city is chosen arbitrarily and data on the retail land use of each block within the central business district (CBD) are collected. From this data, each block surveyed is allocated to one of the five hierarchical types of the Garner model. The distance of each block from the center of the CBD is also measured. We are now in a position to assess the model by testing the hypothesis that each of the five types of block is successively further from the center of the CBD. The average distance from the center for each type of land use block is listed in Table 3.1.

Table 3.1 looks very encouraging for our hypothesis, because the five types of block are in exactly the same order as predicted by the Garner model. We should not simply accept our hypothesis at this point by concluding that the model fits our chosen city. We

Table 3.1 Hierarchical Types of Block and Distance from the Center of the CBD

Type of block	*Average distance from center (yards)*
I	100
II	190
III	370
IV	390
V	580

must admit, for instance, that even if the city's retail pattern had been formed under purely random conditions, and if we had carried out data collection and measurement as described above, we would be sure to produce some ordering of our types of block. We must therefore ask what the probability is that the Garner sequence will occur *even if the CBD's retail structure is simply randomly ordered.* To answer this question, we must translate the problem into probability language.

Let's begin by considering blocks of type I. If we assume for the moment a random pattern, then blocks classified as type I may still be the nearest to the center of the CBD 1 time in 5. The probability that the first part of our results in Table 3.1 will occur is 0.2. If we turn now to the remaining blocks and concentrate on type II, we can see that even in a random situation these may be closest to the center 1 time in 4 (remember that, having disposed of type I, only four types remain at this point). The probability is therefore 0.25. If we repeat this procedure with the remaining blocks, then the probability that type III will be nearest to the center of those remaining is 0.33. This leaves just two types of block, and the probability that these will occur with IV followed by V in terms of distance is 0.5. If the retail pattern was controlled by random forces, then each block type is located independently of each other, and we may apply the multiplication theorem for independent events. The probability of finding the overall sequence I, II, III, IV, and V in a randomly organized CBD can be calculated as $0.2 \cdot 0.25 \cdot 0.33 \cdot 0.5 = 0.0083$. This means simply that, if we carried out this operation in a thousand cities, all of which had random shopping patterns in their CBDs, then in about eight examples we would find a sequence that exactly fits the Garner model prediction.

As a check we can derive the same conclusion by an alternative route. The sequence I, II, III, IV, V is one permutation of 5 events. We know from previous discussion that the total number of permutations of 5 objects taken 5 at a time is

$$P(5, 5) = \frac{5!}{(5-5)!} = \frac{5 \cdot 4 \cdot 3 \cdot 2 \cdot 1}{1} = 120$$

In purely randomly organized CBDs, every one of the 120 possible permutations is equally likely to occur. Thus, the probability that our observed sequence will occur under these conditions is $1/120 = 0.0083$.

What does this mathematical conclusion of $p = 0.0083$ mean in terms ot our geographical research problem? We can interpret it as follows. We do not reject totally the possibility that purely random forces have caused the pattern we have measured in Table 3.1, but we notice that, if this is in fact the case, then we have hit upon a particularly rare instance that occurs approximately only 8 times in 1000. It seems more likely that our data reflect a city that fits the Garner model. If we draw this conclusion, we are in a position to specify explicitly the degree of confidence we can place in this inductive inference. The chance that this inference is incorrect and that the alternative random model is valid is only 0.0083. Our conclusion is probabilistic and not deterministic; there is a 0.0083 chance that we have drawn an incorrect inference from our data in Table 3.1. This is probability theory acting as a very useful tool in decision-making in research.

The example above is a rather idealized and hypothetical research problem chosen for its simplicity. If we assume that it has illustrated the basic principles of applying probability theory in research, we may turn to a research problem that is not hypothetical and that therefore introduces us to some of the problems of actually applying probability theory in practice.

In 1849 and 1850 the town of Nottingham in England experienced a minor cholera epidemic in which twenty-seven people died. The deaths may be divided into two groups: (1) the initial victims of the disease

(a) August-October 1849

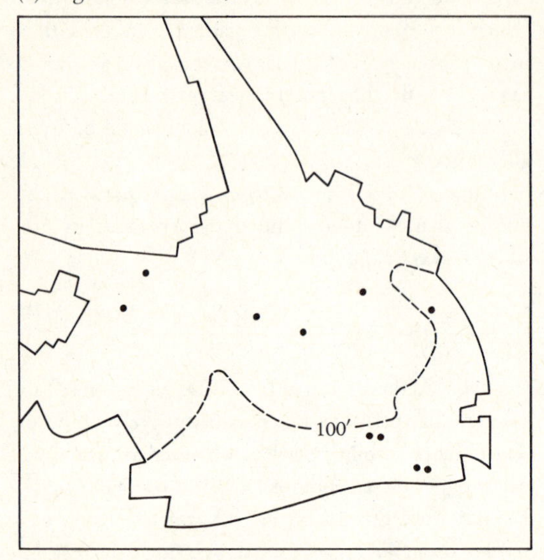

(b) November 1849-September 1850

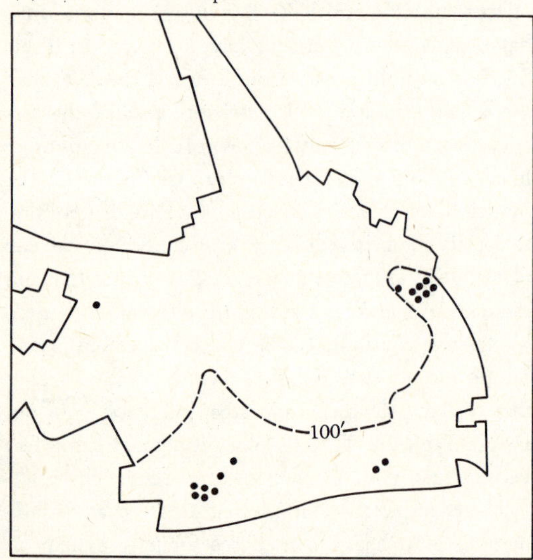

Figure 3.5 Cholera deaths in Nottingham, 1849 and 1850.

when it first visited Nottingham between August and October 1849 and (2) the subsequent victims of the disease as it spread within the town between November 1849 and September 1850. The residences of these two groups are shown in Figures 3.5a and 3.5b. It is hypothesized that the distribution of the cholera deaths reflects the manner in which the disease is transmitted. Cholera is a water-borne disease and mid-nineteenth-century Nottingham was a town with many open sewers. They drained from the higher parts of the town in the north toward the lower land on the southern edge of town. People living in the south of town got everybody else's filth and often their diseases also. Cholera would seem to be a case in point. A one-hundred-foot contour has been included in Figure 3.5 to test this hypothesis. This contour line has been chosen for convenience because it divides Nottingham's population into approximately two equal parts. If the north-south differences in elevation had no influence on the spread of cholera within the town, then we would expect approximately half the cases to lie north of the hundred-foot contour line and half to lie below it. This is in fact true for the initial pattern of victims; there were five deaths on each side of the contour line. This is not surprising when we remember that the initial pattern reflects the locations where the disease was introduced into Nottingham from outside. If our hypothesis is correct, however, once the disease is established within the town, we might expect it to have been influenced by the physical structure of town site as it claimed more victims. The evidence of the deaths after October 1849 is encouraging in this respect. Of seventeen deaths, all but one occurred to residents living in the lower land south of the hundred-foot contour line. But we should not accept the hypothesis with this seemingly clear-cut evidence. We must admit that, even if the relief factor had had no influence on the pattern of deaths, the seventeen victims would still exhibit some pattern of points on our map. What are the chances that our particular pattern will occur if the cholera victims are considered to be unlucky random victims? If we ignore the one death north of the contour line for the moment, we can repeat the basic argument developed in the shopping center sample. We begin with the first victim. The probability that this person lived south of the contour line in the simple random case is 0.5, because half the population of potential victims lived in this area. Similarly, the probability that the second victim lived south of the contour line is 0.5. The probability that both lived south of the contour line is, by the multiplication law, $0.5 \cdot 0.5$, or 0.5^2. If we continue this procedure, we conclude that the probability that sixteen consecutive victims lived south of the contour line is 0.5^{16}. This is such a small probability that we might be tempted to dismiss the chance that our pattern was produced by a random selection process and conclude that Nottingham's relief was in fact influencing the pattern of cholera deaths, as our hypothesis suggests.

Before we draw this conclusion, let's take another look at our map of deaths in Figure 3.5b. The sixteen victims south of the contour line were not spread throughout the area but were concentrated into three compact groups. In fact, the deaths occurred in or adjacent to three small courts. Therefore we may question whether our application of the multiplication law is valid. This "law" is the multiplication theorem for *independent* events. The individual deaths within a single court clearly cannot be considered realistically as independent events, because they almost certainly relate to a single source for the disease. We probably have only *three* independent events under consideration—namely, the clusters of victims around the three separate courts. This line of argument now leads to a very different probability conclusion. The chance that three groups of victims lived south of the contour line, with a simple random selection of groups, is 0.5^3, which equals 0.125. This means that we might expect a result like this one time in every eight, even if the groups of victims were completely uninfluenced by their location within the town.

This probability conclusion is rather disappointing, considering what seemed to be originally conclusive map evidence. When we bring the single death north of the contour line after October 1849 into consideration, we find that we have even poorer grounds for supporting our hypothesis. We may consider this another independent event, so that we have four such events to consider. Now, if these events conform precisely to our hypothesis prediction, then all four will be south of the contour line. This would be the most favorable pattern with respect to our hypothesis. On the other hand, the worst possible evidence would be that all four events occur north of the line. Our evidence falls between these two extremes; we can best consider its worth if we find its rank among all possible patterns, in terms of their compatibility with our research hypothesis. This ranking is presented in Table 3.2. Five patterns are possible, and our particular set of evidence ranks second to best. Not all these patterns are equally likely, however. If we number the events A, B, C, and D, then the two-above-and-two-below pattern can occur as A and B above with C and D below, A and C above with B and D below, and so on. We need not list all possible ways, because we can use our knowledge of combinations.

Table 3.2 Possible Patterns of Four Independent Events Ranked According to Hypothesis Compatibility

Rank	*Events above 100 ft*	*Events below 100 ft*	*Number of ways of occurrence*
1	0	4	1
2	1	3	4
3	2	2	6
4	3	1	4
5	4	0	1

The example now becomes one of how many combinations of two can occur given four events, and this is given by the binomial coefficient

$$\binom{4}{2} = \frac{4!}{2!(4-2)!} = 6$$

On the other hand, there are only $\binom{4}{1} = 4$ ways in which we can obtain a combination of one from four events. All binomial coefficients are listed in Table 3.2, showing sixteen possible combinations in all, five of which are *as good as or better than* the pattern observed with respect to our hypothesis. Therefore, we can expect a pattern as good as ours five times in every sixteen ($p = 0.3125$) in a random distribution of victims. Our seemingly distinctive pattern in Figure 3.5b has become thus even less impressive. If evidence such as ours can be produced 31 percent of the time in which relief had no influence on the pattern of deaths, we are hardly in a position to say that the pattern of cholera deaths in Nottingham between

November 1849 and September 1850 is the result of the structure of the town site and the drainage system. By translating our initial problem into probability language, we conclude that the evidence does not support our hypothesis strongly enough to warrant us drawing the inference that the pattern of deaths from cholera is a consequence of the relief pattern..

This second research application of probability theory illustrates two important points. First, we must always be very careful when we translate a geographical problem into mathematical language. No matter how skillful our mathematical manipulations may be, they will be to no avail if the translation does not take into account the assumptions of the mathematics that is being used. Remember that all deductive conclusions are only as good as their assumptions; if the theorem being employed is concerned with "independent events," we must be sure that this is part of our geographical problem.

The second point is that a set of empirical evidence is not always as clear cut in its implications as may seem from simple inspection. When we view it in the rigorous light of a mathematical language, what has been obvious may become much less certain. In the context of this chapter, the basic advantage of translating empirical evidence into a probability language is that it allows us to specify the degree of confidence we can place in any inferences we make. This is what is meant by using probability theory in research decision-making; the procedure has been formalized into inferential statistics.

INFERENTIAL STATISTICS

The use of probabilistic inferences in scientific research does not normally require return to basic probability theory. Such an approach has become largely unnecessary since generally accepted formal rules involving so-called *statistical tests* have been devised. These techniques define a standardized procedure for drawing probabilistic inferences from a set of data. They consist of many techniques but just one basic procedure. Their method is the method we have described above. The procedure is a simple standardization of this method.

The Standard Procedure for Statistical Tests

In our discussion of sampling in Chapter Two, we distinguished between the *population* we are interested in and the *sample* we draw from it. Our information about the population is derived from the sample. If we wish to compare the population with some hypothetical population, we can do so only indirectly through our sample. Similarly, if we wish to compare two populations, we can only compare their samples. *Drawing inferences from samples about populations is the basis of all problems in inferential statistics.*

Let's consider a very simple example. Suppose we wish to compare two populations of farmers, one group on an offshore island and the other group on an area of the adjacent mainland. Our hypothesis is that the farmers on the mainland are more progressive in their farming practice than their more remote colleagues on the island. There are far too many farmers in both groups to interview, so we take a small sample of each. In a pilot survey, we sample five farmers from the mainland and four from the island. One way in which we can measure progressiveness is by looking at the amount of agricultural literature that each farmer reads. We assume that the more they read the more progressive they are. From careful interviewing, we are able to range farmers on a scale from 0 to 30, and the results are:

Island farmers	0	11	12	20	
Mainland farmers	16	19	22	24	29

Notice first that the samples are different. Let's put this point in perspective. Even if the two samples had been

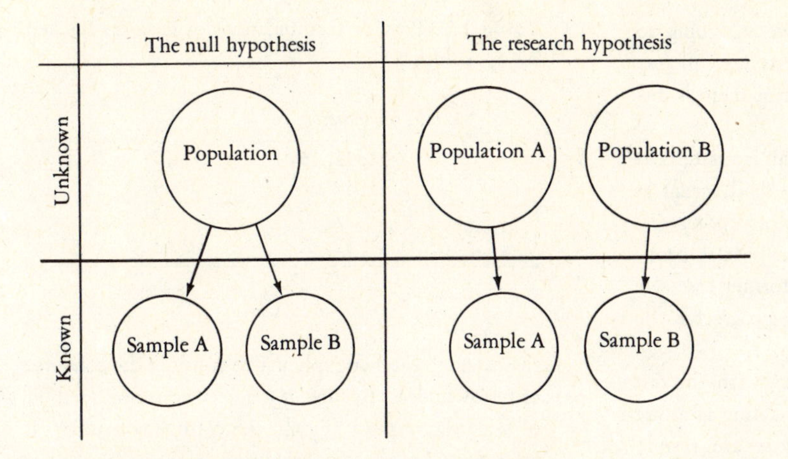

Figure 3.6 The null hypothesis and the research hypothesis.

drawn from the same population—both from the island, for example—we would still expect two different sets of scores. To find two samples of farmers in the same population that are identical on this scale is highly unlikely. Thus, we concede that the two samples in the pilot report are different, as we expect, but we remain interested only in whether populations are different. This we do not know directly, so we use the evidence of the two samples to infer similarity or difference between the populations.

The situation confronting us can be seen in Figure 3.6. We represent our samples as circles, which we know to be different, and which fall below the line. Above the line we have populations A and B as circles, which are largely unknown. There are thus two possibilities: the populations can be the same or they can be different. The second case coincides with our research hypothesis; the converse is usually termed the *null hypothesis*, a hypothesis of no differences between populations. It is the purpose of statistical tests to determine which of these two situations obtains. The conclusion clearly cannot be deterministic—we do not have enough information—so the inference is probabilistic. Thus, we talk of the differences between samples being *statistically significant* at a specified level of probability from which we infer that the populations differ. The procedure for reaching this type of statement is laid out below.

The procedure for comparing populations from sample evidence is basically the same, whatever statistical test we employ. The specification and general acceptance of this procedure has several notable advantages. First and foremost is the intersubjective nature of the research that ensues. Different researchers with the same evidence can come to the same probabilistic conclusion. This allows for ease of replication within a discipline. A second advantage is the equalization of standards between disciplines in their research inferences. This allows for ease of communication between disciplines. A geographer who understands inferential statistics can understand an analysis in, say, psychology. Let us, therefore, consider this procedure in some detail.

How It's Done: The Randomization Test

We can identify six basic steps in a statistical test.

1. We state our null hypothesis (sometimes written H_0), which is normally the opposite of our research hypothesis (H_1). H_0 is typically a statement asserting no differences between populations.

2. We have to choose the test that we are going to employ. There are many different tests we can use, and the way in which we choose among them is discussed below.
3. We specify our significance level. This is a statement of what degree of difference (α) we shall accept as evidence for rejecting H_0 and, hence, accepting H_1. By convention, the $\alpha = 0.05$ and $\alpha = 0.01$ probability levels are used. Thus, in the former case, we expect the probability that H_0 will be correct to be less than one in twenty.
4. We compute the statistic associated with the test we have chosen. Each test involves calculating a statistic from our sample data. These statistics are simply measures of differences between the samples.
5. We find the probability that a statistic of the value we have computed will occur if H_0 is correct. The reason we measure differences by these statistics is that they have been derived by statisticians along with their probabilities under H_0 conditions. The values of the statistics associated with the significance levels of $\alpha = 0.05$ and $\alpha = 0.01$ usually can be directly looked up in tables.
6. We compare the probability that H_0 is incorrect with the significance level we have chosen. If the probability is lower than the significance level, we reject H_0 and accept H_1 at that level of probability. Thus, we make a probabilistic rather than a deterministic inference. If the level of probability that H_0 will be incorrect is greater than our significance level, we conclude that we cannot reject H_0 at that level of probability. Notice that we do not accept H_0; the whole purpose has been to see whether we can reject H_0. In this situation, we are in a sort of limbo—we do not have sufficient evidence to support H_1 while H_0 remains a doubtful proposition.

These then are the six steps in carrying out a statistical test. In themselves they probably seem straightforward enough, if a little formal and long-winded. Some people may feel that the approach is a little strange in terms of its emphasis on the null hypothesis rather than the research hypothesis. Refuting the opposite of a research hypothesis is, in fact, a very rigorous approach to hypothesis testing, in that it forces the researcher explicitly to consider the converse of his own ideas. As such, it is a very common method of scientific inquiry beyond the confines of statistical inference. An example covering each of the steps will help clarify the basis of this approach.

Table 3.3 Possible Combinations of Farmers Ranked by S_R Values

b_i *combinations*	S_R
0, 11, 12, 16	75
0, 11, 12 19	69
0, 11, 12, 20	67
0, 11, 12, 22	63

Let's return to the farmers on the mainland and the island. We test the hypothesis H_1 that the mainland farmers are more progressive than island farmers: the scores for the mainland farmers (a_i's) are greater than the scores for the island farmers (b_i's).

1. H_0 states that there are no differences in progressiveness between the mainland and the island farmers.
2. We choose a test—the randomization test. This test is particularly suited to situations in which we have interval data for two very small samples.
3. The significance level is $\alpha = 0.05$. This means that we shall expect the differences reflected in our samples to occur less than 1 time in 20 under H_0 conditions.
4. We compute statistic S_R. S_R is simply the difference between the sums of the two sample scores, defined as:

$$S_R = \Sigma a_i - \Sigma b_i$$

Thus,

$$S_R = (29 + 24 + 22 + 19 + 16) - (0 + 11 + 12 + 20) = 67$$

5. We find the probability $p(S_R = 67)$. If H_0 is in fact correct and there are no differences in the populations, then the differences observed in the samples simply reflect the fact that two different combinations of 4 and 5 farmers have been allocated to each sample of the 9 chosen. We can begin to assess the probability that this will occur if we find out how many different

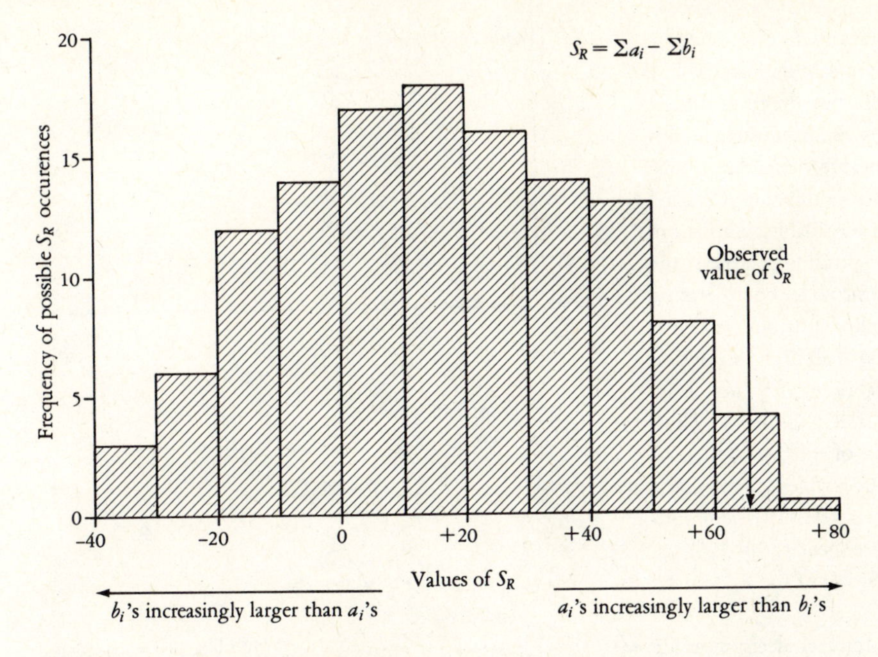

Figure 3.7 The probability distribution of S_R ($S_R = \Sigma a_i - \Sigma b_i$, where the b_i's are *any* combination of four from the set (29, 24, 22, 20, 19, 16, 12, 11, 0) and the a_i's are the remaining five).

combinations of 4 can be obtained from 9. This is given by:

$$\binom{9}{4} = \frac{9!}{4!\,(9-4)!} = 126$$

Each of the 126 possible combinations of 4 farmers has an S_R value associated with it. In fact, we can list them from largest to smallest. The largest S_R value clearly occurs when the a_i's are 29, 24, 22, 20, 19, leaving the four smallest values as b_i's. In this case, $S_R = 75$. This is shown on the top row of Table 3.3. Other combinations of b_i (the associated a_i's follow automatically as the five other observations) are ranked in terms of S_R values. We can now come back to our original problem. What is the probability of obtaining an S_R value as high as ours if H_0 is correct? Our S_R value is the third highest of the 126 possibilities. The probability of an S_R value as high as this if H_0 is correct is, therefore, 3/126. Thus, $p(S_R = 67) = 0.024$.

6. Compare $p(S_R = 67)$ with α. We can see that $p(S_R = 67) < \alpha$, so we reject our null hypothesis. We now accept our research hypothesis: farmers do differ between mainland and island at a significance level of 0.05.

This example shows the strength of this research approach. With only a little evidence, two very small samples, we would normally be very careful about what conclusion we draw. In this case, we can make a quite clear-cut statement concerning our research hypothesis, despite the paucity of information. We have not found a "universal truth," however, and our inference is probabilistic. In fact, the probability level at which our samples are significant is, in its way, a direct measure of how careful we should be in using our finding.

What the Research Decision Means

We have used the randomization test to illustrate the procedure of inferential statistics because it is most like our previous discussion of probabilistic inferences. In particular, we ranked the "best" sets of possible evidence to derive the probability that our evidence could actually occur under the null hypothesis. While this serves as a connection with our previous discussion, the randomization test is quite unusual in this respect. Almost all other statistical tests avoid reference to basic probability theory because this step is carried out simply by using the tables available. This more straightforward approach is possible because the *probability distribution of the statistic* under consideration is known and it is this probability distribution that is presented in the tables. The probability distribution of a statistic is simply the probabilities of values of the statistic occurring under the null hypothesis. Thus, Table 3.3 shows the higher end of the distribution of the 126 possible S_R values in our problem. If we compute all 126 possible S_R values, classify them into classes of 10, and divide the frequencies by 126, we produce the probability distribution of the S_R statistic in Figure 3.7.

We can now reconsider our research decision in the light of the probability distribution. What we have done in Table 3.3 is identify one of the tails of the distribution. Our actual sample evidence lies beyond the line that separates the top 5 percent of the distribution. This area is known sometimes as the *region of rejection* in the test. Because our sample S_R value occurs in the region of rejection, we reject H_0. Notice that we have been concerned only with the top end of the distribution. If our research hypothesis had been that island farmers are more progressive than mainland farmers, we would have been interested in the bottom end of the distribution. In this case, we would have defined a region of rejection in the lower tail. Both of these research situations are known as *one-tailed tests* (Figure 3.8a). On the other hand, if we had been hypothesizing only that mainland and island farmers are *different* in terms of progressiveness, then both high and low S_R values would indicate this. In this situation, we would have conducted a *two-tailed test* by specifying the final 2.5 percent of each tail, to produce a new, total 5 percent region of rejection (Figure 3.8b). Such tests are made when we have little knowledge concerning our samples beyond that contained within them. In our progressive farmers example, we had reason to believe that the more remote farmers (is-

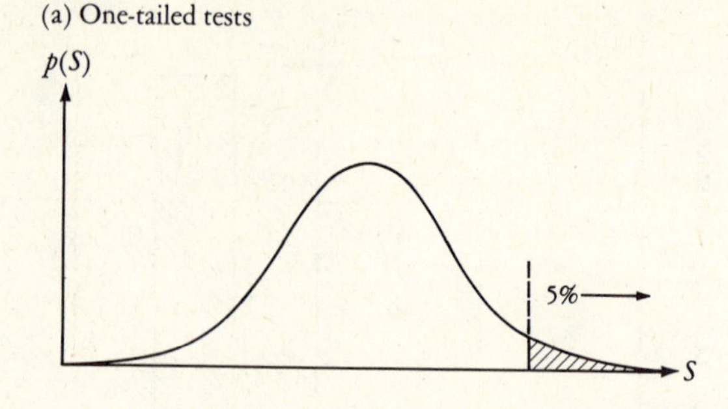

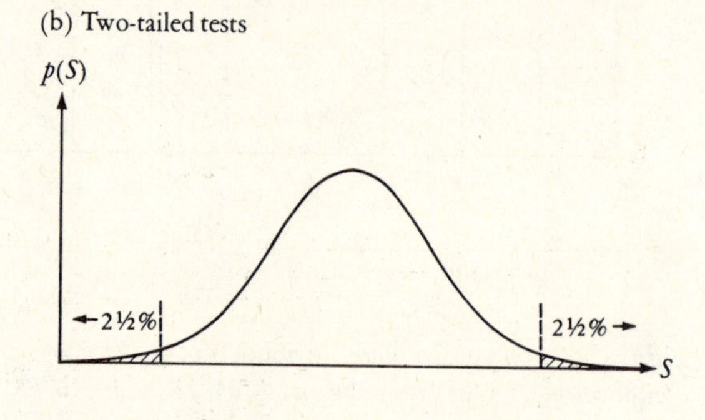

= Region of rejection
$\alpha = 0.05$
S = Any statistic which measures the differences between samples

Figure 3.8 Regions of rejection.

landers) are less progressive, and we specified our research hypothesis accordingly. All tables of probability distributions indicate whether one- or two-tailed cases are specified. We leave consideration of the practical use of such tables until we consider further statistical tests below.

While the availability of tables for statistical testing makes the procedure much easier, it also invites the danger that researchers will forget that it is probability distributions they are dealing with. This problem extends to the whole idea of having a very strict procedure of operation. However straightforward and universal the procedure is, we must not forget that we are dealing with probabilistic inferences. The 5 percent or 1 percent levels of significance must never become so firmly entrenched that they produce inferences that are treated, to all intents and purposes, like deterministic findings. Statisticians clearly have not intended researchers to employ these tools in this manner. In fact, they identify two types of error that can occur in the application of a statistical test.

A *type I error* is to reject H_0 when it is in fact true. This means that the samples have been found to be significantly different when in fact both really do belong to the same or identical populations. If researchers carry out 100 tests on pairs of samples, all from the same population, they are likely to find 4, 5, or 6 significant differences at the 5 percent level. Acceptance of these results would be to commit an error of type I.

A *type II error* is the converse of type I—to fail to reject H_0 when it is in fact false. In this case, our statistical testing is obscuring real differences. If differences between populations are quite small, they will not be discernible with small samples, and thus an error of type II may be committed. Such error can be avoided, however, if we take a broader view of hypothesis verification and avoid a dogmatic and inflexible use of inferential statistics.

Hans Zetterberg (1962), a sociologist, has suggested that we ought to consider several pieces of evidence in the verification of any hypothesis. Five basic questions should be asked in any hypothesis testing:

1. How well do the measurements reflect the concepts under consideration?
2. How good is the fit between the evidence and the hypothesis?
3. How far can the evidence be explained by alternative hypotheses?
4. How well do our samples represent the populations?
5. How far is the hypothesis supported by established laws and theories?

The discussion above has concentrated solely on the second question. Inferential statistics enable us to give a sophisticated answer to this question, but the answer is meaningless if the measurements themselves are biased or if a sample is not representative of a population. Questions 1 and 4 have been dealt with previously and are of continuing relevance for all subsequent research operations. In our progressive farmers example, we might well argue that reading farming literature is a rather academic exercise and that a real measure of progressiveness should relate to what farmers actually do. Furthermore, we have been presented with no information on how the two samples were selected. Thus the testing we have done on our hypothesis is much less satisfactory than it seems from our initial consideration of only question 2. The hypothesis fares much better in terms of questions 3 and 5. The hypothesis seems to have no very obvious rival to explain the significant differences observed and, in fact, the finding is consistent with other work on cultural diffusion and isolation in human geography. Based on these criteria, the hypothesis is much more successful.

In general we can notice that all five criteria are important in a hypothesis-testing situation and that, while inferential statistics remains central to relating evidence to hypothesis, we must always consider the four other questions. If a hypothesis is at variance with much-accepted geographical theory, it should be very stringently tested ($p = 0.001$?) before acceptance. We put priority on avoiding a type I error. On the other hand, if a hypothesis is very well integrated into an accepted geographical theory, to reject it because it fails to make the arbitrary 0.05 probability threshold is normally considered to be unreasonable. In this case, our priority is to avoid a type II error. Inferential statistics brings an intersubjective element into hypothesis testing, but other considerations are involved in the ultimate decision. We have a very useful tool for the research worker, not a replacement.

Choice in Statistical Testing

Once a researcher has understood the general procedure of inferential statistics, then a wide variety of statistical tests become available. They do not on the whole duplicate one another; each is specifically suited to a different research context. The discussion below focuses on the problem of choosing the statistical test that suits a given research context. We can identify two levels of evidence in this decision—general criteria that can be used to draw up a typology of tests and specific criteria that are important considerations in

Table 3.4 A Typology of Statistical Tests

	One-sample case	*Two-sample case*	*k-sample case*
Nominal	χ^2 one-sample test	χ^2 two-sample test	χ^2 k-sample test
Ordinal			
Weak ordering	Kolmogorov-Smirnov one-sample test	Kolmogorov-Smirnov two-sample test	
Strong ordering		Mann-Whitney U test	Kruskal-Wallis one-way analysis of variance
Interval/ratio	Student one-sample t test	Randomization test; student two-sample t test	Snedecor F ratio test (analysis of variance)

Source: Adapted from *Nonparametric Statistics* by S. Siegel. Copyright 1956 by McGraw-Hill, Inc. Used with permission of McGraw-Hill Book Company.

terms of a particular test. Let's begin by describing what has become a very useful typology of statistical tests.

Psychologist Sidney Siegel has identified two major criteria to consider when choosing a statistical test. The first concerns the scale of measurement of data. Thus, tests are available at the main levels of measurement. Second, we must identify the type of research question we are asking in terms of numbers of samples. In our randomization test we had two samples, and we would identify this as a *two-sample research design*. An alternative design is to draw one sample and hypothesize that it comes from a population of known characteristics, such as a truncated rectangular distribution or a normal distribution. This is a *one-sample research design*. It is obvious that we can also consider more than two samples simultaneously, as in a *k-sample research design* ($k > 2$). For instance, we might wish to compare samples of farmers from an outer island, an inner island, and the mainland in a three-sample design.

Siegel used these criteria to produce a three times three table, which is our typology of statistical tests (Table 3.4). Each test is appropriate for research designs and measurement levels as indicated by row and column. If we have understood the discussion, we can add the techniques presented in this typology to our statistical repertoire. All that we require to know is how to calculate their particular statistics and how to find the probability distribution. To get some idea of this learning process, let's consider the first two tests in our table—the chi-square (χ^2) and Kolmogorov-Smirnov tests for the one-sample case.

The Chi-square and Kolmogorov-Smirnov Tests

Both tests deal with data in the form of frequencies, but we can see from the typology that the Kolmogorov-Smirnov test requires ordinal data—that is, ordered classes. We illustrate both tests by comparing agricultural land use sample data with simple expected population distributions. Our research hypothesis is that horticultural farming is located with respect to two factors: market and soil type. A random sample of 100 horticultural holdings is allocated to two classifications relating to our hypothesis as follows:

	Distance to market (mi)			
Class	0–4.9	5–9.9	10–19.4	15+
Frequency	40	30	20	10
Percent land	20	30	30	20

	Soil type			
Class	Clay	Sand	Loam	Limestone
Frequency	30	30	30	10
Percent land	30	40	20	10

Notice that in both cases we have shown the percentage of land for each classification. We use these as our expected distributions under the null hypothesis. Thus, in the first case, if the distance to market has no influence on horticultural location, then we expect frequencies similar to the total land area proportions for each category. Because we are dealing with samples, however, we do not expect exact correspondence, and so we ask whether the differences in the table are significant. The test is as follows.

1. H_0 indicates that distance to market has no influence on horticultural location.
2. We choose a one-sample test for ordered classes; in this case, we choose Kolmogorov-Smirnov one-sample test.
3. The significance level is $\alpha = 0.05$.
4. We compute statistic D. First, we convert observed and expected distributions into cumulative proportions and list the differences:

Observed (O_i)	0.4	0.7	0.9	1.0
Expected (E_i)	0.2	0.5	0.8	1.0
Differences $\lvert O_i - E_i \rvert$	0.2	0.2	0.1	0.0

D is now found, by inspection, as the largest difference, 0.2. We can define D as

$$D = \text{Max}\,\lvert O_i - E_i \rvert$$

5. We find the probability: the probability distribution of D is given in the Appendix (Table A.3), in which critical values of D are given for different sample sizes and different probability levels. We are interested in $\alpha = 0.05$, and our sample size (n) is 100, so the critical value of D is

$$D(\alpha = 0.05) = \frac{1.36}{\sqrt{n}} = 0.136$$

If our observed and expected values are identical, then D will be 0. The critical value of 0.136 means that if the null hypothesis is true, we expect an observed D value as large as this in about 5 percent of the samples. Thus, $D = 0.136$ defines the lower limit of the top 5 percent of the probability distribution of D with a sample of 100. In this case, $D = 0.2$, which is in the top 5 percent because D is greater than the critical D value of 0.136.

6. We compare probabilities. The probability of obtaining a D value of 0.2 is less than $\alpha = 0.05$, and so we reject H_0 and conclude that the samples are significantly different at the level of 5 percent. Thus, we have evidence that the population of horticultural holdings is located with respect to market.

Let's turn now to the table for soil type. The frequencies expected are percentage of land under different soil types. The observed and expected frequencies for soil types differ, so we ask whether these sample differences reflect population differences over soil types.

1. H_0 indicates that soil type has no influence on horticultural location.
2. We choose a one-sample test for nominal data; in this case, we choose chi-square one-sample test.
3. The significance level is $\alpha = 0.05$.
4. We compute statistic χ^2.

$$\chi^2 = \Sigma \frac{(O_i - E_i)^2}{E_i}$$

where O_i is the observed frequency and E_i is the expected frequency. These produce:

$$\chi^2 = \frac{(30-30)^2}{30} + \frac{(30-40)^2}{40} + \frac{(30-20)^2}{20} + \frac{(10-10)^2}{10}$$

$$\chi^2 = 0 + 2.5 + 5 + 0 = 7.5$$

5. We find the probability: the probability distribution of χ^2 is given in the Appendix (Table A.4), in which critical values of χ^2 are given for different probabilities by *degrees of freedom* (*df*). These are defined as

$$df = k - 1$$

where there are k categories. This means simply that, given the total frequency, once the frequencies for all but one category are known the final category frequency automatically follows. There are $k - 1$ degrees of freedom. For this example, $df = 3$, so the critical value is

$\chi^2 (\alpha = 0.05, df = 3) = 7.82$

If the observed and expected samples are identical, χ^2 becomes 0, and 7.82 defines the beginning of the top 5 percent of the χ^2 distribution with $df = 3$. In this case, the computed χ^2 value of 7.25 does not quite make this 5 percent tail.

6. We compare probabilities. Our χ^2 value has a probability of occurring, if H_0 is correct, at greater than $\alpha = 0.05$, so we cannot reject our null hypothesis. We can see that we are in fact very close to the threshold. We certainly would not go ahead on the firm assumption that the research hypothesis is incorrect. This is a situation in which we would probably draw a new, independent, and larger sample to retest our hypothesis.

All the tests in our typology follow the procedure above. We do not go through an example of each in the text because this clearly becomes quite repetitive. We illustrate the tests in Table 3.4 in Work Tables at the end of the chapter using $\alpha = 0.05$ as the significance level in all cases. Consideration of the student's t test and the Snedecor F test is left to later discussion. Examples of actual use of these tests in geographical research are given in the reading notes at the end of the chapter.

Table 3.4 gives us the basis for initial consideration in choosing a statistical test. As we have mentioned, particular criteria relate to some of the tests and sometimes limit their application. The most notable of these is the chi-square test's sensitivity to very small cell frequencies. As a general rule, the test is not to be used when expected frequencies work out at less than five in 20% of the cells. Furthermore, no cell should have an expected frequency of less than one. This is clearly an operational constraint on the use of chi-square tests with small samples. The Kolmogorov-Smirnov test, on the other hand, can accommodate very small samples, because cell frequencies are summed to n and directly taken into account in specifying the critical values of D. From the typology, we can see that this test has the general constraints of not being amenable to analysis of nominal data or of more than two samples. Solutions to all these problems lead researchers to accept some loss of information in data. If it is found that more than the accepted number of cells with low expectations are produced in a chi-square test, the problem can be overcome usually by combining some of the classes. This procedure makes the test acceptable, but the hypothesis is much less interesting. We might, for instance, combine frequencies of forest and moorland to make our data fit the test, even though the distinction between moorland and forest is important to our test. Furthermore, if we have ranked classes (that is, weakly ordered data) with more than two samples, the only test available is the chi-square k-sample test. By using this test, we literally ignore the fact that the classes are ordered and treat them as if they were a nominal scale. Information is sacrificed in order to use the test. Such situations are not completely satisfactory but are necessary if a statistically valid test is to be made. Whether the results are geographically valid given the forced loss of information is a further point of consideration. Let's consider some examples of testing with frequency data that involve making decisions between these two tests.

In a study of perceptions and attitudes toward water management in southwestern Ontario by B. Mitchell (1971), two samples were drawn for comparison: 400 residents of Waterloo County (the public) and 30 water managers (the professionals). The public sample was stratified into urban, rural-urban, and rural. Table 3.5 shows responses to two questions. Since sample sizes differ greatly, direct visual comparison is facilitated by presenting the data as percentages, although it should be emphasized that the tests use frequencies for their computations. From this table we can identify three research designs.

Design A gives nominal data with two samples; only chi-square is appropriate in this case because of the level of measurement. Furthermore, since only 13% of the cells have expected frequencies of less than five, we can go straight ahead with the test. In fact, we can easily reject a null hypothesis of no differences between the public and professional samples, so that it seems that these two groups have different preferences concerning who should make water decisions. (Chi-square two-sample tests are described in Work Table 3.1.)

Design B gives weak ordinal data with two samples; only the Kolmogorov-Smirnov test is directly applicable, because 50% of the cells have expected fre-

Table 3.5 Perceptions and Attitudes toward Water Management in Southwestern Ontario

A Percentage preference for levels at which decisions should be made

	City-township	*County*	*Province*	*Other*
Public	26	38	34	2
Professional	12	37	37	14

B The average person may not know what is best when technical problems are concerned and should rely on professionals

	Agree strongly	*Agree*	*Neutral*	*Disagree*	*Disagree strongly*
Public	9	80	7	4	0
Professional	18	55	8	16	3

C As B above; all samples public

	Agree strongly	*Agree*	*Neutral*	*Disagree*	*Disagree strongly*
Urban	7	84	5	4	0
Rural-urban	19	73	6	2	0
Rural	7	77	9	6	0

Source: Adapted from B. Mitchell, A comparison of χ^2 and Kolmogorov-Smirnov tests, *Area* 4: 239–240, Tables 1 and 2, 1971. With permission of the Institute of British Geographers.

quencies of less than five. We can easily reject a null hypothesis of no differences between public and professional samples, suggesting a real difference in attitudes toward professionals between the two populations. (Kolmogorov-Smirnov two-sample tests are described in Work Table 3.2.)

Design C gives weak ordinal data with three samples; a Kolmogorov-Smirnov test is not applicable because of the three samples, and a chi-square test is not applicable because 28% of the cells have expected frequencies of less than five. We cannot analyze this data in its present form. Two solutions are possible. We could make three paired comparisons using the Kolmogorov-Smirnov test, but this would change the research design from its three-sample status. An alternative strategy is to combine classes and use a chi-square three-sample test. If we combine the categories "agree strongly" and "agree" and the categories "disagree strongly" and "disagree," we produce a three times three contingency table with only 11 percent of the cells having expected frequencies of less than five. A chi-square test on this reorganized data shows that we cannot reject a null hypothesis of no differences in attitude between the three groups at the 5 percent level. We should not place too much weight on this finding. It has became possible only after the loss of much information; not only are classes combined but also their order is ignored. The order of the classes is clearly in many ways the essence of the answers to this question, and so our testing has been particularly disappointing. In the present state of inferential statistics, with no weakly ordinal three-sample test, such unsatisfactory research designs are inevitable. In making our final research decision, we have to draw more heavily than usual on criteria other than inferential statistics.

We conclude this discussion with a description of a fairly typical application of inferential statistics in geographical research by Australian geographer Peter Scott. The problem he was concerned with relates to the hierarchy of settlements in Tasmania. When Tasmanian settlements are divided into hamlets, villages, and towns, we find that there are far more hamlets and villages than would be expected from central-place theory. Scott assumed that this is explicable in terms of Tasmanian violation of central-

Table 3.6 Settlement Hierarchy and Rural Population Density in Tasmania

Density	*Hamlets*	*Hierarchy* *Villages*	*Towns*
44+	46	39	14
18–44	53	47	7
7–17	46	46	3
6–	37	28	5

Source: P. Scott, Areal Variations in Class Structure, *Australian Geographical studies,* October 1964, p. 77.

place theory's isotropic plane. The purpose of the research exercise is to find the characteristics of Tasmania that lead specifically to the overabundance of hamlets and villages. The research strategy is to allocate settlements to samples or subpopulations defined in terms of a particular characteristic. In Table 3.6 we have four subpopulations of varying rural population density, with settlements allocated to the density group by location. We consider the hypothesis that variations in rural population density lead to variations in hamlet and village frequency. Scott constructed eight other such tables by dividing Tasmania further into zones of urban systems, farm-size areas, farm-type areas, altitudinal zones, landform regions, physiographic regions, climatic regions, and soil-type areas. If any of these factors influences the settlement pattern it should be reflected in different patterns of frequencies between regions. We can test whether such differences exist by applying statistical tests to the tables. From Table 3.6 Scott conducted a four-sample test to produce $\chi^2 = 9.52$ with $df = 6$. This does not allow us to reject a null hypothesis of no differences between density areas at $\alpha = 0.05$. This means that the extra towns at high densities, extra villages at medium densities, and extra hamlets at low densities shown in Table 3.6 do not reflect significant differences. (Chi-square *k*-sample tests are described in Work Table 3.1.)

The results from all nine analyses of tables of subpopulations are shown in Table 3.7. The actual probability range in which the χ^2s occur is given. We can present this diagrammatically as a probability scale along which each subpopulation is located (Figure 3.9). From this evidence it seems that we can confidently dismiss physiographic factors and urban regions as irrelevant. These results suggest that we get much the same pattern of hamlets, villages, and towns whether we consider Hobart's urban system or Launceston's urban system and similarly whether we consider the northeastern highlands or the northwestern plateau. Only three subpopulation analyses produce χ^2 values in the top 5 percent of the probability distribution. These are altitude, farm type, and soil type. Nonetheless, all other subpopulation samples occur close to this threshold of $\alpha = 0.05$. We have a set

Table 3.7 Chi-Square Tests on Settlement Hierarchy and Nine Factors for Tasmania

Factor	χ^2	*df*	*Probability*
Urban systems	2.67	4	$0.70 > p > 0.50$
Rural population density	9.52	6	$0.20 > p > 0.10$
Farm size	10.10	6	$0.10 > p > 0.05$
Farm type	21.98	10	$0.05 > p > 0.01$
Altitudinal zones	15.90	4	$0.01 > p > 0.001$
Landform regions	13.88	8	$0.10 > p > 0.05$
Physiographic regions	9.55	14	$0.70 > p > 0.50$
Climatic regions	11.69	8	$0.20 > p > 0.10$
Soil type	15.25	6	$0.05 > p > 0.01$

Source: Adapted from P. Scott, Areal Variations in Class Structure, *Australian Geographical Studies,* October 1964, p. 85.

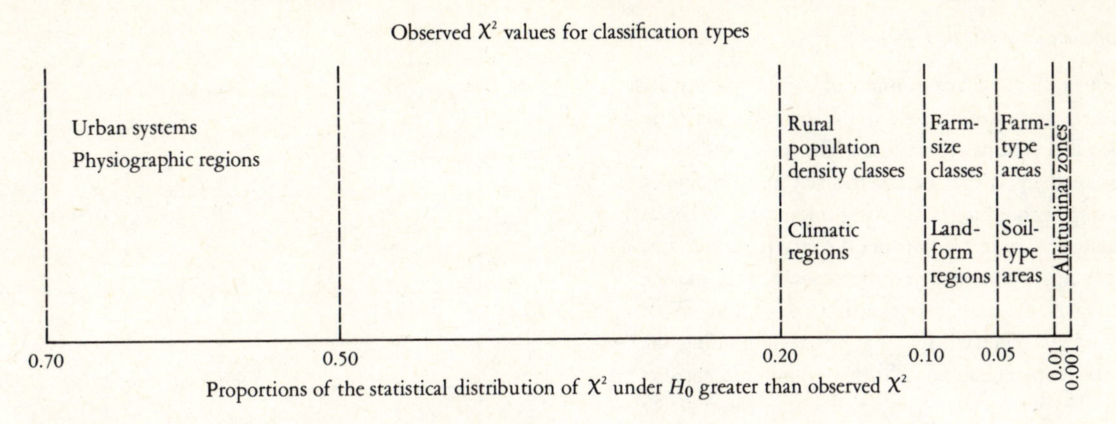

Figure 3.9 Chi-square "probability scale" and central place hierarchy hypotheses.

of closely interrelated factors under the guise of independent alternative hypotheses. Climatic regions clearly relate to farm type, which relates to farm size, which in turn relates to landform regions, soil type being related to several of the preceding factors. Interpretation of these χ^2 results is particularly difficult. We can begin by identifying altitude as the most extreme case in terms of value. We can probably conclude safely that the broken nature of much of the terrain favors village and hamlet settlement at the expense of towns. Farm type probably reflects soil type, but normally we would consider farm type as operatively affecting the settlement pattern through the predominance of intensive farming, favoring a system of many small settlements. Most other factors that do not quite reach the 5 percent level of significance can be viewed as influencing settlement in conjunction with altitude and farm type.

An interesting point is the use of the concept of "sample" in these tests. The researcher began with a population (all settlements in Tasmania) and at no stage did he sample in the way we described sampling previously. Because inferential statistics is concerned with drawing inferences about populations from samples, we can ask legitimately whether the research design above is really suited to statistical testing. The answer is yes, and the justification is as follows. Consider Table 3.6. It was produced by dividing the total population into four density categories. This is one way in which 371 settlements can be allocated to four classes. All the different ways of allocating the settlements to four classes constitute our theoretical population. Each different, arbitrary classification of this population will produce its own frequency patterns of hamlets, villages, and towns, most of which will be slightly different. If the factor underlying our *nonarbitrary* classification affects the settlement hierarchy, then we expect it to produce more distinctively different hierarchies in each class. In the first case, differences between class hierarchies reflect random factors in the interaction of the arbitrary classes and settlement patterns. Such arbitrary classification may well produce quite distinctive hierarchies. Our chi-square tests assess the probability that the nonarbitrary classification under consideration will be produced by sampling one of the numerous arbitrary classifications that could occur. The null hypothesis is that there are no significant differences between classes—that is, that the classification is no different from a purely arbitrary one unrelated to settlement hierarchy. This use of the terms "sample" and "population" is quite different from the strict use of the terms in sampling theory. This wider interpretation allows us to broaden the application of inferential statistics to a very common type of research problem. The important point is that the inferential logic is consistent, so that the resulting probabilities are legitimate tests of the research hypothesis.

Parametric Statistical Tests

All the tests we have considered so far are usually classified as nonparametric statistics. The converse is so-called parametric statistics. The major difference between them is that parametric statistics involves a two-stage procedure in drawing inferences. Whereas the tests above derive statistics directly from the data for testing, parametric statistics concentrates on the parameters of samples and estimated parameters of populations. The two stages are, first, estimating the population parameters and, then, deriving the statistics from the parameters. We see that this entails two inferential decisions.

The use of parametric statistics involves the researcher in checking that his data meet certain conditions. In particular, four conditions are generally stipulated.

1. The observations must be independent. Parametric statistics must employ random sampling procedures and not systematic sampling.
2. The observations must be drawn from a normally distributed population. Nonparametric statistics do not involve this condition and are thus sometimes referred to as distribution free.
3. These populations must have the same variance. Notice that the two latter conditions refer to population properties and *not* directly to sample properties. We have evidence only for our samples; if this were not the case, we would not be interested in inferential statistics. We come to our first necessary sample to population inference. We consider drawing such inferences below.
4. The variables must be measured at least at the interval scale. Parametric tests occur on the bottom row of Table 3.4 along with the randomization test. Notice that this test does not require the three other conditions and is nonparametric.

The student's t test and the Snedecor F test are parametric techniques. These parametric conditions can be considered as further particular requirements of these two tests, above and beyond the general criteria upon which our typology has been based.

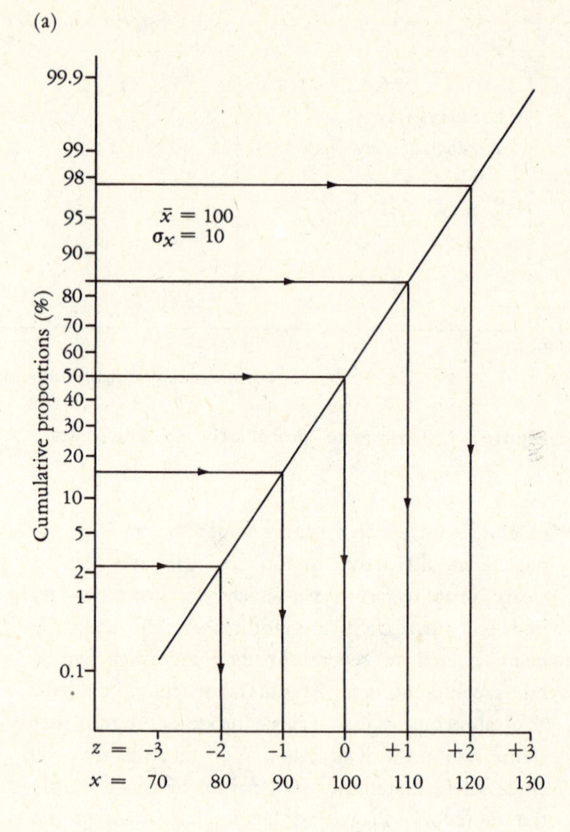

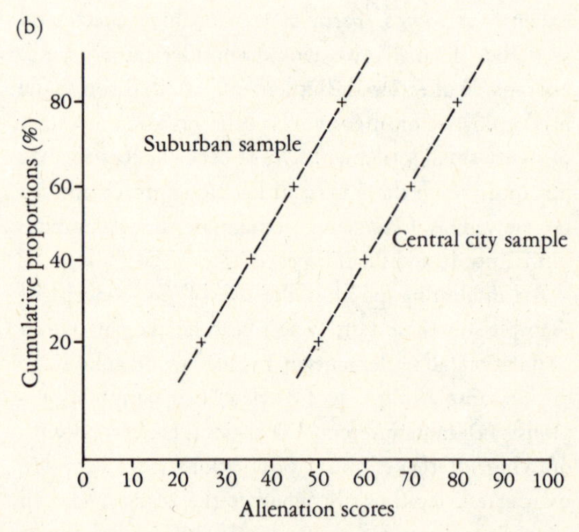

Figure 3.10 The use of probability paper. (a) Normal distribution. (b) Alienation scores.

The assumption of normally distributed populations with equal variances is the first stage of our inference procedure. We have to find out whether it is reasonable to assume that our samples could have been drawn from normal populations. There are two ways of considering this question. In the first we derive a hypothetical normal distribution and then carry out a one-sample test (chi-square or Kolmogorov-Smirnov) to see whether the sample data and theoretical distributions are significantly different. We illustrate this strategy in Work Table 3.5. An alternative and simpler approach is to employ arithmetic probability graph paper. This is really misnamed and would be better termed normal probability paper. We refer to it generally as probability paper.

Probability paper is specially drawn so that plotted *cumulative* normal distributions appear as straight lines. The x axis is the usual interval scale for describing the variable and the y axis is a special scale relating to the properties of the normal distribution (Figure 3.3d). When the calibrated normal distribution from Figure 3.3a is plotted, it defines a straight line (Figure 3.10a). Now consider the following sets of random sample data (Table 3.8). They consist of alienation scores from a questionnaire survey of twenty respondents, each from the central city and suburban zones in a metropolitan area. When we plot this data cumulatively in 20% intervals on probability graph paper (Figure 3.10b), we find that they also define straight lines. This is a visual check on normality. It is not as stringent as a goodness-of-fit test, but for fairly clear-cut cases such as this, the visual check is adequate for us to assume that our sampled populations satisfy the normality condition. When this condition is not satisfied, we can either revert to nonparametric techniques or transform our data to make them normal. The principles of the latter approach are illustrated in Figure 3.11).

Transformation of a single set of data involves simply transferring the measurements from the original scale to a new one. The result is to change the *intervals* between the individual observations on the measurement scale. The overall effect is to change the form of the frequency distribution of the measures. In Figures 3.11a–c, we have three sets of 10 measures (x), each ranging from 1 to 4. In Figure 3.11a, the original measures are slightly biased toward the lower end of the scale, but we can see that by transforming them to their square roots we produce a new scale ranging from 1 to 2, within which the new measures are arranged symmetrically. In Figure 3.11b, we have the opposite situation, with a bias toward the upper end of the scale, made symmetric by squaring. Figure 3.11c reverts to a lower bias, more pronounced than in Figure 3.11a. This is corrected by transforming to common logarithms. These three examples illustrate the three most common transformations: square roots, squares, and logarithms. Many other transformations can be employed, but these cover most empirical needs.

Table 3.8 Alienation Scores for Two Metropolitan Areas

Suburban sample	*Cumulative percentages*	*Central city sample*
12		35
17		41
20		45
25	20	50
30		53
32		56
34		58
36	40	60
39		62
40		65
43		67
45	60	71
46		73
49		76
52		78
55	80	81
59		85
62		89
64		91
70	(100)	95

Let's reconsider the original problem for which the transformations are a solution. We wish to employ parametric statistics but our data are not symmetric

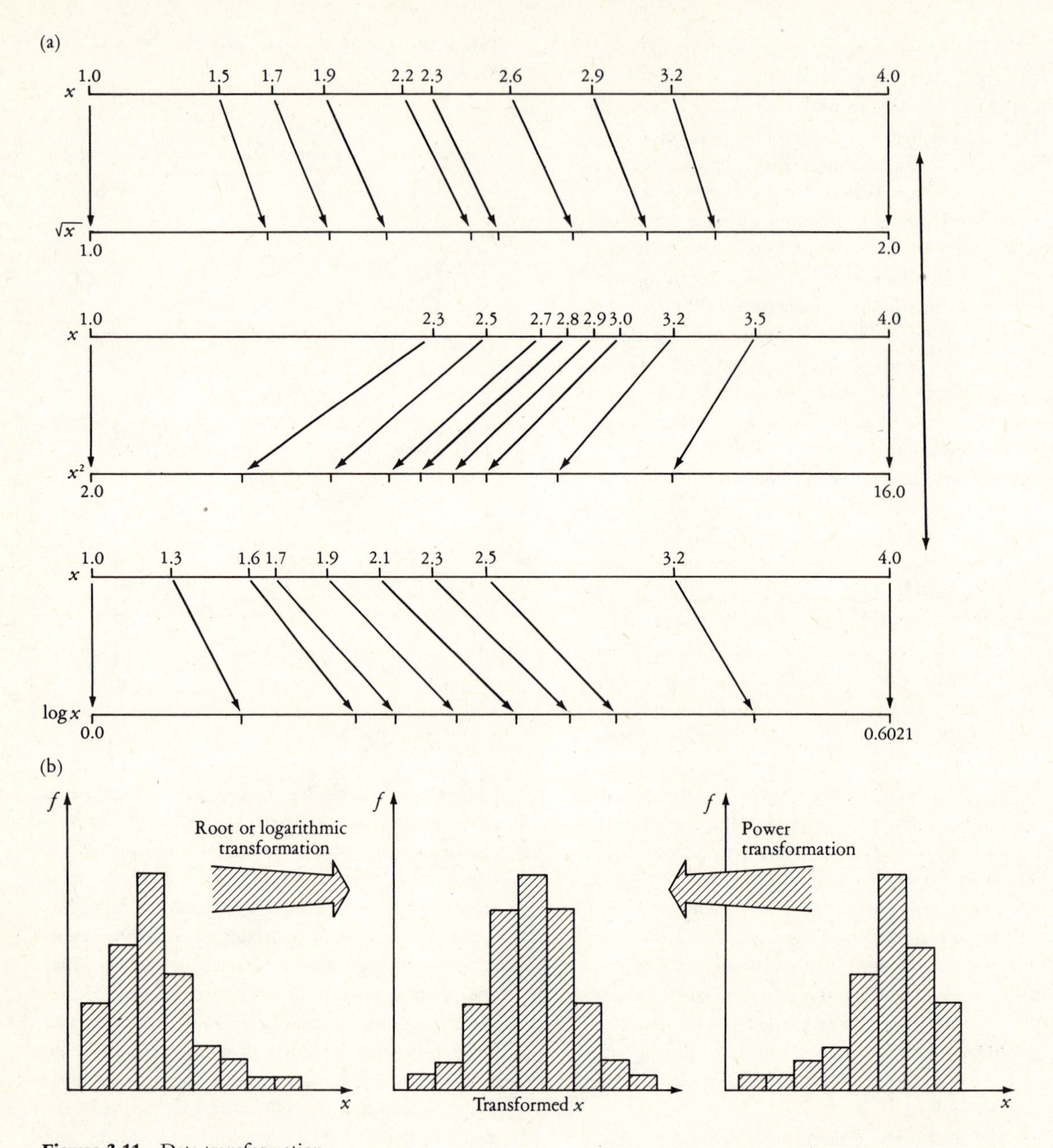

Figure 3.11 Data transformation.

and, hence, do not qualify as normal. By converting the measurement scale, we can produce an approximately normal, symmetric distribution. Parametric statistics can now be employed on this transformed data. If the data are positively skewed (there is a long tail to the right), a square root transformation may be the answer. With more extreme positive skewness, a logarithmic transformation may be employed. For even more extreme positive skewness, higher roots may be employed (cube root, fourth root, and so on). On the other hand, for negative skewness (the long tail is on the left), various power transformations are appropriate, starting with x^2, and with x^3 or x^4 for more extremely skewed data. This procedure is summarized in Figure 3.11d. In practice, we can experiment with various transformations until the data define a straight line on the probability paper.

Once the data have been plotted satisfactorily on probability paper, we can read properties of the normally distributed population that is being approximated. Where the line crosses $p = 0.5$ on the y axis defines the mean on the x axis. This is beause a normal distribution is symmetric, with half of its frequencies on one side of the mean and half on the other. In our normal distribution in Figure 3.10a, we can see that the mean can be read from the x axis as 100. Furthermore, we know that one standard deviation on each side of the mean defines the cumulative distribution between $p = 0.16$ and $p = 0.84$. If we find the range that these y values define on the x axis, we have an estimate of two standard deviations for the population. If we halve this range and square the answer, we have an estimate of the population variance. This is shown for the normal distribution in Figure 3.10a. In fact, we can check for equal variances by seeing whether the slopes of the cumulative distributions are approximately the same.

In Figure 3.10b, the two slopes for the alienation data seem to be very close and we may consider the assumption of equal variance as fulfilled. Because the data are interval measures from two random samples, it follows that we have met all four parametric conditions, and we can employ parametric techniques on the data.

In the application of parametric statistics, we must have more precise estimates of population parameters than are read from probability paper. If we begin with the standard deviation, we find that a random sample does not give an unbiased estimate of the population standard deviation. In fact, small samples tend to underestimate the true population standard deviation. This is fortunately a case in which the bias is well known and can be corrected easily by multiplying by $\sqrt{n/(n-1)}$. This gives us what is termed the *best estimate of the population standard deviation* $\hat{\sigma}$. Thus

$$\hat{\sigma} = s\sqrt{n/(n-1)}$$

1.15 √25(24) σ1

σ2

in which s is the sample standard deviation.

In our two samples in Table 3.8, the best estimates of the population standard deviations are derived from sample deviations of 15.7 and 16.7, respectively, giving

$$\hat{\sigma} = 16.1 \qquad \hat{\sigma} = 17.1$$

We can see from these results that our visual check on variances from the probability graphs is reasonable.

The sample mean is an unbiased estimate of the true population mean. This is not to suggest that it is exactly the same as the population mean; in fact, this is highly unlikely, because, we have only one of an infinite number of possible random samples. We can measure the degree of precision of our estimated population (sample) mean. It is clear that this will depend on sample size (n) and the spread of results in the population ($\hat{\sigma}$), so that we define the *sampling error*, or *standard error of the mean*, as

$$\text{SE}\ \bar{x} = \frac{\hat{\sigma}}{\sqrt{n}}$$

We can envisage what the standard error actually represents as follows. Suppose we draw a very large number of samples of size n from the same population. All will tend to differ and, hence, produce slightly different sample means. Let's now array these various mean values as a frequency distribution. The standard error of the mean can be interpreted as the standard deviation of this frequency distribution of sample means. Even though in practice we draw only one or two samples, we are able to define the frequency distribution to which the mean values belong. The standard error of the mean describes the spread of this frequency distribution: it is a direct measure of the precision of our sample mean.

In our example, average sample alienation scores of 41.5 and 66.6 can be treated as unbiased estimates of the population means, although they have the following standard errors:

$$\text{SE}\ \bar{x}_1 = 3.6$$

$$\text{SE}\ \bar{x}_2 = 3.8$$

We are at last in a position to begin testing whether our two samples reflect significant differences between their respective populations. In fact, we test whether the means are significantly different. Thus, we focus on the difference between the means $|\bar{x}_1 - \bar{x}_2|$. Because we know that each sample mean value is a relatively imprecise estimate of the population mean, it follows that this sample difference is similarly imprecise. The imprecision in the samples' estimate of the population difference between means is a simple function of the two sample mean standard errors:

$$SE|\bar{x}_1 - \bar{x}_2| = \sqrt{(SE\,\bar{x}_1)^2 + (SE\,\bar{x}_2)^2}$$
$$= \sqrt{\frac{\hat{\sigma}_1^2}{n_1} + \frac{\hat{\sigma}_2^2}{n_2}}$$

In order for differences between sample means to be significant, it is obvious that they be larger than their own level of imprecision—the standard error of the differences. The ratio of the difference to its standard error is a direct measure of the degree to which the observed difference is greater than its level of imprecision. This ratio is known as *student's t* and is the statistic employed in the student's t test. Hence:

$$t = \frac{|\bar{x}_1 - \bar{x}_2|}{SE|\bar{x}_1 - \bar{x}_2|}$$

The probability distribution associated with this is the student's t distribution, given in the Appendix (Table A.8) by degrees of freedom where $df = (n_1 - 1) + (n_2 - 1)$.

If we return to the alienation samples, we find that the standard error of the difference is 5.235 for the means $\bar{x}_1 = 41.5$ and $\bar{x}_2 = 66.6$, so that

$$t = \frac{25.1}{5.235} = 4.8$$

and

$$df = (20 - 1) + (20 - 1) = 38$$

Our null hypothesis is that there are no differences between populations in terms of alienation

$$t(\alpha = 0.05, df = 38) = 2.02$$

Thus we can easily reject H_0 and proceed on the assumption that there are real differences in levels of alienation between the central city and the suburban zones.

A one-sample student's t test is illustrated in Work Table 3.6. The Snedecor F test is used for parametric k-sample situations and is associated with the powerful research design known as *analysis of variance*. We illustrate and discuss this design in Chapter Six.

The application of a parametric statistical test is much more complicated than the nonparametric counterparts. The advantages of the parametric approach are that, if the sample data conform to the assumptions, then the resulting test utilizes a maximum amount of information in the data. In such a situation, parametric statistics are said to have greater *power efficiency*. Despite this admitted advantage of parametric tests, the trend in social science has been toward increasing use of nonparametric techniques, even to the extent of converting interval or ratio data into ranks. The innate simplicity of many nonparametric tests certainly appeals to less-numerate social scientists. Reluctance to use parametric tests may, unfortunately, reflect lack of information concerning the stringency of parametric conditions. We have been careful to list the parametric conditions, but statisticians have long shown that many parametric tests are quite *robust*—that is, they perform well under varying conditions outside their specific assumptions. The full implications of this robustness seems not to have been fully explored for either social science or geographical research. Some statisticians would agree that, with the exception of very small data sets, there is hardly any need for distribution-free statistics because of the robustness of the parametric statistics. If this is true, the major usefulness of nonparametric statistics is in dealing with data that can be measured only at less than an interval scale.

Inferences About Maps: The Case of Spatial Autocorrelation

Our general discussion of inferential statistics is applicable to nonspatial studies and spatial analysis, but our examples have all had to do with areas or regions. Tasmania was divided into numerous sets of areas that were then compared to each other by settlement type; progressive and nonprogressive farmers were divided into mainland and island groups; alienation scores were calculated for a central city and a suburb. In each case, samples were related directly to areas or regions in the hypothesis, so the ultimate purpose of the inferential statistics has been to show differences between areas. When we view the application of statistical tests in this way, we can see that they fit very easily into the traditional geographical pursuit of areal differentiation. The search for areal differentiation, in the narrow sense of showing differences between areas, is only part of what geographers are interested in when they view a map—they are also interested in similarities between areas. Much recent research in spatial analysis can be described as dealing with the question of whether there is any order in the pattern of similarities and differences between areas as we portray them on maps. In this section, we consider the identification of spatial autocorrelation among data given for a set of areal units.

We came across spatial autocorrelation in Chapter Two when we were discussing point sampling. We suggested in that discussion that areal stratified sampling is usually successful because most maps contain data that is spatially autocorrelated. We can refine our definition of spatial autocorrelation. The autocorrelation referred to in Chapter Two can be designated as *positive spatial autocorrelation:* an area is likely to be more similar to neighboring areas than to areas further afield. Alternatively, if areas are unlike their neighbors but similar to more distant areas, the pattern is *negatively spatially autocorrelated.* Most attention has been focused on positive autocorrelation because of its common occurrence in physical and human areal data. We normally expect patterns of rainfall and the distribution of social classes in a city to be positively autocorrelated, for example. Our purpose is to test such assertions using inferential statistics.

The research strategy is as follows. We begin by assuming that there is no pattern in the map. Every observation is unrelated to every other observation; one areal unit may or may not be like its neighbor. This is an assumption of independence with a purely random allocation of values to a set of areal units. We test whether the pattern we observe in our particular set of areal units is significantly different from the random case of no autocorrelation. Thus, we deal with a special case of a one-sample test specially designed to make inferences about map patterns. The justification for using inferential statistics in this context is similar to our argument in the Tasmanian central-place study. Although we do not actually select a sample, the areal units on our map represent one possible arrangement of areal units in an infinite population of ways to divide the map into areal units. We ask the question: "Is the observed spatial autocorrelation in this particular set of areal units based on an essentially random pattern?" Let's see how this research design operates in practice. We illustrate its application for simple nominal dichotomous data (two-color maps) and interval/ratio data (k-color maps).

Testing for Pattern in Two-Color Maps

Figure 3.12 shows a hypothetical study region divided into six areal units classified as either black or white. These two categories represent nominal data, although higher-level measurements can be reduced to two classes for analysis purposes. A common dichotomy of areal data in electoral geography, for instance, is areas won by party A and areas won by party B in a two-party contest. The most common form of dichotomizing interval or ratio measures is about their mean values. The black and white dichotomy on Figure 3.12 may be interpreted as the result of dichotomizing the reported x values about their mean value of 7. In any of these cases, we proceed by employing only the information contained in the dichotomous classification.

We concentrate on the contiguous pairs of areas and count the number of white areas touching white areas (WW), white areas touching black areas (WB), and black areas touching black areas (BB). Thus, in Figure 3.12, there are 2 WW joins (4–5, 5–6), 4 WB joins (2–4, 2–5, 2–6, 3–4), and 3 BB joins (1–2, 1–3, 2–3). It is clear that the more WW and BB joins there are, the more *positively* autocorrelated will be the map pattern, whereas the more pairs of different neighbors (WB joins) there are, the more *negatively* autocorrelated will be the pattern. We can assess these possibilities by comparing the number of different types of joins we have recorded with the number that would be expected if the areal pattern were random (not spatially autocorrelated). These are given by the following:

(a)

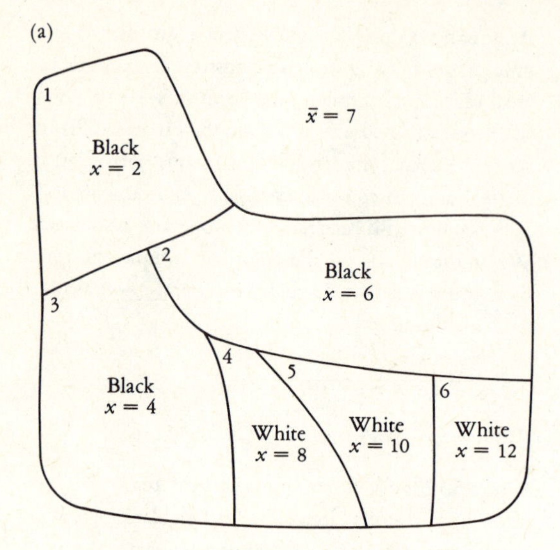

(b)

Areas	1	2	3	4	5	6
1	0	1	1	0	0	0
2	1	0	1	1	1	1
3	1	1	0	1	0	0
4	0	1	1	0	1	0
5	0	1	0	1	0	1
6	0	1	0	0	1	0

$\delta_{ij} = 1$, where i and j are contiguous
$\delta_{ij} = 0$, where i and j are noncontiguous or $i = j$

Figure 3.12 A hypothetical map pattern to test for spatial autocorrelation. (a) Map pattern. (b) Contiguity matrix, Δ.

$$E(WW) = p^2 l$$
$$E(WB) = 2pql$$
$$E(BB) = q^2 l$$

in which E signifies expectations under random allocation, l is the total number of joins, and p and q are the probabilities that areas will be white and black, respectively:

$$p = \frac{\text{number of white areas}}{\text{total number of areas}} = \frac{3}{6} = 0.5$$

$$q = \frac{\text{number of black areas}}{\text{total number of areas}} = \frac{3}{6} = 0.5$$

Because this is a two-color map, probabilities are given by the binomial expression. In our example, in which $l = 9$, the expectations are 2.25 for both WW and BB joins and 4.5 for WB joins. Of course, these expectations will not occur in every random areal distribution, and their variances are given by

$$\text{Var}(WW) = p^2 l + p^3 k - p^4(l + k)$$
$$\text{Var}(WB) = 2pql + pqk - 4p^2q^2(l + k)$$

and

$$\text{Var}(BB) = q^2 l + q^3 k - q^4(l + k)$$

k is given by

$$k = \sum_{j=1}^{n} l_j(l_j - 1)$$

in which l_j is the number of contiguous links for the jth of the n areas. In our example, the numbers of links are 2, 5, 3, 3, 3, and 2, so that $k = 42$. Thus, the variance for WW and BB is $2.25 + 5.25 - 3.1875 = 4.3125$, since $p = q = 0.5$. The variance for WB, on the other hand, is $4.5 + 10.5 - 12.75 = 2.25$. These variances merely indicate the spread of results that could be expected for different random patterns. Our purpose is to compare these random expectations with our actually observed contiguous pairs. Therefore, we define three statistics z_{WW}, z_{WB}, and z_{BB}, which relate the difference between observed and expected patterns of contiguous pairs to the standard deviation of the expected values

$$z_{WW} = \frac{WW - E(WW)}{\sqrt{\text{Var}(WW)}}$$

$$z_{WB} = \frac{WB - E(WB)}{\sqrt{\text{Var}(WB)}}$$

and

$$z_{BB} = \frac{BB - E(BB)}{\sqrt{\text{Var}(BB)}}$$

These z statistics are known to be standardized normal deviates, so we can compare our computed values with a normal distribution to assess the significance of the observed pattern. We carry out a statistical test in the normal way. (This example is laid out in full in Work Table 3.7.) The results show that the probabilities associated with our z values are $p(z_{WW} = 0.12) = 0.4522$; $p(z_{WB} = 0.33) = 0.3707$; and $p(z_{BB} = 0.36) = 0.3594$. This means simply that z values such as ours frequently occur in simple random map patterns with 9 links, so that we cannot reject a null hypothesis from randomness at any usual significant level. Our research conclusion based on a two-color map analysis of Figure 3.12 is that the pattern is not spatially autocorrelated.

This conclusion may seem a little unexpected, given the apparent visual autocorrelation in the map with its clear north–south trend (Figure 3.12). This is the result largely of area 2, which is contiguous with every other area and, hence, contributes to 3 of the 4 WB links. This situation is doubly important if we interpret our results as reflecting the pattern of the dichotomized x variable. Area 2 has a numerical score of only one less than the average, but in the two-color map analysis we treat it the same as area 1, with its x value of 5 less than the average. This suggests that failure to find autocorrelation in this case may be the result of the information we have lost in treating the x variable as a two-class nominal scale. There is clearly a need for techniques to test k-color maps that incorporate interval and ratio data.

Testing for Pattern in k-color Maps

Statisticians have developed two methods of measuring spatial autocorrelation for k-color maps. In each the basic strategy is the same. The differences between neighboring areal units are related to differences in the set of areal units as a whole. The two methods differ in terms of how these differences are assessed.

Let's begin by constructing a connection matrix that shows contiguity between areas. This is given for the example in the six-times-six matrix, Δ, in Figure 3.12b. We refer to entries in this matrix as δ_{ij}, for which $\delta_{ij} = 1$ when area i and area j are contiguous and $\delta_{ij} = 0$ when they are noncontiguous. We define $\delta_{ii} = 0$. Now, consider the following expression:

$$c = \frac{n - 1 \sum\sum(\delta_{ij}(x_i - x_j)^2)}{4l \sum(x_i - \bar{x})^2}$$

(The double sigma notation instructs us to add rows and columns in a table.) n and l are numbers of areas and links as before; $n - 1$ and $4l$ are basically scaling factors so that with a random pattern the expectation is $c = 1$. Statistic c is known as *Geary's contiguity ratio*, after the statistician who devised it. It is a contiguity ratio, because it is the ratio of the sum of *squared differences* between contiguous neighbors and the sum of squared differences of all observations from the mean. Differences between only contiguous neighbors are ensured by having δ_{ij} as a multiplier so that differences between noncontiguous pairs of observations disappear (δ_{ij} is 0). The data in Figure 3.12a are analyzed in this way in Work Table 3.8 to produce a c ratio of 0.413, compared to a random expectation of 1. We still have to consider the variance of the c values under random conditions; exactly 1 certainly will not occur for every random areal distribution. This variance is given by

$$\text{Var}(c) = \frac{(2l + k)(n - 1) - 2l^2}{(n + 1)l^2}$$

which in our example comes to 0.243 (Work Table 3.8). We now define

$$z_c = \frac{E(c) - c}{\sqrt{\text{Var}(c)}}$$

The distribution of z_c is known to be normal, and we can again use normal distribution probabilities. From Work Table 3.8 $p(z_c = 1.19) = 0.1170$. Thus, although this z value is much rarer than the z values derived from the two-color map analysis, we cannot reject a null hypothesis that there are no differences from random expectations at $\alpha = 0.05$. We conclude that there is not sufficient evidence to support a hypothesis of autocorrelation in the pattern.

Let's consider the alternative measure of autocorrelation, associated with statistician Moran (1948). He proposed

$$r_a = \left(\frac{n}{l}\right) \frac{\sum\sum(x_i - \bar{x})(x_j - \bar{x})\delta_{ij}}{\sum(x_i - \bar{x})^2}$$

In this case, we have the *cross-products of deviations from the mean* for contiguous pairs as a proportion of the total sum of squares multiplied by n/l. Notice that cross-products of contiguous pairs are ensured by multiplying by δ_{ij} again. In Work Table 3.9 the Figure 3.12 data produce $r_a = 0.56$. The statistic r_a may be interpreted as a measure of autocorrelation in which positive autocorrelation produces a positive score to a maximum of +1, and negative autocorrelation produces a negative score to a maximum of −1. For inferential purposes, however, interest is generally focused on the second half of the formula, denoted by I in

$$r_a = \frac{n}{l} I$$

In this case, $I = 0.83$. We assess I for randomness. The expected value of I under randomness is given by

$$E(I) = \frac{-l}{n(n-1)}$$

which is $\frac{-9}{30} = -0.3$

In order to define Var(I) under randomness, we must define

$$\rho = -(n-1)^{-1}$$

and

$$E(I)^2 = \frac{l(1 + 2\rho^2) + 2k(\rho + 2\rho^2) + 3l(l-1) - 2k\rho^2(n-1)}{n^2(n+1)}$$

We can define Var(I) as

$$\mathrm{Var}(I) = E(I^2) - (E(I))^2$$

Our standard normal deviate statistic is

$$z_I = \frac{I - E(I)}{\sqrt{\mathrm{Var}(I)}}$$

These formulas look a little forbidding at first sight, but in fact they involve only substitution of a few known values and a little arithmetic. The data in Figure 3.12 is analyzed in this way with the result

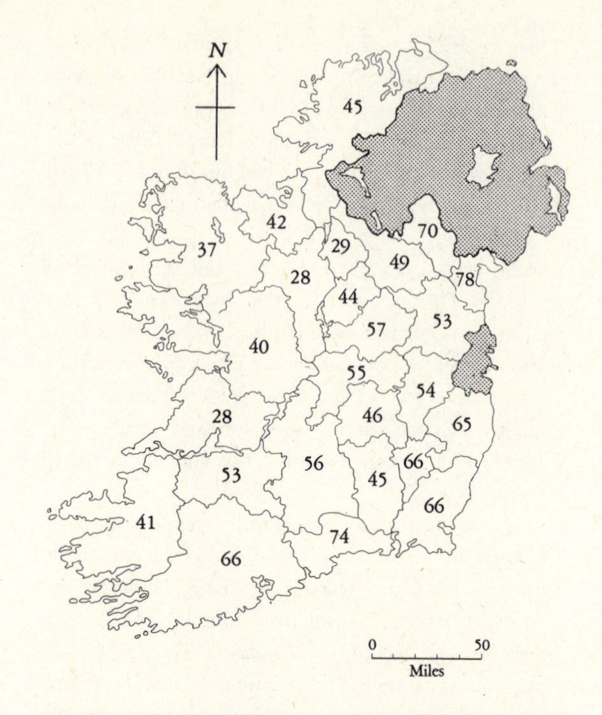

Figure 3.13 Per capita retail sales for Irish counties, 1951.

that $p(z_I = 1.35) = 0.0885$. Thus, z_I is not a very rare normal deviate but may be equaled or exceeded in nearly 10 percent of all simple random map patterns. Once again, we cannot reject a null hypothesis that there are no differences from a random pattern. We conclude that there is no autocorrelation in our hypothetical map pattern as measured by Moran's technique.

Notice that the normal deviates computed from Geary's and Moran's techniques are not the same, despite the fact that both used interval level data from the same map pattern. This is, of course, because they define spatial autocorrelation in different ways. One is based on sums of differences; the other is based on products of deviations from means. They will always produce slightly different results although they tend to be consistent with one another. Because we work with a threshold level in inferential statistics, usually $\alpha = 0.05$, it may happen that we reject H_0 in using one approach and accept H_0 in using the alternative approach. In this situation, we must decide what we mean by spatial autocorrelation in a particular problem. In general, however, we have no real geographical reasons, theoretical or empirical, for preferring a cross-products or differences definition of autocorrelation. This is another overidentified spatial concept, and we should be especially careful in making a research decision from inferential procedures.

We can illustrate the application of the Geary and Moran statistics by using some of Geary's original data for Irish counties, excluding Dublin. The variable we consider is retail sales per person in 1951. This variable reflects income and, hence, economic prosperity; we expect it to be positively spatially autocorrelated. The pattern of values is shown in Figure 3.13, which suggests that our hypothesis is reasonable. When we test it by using Geary's approach, we obtain $z_c = 3.1303$; with Moran's statistics we obtain $z_I = 3.4985$. In both cases, the statistics are very rare normal deviates and, hence, we can reject a null hypothesis that there are no differences from randomness when using either approach. The two approaches do not give identical results, but they do produce consistent inferences. With most data sets, this is normally true, and the definitional problem alluded to above is of more theoretical than practical importance.

The main criticism of such measures by researchers has been that they are topologically limited in their search for autocorrelation in map patterns. The original general concept of spatial autocorrelation involves more than simple contiguity relations. This limitation has been studied in recent years by British spatial analysts Cliff and Ord (1969 and 1970), who suggest replacing δ_{ij} (whether areas are contiguous) by more general weights (w_{ij}), which may take any form. If the weights are taken as distances, d_{ij}, between areas, the definition of pattern goes beyond neighbors to produce Euclidean measures of spatial autocorrelation.

These measures of spatial autocorrelation and their associated z tests illustrate how we can draw inferences from map evidence in geographical research. They represent only one special case of a very general strategy for making inferences about map patterns, however. This strategy is most developed with respect to simple dot maps, which we discuss in Chapter Four.

Further Reading

The relation between geographical research and statistics is briefly described by Berry and Marble (1968) and the French connection of the possibilist school is postulated by Lukerman (1965). For a more general discussion of the origins and development of probability theory, reference should be made to the early chapters of Hogben (1957).

Topics such as permutations and combinations and the pure probability theory we present in this chapter are covered in numerous probability textbooks. A good simple introduction is given by Lipschutz (1968), and a more advanced exposition can be found in Feller (1957). The applied probability approach we develop is based largely on the way in which Langley (1968) treats the subject.

Our typology of statistical tests is based on Siegel (1956), who gives a good coverage of nonparametric statistics in general. French (1971) considers the utility of nonparametric statistics in geography, and Gregory (1968) and Hammond and McCullagh (1974) both cover nonparametric and parametric tests in a geographical context. It is always useful to be able to see techniques working in real research applications. Examples of the various tests in geographical research can be found in the following.

1. Chi-square tests: Mitchell (1971) and Scott (1964). Zobler (1957, 1959), Mackay (1959), and Berry (1959) conduct a debate on the importance of frequency data for this test.

2. Kolmogorov-Smirnov tests: Mitchell (1971). Further one-sample examples are given in the next chapter.
3. Mann-Whitney test: Driscol and Hopley (1968)
4. Kruskal-Wallis test: Johnston (1965)
5. Student's t test: Rushton, Golledge, and Clark (1967)
6. Snedecor F test is discussed in Chapter Six

Formal presentation of inferential statistics can be found in many textbooks: Dixon and Massey (1969) present a rigorous introduction, and Roscoe (1969) pursues a practical scientific approach with a comprehensive coverage of tests. In subjects closely related to geography, there are two excellent expositions: Krumbein and Graybill (1965) in geology and Blalock (1960) in sociology.

For advanced students, the general debate concerning the utility of the research strategy of inferential statistics is introduced into geography by Gould (1970), who suggests that this approach is probably not appropriate for most geographical research. His argument is supported by Meyer (1972). A counter-argument can be found in Gudgin and Thornes (1974). The debate outside geography has been led by Hogben (1957), parts of whose arguments have been reprinted in Morrison and Henkel (1970). This last reference is a set of readings that relate to the statistical test controversy, largely in sociology. This supplementary reading can be extended usefully by referring to the readings in Steger (1971), where several key articles on robustness can be found. A very useful general discussion of hypothesis testing is presented in Zetterberg (1962).

The original work on spatial autocorrelation by statisticians can be found in Moran (1948) and Geary (1954) and was initially introduced into geography by Dacey (1968b). This work has subsequently been built upon by a small group of British geographers who have been concerned both with measuring and testing for spatial autocorrelation, and with assessing its influence on standard statistical procedures. A part of this work can be found in Cliff and Ord (1969, 1970, 1973, 1975a and 1975b), Cliff, Martin, and Ord (1975), and Cliff *et al.* (1975). Cliff and Ord (1975b) specifically deal with the effects of spatial autocorrelation on the student's t test. We introduce the topic of spatial autocorrelation effects on statistics in Chapter Five where additional references are given.

Work Table 3.1 Chi-Square Two- and k-Sample Tests

Data

The table shows frequencies of land use by field.

Samples	*Land Use Classes*				
	Arable	*Pasture*	*Forest*	*Moorland*	*Total*
Upland	5	9	10	15	39
Valley	11	11	5	0	27
Hillside	9	5	6	2	22
Total	25	25	21	17	88

This array of frequencies is known as a *contingency table.*

H_0

Two-sample case: there are no differences between upland and valley in terms of land use type.

Three-sample case: there are no differences between upland, valley, and hillside in terms of land use type.

Statistic

$$\chi^2 = \sum_{i=1}^{r} \sum_{j=1}^{k} \frac{(O_{ij} - E_{ij})^2}{E_{ij}}$$

There are r rows and k columns in the contingency table whose members are O_{ij}'s.

$$E_{ij} = \frac{\Sigma O_i \, \Sigma O_j}{\Sigma\Sigma O_{ij}}$$

which produces the expected frequency table from: row total · column total divided by overall total, thus

	A	P	F	M
Upland	11.1	11.1	9.3	7.5
Valley	7.7	7.7	6.4	5.2
Hillside	6.3	6.3	5.3	4.3

For example, $E_{11} = \dfrac{39 \cdot 25}{88} = 11.1$

If we consider only two samples (upland and valley), the expected frequencies are slightly revised:

$$E_{11} = \frac{39 \cdot 16}{66} = 9.5$$

The complete set of two-sample expectations is:

	A	P	F	M
Upland	9.5	11.8	8.9	8.9
Valley	6.5	8.2	6.1	6.1

In the two-sample example:

$$\chi^2 = \frac{(5-9.5)^2}{9.5} + \frac{(9-11.8)^2}{11.8} + \frac{(10-8.9)^2}{8.9} + \frac{(15-8.9)^2}{8.9} + \frac{(11-6.5)^2}{6.5} + \frac{(11-8.2)^2}{8.2} + \frac{(5-6.1)^2}{6.1} + \frac{(0-6.1)^2}{6.1} = 17.5$$

$df = k - 1 = 4 - 1 = 3$ (there are k classes)

In the three-sample example:

$$\chi^2 = \frac{(5-11.1)^2}{11.1} + \frac{(9-11.1)^2}{11.1} + \frac{(10-9.3)^2}{9.3} + \frac{(15-7.5)^2}{7.5} + \frac{(11-7.7)^2}{7.7} + \frac{(11-7.7)^2}{7.7}$$
$$+ \frac{(5-6.4)^2}{6.4} + \frac{(0-5.2)^2}{5.2} + \frac{(9-6.3)^2}{6.3} + \frac{(5-6.3)^2}{6.3} + \frac{(6-5.3)^2}{5.3} + \frac{(2-4.3)^2}{4.3}$$
$$= 22.4$$

$df = (r-1)(k-1) = (3-1)\cdot(4-1) = 6$
(there are r samples)

Probability

From Table A.4:

$\chi^2(\alpha = 0.05, df = 3) = 7.82$

$\chi^2(\alpha = 0.05, df = 6) = 12.59$

Decision

In both cases, we can clearly reject H_0, and we conclude that there are significant differences between upland and valley and between upland, valley, and hillside in terms of land use types.

Exercise

We have shown differences between all three areas taken together and between upland and valley. We have not shown that two-sample tests for the hillside sample against upland and valley, respectively, show significant differences. State a null hypothesis, compute χ^2, and test these two outstanding hypotheses from the same data. Do we continue to find significant differences between samples at $\alpha = 0.05$?

Work Table 3.2 Kolmogorov-Smirnov Two-Sample Test

Data

The table shows frequencies of commuting distance in miles.

Samples	*Distance classes (mi)*					
	0–0.9	*1–1.9*	*2–4.9*	*5–9.9*	*10 +*	*Total*
Working class (S_1)	10	20	10	5	0	$n_1 = 45$
Middle class (S_2)	5	10	20	20	5	$n_2 = 60$

H_0

One-tail test: middle-class commuters do not travel longer distances than working-class commuters.

Two-tail test: there are no differences between middle-class and working-class commuters in terms of distance.

Statistics

One-tail case: $D = \max(c_1 - c_2)$

Two-tail case: $D = \max|c_1 - c_2|$

c_1 and c_2 are cumulative proportional distributions of S_1 and S_2, respectively:

c_1	0.22	0.67	0.89	1.00	1.00
c_2	0.08	0.25	0.58	0.92	1.00
Difference	0.14	0.42	0.31	0.08	0.00
$(c_1 - c_2)$	0.14	0.42	0.31	0.08	0.00

$D = 0.42$ for one- and two-tail cases. In the one-tail case, convert to $\chi^2 = 4D^2 \dfrac{n_1 n_2}{n_1 + n_2} = 18.1$ (with $df = 2$.).

Probability

One-tail case from Table A.4:

$\chi^2(\alpha = 0.05, df = 2) = 5.99$

Two-tail case from Table A.5:

$D(\alpha = 0.05) = 1.36 \cdot \sqrt{\dfrac{n_1 + n_2}{n_1 n_2}} = 0.27$

When $n_1 = n_2 < 40$, then Table A.5 can be used.

Decision

We can easily reject H_0 at $\alpha = 0.05$ for either a one-tail or a two-tail research design. It would seem that there are real differences in distance traveled by working-class commuters and middle-class commuters. The middle class tends to travel further.

Exercise

In a sample taken ten years before the one reported above, the distances traveled by middle-class commuters in miles had been found to be as follows:

Miles	0–0.9	1–1.9	2–4.9	5–9.9	10+
Commuters	5	20	20	10	5

Visual comparison with the middle-class commuters in the data table suggests slight differences. Test these differences with the Kolmogorov-Smirnov D statistic at $\alpha = 0.05$, using both one-tail and two-tail strategies. Do we have sufficient evidence to suggest differences in middle-class commuting habits between the two periods?

Work Table 3.3 Mann-Whitney *U* Test

Data

The preference rankings of cities in terms of residential desirability treated as two samples are:

Northern cities	3	6	8	9		$(n_1 = 4)$
Southern cities	1	2	4	5	7	$(n_2 = 5)$

H_0

One-tail test: there is no southern bias in preference rankings.

Two-tail test: there are no differences in ranking between northern and southern cities.

Statistic

Order ranks but maintain sample identification:

S S N S S N S N N

S is a southern city and N is a northern city. Now count the number of lower ranked sample cases (the

N's) before each higher ranked sample case (the S's) and sum:

$U = 0 + 0 + 1 + 1 + 2 = 4$

Notice total difference between samples is reflected in $U = 0$. This means that U is a direct measure of *similarity* between samples, whereas most other statistics relate to differences between samples.

Probability

One-tail test (from Table A.6):

$U\ (\alpha = 0.05,\ n_1 = 4,\ n_2 = 5) = 2$

Two-tail test (from Table A.6):

$U\ (\alpha = 0.05,\ n_1 = 4,\ n_2 = 5) = 0$

Decision

In neither case is the U value *low enough* to reject H_0, and we must conclude that we do not have sufficient evidence to suggest simple differences or a southern bias in the preference rankings.

Exercise

We might suspect that the small sample size prevented us from showing real differences in the analysis—that is, we committed a type II error. We can consider this assertion by obtaining preference rankings for a larger number of cities with the following results:

Northern cities	3	5	10	11	13	14	15	16
Southern cities	1	2	4	6	7	8	9	12

Carry out Mann-Whitney one-tail and two-tail tests on these new data. Does this larger data set enable us to reject our null hypotheses on this occasion and, hence, suggest that our previous analysis involved a type II error?

Notice that when tied ranks occur, each should be allocated the average position it would have occupied had it not been tied. Thus, 1, 2, 2, and 4 becomes 1, 2.5, 2.5, and 4, because the two objects ranked 2 take positions 2 and 3 in the ranking order. With such tied ranks, the counting method used above cannot operate, so we define

$$U = n_1 n_2 + \frac{n_1(n_1 + 1)}{2} - r_1$$

in which r_1 is the sum of the ranks of the sample with the lower ranks, n_1 is the number of ranks in this sample, and n_2 is the number of ranks in the other sample.

Work Table 3.4 Kruskal-Wallis One-Way Analysis of Variance

Data

The preference rankings of cities in terms of residential desirability for three samples are:

Northern cities	4	8	10	14	15
Central cities	3	6	7	11	13
Southern cities	1	2	5	9	12

H_0

There are no differences between regions in terms of preference rankings.

Statistic

$$H = \frac{12}{n(n + 1)} \sum_{j=1}^{k} \left(\frac{r_j^2}{n_j}\right) - 3(n + 1)$$

There are k samples; n_j is the number of ranks in the jth sample; $n = \Sigma n_j$; r_j is the sum of the ranks in the jth sample.

$$H = \frac{12}{15 \cdot 16}(520.2 + 320 + 168.2) - (3 \cdot 16)$$

$$= (0.05 \cdot 1008.4) - 48$$

$$= 50.42 - 48 = 2.42$$

Probability

When $n_j \leqslant 5$, we can use Table A.7 for three sample tests. Hence, from Table A.7:

$H(\alpha = 0.05, k = 3, n_j\text{'s} = 5) = 5.78$

Decision

We are clearly unable to reject H_0 on this evidence, and we cannot conclude that we have observed significant differences between southern, central, and northern cities in terms of preference rankings.

Exercise

Just as in Work Table 3.3, we might argue that our samples are a little on the small size and, hence, are a very limited test of our research hypothesis. With a new larger sample of cities, we obtain the following rankings:

Northern cities	4	10	11	17	18	21	23	24
Central cities	5	6	9	12	15	16	20	22
Southern cities	1	2	3	7	8	13	14	19

Use a Kruskal-Wallis test to see whether we can find significant differences with this new evidence.

Notice that $n_j > 5$ and $k > 3$, H approximates χ^2 and Table A.4 should be used. Notice also that tied ranks should be treated as described in Work Table 3.3. A small adjustment to H then is required by dividing by

$$1 - \frac{\Sigma T_i}{n^3 - n}$$

$T_i = t_i^3 - t_i$, with t_i being the number of tied observations in the ith group of scores.

Work Table 3.5 A Goodness-of-Fit Test for Normality

Data

These are alienation scores for the suburban sample in Table 3.8 grouped into classes of standard deviation ($\sigma = 1.61$) about the mean $\bar{x} = 41.5$

$(< -2\sigma)$	$(-2\sigma$ *to* $-\sigma)$	$(-\sigma$ *to* $\bar{x})$
0	4	6
$(\bar{x}$ *to* $+\sigma)$	$(+\sigma$ *to* $+2\sigma)$	$(> +2\sigma)$
6	4	0

A normal distribution with the same parameters arranged in the same way (from Table A.2) gives the following probabilities:

$(< -2\sigma)$	$(-2\sigma$ *to* $-\sigma)$	$(-\sigma$ *to* $\bar{x})$
0.0228	0.1359	0.3413
$(\bar{x}$ *to* $+\sigma)$	$(+\sigma$ *to* $+2\sigma)$	$(> +2\sigma)$
0.3413	0.1359	0.0228

H_0

There is no difference between the population of alienation scores and a normal distribution.

Statistic

Because our sample size is so small, we employ a Kolmogorov-Smirnov test.

$$D = \max |c_T - c_S|$$

c_T is the cumulative proportional theoretical distribution, and c_S is the cumulative proportional sample distribution which are laid out below to produce:

$$D = 0.0413$$

	$(< -2\sigma)$	$(-2\sigma$ *to* $-\sigma)$	$(-\sigma$ *to* $\bar{x})$	$(\bar{x}$ *to* $+\sigma)$	$(+\sigma$ *to* $+2\sigma)$	$(> +2\sigma)$
c_T	0.0228	0.1587	0.5000	0.8413	0.9772	1.0000
c_S	0.0000	0.2000	0.5000	0.8000	1.0000	1.0000
Difference	0.0228	0.0413	0.0000	0.0413	0.0228	1.0000

Probability

Special tables have been produced for testing normality using the D statistic. These are given in Table A.3:

$D(\alpha = 0.05, n = 20) = 0.190$

and

$D(\alpha = 0.20, n = 20) = 0.160$

Decision

Our D value is clearly extremely low, and we cannot begin to consider rejecting H_0. We conclude that our alienation scores have a distribution that can be described as reasonably normal.

Exercise

Repeat the analysis above for the central city sample in Table 3.8. Are these alienation scores reasonably normal?

Work Table 3.6 Student's *t* Test with One Sample

Data

SAMPLE Epsilon measures of the length of the Spanish–Portuguese boundary for $\varepsilon = 12.8$ km (from Table 2.2) are: 919, 876, 1029, 946, 946, 927, 998, 927, 1030, 964, and 997. $n = 11$; $\bar{x} = 960$; and the sample standard deviation is given by

$$\hat{\sigma} = \sqrt{\frac{\Sigma(x - \bar{x})^2}{(n - 1)}} = 48.7$$

POPULATION Spanish estimate of boundary = 987 km. We are testing whether our sample results for the scale $\varepsilon = 12.8$ km could have been drawn from a population represented by the Spanish estimate (population mean, M = 987).

H_0

There is no difference between our sample scores and the population represented by the Spanish estimate.

Statistic

$$t = \frac{\sqrt{n}\,|\bar{x} - \mathrm{M}|}{\hat{\sigma}} = \frac{\sqrt{11} \cdot |960 - 987|}{48.7} = 1.8$$

$df = n - 1 = 10$

Probability

From Table A.8:

$t(\alpha = 0.05, df = 10) = 2.23$

Decision

We cannot reject H_0. We have not been able to show that the sample at this degree of resolution differs significantly from the Spanish estimate at $\alpha = 0.05$.

Exercise

Repeat the exercise above by comparing the results obtained for $\varepsilon = 6.4$ km (from Table 2.2) with the Portuguese estimate of the boundary of 1214 km. Could this estimate have been derived reasonably from a degree of resolution represented by $\varepsilon = 6.4$ km?

Work Table 3.7 *z* Tests for Spatial Autocorrelation in Two-Color Maps

Data

From Figure 3.12, the joins are WW = 2, WB = 4, and BB = 3.

H_0

The number of WW joins is not different from random expectations: E(WW).

The number of WB joins is not different from random expectations: $E(\text{WB})$.

The number of BB joins is not different from random expectations: $E(\text{BB})$.

Statistics

$$z_{\text{WW}} = \frac{\text{WW} - E(\text{WW})}{\sqrt{\text{Var(WW)}}}$$

$$z_{\text{WB}} = \frac{\text{WB} - E(\text{WB})}{\sqrt{\text{Var(WB)}}}$$

$$z_{\text{BB}} = \frac{\text{BB} - E(\text{BB})}{\sqrt{\text{Var(BB)}}}$$

Let p = white areas/total areas = 3/6 = 0.5. Let q = black areas/total areas = 3/6 = 0.5. Let l = total number of joins = 9.

$$E(\text{WW}) = p^2 l = 0.25 \cdot 9 = 2.25$$

$$E(\text{WB}) = 2pql = 2 \cdot 0.25 \cdot 9 = 4.5$$

$$E(\text{BB}) = q^2 l = 0.25 \cdot 9 = 2.25$$

Let $k = \sum_{j=1}^{j=n} (l_j(l_j - 1))$, in which l_j is the number of joins for the jth area (2, 5, 3, 3, 3, 2), so that

$$k = (2 \cdot 1) + (5 \cdot 4) + (3 \cdot 2) + (3 \cdot 2) + (3 \cdot 2) + (2 \cdot 1) = 42$$

$$\begin{aligned}\text{Var (WW)} &= p^2 l + p^3 k - p^4(l + k) \\ &= (0.25 \cdot 9) + (0.125 \cdot 42) - (0.0625 \cdot 51) \\ &= 4.3125\end{aligned}$$

$$\begin{aligned}\text{Var(WB)} &= 2pql + pqk - 4p^2q^2(l + k) \\ &= (2 \cdot 0.25 \cdot 9) + (0.25 \cdot 42) - (4 \cdot 0.25 \cdot 0.25 \cdot 51) \\ &= 2.25\end{aligned}$$

$$\begin{aligned}\text{Var(BB)} &= q^2 l + q^3 k - q^4(l + k) \\ &= (0.25 \cdot 9) + (0.125 \cdot 42) - (0.0625 \cdot 51) \\ &= 4.3125\end{aligned}$$

Now, we find that

$$z_{\text{WW}} = \frac{2 - 2.25}{\sqrt{4.3125}} = \frac{-0.25}{2.08} = -0.1202$$

$$z_{\text{WB}} = \frac{4 - 4.5}{\sqrt{2.25}} = \frac{-0.5}{1.5} = -0.3333$$

$$z_{\text{BB}} = \frac{3 - 2.25}{\sqrt{4.3125}} = \frac{0.75}{2.08} = +0.3606$$

Probabilities

The statistical distribution of the z values follows the normal distribution, so that we use Table A.2 to test the significance of z_{WW}, z_{WB}, and z_{BB}:

$$p(z_{\text{WW}} = 0.12) = 0.4522$$

$$p(z_{\text{WB}} = 0.33) = 0.3707$$

$$p(z_{\text{BB}} = 0.36) = 0.3594$$

Decisions

In no case is the z value particularly rare and it does not reach $\alpha = 0.05$ or any other usual significance level. In all three tests, we cannot reject H_0, and we conclude that we have no evidence of spatial autocorrelation in the two-color map.

Exercise

Add a new area 7 to the map in Figure 3.12 that is contiguous only to areas 1 and 3 and colored black. What effect does this have on all three z values? Are there now any significant differences from random expectations with $\alpha = 0.05$?

Work Table 3.8 *z* Tests and Geary's Contiguity Ratio

Data

These are from Figure 3.12: a k-color map with x as indicated.

H_0

There is no autocorrelation between contiguity relations in the map pattern.

Statistic

$$z_c = \frac{E(c) - c}{\sqrt{\mathrm{Var}(c)}}$$

in which

$$c = \frac{n - 1\, \Sigma\Sigma\, (\delta_{ij}(x_i - x_j)^2)}{4l\, \Sigma\, (x_i - \bar{x})^2}$$

n and l are numbers of areas and links, respectively, and δ_{ij} is a member of the contiguity matrix Δ (Figure 3.12b). In Δ in Figure 3.12b, there are 18 cells for which $\delta_{ij} = 1$, which give:

$16 + 4 + 16 + 4 + 4 + 16 + 36 + 4 + 4 + 16 + 4 + 16 + 4 + 16 + 4 + 4 + 36 + 4 = 208$

There are six areas, so that squared deviations from the mean of 7 give:

$25 + 1 + 9 + 1 + 9 + 25 = 70$

$$c = \frac{(6 - 1) \cdot 208}{(4 \cdot 9) \cdot 70} = 0.4127$$

$E(c) = 1$

$$\mathrm{Var}(c) = \frac{(2l + k)(n - 1) - 2l^2}{(n + 1)l^2}$$

in which $k = 42$, as defined in Work Table 3.7.

$$\mathrm{Var}(c) = \frac{(18 + 42) \cdot 5 - (2 \cdot 81)}{7 \cdot 81}$$

$$= \frac{(300 - 162)}{567} = 0.2434$$

$$z_c = \frac{1 - 0.4127}{\sqrt{0.2434}} = 1.1903$$

Probability

As in Work Table 3.7, the z value is normally distributed, so that from Table A.2 $p(z_c = 1.1903) = 0.1170$.

Decision

$p(z_c = 1.1903) = 0.1170$ is much larger than $\alpha = 0.05$, so that we cannot reject H_0.

Exercise

Recompute c and test for spatial autocorrelation using z_c when, in area 1, $x = 4$ and in area 6, $x = 10$. This change in x values makes two pairs of neighbors identical. Does it also produce significant spatial autocorrelation at $\alpha = 0.05$?

Work Table 3.9 *z* Tests and Moran's Autocorrelation Statistic

Data

These are from Figure 3.12: a *k*-color map.

H_0

There is no autocorrelation between contiguity relations in the map pattern.

Statistic

$$r_a = \frac{n}{l} I$$

n and l are defined as before and

$$I = \frac{\Sigma\Sigma((x_i - \bar{x})(x_j - \bar{x})\, \delta_{ij})}{\Sigma(x_i - \bar{x})^2}$$

δ_{ij} is from Δ as before (Figure 3.12b). The 18 cells with $\delta_{ij} = 1$ in Figure 3.12b, give

$+ 5 + 15 + 5 + 3 - 1 - 3 - 5 + 15 + 3 - 3 - 1 - 3 + 3 - 3 + 3 + 15 - 5 + 15 = 58$

and from Work Table 3.8

$\Sigma\, (x_i - \bar{x})^2 = 70$

so that

$$I = \frac{58}{70} = 0.8286$$

and

$$r_a = \frac{6}{9} \cdot 0.8286 = 0.5524$$

$$z_I = \frac{I - E(I)}{\sqrt{\text{Var}(I)}}$$

where

$$E(I) = \frac{-l}{n(n-1)} = \frac{-9}{30} = -0.3$$

Let

$$\rho = -(n-1)^{-1} = \frac{-1}{5} = -0.2$$

and

$$E(I^2)$$
$$= \frac{l(1+2\rho^2) + 2k(\rho+2\rho^2) + 3l(l-1) - 2k\rho^2(n-1)}{n^2(n+1)}$$
$$= \frac{(9 \cdot 1.08) + 84(-0.12) + (27 \cdot 8) - (84 \cdot 0.04 \cdot 5)}{36 \cdot 7}$$
$$= \frac{9.72 - 10.08 + 216 - 16.8}{252}$$
$$= 0.7890$$

so that

$$\text{Var}(I) = E(I^2) - (E(I))^2 = 0.7890 - 0.09 = 0.699$$

Now, therefore,

$$z_I = \frac{0.8286 - (-0.3)}{\sqrt{0.699}} = 1.3499$$

Probability

As in Work Tables 3.7 and 3.8, the z value is normally distributed, so that, from Table A.2, $p(z_I = 1.3499) = 0.0885$.

Decision

$p(z_I = 1.3499) = 0.0885$ is larger than $\alpha = 0.05$, so we cannot reject H_0.

Exercise

Recompute r_a and z_I for the modified data given in the exercise for Work Table 3.8. Do we obtain significant spatial autocorrelation with this new map pattern?

IV

POINT PATTERN ANALYSIS

The dot map has long been one of the most popular cartographic tools of geographers. Its widespread use is based on the simplicity and clarity with which it displays spatial distributions. This arises from its abstraction of some discrete object from a real world situation into a framework involving locational coordinates. At a global scale, millionaire cities may be represented as single dots; at a local scale, individual incidences of disease may be plotted similarly. Such abstraction and portrayal is, of course, only a first stage in geographical analysis. The next step is to see whether the resulting spatial distribution exhibits recognizable patterns. Search for patterns should be directed ideally by hypotheses generated by previous studies, particularly studies that have developed theories and models. In this way, search for patterns can be structured, enabling geographers to employ specific techniques.

In this chapter, we consider what has come to be known as point pattern analysis, which consists of a set of techniques designed specifically to search for patterns in dot maps. This approach has been developed largely by plant ecologists in the last half-century as part of their studies of plant distributions. The concepts and techniques they have generated have been shown recently to be relevant in several geographical contexts. In fact, similar types of analyses using dot maps depicting house and farm patterns in the Tonami Plain, Japan, were being carried out forty years ago by a Japanese geographer, Isamu Matui. His work only recently has been fully appreciated within geography. Geographers, with increasing awareness of and interest in point pattern analysis, have begun to make their own particular contributions. The pioneer in this development has been American mathematical geographer Michael Dacey, and we describe some of his work together with geographical applications of traditional point pattern analysis.

The first research principle in any search for patterns must be that *we must never forget that every pattern is the result of some process at a given point in time and space.* Analysis of a point pattern ought to incorporate some idea of how the pattern evolved. We deal with the static evidence of point patterns, but we approach their description with concepts derived from processes over space. After all, patterns are only abstractions because, in reality, time never stops; it is a continuous on-going process. There are no patterns; there are only processes. It is imperative that we view our point patterns as simply the visual expression at one point in time of processes operating continuously over space. The key step in this approach is the interpretation of the occurrence of each dot on the map as an event in the context of the probability calculus described in Chapter Three. The processes we deal with are modeled in relation to probability distributions. In technical terms, two distinct approaches have evolved: quadrat analysis and spacing. These are used in most map analyses in geographical research, and we treat of each in turn.

QUADRAT ANALYSIS

Consider Figure 4.1. What sort of pattern exists on this map? One feature often visibly apparent on a dot map is that the density of points varies. In Figure 4.1, the top lefthand corner of the map is empty whereas several points occur in the bottom lefthand corner. Such visual impressions can be converted into quantitative expressions of pattern by sampling the density or frequency of points at different locations. This is the strategy for seeking patterns in quadrat analysis.

Randomness in Point Patterns

Anybody would be forgiven for dismissing the pattern in Figure 4.1 as chaotic or random. Surely there is no pattern in this map. In a very real sense, this response is correct. We have already argued that we should view patterns through the processes that produce them, so that we must ask, "How did Figure 4.1 evolve.?" The pattern was generated by drawing twenty pairs of random numbers from Table A.1 to represent point coordinates, as in simple random sampling. The allocation of points to the map is, therefore, random and we can term this procedure a random process. A process like this may at first seem rather uninteresting to geographers, but we can use it as a link between empirical map patterns and probability theory.

Randomness and the Poisson Distribution

What are the essential characteristics of the random process of allocating points to a map? From our discussion of random numbers in Chapter Two, we know that each allocation of a point is independent of all previous and future allocation. The fact that a point occurs in the top righthand corner of the map neither attracts nor repels other points. We can model this process as follows. Consider the small area Δ (depicted in Figure 4.1); it has an area of 1 square unit. The total map is $10 \cdot 10 = 100$ square units, so that the probability of finding 1 of the 20 observed points in an area of 1 square unit is $20/100 = 0.2$. We term this the expectation and denote it λ. We can now return to the process that produced the pattern. Initially there will be no points on the map; Δ and all other small areas will have no points within their areas. By the time we observe the process on Figure 4.1, 20 events have occurred, each producing a dot on the map. Thus, the probability that

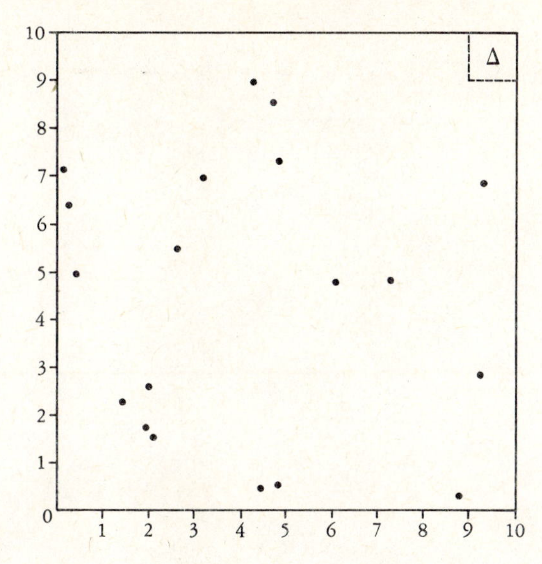

Figure 4.1 A random distribution of points.

our small area Δ will experience one of these events is 20/100 = the expectation. The expectation λ is the transition rate for area Δ having changed from the state of having no point to the state of having 1 point. This is shown diagrammatically in Figure 4.2a. Because points are allocated independently in a random process, it follows that Δ receiving a first point in no way affects its chances of obtaining a second point. The transition rate for having moved from the state of having 1 point to that of having 2 points in Δ remains λ. We can extend this argument to cover all transitions. This model of a constant transition rate defines a *Poisson process*. The Poisson distribution is the frequency distribution of small areas like Δ among the various states—the proportion of areas with no point (state 0), the proportion of areas with 1 point (state 1), the proportion with 2 points (state 2), and so on. The distribution is discrete; there are no fractions of points in an area. An example of the discrete probability distribution for $\lambda = 0.2$ is given in Figure 4.2b. Let's consider how the specific proportions in the different states are found.

The Poisson probability function is given by

$$p(x) = \frac{e^{-\lambda}\,\lambda^x}{x\,!}$$

$p(x)$ is the probability that a small area Δ will contain x points; e is the exponential constant. The probability that Δ will contain no points in the pattern depicted in Figure 4.1a is found by substituting $x = 0$, $\lambda = 0.2$, and $e = 2.7183$ in the equation, so that

$$p(0) = \frac{2.7183^{-0.2} \cdot 0.2^0}{0\,!} = \frac{1}{2.7183^{0.2}} = 0.82$$

Similarly, the probability that there will be 1 point is

$$p(1) = \frac{2.7183^{-0.2} \cdot 0.2^1}{1\,!} = \frac{0.2}{2.7183^{0.2}} = 0.16$$

and so on. These Poisson probabilities are shown in Figure 4.2b and also given in the second row of Table 4.1.

Because the Poisson distribution depends solely on λ, it has been tabulated for various values of λ to avoid repetitive calculation by many researchers in many disciplines. The tables are available in special books on the Poisson distribution; a part of them is reproduced in the Appendix in Table A.9. The probabilities given to four decimal places in Table 4.1 are from this source and, henceforth, we use these tables for finding Poisson probabilities, remembering that they are calculated as above. These represent our theoretical predictions of the probabilities of points over small areas for a random process. Notice that area Δ in Figure 4.1 has no points,

Table 4.1 Poisson Probabilities and Frequencies for $\lambda = 0.2$ and $n = 100$

State x	0	1	2	3	4
$p(x)$	0.8187	0.1637	0.0164	0.0011	0.0001
Observed frequencies	82.	16.	2.	0.	0
Expected frequencies ($p(x) \cdot n$)	81.87	16.37	1.64	0.11	0.01

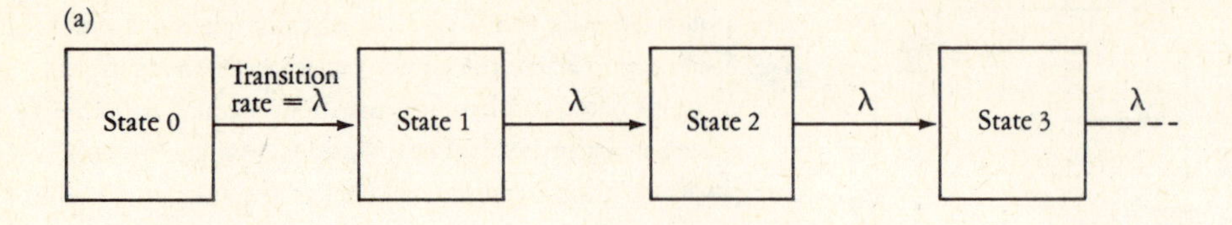

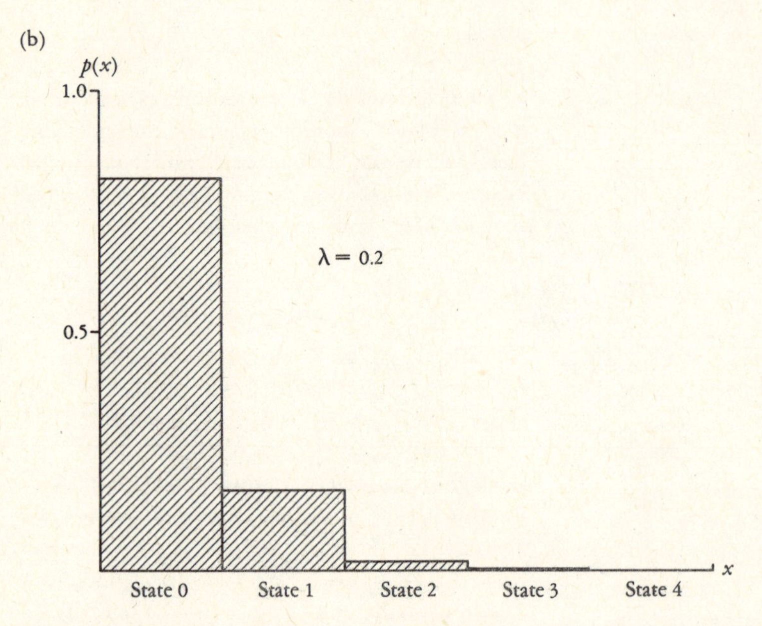

Figure 4.2 The Poisson model: (a) the Poisson process, (b) the Poisson distribution.

which we can now see to be very likely for a random process in which $\lambda = 0.2$ In fact, we expect that more than 8 in every 10 such small areas will record no points.

How can we apply this knowledge of the Poisson process to the analysis of point patterns? All we have to do is give an empirical definition of the small area Δ to which the events in the model occur. Geographers have defined Δ empirically in two different ways. The simplest is to divide the map area by a fine grid and to treat each grid square as Δ in the theoretical statements. The alternative approach is areal random sampling. We represent Δ by a small areal sampling unit (quadrat) that is laid upon the map at a sample of locations. The quadrat is normally a small square assigned randomly by using pairs of random numbers to allocate one of its corners or center. Notice that a square is used merely by convention, because our theoretical arguments are not concerned with the shape of Δ. This method produces random quadrat sampling, and we can consider the former method as an alternative sampling procedure, or systematic quadrat sampling.

Let's consider systematic quadrat sampling in the map in Figure 4.1. In Figure 4.3a, we have laid a grid of one-unit squares (quadrats) over this area and noted the number of points that fall in each area. In the top righthand cell (the original Δ) there are no points, next along the top row there are no points, and so on. This gives us 100 frequencies, one for each cell. We can aggregate them into a frequency distribution, as given in the third row of Table 4.1 and illustrated as a histogram in Figure 4.3b. We converted the two-dimen-

(a)

0	0	0	0	1	0	0	0	0	0
0	0	0	0	1	0	0	0	0	0
1	0	0	1	1	0	0	0	0	1
1	0	0	0	0	0	0	0	0	0
0	0	1	0	0	0	0	0	0	0
1	0	0	0	0	0	1	1	0	0
0	0	0	0	0	0	0	0	0	0
0	1	1	0	0	0	0	0	0	1
0	0	2	0	0	0	0	0	0	0
0	0	0	0	2	0	1	0	1	0

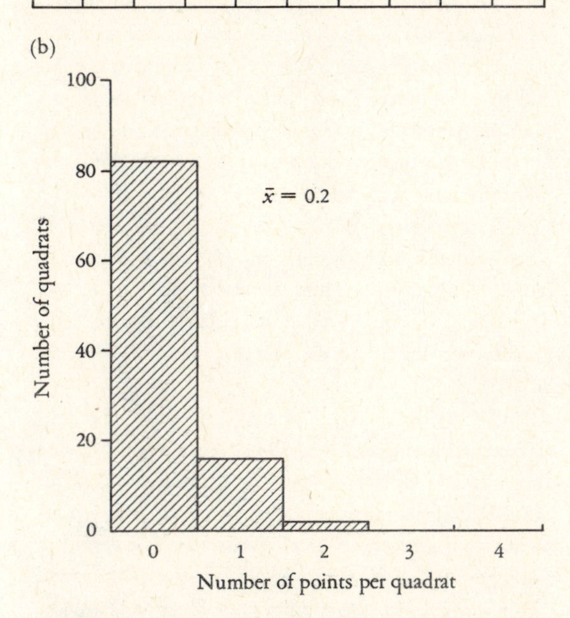

Figure 4.3 Quadrat sampling by grid: (a) quadrat counts, (b) frequency distribution.

sional point pattern into a one-dimensional frequency distribution. The frequencies represent the number of small areas, Δ, in each of the states of the theoretical model. If the point pattern is generated by a process like the Poisson process (constant transition rates), then the empirical frequency distribution should have a shape similar to that of the equivalent Poisson distribution. Figure 4.3b certainly looks like Figure 4.2b. Let's consider how we can make more rigorous comparisons in this context.

Goodness-of-Fit Tests

Table 4.1 displays probabilities of the Poisson function with an expectation of $\lambda = 0.2$. This expectation was in fact originally derived from Figure 4.1 with its 20 points in 100 square units. With quadrats of 1 square unit, we should expect 0.2 points per quadrat. Put another way, λ is the average number of points per quadrat, or

$$\lambda = \bar{x} = \frac{\Sigma(x_i f_i)}{\Sigma f_i}$$

in which x_i is the number of points per quadrat and f_i the frequency in each class. As we have seen, $\Sigma(x_i f_i) = 20$ and $\Sigma f_i = 100$, so that $\bar{x} = \lambda = 0.2$.

Consider now the frequencies in the third row of Table 4.2. Here we deal with empirically derived frequencies. These can be compared with the theoretical predictions of Table 4.1 by multiplying the latter by n to produce expected frequencies. We expect $0.8187 \cdot 100$ of the cells to have no points in them. Similarly, we expect $0.1637 \cdot 100$ of the cells to have 1 point in them, and so on. These expected frequencies are given in the fourth row of Table 4.1. The problem of testing the probability model reduces now to comparing the observed and the expected frequencies. This is a well known problem in inferential statistics, and it is known as goodness-of-fit testing.

Our example is rather artificial. We know that the process is random, because the points were allocated simply by using random number coordinates. We can use the results of this exercise to illustrate goodness-of-fit tests, however. Our empirical distribution should differ from the expected frequencies 1 time in 20 if we use $\alpha = 0.05$ as the significance level. From Table 4.1 it seems highly unlikely that our example is in fact the 1 of every 20 analyses that proves significantly different. Our observed frequencies are actually as close as they possibly could be to prediction, given the fact

Table 4.2 Contingency Table from Table 4.1

x	0	1	$\geqslant 2$
Observed (O)	82	16	2
Expected (E)	81.87	16.37	1.76

that the observed frequencies are counted in integer numbers. Let's see what sort of probability level relates to this evidence.

The data are in frequencies so that we can employ either a χ^2 or a Kolmogorov-Smirnov one-sample test. Problems arise in employing either. With the χ^2 test, we come across small class frequencies. Remember that χ^2 tests require a contingency or frequency table with no more than 20% of classes having expected frequencies of less than 5. In Table 4.2, 33% (that is, 1 out of 3) of the classes have expected frequencies of less than 5. Therefore, we have to combine them to overcome this problem. In order to achieve this successfully, we are forced to have two classes: $x = 0$ and $x \geqslant 1$. When we do this, a further problem arises. In Chapter Three we defined degrees of freedom for this type of one-sample test as $k - 1$, where there are k classes. When we derive the parameters of a theoretical distribution from the same empirical evidence as observed frequencies, as we usually do, further degrees of freedom are lost. In general,

$$df = k - 1 - c$$

when c parameters are estimated for the theoretical distribution. The Poisson distribution has the one parameter λ, so that, with 2 classes, degrees of freedom become

$$df = 2 - 1 - 1 = 0$$

which makes a χ^2 assessment impossible. In order to carry out a χ^2 test, we have to violate the rules attached to its procedure. We maintain 3 classes, one of which (33%) has an expectation of fewer than 5. From Table 4.2 we calculate χ^2.

$$\chi^2 = \Sigma \frac{(O - E)^2}{E} = \frac{0.13^2}{81.87} + \frac{0.37^2}{16.37} + \frac{0.24^2}{1.76}$$

$$= 0.0413$$

$$df = 3 - 1 - 1 = 1$$

The effect of violation of the 20% rule on this computed χ^2 value is not known exactly. We proceed on the assumption that the rule violation is small and that the value of 0.0413 is a good estimate of χ^2.

What, therefore, does $\chi^2 = 0.0413$ represent in this case? Table A.4 for χ^2 in the Appendix presents the whole range of probabilities from $p = 0.99$ to $p = 0.001$. For every column, we are given χ^2 values for different degrees of freedom that we can expect at the column probability level. Hence, at $p = 0.99$ (99 times in 100), a χ^2 value with $df = 1$ will be larger than 0.00016. From Chapter Three we remember that this means that 0.00016 defines the lefthand end of the statistical distribution of χ^2 such that 99% of the distribution is beyond this value. Here, $\chi^2 = 0.0413$, which falls between columns 4 and 5, so the probability of a value this large is between $p = 0.9$ and $p = 0.8$. Under a null hypothesis of no differences, a χ^2 value such as ours will occur between 80 percent and 90 percent of the time. We clearly cannot reject H_0 and so our observed and expected frequencies have a high probability of belonging to the same population.

Notice that in a goodness-of-fit test we do not automatically look at the standard $\alpha = 0.001$ or $\alpha = 0.05$ significance levels. We are not usually interested in proving differences, in rejecting H_0, but in trying to test for similarities. Thus, we refer directly to the probability that our χ^2 value will occur under H_0. This is why we stated above that for $\chi^2 = 0.0413$ and $df = 1$;

$$0.8 < p < 0.9$$

rather than stating that χ^2 does not reach a critical value at $\alpha = 0.05$. Goodness-of-fit tests often differ from other inferential statistics in that the research hypothesis and null hypothesis usually coincide. The resulting testing procedure is less standardized, as we have shown.

The comments above apply to all goodness-of-fit tests. If we use a Kolmogorov-Smirnov test on our data, several advantages accrue. First of all, we avoid problems associated with small frequencies. We do not lose information from combining classes. Furthermore, this test also uses more information than a χ^2 test because it takes into account the ordering of the classes. It is for this reason that the Kolmogorov-Smirnov test has been preferred to the χ^2 test in many goodness-of-

Table 4.3 Table for Kolmogorov-Smirnov Test from Table 4.1

x	0	1	2	3	4
Cumulative observed proportions	0.82	0.98	1.00	1.00	1.00
Cumulative expected proportions	0.8183	0.9824	0.9988	0.9999	1.0000
Differences	0.0013	0.0024	0.0012	0.0001	0.0000

fit situations in geographical research. The case for using Kolmogorov-Smirnov tests is not as clear cut as it seems, however. The tables of D such as those in our appendix (Table A.3a) are produced on the assumption that the theoretical distribution is completely specified before data analysis. This means simply that parameters should not be derived from the same evidence as the observed frequencies. As we have seen, this is not usually the case in goodness-of-fit testing. This can be taken into account with χ^2 by adjusting the degrees of freedom. No such adjustment is possible with the Kolmogorov-Smirnov test because degrees of freedom are not employed.

We can compute a D value, but we have no appropriate table in which to assess its probability. In recent years, statisticians have been producing some alternative tables for some theoretical distributions, including the normal distribution, but none is available for testing against a Poisson distribution. The resulting differences in critical D values are fortunately quite small, and we can use the usual tables for D, because we are not exactly specifying the probability level. To calculate the D statistic, we compute proportions of observed and expected, we cumulate, and we find the largest difference, as shown in Table 4.3. We see that $D = 0.0024$, which is very small indeed. From the tables of D we find we do not have a complete range of probabilities available; at the relatively high level of probability of $p = 0.2$, the critical D value is given by $1.07/\sqrt{n} = 0.107$. Our observed D is much smaller than this, so we can say that the probability of obtaining $D = 0.0024$ under the null hypothesis of no differences between observed and expected frequencies is much greater than $p = 0.2$. Clearly, as in our χ^2 test, we are not able to begin to consider rejecting H_0.

The tests above will enable us to decide whether it is reasonable to assume that our point pattern could have been generated randomly. In the example we have considered, the pattern proved to be compatible with a Poisson interpretation, and this is quite comforting because we know the process to have been random. In most research contexts, however, we are not so certain of the process. If the goodness-of-fit tests suggest that the Poisson distribution is not a good description of our data, then we must turn our attention to the other nonrandom types of process.

Divergence from Randomness

We have suggested that random processes producing random point patterns do not hold much intrinsic interest for the geographer. Here we treat of them as a starting point for more interesting analyses. The random distribution and its properties are a standard against which we can compare other distributions that we suspect are produced by nonrandom forces. In a sense, the Poisson distribution is as basic to probability process models as the normal distribution is to inferential statistics. Let's consider, in a general way, what sort of nonrandom processes geographers deal with in their research.

Generalized Types of Pattern and Process

Consider the situation in which settlers are moving into a new area. At certain points where routes meet, small supply stores may arise to serve the settlers with provisions and goods they cannot themselves grow or make. Stores may be haphazardly located, although very soon it will become apparent that some stores are doing much better than others. Those that are some distance from their rivals enjoy a greater share of the potential market and therefore prosper. On the other hand, where there are two or three stores located close to one another, there probably will not be enough settlers to support each store as a separate entity. In time, one or two will be forced to close, leaving a single store to serve that portion of the market. If the settlers are spread fairly evenly over the land, this competition for markets is a competition for space, and such processes may be termed *competitive* processes over space. In abstract terms, such processes can be viewed as objects locating themselves in space in such a way as to be as far away from other objects as possible. The result is an evenly spread distribution of points, and this has been termed a *dispersed, uniform,* or *regular pattern.* We use the former term here, reserving the others for specific, limiting cases. The classic example in geographical theory is in central-place theory according to which competition among service centers results in a regular triangular pattern of central places.

This process certainly is not random, because the allocation of points is anything but independent of the location of existing points. This is not the only way in which the Poisson independence assumption can be violated, however. From a completely opposite view, we can consider a situation in which, instead of existing points repelling new points, existing points attract further points. Such processes are termed *contagious* processes over space. We use this term in essentially the same sense as it is when commonly used to describe diseases passed from one person to another by some form of contact. Contagious processes over space are, therefore, those in which some property is transmitted between objects in such a way that transmission is facilitated by proximity between objects. The resulting distribution is one in which points tend to be located close to one another and has been termed an *aggregated* or a *clustered pattern.* We use the latter term here. In cultural geography, it has been suggested that information about an innovation spreads between neighbors in a farming community by word of mouth so that it may be termed a contagious process.

With the random case, we have three basic processes producing three distinct patterns (Figure 4.4), two of which are directly predicted by two of geography's most respected models. These are summarized in Table 4.4. This gives us a conceptual background upon which divergences from randomness can be considered. How will these two general divergences from randomness be reflected empirically in our quadrat analysis?

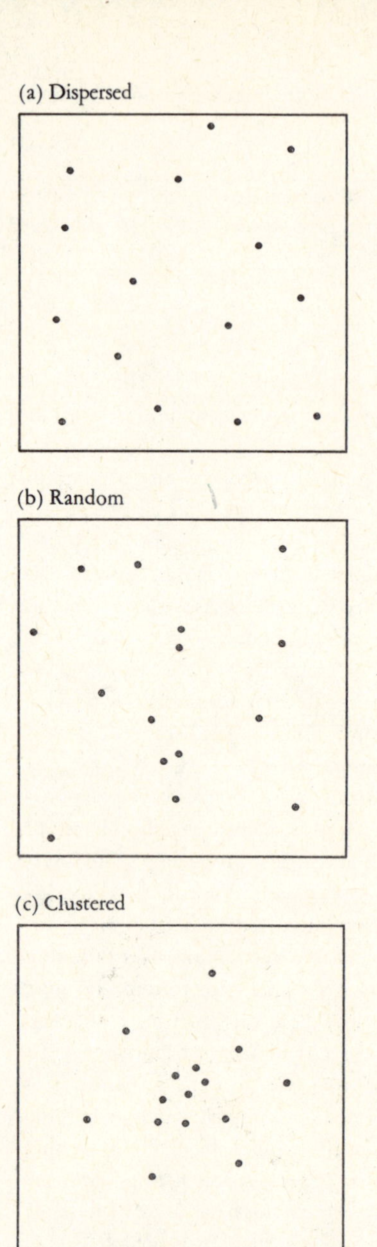

Figure 4.4 Three types of point pattern.

Table 4.4 Processes and Patterns

Process	*Pattern*	*Geographical example*	*Hypothesized process*	*Predicted distribution*
Competitive	Dispersed	Central-place theory	Market (space) competition	Regular triangular lattice
Random	Random			
Contagious	Clustered	Spatial diffusion model	Face-to-face contact (proximity)	Neighborhood effect

Consider a highly clustered distribution. It has many quadrats with no points in them and one or two with many points. The resulting frequency distribution of quadrats has a relatively large variance. On the other hand, a highly dispersed distribution tends to have approximately the same number of points in each quadrat. The resulting frequency distribution has a relatively small variance. Plant ecologists have long used the variance of the frequency quadrat count distribution to indicate pattern. We report on their use of this in experiments relating to scale effects in point patterns below. If we wish to compare general patterns using this approach, we must relate the variance to the mean. Thus, ratios of variance/mean are used to determine the type of divergence in a given pattern. This measure is very convenient because of the property of the Poisson distribution that the mean and variance are equal. A random process producing a Poisson distribution has an expected variance over mean ratio of unity. If the ratio is more than one, there is a relatively large variance and, hence, a divergence from randomness in a clustered direction. Conversely, a variance/mean ratio of less than one results from a low variance reflecting a dispersed divergence from a random pattern. Thus, we have a simple tool for identifying and measuring the two types of divergence from randomness.

We illustrate this strategy of assessing point patterns by using the study of Arthus Getis on grocery store locations in Lansing, Michigan, for ten-year periods from 1900 through 1960. He hypothesized that changes in the pattern of grocery stores would reflect changes in the population densities and transport technology. Before the major growth of the town in 1900, the pattern would simply reflect the varying population densities existing at that time. During the town's growth early in the twentieth century, the very uneven population distribution would have led to clustering in the pattern as small groups of grocery stores served local concentrations of population. In more recent years, this tendency would have reversed as the automobile led to more even population densities and allowed the retail trade to reorganize its grocery shops into fewer but larger outlets that no longer required to be clustered because they served a more mobile population. The frequency distributions and analyses using grid cells of 542 square feet are shown in Table 4.5. The first step is to ascertain whether these frequencies differ from a simple Poisson expectation. (Notice that in this particular goodness-of-fit test the null hypothesis and research hypothesis do not coincide.) The probabilities from the χ^2 test indicate that we can reject the null hypothesis that there are no differences between observed and Poisson predictions at a very high level from 1910 through 1950. The variance/mean ratios then show that these significant divergences from randomness are in a clustered direction, and Getis' hypothesis is confirmed.

The variance/mean ratio allows us to specify the pattern divergence from the Poisson predictions with its ratio of unity. This reference to the Poisson standard is a relatively weak research strategy for studying clustered and dispersed patterns. Instead of considering pattern divergences from random, often we are better served by considering *process* divergences from the Poisson. This means using contagious and competitive probability models. We consider one example of each.

Dacey's County Seat Model

Central-place theory predicts a regular point pattern of settlements on its isotropic plane. In real world situations, we do not expect exact regularity, but in the absence of major distortions we expect a dispersed

Table 4.5 Quadrat Analysis of Grocery Store Locations in Lansing, Michigan (1900–1960)

	Frequency points per cell(x)						*Total points*	*Total cells*	$p(\chi^2)$	*Variance/mean*
	0	1	2	3	4	5				
1900	89	13	3	0	0	0	19	105	0.16	1.13
1910	112	20	4	3	0	0	37	139	0.0005	1.44
1920	261	43	12	2	3	1	110	322	0.0005	1.31
1930	315	66	18	5	3	0	129	407	0.0005	1.47
1940	389	82	15	5	2	2	145	495	0.0005	1.56
1950	433	76	15	5	1	0	125	530	0.0005	1.34
1960	598	64	4	0	0	0	72	657	0.50	1.00

Source: A. Getis, Temporal land use pattern analysis. Reproduced by permission from the *Annals* of the Association of American Geographers, vol. 54, 1964, Tables 1 and 2.

settlement pattern. The variance/mean ratio should be less than 1. Mathematical geographer Michael Dacey has devised a probability law that specifies a process that generates such a pattern. It is referred to usually as Dacey's county seat model.

The county seat model is a modified Poisson distribution that combines a purely random process with a specified competitive process. It is easiest to understand in terms of the settlement context that it is designed to model. We begin by dividing urban places into county seats and noncounty seats. These are administrative centers (in England, they are county towns) and other urban centers. The study region is assumed to be divided into counties of equal size to which county seats are randomly assigned. The process is made competitive by limiting the number of county seats to one per county. Once a county has been allocated a county seat, it cannot be allocated another. The noncounty seats are allocated randomly without this restriction and, thus, conform to the Poisson law. These two processes are combined to form a modified Poisson model that produces a pattern more regular than random. The probability function is

$$p(x) = \frac{q\gamma^x e^{-\gamma}}{x!} + \frac{p\gamma^{x-1} e^{-\gamma}}{x!}$$

$x = 0, 1, 2, \cdots$, and $q = 1 - p$. The distribution depends on two parameters, γ and p. These can be estimated from the mean and variance of the observed frequencies such that

$$\hat{p} = \sqrt{\bar{x} - \mathrm{Var}(x)}$$

and

$$\hat{\gamma} = \bar{x} - \hat{p}$$

Both parameters vary from 0 to 1. Parameter p is interpreted as indicating the degree of regularity: when $p = 0$, $\gamma = \bar{x}$ and the process is purely Poisson, but when $p = 1$ there is a seat in every county. In most applications p is between 0 and 1 because the point pattern of settlements is usually defined to include all towns above a certain size that include some but not all county seats. A simple example of fitting this model is described in Work Table 4.1.

Dacey's county seat model is clearly a probability law of particular interest to geographers in their studies of spatial regularity. It is not a probabilistic translation of Christaller's deterministic central-place theory. Neither is it a very general model of competitive processes as described at the beginning of the chapter. We have a very specific process being modelled. It is a random allocation of towns with an administrative or political addition. This process is not unlike that reported for the development of the central place system in the American Midwest. In the competition between towns in this area, the county seats certainly prosper

Table 4.6 The County Seat Model and the 99 Largest Towns Among Iowa Counties

Number of cities per county (x)	*0*	*1*	*2*	*3*
Observed frequency	18	65	14	2
Predicted frequency	18	65	14	2

Source: M.F. Dacey, Modified Poisson probability law. Reproduced by permission from the *Annals* of the Association of American Geographers, vol. 54, 1964, Table 5.

often at the expense of other towns. It is rather appropriate, therefore, that the model was originally tested by using Iowa's settlement pattern.

The simplest test is to take the 99 largest towns in Iowa as the point pattern and the actual 99 counties as the counties in the model. The distribution of these largest towns observed among the counties is given in the second row of Table 4.6. From this data $p = 0.77$ and $\gamma = 0.23$, which produce the predictions in row 3 of Table 4.6. We have a perfect fit, an empirical scientist's dream. The county seat model certainly seems to fit Iowa's settlement pattern well.

Of course, the finding could be a fluke. In order to show that this was not the case, Dacey used the model to predict settlement frequencies per county for towns of over 2,500 for every 10 years from 1840 through 1960. The predictions were not always exactly right, but they were very close to the observed pattern at each date. The increasing regularity of the pattern as more county seats gain population of over 2,500 is shown by the increasing value of p from 0.08 in 1840 to 0.83 in 1960.

The initial empirical assessment of this model is clearly encouraging. The process is realistic in the sense that something akin to it has been observed by historical research on Midwestern settlement patterns. There are some objections, however, concerning both the analysis and the model. First of all, the model assumes counties of equal size whereas Iowa's counties differ in size. Second, the model assumes equal propensity to produce towns whereas varying agricultural potential in Iowa leads to different rural population densities and, hence, variations in propensity for town growth. These two objections were admitted by Dacey, but he assumed that their combined effect is negligible. This seems reasonable because Iowa's counties probably fit these assumptions as well as any other area in the world. The suitability of Iowa as a central place laboratory is certainly attested by its having been a favorite testing ground for central place researchers as far back as the original studies of August Losch.

A more technical objection concerns the independence of events assumed by the Poisson aspect of the model. In a spatial sense, this means that if a county receives a noncounty seat, this should not affect the probability that this county will receive another noncounty seat or, in fact, that adjacent counties will subsequently receive noncounty seats. In many parts of the world, particularly in industrial regions, we are used to the grouping of some towns, and in them this assumption of Dacey's model would be violated. Is this true for his Iowa data? It is clearly a problem of spatial autocorrelation: we have a map of areal units (counties) with values (number of noncounty seats) attached to each and we wish to see whether adjacent area values are correlated. To his data Dacey applied a spatial autocorrelation analysis (as we described it in Chapter Three) and was able to show that the county pattern was insignificantly different from random expectations. Thus, the assumption of independence of events seems to hold for Iowa counties.

Dacey's analysis seems to stand up to his own rigorous scrutiny. We may still ask how useful this model is. We have already noticed that it is a very particular process that seems to describe Midwestern central place development. What is unanswered is how well the model performs outside the ideal conditions of the Midwest. We consider this problem and propose an alternative research strategy for difficult study regions below.

The Negative Binomial Model

When we come to consider contagious processes, we find that we have many probability models to choose from because contagious processes are common in the biological sciences and this fact has stimulated the development of several different models.

Several have been employed in geographical research, although it has been argued that the negative binomial distribution is the contagious probability law most appropriate to geographical needs, particularly for diffusion research. We describe this probability model as our example of a contagious process model for describing point patterns.

From a geographical viewpoint, the most suggestive physical interpretation of the negative binomial model is in terms of randomly distributed colonies of individuals. Random colonies or clusters have a Poisson distribution; the numbers of individuals in colonies are independently distributed logarithmically. This fits a two-stage diffusion process of opinion-leaders and followers. Consider an area in which some information is available. It is received and accepted by any one individual in the area. She is the first opinion-leader. In subsequent time periods, this opinion-leader passes on the information at a logarithmic rate to adjacent followers. Meanwhile, other opinion-leaders emerge, independently of one another, and also begin spreading the information at a logarithmic rate. If points represent acceptors (either opinion-leader or follower), then the process leads to a pattern of distinct clusters. The process and pattern are clearly sufficiently similar to Hägerstrand's (1968) diffusion studies to warrant further investigation.

The negative binomial probability function can be given by

$$p(x) = \frac{(k + x - 1)!}{x!(k - 1)!} p^k(1-p)^x$$

p and k are two parameters that have to be estimated. Of the several methods of estimating these constants, the simplest is the so-called method of moments, which employs the mean ($\bar{x}$) and variance (Var(x)) of the observed frequency distribution. Thus,

$$p = \frac{\bar{x}}{\text{Var}(x)}$$

and

$$k = \frac{\bar{x}p}{(1 - p)}$$

Because this is a frequently used distribution, tables of negative binomial probabilities are available for direct consultation, eliminating individual need for computation. Because the tables have to accommodate two parameters, however, the values of the parameters that are used to define the probability distribution are not as comprehensive as in the case of the Poisson with its single parameter. Hence, interpolation is required for many situations in which computed p and k values fall between the p and k values for which probabilities have been tabulated. The researcher may prefer to calculate probabilities directly, using computed p and k values. A fully worked example of fitting a negative binomial model in this way is illustrated in Work Table 4.2.

The negative binomial model has been used to test Hägerstrand's ideas on the diffusion of innovations. David Harvey has analyzed the empirical patterns of acceptors on Hägerstrand's maps and some of the point patterns generated by his simulation model. The former were sampled systematically from a grid overlay; the latter were sampled by using random allocation of quadrats. In both cases, the quadrats were 5 kilometer squares. Some of Harvey's results in terms of observed and negative binomial expected frequencies are shown in Table 4.7.

Simple visual inspection of Table 4.7 indicates the close fit between the negative binomial model and Hägerstrand's empirical and simulated maps. This is confirmed by the low χ^2 values, all indicating that the observed levels of difference are not inconsistent with a null hypothesis of no differences. We have argued previously that research inferences should be something more than statistical inference. We can begin by asking two sets of questions, one relating to how well this application fits accepted geographical ideas and the other asking how well it fits the mathematical assumptions.

From our initial description of the process behind the negative binomial distribution, it seems that, superficially at least, the model fits our ideas on diffusion of innovation rather well. We can, in fact, develop these links a little further. For example, we can argue that placing the quadrat over the map is not unrelated to the placement of the mean information field (MIF) in the original simulation model. In MIF placement, we generate a diffusion map; in quadrat placement, we

Table 4.7 The Negative Binomial Distribution Fitted to Diffusion Maps

x	Map A		Map B		Map C		Map D	
	Observed	*Expected*	*Observed*	*Expected*	*Observed*	*Expected*	*Observed*	*Expected*
0	27	27.1	32	32.2	45	44.8	32	32.3
1	15	13.4	15	16.0	13	14.1	17	14.7
2	9	9.3	9	10.7	9	7.6	10	9.2
3	4	7.0	7	7.6	3	4.7	5	6.3
4	3	5.5	8	5.5	3	3.2	3	4.4
5	4	4.4	5	4.1	3	2.0	4	3.2
6	4	3.6	2	3.1	2	1.4	2	2.4
7	2	3.0	3	2.3	1	1.0	3	1.9
8	3	2.5	1	1.8				
9	3	2.1	1	1.4				
10	3	1.8	2	1.1				
11	2	1.5	1	0.8	2	2.2	5	5.2
12	4	1.4	1	0.6				
13	1	1.1	3	2.8				
⩾14	6	6.3						
χ^2	6.67		2.11		1.10		0.84	
df	6.		5.		3.		4.	
Probability	0.3		0.8		0.7		0.9	

Map A shows acceptors of pasture improvement grants (1928–1933).
Map B shows acceptors of milking machines (1944).
Map C shows simulation on anisotropic plane.
Map D shows simulation on anisotropic plane.
Source: D.W. Harvey. *The Analysis of Point Patterns.* Transactions No. 40, 1966, Table VI, p. 92. With permission of the Institute of British Geographers.

convert it to a frequency distribution for analysis. Harvey's quadrats are the same size as Hägerstrand's original MIFs. Furthermore, the *k* parameter of the model can be interpreted as the diffusion rate, because higher *k*'s produce more quadrats with high frequency counts. It is clear that at any point in time, the more high frequency counts there are, the more rapidly the diffusion has spread. On both these grounds, this model is consistent with previous ideas on diffusion of innovation.

Diffusion, like all processes, occurs in space and time. Point pattern analysis leads us to emphasize the spatial aspect but we should not forget the temporal dimension. In the negative binomial model, the followers of opinion-leaders are generated at a logarithmic rate (increasing proportionately 2, 4, 8, 16, 32). In any real world situation, there is a maximum number of possible acceptors. Followers cannot continue to grow at a logarithmic rate because the population soon becomes saturated. In fact, it is well known that diffusion processes fit a *logistic* trend in which the diffusion rate slows down after half the population have become acceptors. After the 50% level of acceptance has been reached, the negative binomial model is quite inappropriate. In the early stages of diffusion, however, logarithmic and logistic trends need not be so dissimilar and the negative binomial model gives an approximation adequate to the process in the time dimension. Harvey's analyses of Hägerstrand's maps (Table 4.7) are restricted to early stages of the diffusion process. The fact nonetheless remains that the negative binomial distribution cannot adequately model dif-

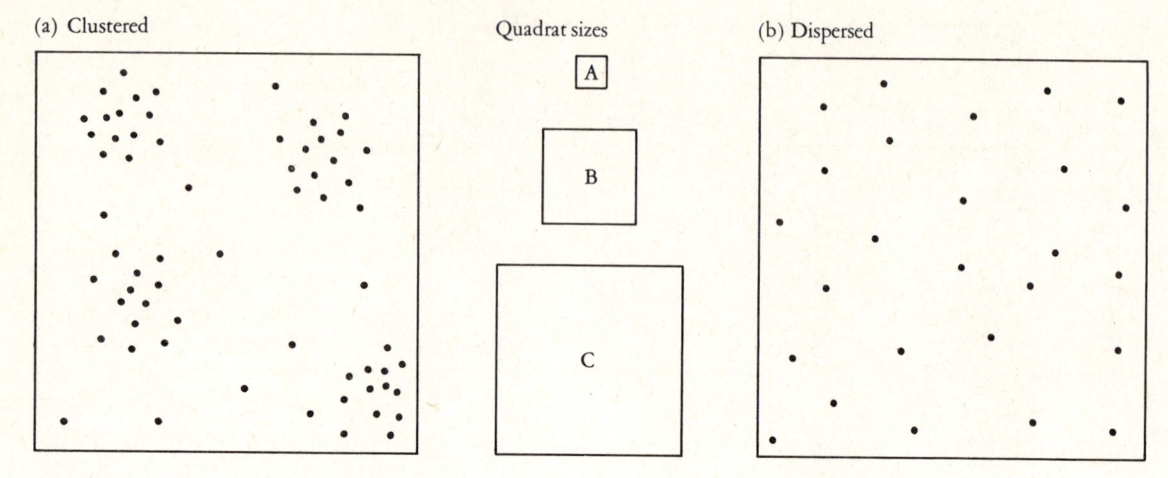

Figure 4.5 Pattern and quadrat size.

fusion process to saturation level. It is for this reason that the model cannot be regarded as a mathematical replacement for Hägerstrand's Monte Carlo simulation model.

Problems in the Use of Quadrat Analysis

We have considered problems of particular models as we have dealt with the models themselves. There are other, more general problems of quadrat analysis that occur irrespective of the probability model being used. These problems are our concern in this section.

Quadrat Size

Let's begin by considering an important technical problem. In all our discussion, we have assumed the quadrat size is given. It can easily be shown that changing the quadrat size produces changes in the observed frequency distribution that can lead to different process inferences. This problem warrants careful consideration.

Consider Figure 4.5a. Our visual impression is of a clustered distribution but it is not automatically confirmed by a quadrat analysis. If we use the smallest quadrat size, A in Figure 4.5, we are likely to produce a frequency distribution of many 0 and 1 counts, producing a frequency count with a variance not unlike Poisson predictions. If we increase quadrat size to cluster size (quadrat B), we produce quadrats with many points (when the quadrat falls on a cluster) and some with few points (when the quadrat falls between clusters). Such a frequency distribution has a very high variance so that the variance/mean ratio is greater than one, indicating clustering. This result would be consistent with our visual impression. As we increase quadrat size further (quadrat C), however, very large quadrats cause most quadrats to have similar numbers of points, and the variance of the frequency distribution is relatively low. The resulting variance/mean ratio may, therefore, fall below unity to indicate a pattern tending toward dispersion. One point pattern may produce results suggesting all three general pattern types.

A similar conclusion can be arrived at by starting with what seems visually to be a dispersed point pattern (Figure 4.5b). With very small quadrats of size A, the numerous 0 quadrat counts may mean that the pattern cannot be distinguished from a random pattern. In fact, it is found generally that small quadrats produce Poisson-like frequency distributions almost independently of the actual character of the point pattern. In the case of a dispersed pattern, however,

Table 4.8 Quadrat Size and Frequency Distribution Parameters: Artificial Data Experiments

Quadrat size (cm)	*Mean*	*Variance*	*Variance/mean*
10	0.07	0.066	0.9394
15	0.18	0.190	0.9527
20	0.29	0.289	0.9958
25	0.39	0.443	1.1342
30	0.68	0.947	1.3928
35	0.79	1.238	1.5674
40	0.97	1.585	1.6340

Source: Adapted from P. Greig-Smith. The use of random and contiguous quadrats in the study of the structure of plant communities. *Annals of Botany* (London), New Series, 16:312, 1952, Table V. By permission of The Clarendon Press, Oxford.

dispersion will be detected as quadrat size is increased, and this finding continues with very large quadrats for the same reason as for clustered patterns. At first sight, these complementary sets of results cast severe doubt on the utility of quadrat analysis. If we can produce any pattern by judicious choice of quadrat size, why not simply pronounce a preferred pattern at the outset and save time on the analysis? A less cynical attitude is to investigate this rather interesting phenomena. Several plant ecologists have done this in their search for an optimum quadrat size.

Greig-Smith employed varying quadrat sizes to analyze artificial point patterns. We show his results in Table 4.8 for a visually clustered pattern with random sampling using quadrats of seven different sizes. As we would expect, the mean count per quadrat increases with quadrat size, although not as rapidly as the variance. This finding is consistent with our arguments above. From the variance/mean ratio, we see that the pattern tends to become more clustered as quadrat size increases. In this experiment we seem to reach no quadrat size large enough to show the reverse of this trend. We can use this sort of experimental evidence to define the optimum quadrat size for studies in which we attempt to detect nonrandomness. In general, we must specify a quadrat size that approximates the cluster size. In clearly clustered distributions, such as in Figure 4.5a, this can be done by visual inspection but with more complicated patterns we have to rely on analyses such as Greig-Smith's study.

An alternative approach by ecologists to the notion of optimum quadrat size has been to define this size as $2A/n$, in which A is the area of the map and n is the number of points. Thus quadrat size is defined as twice the average area around each point and is much larger than quadrats as usually employed in geography. The effect is to ensure that there are few quadrats recording zeros or very low numbers of points. The resulting frequency distributions are less markedly skew, and such symmetric distributions facilitate comparison between study regions based on parametric inferential statistics. This is presumably an important consequence in ecological research. In geography, however, we are generally concerned with comparing the observed frequency distribution with a probability law using nonparametric statistics. Thus, the symmetric property of this large quadrat size is not generally important. This approach to defining quadrat size may be of relevance nonetheless to studies of dispersed patterns. If these are assumed to involve competition for space, and if we employ a size defined by A/n, we have the mean area about each point as our area. This is the area we would expect each point to control if the competition between points is equal. On Christaller's isotropic plane, these are the sizes of the market areas about the central places. We return below to the use of this method of defining quadrat size.

Our discussion of the quadrat size problem has assumed that we want to analyze a point pattern at just one scale and, hence, we have explored the concept of a single optimum quadrat size. An alternative strategy is to employ quadrats of various sizes to search for different patterns at different scales. In this research design, the question of quadrat size is not a problem but is used to learn more about scale effects in spatial patterns. The strategy requires some justification. Although we are used to considering a point pattern visually as a single pattern type, when we consider process we are quite used to the idea that different processes are important at different scales and

(a)

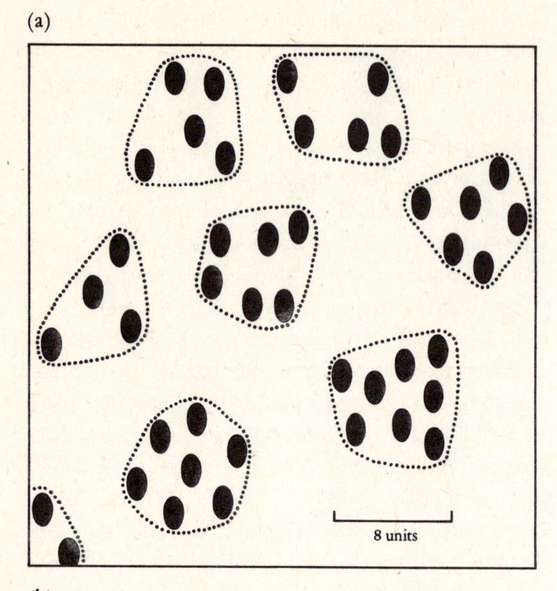

(b)

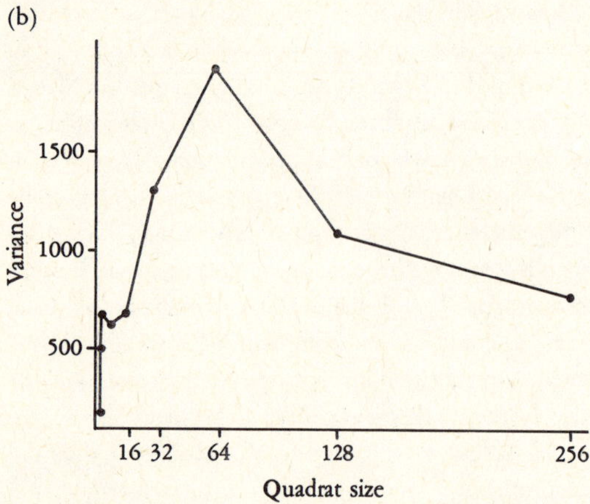

Figure 4.6 An experiment on pattern and scale: (a) an artificial pattern on two scales, (b) the variance-scale relationship.

From K.A. Kershaw. *Quantitative and Dynamic Ecology*, London: Edward Arnold, 1964.

produce patterns that vary by scale. From a general perspective, many simple examples can be thought of for which we normally admit that there are different patterns at two separate scales. Any map of the United States showing the distribution of off-farm facilities in the dairy industry, such as milk collection centers, will show an extreme degree of concentration in Wisconsin. A map of the same facilities for the Wisconsin dairy region alone, however, will show a more dispersed pattern, as the collection centers each serve a separate local area. Similarly, the distribution of collieries in Great Britain has the clustered pattern identified as the coalfields at a national scale, although within any single coalfield the shafts are sunk at fairly regular intervals to produce separate underground workings, giving a dispersed pattern. It would seem that scale is more than a problem of analysis and should be studied as a feature in its own right. David Harvey has noticed that this can be achieved by varying the size of the quadrat in order to observe what patterns are revealed. Such a research design has the advantage of going beyond what we can usually visualize from a dot map. We can quantify explicitly our description of pattern at various scales and thus suggest what processes dominate at particular scales.

Harvey's suggestion has not been followed up in geography, but it can be illustrated from ecological work by K.A. Kershaw with artificial data, part of which is shown in Figure 4.6a. These data were specially prepared to show two levels of clusters: small ones at the scale of 2 units and large ones at the scale of 64 units. The whole map was divided into a grid of 1-unit quadrats representing a systematic sample. Quadrat counts were made at this scale and all higher scales with the grid cells combined in 2s, 4s, 8s, and up to quadrats of 256 units. The resulting levels of variance are shown in Figure 4.6b. Remembering that high variances result from quadrat counts of high and low frequencies, we can interpret peaks on this graph as indicating the existence of clustering at different scales. The actual peaks at scales of 2 and 64 units confirm this argument. Ecologists on many occasions have used such size variance quadrat diagrams to study scales of pattern. Their work emphasizes that this method can detect clustering at different scales in complicated patterns in which clustering is not visually apparent. An approach like this that goes beyond our visual impression of maps obviously should command the attention of geographers.

The Basic Inferential Problem

These analyses of patterns at different scales may produce quite interesting results, but we should be careful never to overstate our conclusions. We quite simply can never infer conclusively a process from the evidence of pattern. This is the basic inferential problem of quadrat analysis and all pattern analysis. If we get a perfect fit to our model from a very large sample of quadrats, we still cannot be sure that the model's process is operating in the real world. *Inferences concerning processes from map patterns must always remain conjectural.* In terms of contagious processes, this problem emerges in practice as concerning *overidentified patterns* and *overidentified models.* Overidentified patterns result from our being able to fit several distinct, contagious probability laws to a single set of data. In the Harvey experiments, other contagious process distributions were tested and, on some occasions, these gave better fits than the negative binomial model. In many cases, the levels of similarity with the empirical evidence were comparable. In this situation, we must look to other evidence to discriminate between models. It was partly on the basis of the similarities between the negative binomial contagious process and previous ideas on diffusion that led Harvey to favor this model.

The problem of the overidentified model is a much more basic issue. Most of the probability laws that predict clustered patterns have *two distinct* process interpretations. There is the contagious process interpretation, which we have emphasized and described for the negative binomial model. An alternative generalized process that leads to the same form of probability models is a *random process on a population of varying densities.* Because population varies in density, the Poisson expectation is different in different parts of the plane. The result is a clustered pattern from a random process. These two interpretations of probability models leading to clustered patterns may be termed contagious and the heterogeneous, respectively. In the particular case of the negative binomial probability law, these two generalized process interpretations have been broken down further into at least six distinct processes, all of which lead to a negative binomial distribution. This model clearly demonstrates the principle that *different processes can lead to the same pattern.* With only empirical evidence of pattern, there is no definite way in which process can be determined.

Let's consider the implications of overidentification for our arguments concerning diffusion processes. What we are saying basically is that, if we fit a negative binomial model to our data, there are two general interpretations we can make. We can suggest a contagious process, as we have argued above. Alternatively, we can interpret the situation as a varying population distribution of potential acceptors that receive the innovation randomly. Because the population varies in density, places with high densities have more chance to receive the information, even though the process is random. Areas of high population density will be represented as clusters of adopters on the diffusion maps. This second interpretation is much less appealing on theoretical grounds but it may nonetheless contain a certain degree of truth in any real world diffusion situation. There are quite simply many environmental forces that tend to promote clustering of rural settlements in more favored areas. If this statement is generally acceptable, then the heterogeneous interpretation of the negative binomial model is likewise acceptable.

From our previous arguments, it seems that we have two acceptable interpretations of why the negative binomial distribution produces a good fit to quadrat counts on diffusion maps. It seems clearly reasonable to infer that both factors, contagion and heterogeneity, are operating in any single situation. The obvious strategy would therefore seem to be to attempt to model two processes: the underlying population, as in settlement colonization models, and the diffusion process. Such double probability models have not been attempted in geographical research.

Thus far we have discussed process inference in terms of contagious models. All the general statements concerning pattern and process apply equally to Dacey's county seat model. This very specific model has been interpreted only in terms of the original process described above. We have no reason to believe that this is the only process possible to generate this probability law. In any case, close fits of this model, such as we presented above, leave inference of process, as always, purely conjectural.

Map Transformation Research Designs

In the inferential problem discussed above, the alternative explanations of the pattern are not usually of equal interest to the geographer. For a clustered pattern, the contagious process hypothesis is usually considered theoretically more pleasing than a random process on an uneven population density hypothesis. In fact, in deductive theoretical arguments, one of the first assumptions is even distribution of population. This constitutes a major characteristic of the theoretical geographer's isotropic plane. Central place theory assumes an even rural population as the demand surface for central goods. Unfortunately, the real world is not quite so accommodating in the matter. Some geographers believe that central place theory is only applicable to a few areas of the Earth's surface where the population is evenly distributed as in southern Germany and Iowa. Geographers in countries such as Great Britain, where the population density is anything but even, have tended to pay only scant attention to the theory. One early result of the quantitative approach is that it produced methods for studying point patterns in areas of uneven population density without violating the even population assumption. The method is simply to produce an even population upon which to test the theory, even where one does not exist in reality. Such distortion of real space is known as *map transformation*. The product is usually termed a *topological cartogram*. Let's consider how one is produced.

The principles of map transformation can be illustrated by the simplified example in Figure 4.7. In Figure 4.7a, a rectangular grid is drawn over a set of equal population administrative areas. The example is simplified because each area has only six neighbors. We can transform the pattern of administrative areas into a regular hexagonal lattice, as in Figure 4.7b. Because we know that all administrative areas have the same population, the resulting map transformation

(a)

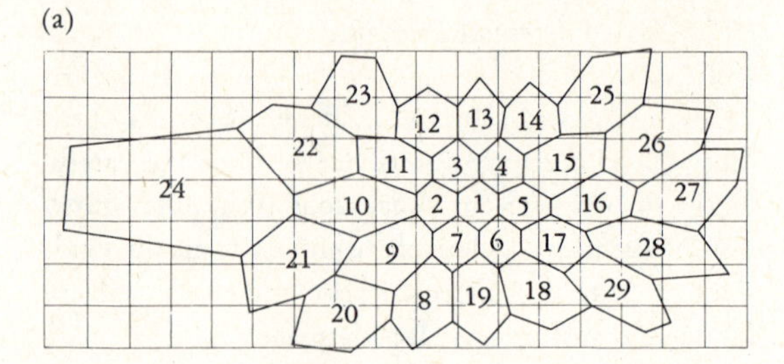

(b)

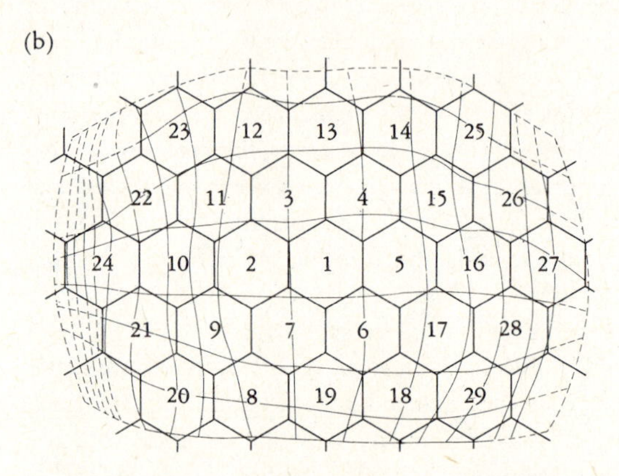

Figure 4.7 A unique map transformation: (a) the original map, (b) a population space.
From W. Bunge. Theoretical geography. *Lund Studies in Geography* Series C, No. 1, 2d ed., 1966.

defines a population space. The resulting cartogram has equal population density at the expense of distortions of the original grid coordinates that relate to the physical basis.

This example resolves into a nice, unique solution. This is possible only because of the highly unusual property that every area has exactly six neighbors and exactly the same population. Neither of these conditions normally obtains. In practice, map transformations begin with a map of areal units for which we have some variable quantities. We continue to assume that these are population totals, but instead of appealing to geometric theorems about regular hexagons we proceed with an empirical trial and error procedure, as follows.

Some population unit scale is chosen so that the

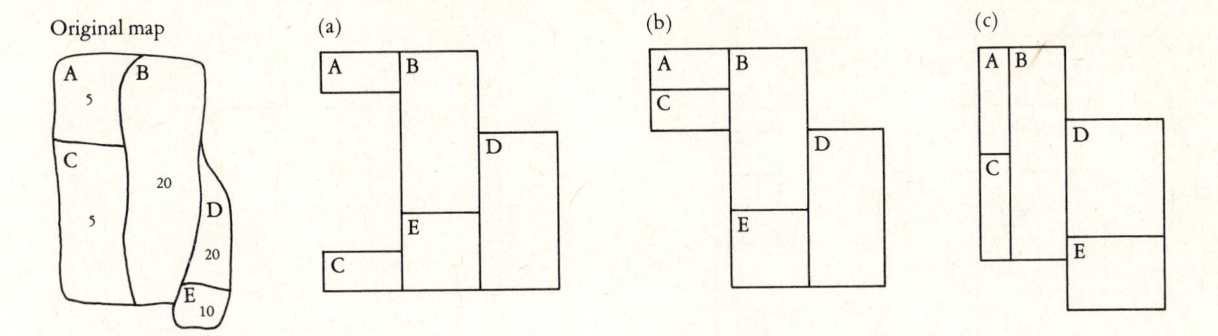

Figure 4.8 An original map (figures refer to population totals) and alternative map transformations.

resulting map is the appropriate size. We might make a square of 0.1 inches equal to 100 people at an urban scale or 1 million people at a world scale. Next we decide how large each individual areal unit will be on the new map. With this information, we can construct a diagram. It is normally helpful to start with the largest area (the largest population unit) and build around it. At this stage, explicit rules end, and we must begin to follow rules of thumb. This is clearly a jigsaw puzzle in which the sizes but not the shapes of the pieces have been decided and so a wide variety of solutions is available. Figure 4.8 shows three possible solutions to a simple five-areal unit map. The three cartograms are not equally desirable. In solution a, for instance, area A is not contiguous with area C, whereas on the original map these two areas are contiguous. The contiguity discrepancy is easily overcome by pushing up area C to touch area A, as in solution b. This transformation is topologically correct. Normally we would prefer solution c, which not only maintains the correct pattern of contiguity but also maintains as far as possible the shape of the study area as a whole and the individual areal units. This argument suggests that we can specify three criteria for constructing cartograms:

1. Maintain contiguity relations,
2. Maintain external shape of the study area,
3. Maintain internal shapes of the areal units.

These criteria are not normally compatible, and the order in which we list them usually agrees with their respective priorities. If the first criteria is violated, we no longer have a topological transformation. Cartograms have been published in which this has been sacrificed in order to make a recognizably shaped diagram. There are no strict rules in this matter, largely because the criteria do not define a unique solution to a transformation problem anyway. Although all Mercator projections of the Earth's surface are identical, all world population cartograms are by no means the same. The development of an acceptable method for producing general solutions to map transformation problems is an unsolved problem in spatial analysis.

Of what value are such distorted maps to theoretical geographers? Despite the problem of nonuniqueness of solutions, such population spaces represents a major aspect of the geographer's isotropic plane. It can form an important link between abstract deductive theory and empirical testing in the real world. The importance of this link was realized and developed by Waldo Tobler and the research strategy is now usually associated with his name. We illustrate two such map transformation research designs.

Constructing a Theoretical Point Pattern for an Anisotropic Plane

The research design we describe first presents a simple method for producing a point pattern that will serve an uneven demand surface, or anisotropic plane. Consider Figure 4.9a. It shows consumption expenditures available for groceries in a part of Tacoma, Washington.

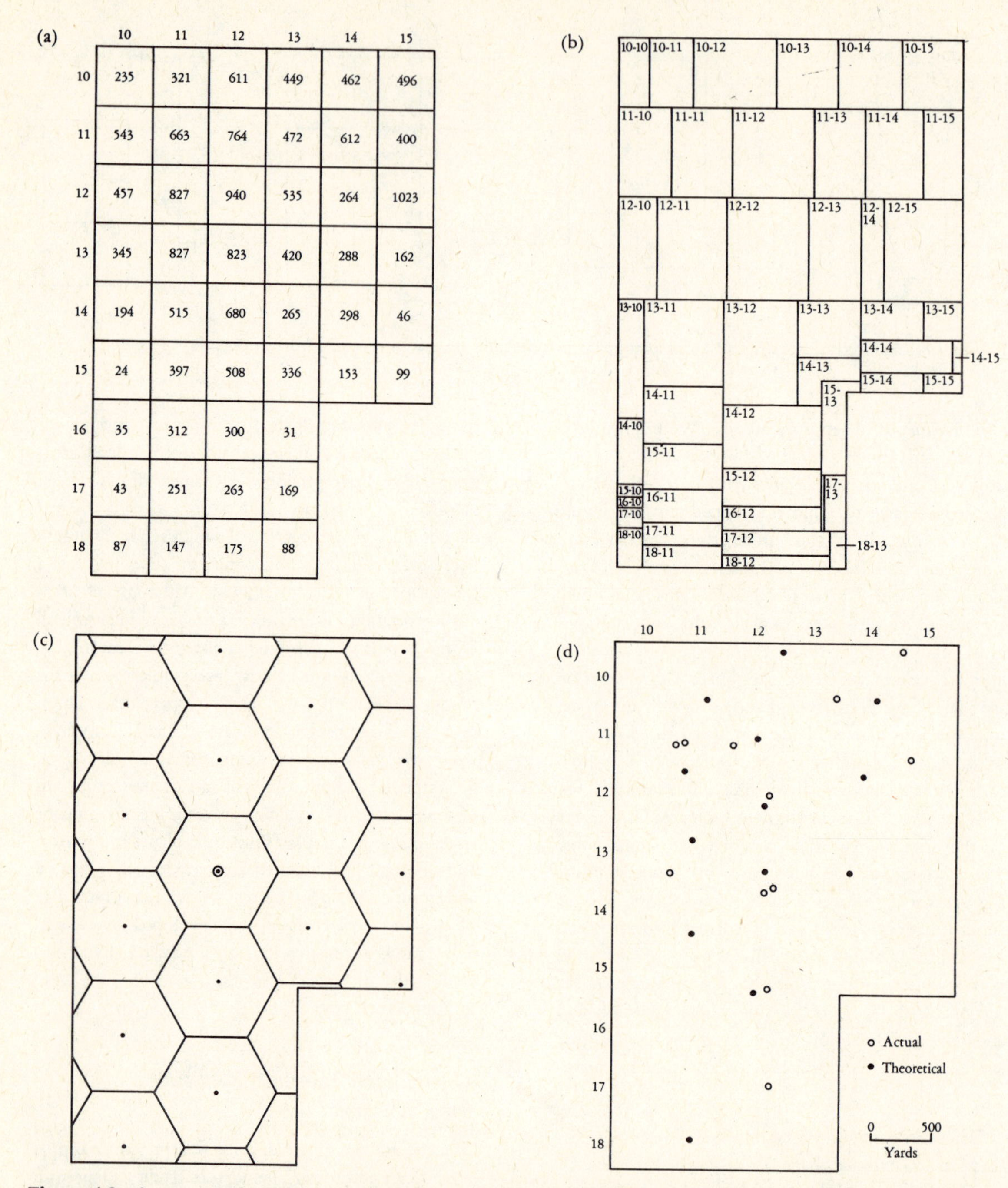

Figure 4.9 A map transformation research design applied to part of Tacoma, Washington. a. Consumption expenditures available for groceries. b. Map transformation to an equal-demand space. c. Determining theoretical market areas. d. Theoretical and actual location of supermarkets.

From Arthur Getis. The determination of the location of retail activities with the use of a map transformation. *Economic Geography* 39:18–19, January 1963.

Demand for groceries is concentrated clearly in the north half of the map. We would not expect a regular pattern of grocery stores to satisfy this demand. Arthur Getis has suggested that with a map transformation utilizing these consumption expenditures we can find a more realistic theoretical pattern. Using as his base the distorted map in Figure 4.9b, Getis arranged a pattern of 12 hexagons about the center of the map, such that they enclose most of the area and, hence, the demand (Figure 4.9c). The centers of the hexagons define a theoretically ideal point pattern to satisfy the demand on the cartogram. The points are transferred to the original map by maintaining their relative position within the transformation grid cells, shown in Figure 4.9d along with the actual locations of the 12 largest supermarkets in the area. The correspondence between the two point patterns is quite reasonable. The average distance of the theoretical locations from the nearest actual supermarket is 345 yards in an area larger than 12 square miles. Of course, the actual supermarkets could not be located as freely as we have located the theoretical points. The most explicit limitation on their locations is zoning policy. If we adjust our theoretical points to the nearest commercially zoned land, the average distances between the theoretical points and the nearest actual locations reduces to 282 yards. An error of the order of 2 blocks in an area of 12 square miles seems to be a satisfying conclusion.

We must mention indeterminacy in locating the theoretical points by this method. There are several placements of 12 hexagons that fill the transformed space. We can, for instance, rotate the existing mesh and produce a different solution. Furthermore, we might arrange the hexagons around an alternative cartogram center. One suggestion is to start with a hexagon about the original supermarket in the area, because all subsequent decisions to locate supermarkets were made with the knowledge that this location was occupied. This procedure might well give a better fit.

The lack of a single, determined theoretical solution is a severe limitation. In fact, it casts doubt on the usefulness of direct comparison with actual locations. Perhaps a better interpretation of the results is simply to use them as an illustration of the distortion that modifies competitive process patterns when they occur on an anisotropic plane.

Testing a Point Pattern for an Anisotropic Plane

Our second map transformation research design is in some ways the reverse of the one given above. Instead of locating theoretical points on a cartogram, we find the actual locations on the diagram and then carry out our analysis in the transformed space. We attempt this strategy by using the central place pattern in England and Wales. It is commonly assumed that central place theory does not work in this area, and this presents the research design with a severe challenge.

Figure 4.10a shows the pattern of major central places in England and Wales, classified as 1, 2, and 3A centers by R.D.P. Smith late in the 1960s. This map includes all the cities and major towns plus the main shopping centers in Greater London. The pattern clearly seems highly clustered whereas central place theory predicts regularity. We know that the population distribution in England and Wales is very uneven, and we would not conclude that the pattern is irrational in any sense or unpredictable given the population pattern. A quadrat analysis with a sample of 30 using a quadrat size defined by $2A/n$ is carried out in Figure 4.10a. Notice that this quadrat size, measuring twice the mean area about each point, is particularly relevant because, with a perfectly regular pattern, each quadrat always includes one or two points. In this particular case, we find five quadrats with five or more points and three with no point at all. The resulting analysis confirms all our suspicions with a variance/mean ratio of 1.5, suggesting a pattern more clustered than random.

If we accept this evidence, it is of little use to go one stage further and attempt to fit some contagious probability model to the point pattern, because we have no theoretical justification for postulating such a process. If we had defined the pattern of industrial towns instead of central places, we might be able to argue for some concentration process, but this does not fit our research context. What we seem to have is a competitive process on an anisotropic plane and the point pattern reflects demand, as central place predicts, but thereby produces a clustered pattern. If we make demand evenly distributed, central place theory predicts a regular pattern of central places. In Figure 4.10b, England and Wales are depicted as a population space,

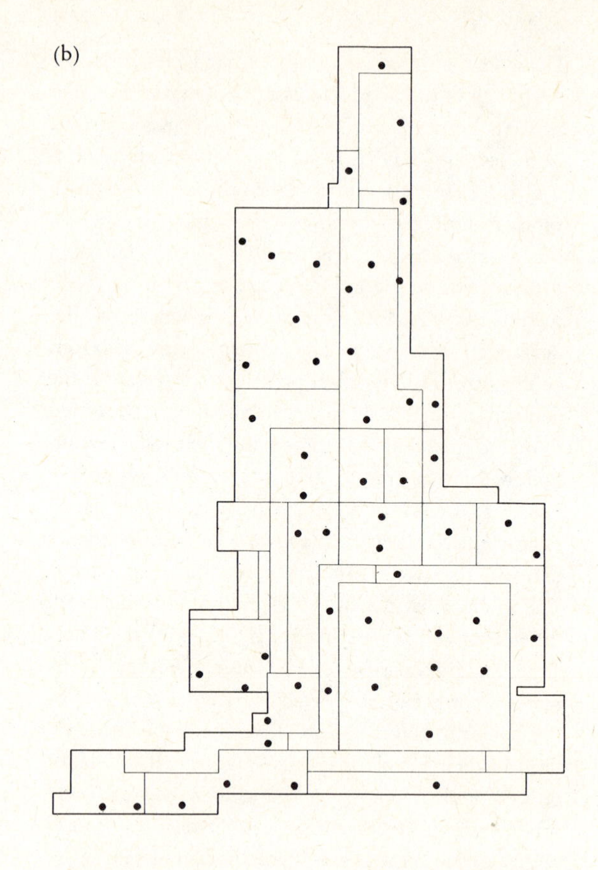

Figure 4.10 A map transformation research design applied to the central place pattern in England and Wales: (a) central place pattern, (b) population space.

using population figures for the 30 areas shown in Figure 4.10a. The map was constructed along principles outlined above; the shape and contiguity are maintained as far as possible. (Two line contiguities have been converted to point contiguities.) The effect is to collapse the rural areas into a small periphery around London, the Midlands, and the North. The original 54 central places are relocated on the new map in their respective areas in the same relative positions as their original locations. This part of the process is rather subjective but it should not prejudice the subsequent analysis. The result is illustrated in Figure 4.10b, which gives a visual impression of a reasonably regular pattern as predicted. A quadrat analysis will test this.

Two random samples of 30 quadrats each were used. The size defined by $2A/n$ was used, for reasons previously stated. It was decided also to use quadrats defined by A/n—that is, the mean area around any point that, on the regular pattern of central-place theory, represents the size of actual market areas served by each town. We refer to these two analyses as B and A, respectively. The resulting distributions are shown in Table 4.9, and both produce very similar variance/mean ratios of 0.49 for analysis A and 0.54 for B. Both confirm our hypothesis by suggesting a pattern much more dispersed than random.

Having confirmed a tendency toward dispersion with both quadrat sizes, we can consider the process. The dispersed pattern suggests the competitive process of central place theory although pure regularity is by no means achieved. We can turn to Dacey's modified Poisson process, which produces a pattern more dispersed than random. In fact, the Dacey model is not far removed from the evolution of the central place pattern in England and Wales. First, there are the sur-

Table 4.9 Expected and Observed Frequencies in England and Wales Population Space

x	A quadrat size A/n			B quadrat size $2A/n$		
	Modified Poisson probabilities	*Modified Poisson expected frequencies*	*Observed frequencies*	*Modified Poisson probabilities*	*Modified Poisson expected frequencies*	*Observed frequencies*
0	0.2858	8.6	9	0.0000	0.0	1
1	0.5591	16.8	15	0.3073	9.2	8
2	0.1439	4.3	6	0.3625	10.9	9
3	0.0194	0.6	0	0.2124	6.4	9
4	0.0018	0.0	0	0.0828	2.5	3
5	0.0001	0.0	0	0.0242	0.07	0

This table was calculated by the author; $p = 0.631$ and $\gamma = 0.28$ for A and $p = 1.0$ and $\gamma = 1.17$ for B.

viving county towns depicted on our map, which dominate their own counties much as the county seats dominate in the American Midwest. Typical examples are Derby, Nottingham, Leicester, Northampton, Norwich, Cambridge, Oxford, and Reading. Second, there are the products of the nineteenth-century industrial revolution that have become important central places, ranging in size from as large as Birmingham and Liverpool to as small as Sunderland and Newport. Their distribution has depended on the particular needs of trade and industry in the past two hundred years and it is expected that this pattern is more random than regular. We may postulate that we have competitive and random processes combined, not unlike Dacey's county seat model. We can test this by calibrating Dacey's model after estimating the two parameters p and γ and calculating the expected probabilities for various numbers of quadrat counts. The results are shown in Table 4.9 along with expected and observed frequencies. Notice the high values of p, indicating the high degree of regularity. The model clearly gives a very close fit to the data, and Kolmogorov-Smirnov tests confirm this. In A, the D statistic is only 0.04 and in B it rises to 0.13. For a sample size of 30, the critical value for D with a low probability level of $p = 0.2$ is 0.19. Thus, we cannot even begin to consider rejecting hypotheses of no difference between observed and theoretical distributions. We are able to conclude, therefore, that Dacey's county seat model gives a very reasonable description of the central place pattern in England and Wales on a population space.

This is a very encouraging result in what is usually considered an unsuitable laboratory for central place studies. The ability to identify a competitive process in the originally clustered pattern is wholly a product of our use of a map transformation. Certainly this approach makes point pattern analysis much more flexible and of more general use in suggesting the processes involved in producing the patterns we observe on all parts of the Earth's surface. It is not surprising that we find theoretical geographer William Bunge particularly enthusiastic about this method. He suggests that at long last geographers can bend space to meet their own need to control experiments to a degree similar to that achieved by physical scientists in their laboratories. This overstates the case, because such space has rather important limitations. For instance, distances are not meaningful in the new space, and this has technical implications relating to techniques we describe in the next section. Furthermore, any sophisticated central place theory must consider patterns of movement of people in the system and this cannot be realistically carried out in a population space. Despite these limitations, we must agree partially with Bunge's enthusiasm. At a methodological level, this research design has illustrated a particularly spatial way of holding a disturbing variable (population density) constant while we concentrate our analysis on the main topic of interest, the pattern of central places.

THE SPACING APPROACH

Visual appraisal of dot maps consists of assessing the variations in point densities over a map. Quadrat sampling is in many ways an objective method of simulating this subjective approach. There is an alternative method whose roots can be found in early theoretical geography. It concentrates on the spacing between points. The notion of points per area (density) is replaced by its reciprocal: area per point. If the points are central places, then the area per point is the average market area for the central places. It is in this area of geographical theory that the spacing approach is mainly to be found. From simple Euclidean geometry, it can be shown that the distance between central places derived from Christaller's hexagonal lattice is given by $1.075 \cdot \sqrt{A/n}$, in which A is the area of the region and n is the number of central places it contains. This finding is of little interest to empirical researchers who are resigned to dealing with central place patterns that are not hexagonal. But we can measure distances between towns in the real world and compare them with this ideal hexagonal standard. Measurements of distance between towns and their nearest neighbors were originally conducted by the central place theorist August Losch. His tables of nearest neighbor distances between towns with populations of between 20,000 and 100,000 for part of the American Midwest and England are reproduced in Table 4.10. This table illustrates the major differences in terms of town density between the two areas. This difference is so much the overwhelming feature of this and other tables of Losch that it has received much comment. In this chapter, we are interested in pattern rather than density. The breakthrough in terms of analyzing spacing for pattern has come with the linking of nearest neighbor distances to the Poisson probability function. Distances between points resulting from a random process may be computed. It is this random standard that is employed in nearest neighbor analysis. This method was pioneered and developed by plant ecologists; later modiffcations were made by Michael Dacey.

Nearest Neighbor Analysis

Let's assume that we have measured the distance r between every point in a given pattern and its nearest neighbor. If we take the average of all these distances, we produce a value $\bar{r}_a$, which we can use to compare this particular pattern with other patterns. The value of $\bar{r}_a$ depends partly on the size of the area under study and the scale of measurement units used. In order to make any general comparisons between this particular distribution and any other pattern, we need a standard to which each can be related. The method we use to achieve this is to calculate what the average spacing distances would be if our observed distribution were random. We know from our previous discussions that

Table 4.10 Frequency Distributions of Nearest Neighbor Distances Between Towns of Population 20,000–100,000

	Illinois, Indiana, Ohio		*England*	
Distance (mi)	*Frequency*	*Percent*	*Frequency*	*Percent*
0–4	8	13.1	196	83.1
4–8	20	32.8	34	14.4
8–12	23	37.7	6	2.5
12–16	9	14.8	0	0.0
16–20	1	1.6	0	0.0

Source: Adapted from A. Losch. *The Economics of Location.* Stuttgart: Gustav Fischer Verlag, 1954, Tables 22 and 23.

a random process is associated with the Poisson probability function. We can use this distribution to derive expected average nearest neighbor distances for a randomly generated pattern. It is known that this expected average distance for a randomness assumption is given by

$$\bar{r}_e = \frac{1}{2\sqrt{n/A}}$$

n is the number of points and A is the area of the study region.

Divergence from Randomness: The R *Scale*

For any nearest neighbor analysis, we have an empirical or actual average distance $\bar{r}_a$, and we can compute an expected average distance under random expectations. The differences between these two values will measure clearly the divergence of the actual point pattern from randomness. This divergence can be expressed as a simple ratio:

$$R = \frac{\bar{r}_a}{\bar{r}_e}$$

which simply tells us how more or less spaced the observed distribution is than a random one. The advantages of this simple index is that we can now place our observed point pattern on a scale ranging from clustered through random to regular. This R *scale* ranges from 0 to 2.149. A score of 0 represents a situation in which $\bar{r}_a = 0$. All distances between points are 0 (there is no spacing, so that all points lie on the same location), and we have the limiting case of the clustered pattern. When $R = 1$, it follows that $\bar{r}_a = \bar{r}_e$, and a random pattern is indicated. High R scores represent various degrees of dispersion, with $R = 2$ resulting from a square regular pattern and $R = 2.149$ resulting from the limiting regular pattern based on a triangular lattice. Point patterns illustrating each of the four special cases are shown in Figure 4.11. In general, R values of less than 1 indicate distributions tending toward a clustered pattern and R values greater than 1 indicate patterns tending toward dispersion. Various degrees of divergence from the random position of

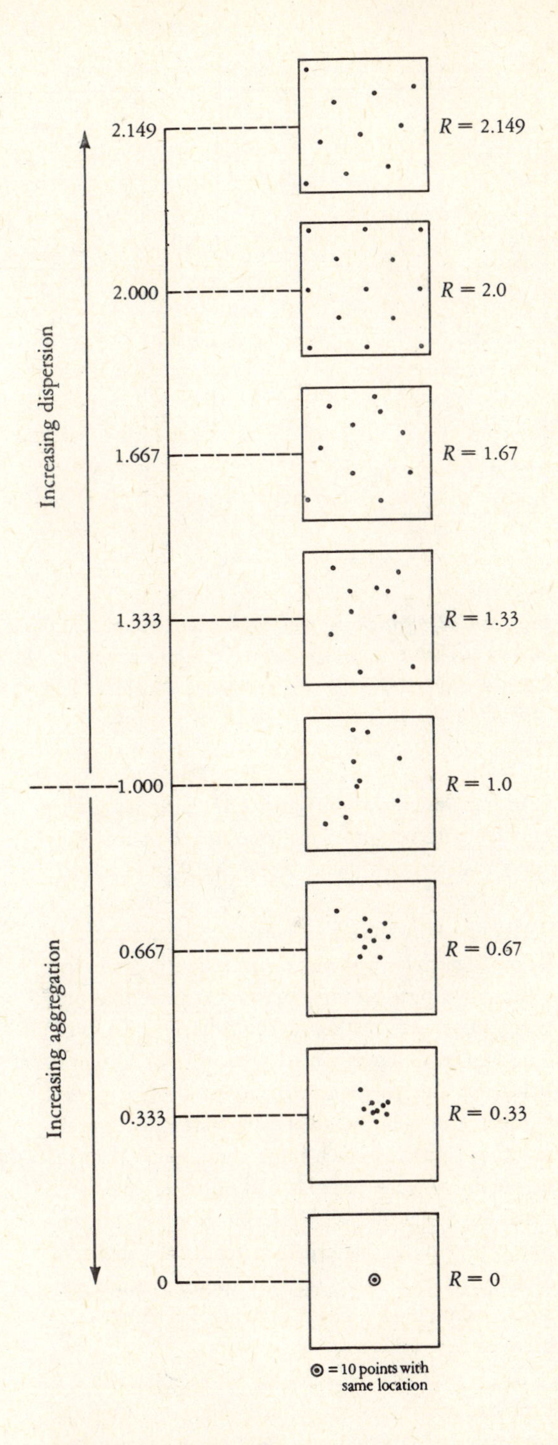

Figure 4.11 The R scale.

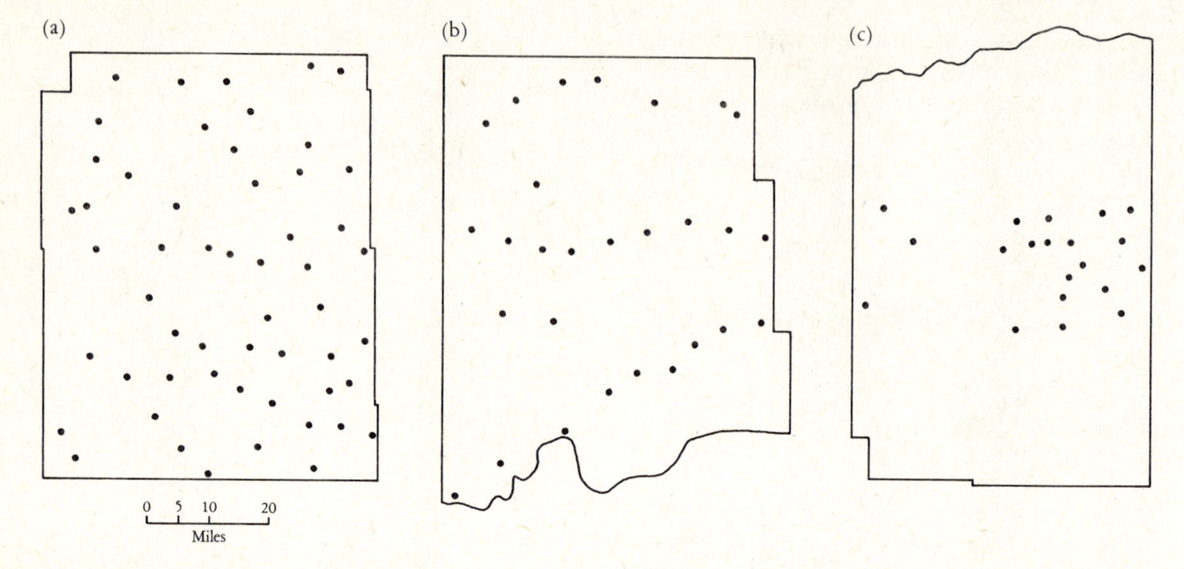

Figure 4.12 Settlement patterns for selected areas in the United States: (a) part of Minnesota, (b) part of North Dakota, (c) part of Utah.
From L.J. King, A quantitative expression of the pattern of urban settlement. *Tijdschrift voor Economische en Sociale Geografie*, 1960, vol. 51, p. 318.

$R = 1$ are also shown in Figure 4.11. Each pattern has only ten points covering the same area so that we may assess visually what values along the R scale actually represent.

Figure 4.11 has a further purpose. The argument so far has been concerned with extreme limiting cases that result from purely contagious and competitive processes. In practice, neither process is likely to be found working away simply on its own in the real world. Whatever the phenomena being studied, processes producing an empirical point pattern are many, and the resulting pattern is much more complicated than the simple limiting case at either end of the R scale. Even when the dominant process is contagious or competitive, the resulting pattern may tend only toward a limiting position along the R scale; in practice, *empirical* examples of point patterns fall somewhere between $R = 0.33$ and $R = 1.67$.

This is well illustrated in a study by Leslie King. A sample of areas across the United States was taken and their respective settlement patterns were analyzed. We have already noticed that central place patterns theoretically show regularity. Even if we agree that the competitive process is dominant, however, many other factors affect a particular empirical distribution. Some settlements may not be primarily service centers. Others may be located to combine an advantageous market location with some other function, such as access to a river. The policy of a railway company may distort the settlement pattern. All these and many other factors may influence a single pattern, and a distribution exhibiting some competitive influences, does not normally even approximate the limiting cases of $R = 2$ and 2.149. In fact, in all his 20 sample areas, King's highest R value was only 1.38, for an area in Minnesota (Figure 4.12a). This is in spite of the fact that many factors favored a regular pattern, such as the uniform relief and the rectangular land division and road system. More typical of King's findings is that for North Dakota, where $R = 1.11$ and factors other than competition helped produce a pattern of settlement little different from random (Figure 4.12b). The competitive process can be almost wholly overshadowed by other factors, notably relief, producing a

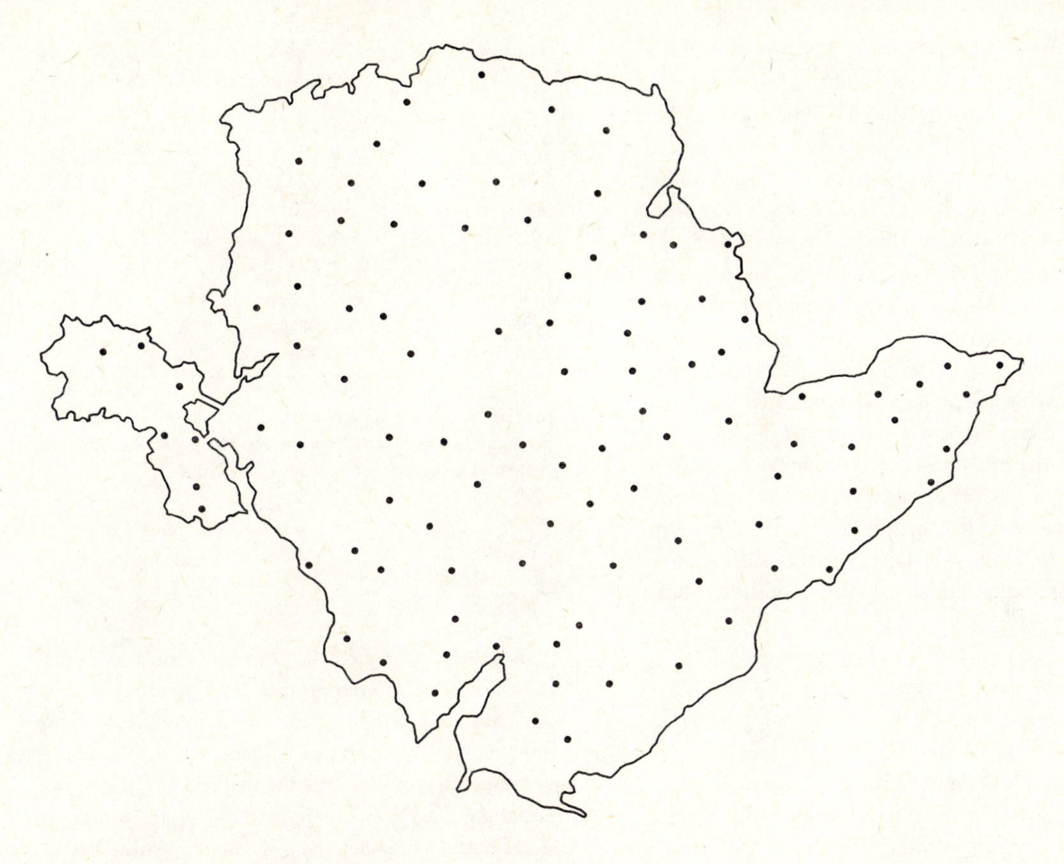

Figure 4.13 Post offices and public telephone booths on Anglesey.

pattern of towns that are clustered in areas favorable to settlement. This is the case in Utah (Figure 4.12c), which has an R value of only 0.70.

Even in situations in which a service is planned to serve a reasonably evenly distributed population, extremely high R values are not produced. Figure 4.13 shows the distribution of post offices and public telephones on the island of Anglesey. The post offices, which include a public telephone, are located at most of the settlements, and separate public telephone booths are located in the areas furthest from post offices. We have a pattern of the very lowest order of service function specifically designed to serve a rural population. The pattern looks very dispersed, but a nearest neighbor analysis produces an R value of only 1.6. This clearly emphasizes our point that the range of R values that we can reasonably expect from observed point patterns is far less than that specified by abstract limiting cases.

Testing Pattern Hypotheses

So far the technique seems to be straightforward. A little thought about our standard Poisson distribution generated by a random process causes concern, however. Whereas the extreme cases of clustering and regularity are deterministic—only one value of average distance and, hence, of R can occur—this is certainly not the case with a randomly produced distribution. By the independence assumption, one random point pattern is very likely to be different from all other random patterns. Will they have exactly the same nearest neighbor distances and the same R values of 1? Let's try the only example of a randomly generated distribution that we have produced (Figure 4.1). If we measure the distances from each of the 20 points to its nearest neighbor, we produce $\bar{r}_a = 0.455$ units. With $n = 20$ and $A = 16$, $\bar{r}_e = 0.559$. Therefore, $R = 0.455/0.559 = 0.81$, which suggests a tendency toward clustering. In this case, however, we know that the process is random. It is clear that random processes do not always give patterns that produce an R value of 1. On average, many randomly generated processes give an R value of 1, but any individual random pattern may diverge slightly from this expectation.

What is the implication of this finding for nearest neighbor analysis? It means that we must take into account the variation that may occur in random processes. This brings us once again into the realm of inferential statistics. The standard error of the $\bar{r}_e$ values is given by

$$\text{SE}\bar{r}_e = \frac{0.26136}{\sqrt{\dfrac{n^2}{A}}}$$

For our example in Figure 4.1,

$$\text{SE}\bar{r}_e = \frac{0.26136}{\sqrt{\dfrac{400}{16}}} = 0.052$$

This standard error is interpreted like any other standard error, with approximately 95% of random pattern $\bar{r}_e$ values within $\pm$ 2SE$\bar{r}_e$ of the computed value. Thus, with computed $\bar{r}_e = 0.559$, we can expect 95% of $\bar{r}_e$ values in random distributions to lie between $0.559 \pm 2(0.052)$, which gives a range of from 0.455 to 0.663. We find that the actual value for a randomly generated pattern $\bar{r}_a$ comes on the borderline for the 95 percent central portion of the statistical distribution of $\bar{r}_e$. We can take this analysis a little bit further by computing a z variate, which we can test in the usual manner. In this situation

$$z_R = \frac{|\bar{r}_e - \bar{r}_a|}{\text{SE}\bar{r}_e}$$

which gives

Table 4.11 Nearest Neighbor Analyses of Grocery Store Locations in Lansing, Michigan (1900–1960)

Year	n	R	z_R	R_3
1900	20	1.074	0.63	1.12
1910	33	0.673	3.59	1.08
1920	94	0.658	6.35	1.15
1930	124	0.772	4.93	1.02
1940	133	0.792	4.59	1.08
1950	117	0.841	3.30	1.02
1960	68	0.998	0.04	1.01

Source: A. Getis, Temporal land use pattern analysis. Reproduced by permission from the *Annals* of the Association of American Geographers, vol. 54, 1964, Table 3.

$$z_R = \frac{0.559 - 0.455}{0.052} = \frac{0.104}{0.052} = 2$$

This value of z_R yields a probability of $p = 0.0228$ (Table A.2). If the significance level were set at $\alpha = 0.05$, then we would reject a null hypothesis of no differences between average distances in the randomly generated distribution and the theoretical random expectation. The conclusion that the point pattern is significantly different from a random distribution would be an example of a type I error, or rejection of H_0 when it is in fact true. We know H_0 is true because we constructed Figure 4.1 from random number tables. Thus, we have an example of an error of type I knowingly made. Such cases are rarely ever known in practice and are hardly ever reported; this example serves to emphasize the basic feature of inferential statistics, their known and accepted uncertainty. The known uncertainty is given by α, in this case 0.05, so that in one example in twenty we will reject a randomly generated pattern as nonrandom. Our example is one such example.

We illustrate the use of these inferential procedures in an actual research application by returning to Arthur Getis' study of grocery stores in Lansing, Michigan. The quadrat sampling reported above was supplemented by a nearest neighbor analysis of the patterns. The results are shown in Table 4.11. The trend toward

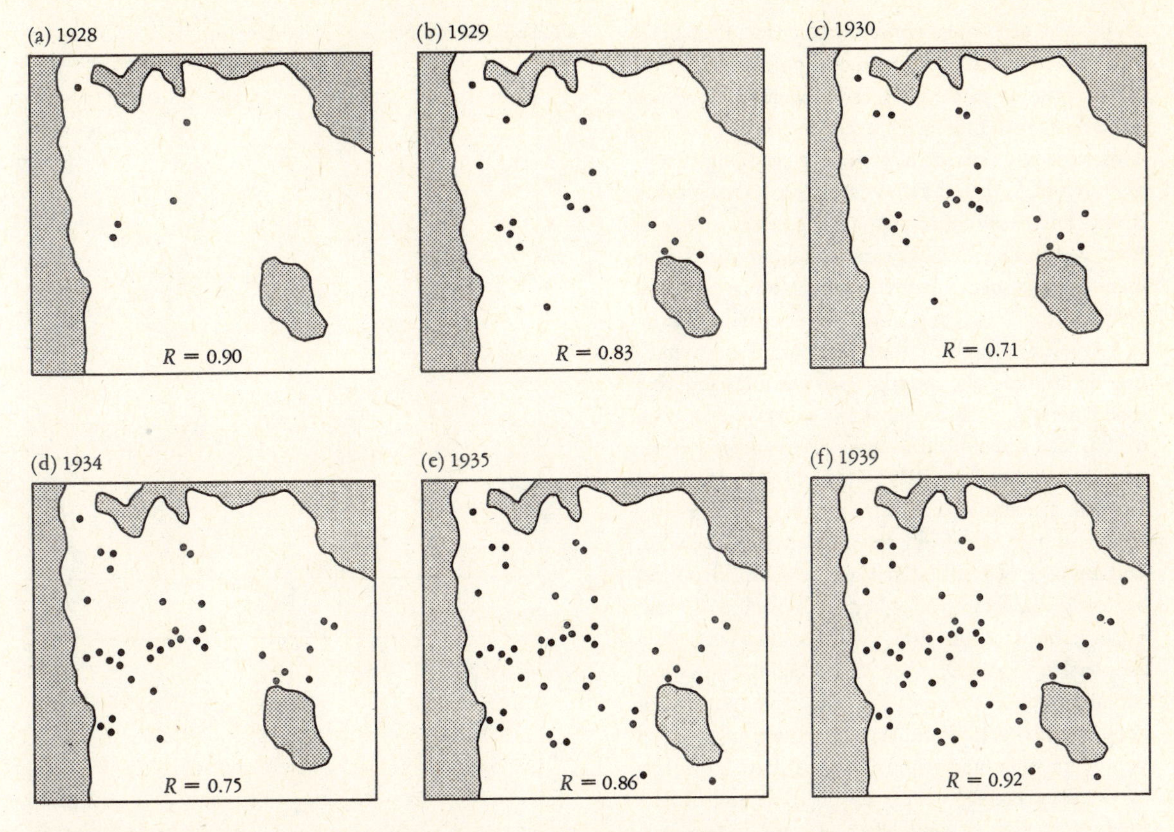

Figure 4.14 The spread of tuberculosis controls for dairy herds in part of central Sweden, 1928–1939. From T. Hägersrtand. *Innovative Diffusion as a Spatial Process.* pp. 160–161. © 1967 by Torsten Hägerstrand.

clustering after 1900 and the return to a more random pattern by 1960 are illustrated by the R values. The z_R statistics allow us to infer whether observed differences from 1910 to 1950 are in fact statistically significant. The large z_R values for these dates enable us to reject a null hypothesis between the observed patterns and the random expectation at less than $\alpha = 0.01$. These results closely support the findings of the quadrat analysis (Table 4.5).

This z test is available for all nearest neighbor analyses. In geographical research it seems to have been used rarely. Geographers have used the R scale as a descriptive tool for measuring pattern. In the real world, complex arrays of counteracting forces tend to produce R values within the central range of the R scale. The frequent result is large numbers of R values showing relatively small divergences from random expectation. This is illustrated in King's studies of settlements in the United States. Notice that these findings do not necessarily justify an assumption that a random process is operating. All we have is a pattern description; we have no direct information about process on a dot map. A statistical test on differences from random expectation also tells us nothing of the processes operating. Geographers have tended not to use the rigor of inferential statistics in this context.

Although geographers have stayed largely at the purely descriptive level of nearest neighbor analysis, they have been able nonetheless to incorporate R values into research designs so that process can be inferred. This is particularly true for cases in which pattern has been monitored over time. If processes are considered in specifying hypotheses concerning changing levels of R, then the study can claim to make quite strong inferences concerning the processes in operation. We illustrate use of nearest neighbor analysis for monitoring changes in pattern in the familiar diffusion context.

A typical example from one of Hägerstrand's original studies is the spread of dairy farmers adopting tuberculosis controls. Figure 4.14a shows the location of farmers accepting this innovation at five time periods from 1919 to 1939 in a small part of central Sweden. The distribution certainly indicates a gradual spread of the idea from the northwest southward. Our purpose is quantitatively to illustrate this spread by using the nearest neighbor approach. We can hypothesize that in 1919, when there were two adopters, and in 1939, when all but eight farmers had adopted the innovation, the respective pattern of adopters closely approximates the pattern of the total population of farmers whose R value is 1.10. However, as the innovation spread, the Hägerstrand model predicts a neighborhood effect that we hypothesize will be reflected in a much more clustered pattern of innovators. We expect a U-shaped pattern of R values between 1919 and 1939. The results in Figure 4.14 confirm this hypothesis. This quantitative description of pattern is quite consistent with Hägerstrand's hypothesized diffusion process.

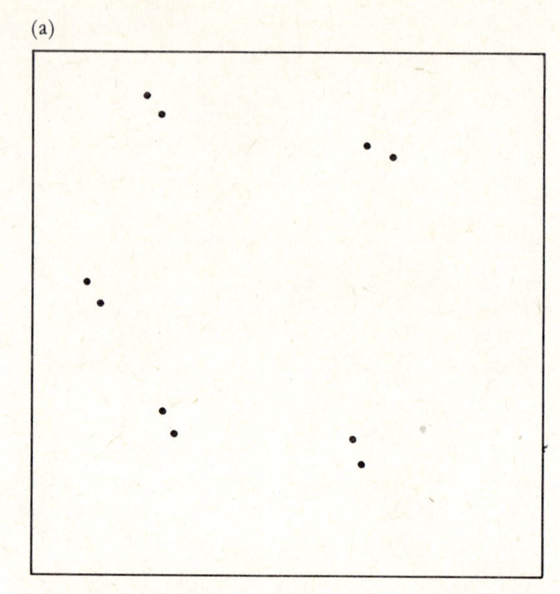

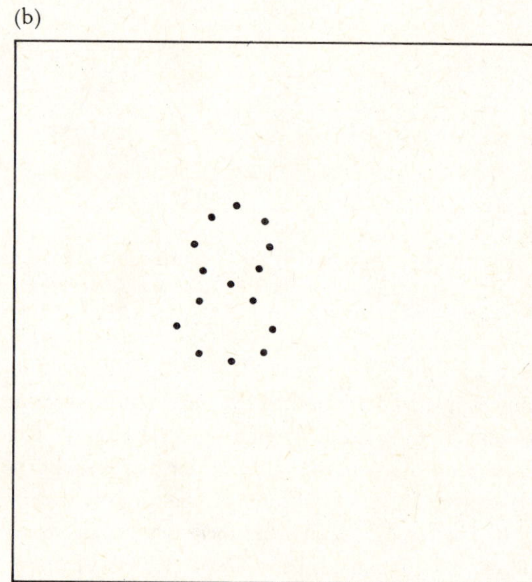

Figure 4.15 Special cases of low R values.

Problems Associated with the Spacing Approach

A well-known problem associated with nearest neighbor analysis is that it is relatively easy to produce intuitively implausible results. For instance, both maps in Figures 4.15a and 4.15b produce very low R values of 0.32 and 0.39 indicating high degrees of clustering. These patterns differ markedly, however, from our idea of the limiting case of clustering as shown on the R scale in Figure 4.11. Furthermore, notice that, if we allow the dots in each pair in Figure 4.15a to come closer together, then when they occupy the same location, all nearest neighbor distances are reduced to 0 and we have an R value of 0. Thus, we have an alternative pattern to our limiting case of $R = 0$. We

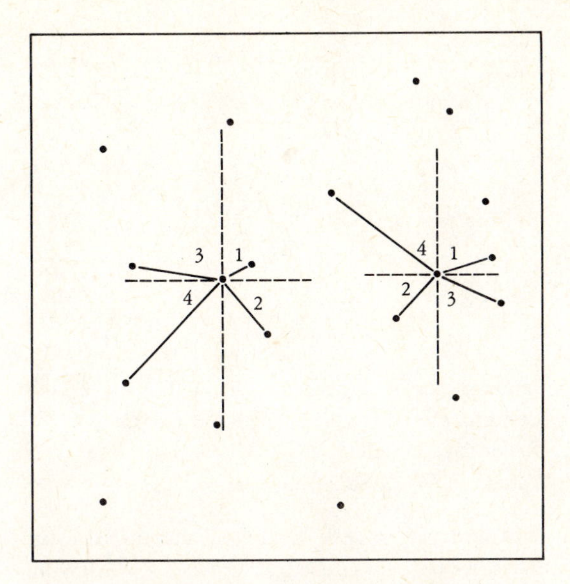

Figure 4.16 Distance relationships in sectors. Sectors are numbered in terms of the ranked order of nearest neighbors in the sectors.

do not normally think of such a pattern as being maximally clustered because of the remaining distances between the separate pairs of points. The problem is clearly that the nearest neighbor approach uses only a small part of the information available on any dot map. In certain patterns, a measure incorporating only nearest neighbor distances can give a misleading numerical impression. Hence, we have the results in Figures 4.15a and 4.15b.

Further Distance Relations in Sectors

The solution to this important limitation of nearest neighbor analysis is fortunately relatively straightforward. Consider a point on a dot map. If we draw two lines at right angles through the point, we produce 4 sectors about the point (Figure 4.16). We can now identify nearest neighbors in each sector. One of these will be the original nearest neighbor we used previously but now we also have three other neighbors to consider. If we measure distances to all 4 neighbors we will have more complete information about the point pattern. If we use 4 such neighbors for every point, we can produce a new observed average neighbor distance $\bar{r}_{a4}$ in which the subscript *a*4 denotes *actual* distances in *four* sectors.

Expected neighbor distances in sectors about points are known for randomly generated patterns from the Poisson probability law. Ecologists Clark and Evans have shown that generally

$$\bar{r}_{ek} = \frac{\sqrt{k}}{2\sqrt{n/A}}$$

in which k is the number of sectors. Thus, with 4 sectors,

$$\bar{r}_{e4} = \frac{\sqrt{4}}{2\sqrt{n/A}}$$

Notice that the previously simple, nearest neighbor approach is incorporated in this equation, because this can be interpreted as employing 1 sector, so that

$$\bar{r}_{e1} = \frac{\sqrt{1}}{2\sqrt{n/A}} = \frac{1}{2\sqrt{n/A}}$$

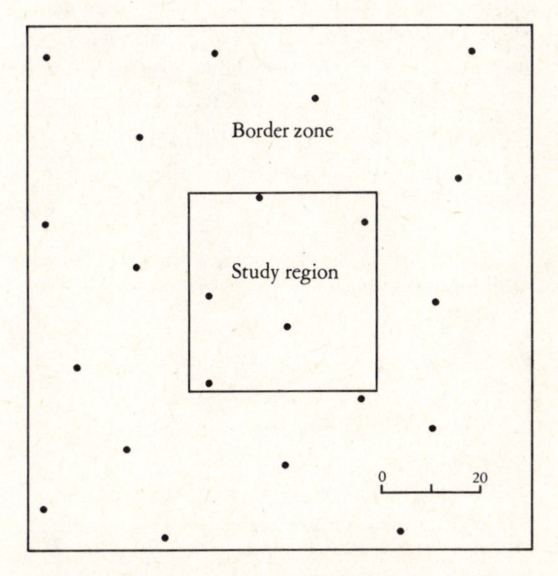

Figure 4.17 A point pattern for study region and border zone.

Table 4.12 Six Sectors Ranked by Neighbor Distance for Figure 4.17

Average Distances:	*Ranked Sectors*					
	1	*2*	*3*	*4*	*5*	*6*
Observed from Work Table 4.3	18.6	19.6	20.8	24.2	27.2	33.2
Random expectation from Work Table 4.3	8.4	13.3	17.6	22.0	27.3	35.5
Triangular lattice	18.3	18.3	18.3	18.3	18.3	18.3

as before. Once we have computed $\bar{r}_{ak}$ and $\bar{r}_{ek}$, we can define a new measure of pattern

$$R_k = \frac{\bar{r}_{ak}}{\bar{r}_{ek}}$$

which has a value of unity for a randomly generated pattern. An example of such analysis is laid out in Work Table 4.3, using a 6-sector method of defining neighbors for the study region in Figure 4.17. In this case, we find that $R_6 = 1.20$. Interpretation of this result is unfortunately not as simple as in single nearest neighbor analysis. Let's consider it in some detail.

In interpreting any numerical measure, we find it useful if we can specify the range of the measure. In general, the lower limiting cluster case for any R_k value is 0. When all points lie on one location, R_k is 0 for any value of k. The other limiting pattern case is not as straightforward. With k sectors, the perfectly even triangular lattice of points yields a value of

$$R_k = \frac{2.1491}{\sqrt{k}}$$

in which $k \leqslant 6$. As we have seen, when $k = 1$ (simple nearest neighbor analysis)

$$R = \frac{2.1491}{\sqrt{1}} = 2.149$$

However, when $k = 6$,

$$R_6 = \frac{2.1491}{\sqrt{6}}$$

which is less than unity, the expected R_6 value for a randomly generated pattern. Thus, we are left with the following situation:

Limiting case of clustering	$R_6 = 0$
Limiting case of regularity	$R_6 = 0.8775$
Random expectation	$R_6 = 1$
Pattern in Work Table 4.3	$R_6 = 1.20$

Our observed value is beyond all our special cases. Thus, we have no simple scale as when $k = 1$ (Figure 4.11). Let's consider why this is.

The average nearest neighbor values for the 6 sectors are given in the top row of Table 4.12, where sectors are ranked in order of smallest neighbor distance to each point (Figure 4.16). Table 4.13 shows the expected average neighbor distances for from 1 to 6 sectors for a randomly generated pattern. The order of neighbor distance predictions for this example has been calculated and added to Table 4.12. Finally, the expected average distance for a triangular point pattern is given by

$$T_t = 1.075\sqrt{A/n}$$

which, in this example, is 18.2687 and is also added as the final row in Table 4.12. In this comparison, the observed case seems to fall between the triangular lattice's constant distance relations and the rapidly increasing distances with ranked sectors for the random pattern. This is different, and intuitively more appealing, than the simpler R_6 finding. We can also now see why the limiting regularity case leads to $R_6 < 1$. With the triangular lattice, neighbor distances do not get larger as we measure to higher order neighbors up to the sixth neighbor. This results in the average neighbor distance over all 6 neighbors of less than random expectation and, hence, $R_6 < 1$. Similarly, $R_5 < 1$, although other R_k limiting regularity cases are greater

Table 4.13 Expected Average Order Neighbor Distances for Six Sectors Under Poisson Assumptions ($d = \sqrt{n/A}$)

$k =$	1	2	3	4	5	6
$\bar{r}_{(1)} =$	$\frac{0.5}{d}$	$\frac{0.5}{d}$	$\frac{0.5}{d}$	$\frac{0.5}{d}$	$\frac{0.5}{d}$	$\frac{0.5}{d}$
$\bar{r}_{(2)} =$		$\frac{0.9142}{d}$	$\frac{0.8371}{d}$	$\frac{0.8095}{d}$	$\frac{0.7951}{d}$	$\frac{0.7863}{d}$
$\bar{r}_{(3)} =$			$\frac{1.2610}{d}$	$\frac{1.1237}{d}$	$\frac{1.0698}{d}$	$\frac{1.0403}{d}$
$\bar{r}_{(4)} =$				$\frac{1.5669}{d}$	$\frac{1.3809}{d}$	$\frac{1.3034}{d}$
$\bar{r}_{(5)} =$					$\frac{1.8444}{d}$	$\frac{1.6175}{d}$
$\bar{r}_{(6)} =$						$\frac{2.1010}{d}$

Source: M.F. Dacey, Analysis of central places. *Lund Studies in Geography*, Series B., No. 24, 1962, pp. 55–76.

than 1, up to the maximum $R = 2.1491$ for $k = 1$. Thus, where k is relatively large, certainly when $k =$ 5 or 6, it is more useful to construct a table of ranked average distances, such as Table 4.12, rather than to compute an uninterpretable R_k value. In order to facilitate interpretation of this table, we can compute the sum of the differences between theoretically expected average neighbor distances and the observed average neighbor distances for numerical comparison. In our example, the sum of differences between observed and random is 24.3, whereas between observed and regular the sum is 45.5. Hence, we may conclude that the neighbor distances in our pattern are more like a random pattern than the triangular regular extreme, although interpretation of the results for this case must remain guarded because of the severe boundary problems (briefly discussed below and in Work Table 4.3).

This extension of the simple nearest neighbor approach is obviously important for more complete assessment of point patterns using the spacing approach. Both R_k and the separate sector distances methods have been employed in geographical research. Michael Dacey employed this second approach with $k = 6$ on the central-place pattern in part of Wisconsin. Previous research had suggested that the pattern is dispersed. Dacey's results are portrayed in Table 4.14, and once again we see that the observed average neighbor distances are between the two theoretical patterns in the table. In this case, however, the observed values are clearly more like the random expectations, as shown by the low difference of 8.25 compared with a difference of 18.47 between observed values and the regular expectations. Hence, Dacey was able to contradict the previous conclusion concerning the settlement pattern and suggest that the pattern is more random than regular.

We have already reported on the use of nearest neighbor analysis in the Lansing, Michigan, study: the trend toward clustering and then randomness in the pattern is confirmed in a simple nearest neighbor analysis (Table 4.11). Getis analyzed his data further by using a 3-sector approach. The R_3 results are also included on Table 4.11. It is clear that we are picking up a rather different story of changing pattern with R_3. The hypothesized U-shaped pattern for R is replaced by a rather irregular trend for R_3. In every case, R_3 suggests more dispersal than R. The interpretation that Getis gave for this result is that the tendency toward clustering that he hypothesized is at a very local scale and, at the broader scale reflected by R_3 values, the pattern remains fairly random throughout the period. This is the extended neighbor distance method identifying scale differences.

Table 4.14 Six Sectors Ranked by Neighbor Distance for Central Places in Wisconsin

Average Distances:	*Ranked Sectors*					
	1	*2*	*3*	*4*	*5*	*6*
Observed	3.4	4.5	5.5	6.6	7.6	9.8
Random expectation	2.6	4.1	5.5	6.8	8.5	11.0
Triangular lattice	5.6	5.6	5.6	5.6	5.6	5.6

Source: M.F. Dacey. Analysis of central places. *Lund Studies in Geography*, Series B., No. 24, 1962, pp. 55–76.

Boundary Problems

If we are interested in studying pattern at various scales, the quadrat sampling method is superior to neighbor distance. This is because scale is clearly specified by quadrat area whereas neighbor distances reflect different scales at different point densities because they are based on ordinal (that is, nearest) information only. We can use the neighbor distance approach to illustrate what is a scale problem little appreciated in point pattern analysis.

The various parameters for theoretically special cases that we use in analyses of spacing all assume that we deal with an infinite plane. Area is brought into the analysis only when we must define point density and area per point. In practice, we always deal with a bounded plane. The result is that the processes and patterns we define in theory include no consideration of boundaries. Nonetheless, in our empirical research, boundaries and their effects cannot be avoided. This is a classic example of translation difficulty as we move from geographical problem to mathematical language. One consequence is that we may have in our study area a point whose nearest neighbor is across the boundary outside our area of interest. If we consider only points within our boundary, we can see that we bias results toward longer neighbor distances and, hence, toward dispersed pattern. If, on the other hand, we consider neighbor distances across the boundary, our definition of point density becomes suspect, because we are using points not incorporated in its calculation. This effect seems not to lead to systematic bias and, hence, is the solution usually employed. In most spacing analyses, points in a zone around the boundary are recorded because they may well turn out to be nearest neighbors. We employ this approach in our example in Work Table 4.3, where we employ the border zone of Figure 4.17. In our diffusion example in Figure 4.14, the study area is defined by a natural boundary (the lake shore on the east) and by arbitrary straightline boundaries selected to run through empty sections of the distribution. This definition of study area leads to all points having nearest neighbors within the study area.

A much less recognized problem associated with the definition of a study area has to do with its size and shape. Enclosing a point pattern in an artificially large frame can drastically influence our resulting R value because area enters the formula for $\bar{r}_e$. If a large study area is defined around a small, regularly spaced pattern of points, the resulting R value will suggest a clustered pattern. In the absence of good theoretical justification for a large study area, such result is merely the product of study area definition. We can therefore state the study area must be carefully defined in the context of the problem under investigation. We still have to be careful concerning the shape of the study area, which must remain reasonably compact in the sense that it should not be greatly indented. If we begin with any point pattern, it is an easy matter to draw a very irregular boundary in and around the points so that the pattern is enclosed in a very small area. This has the opposite effect of extending the area to produce clustering. Because we can make the area under study almost as small as we wish, we may even have results in which R values exceed 2.149, the limiting case in

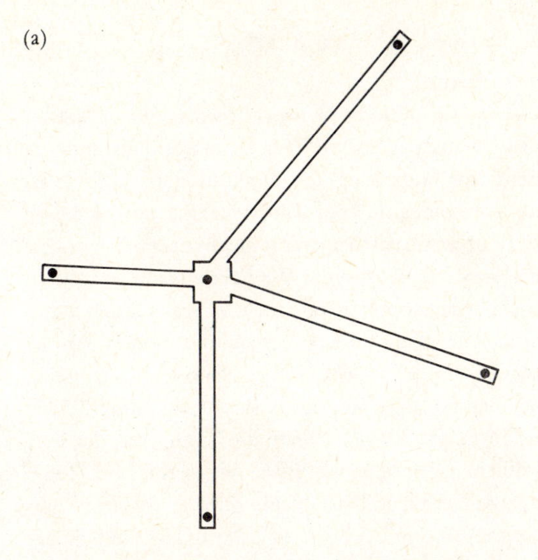

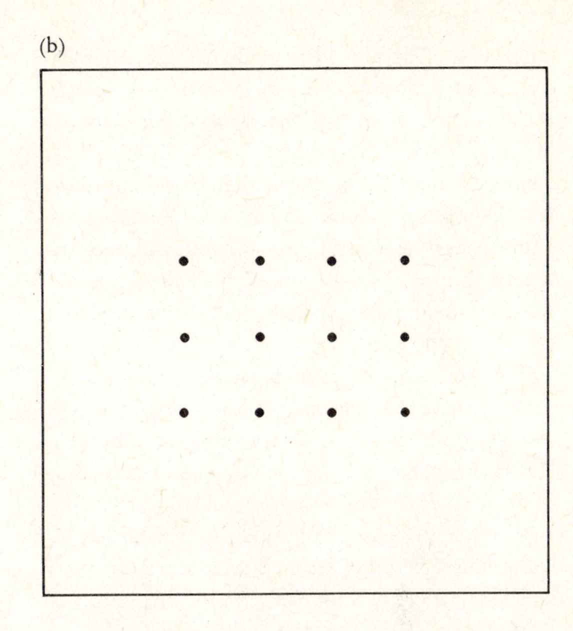

Figure 4.18 Special cases where the boundary affects the *R* value.

Figure 4.11. We have clearly assumed that we are dealing with reasonably compact areas and, in fact, in most situations this is the case. When study areas are defined by such features as the built-up area of small settlements, however, highly irregular shapes can occur and the limitation above must be borne in mind. The best solution is probably that employed by Arthur Getis in his Lansing study, in which study regions were defined as circles enclosing most of the built-up area at each date but avoiding the irregular boundary of the town.

Two examples of the problem above are shown in Figures 4.18a and 4.18b. In the first case, the irregular, small area leads to a ridiculously high *R* value of 7.07 that has no counterpart in our theoretical examples in their boundless planes. The second example is a case of numerical compromise. We clearly have two scales of pattern. At the scale of the study region, the pattern is reasonably clustered; at the scale that encloses the points, the pattern is clearly regular. The *R* value turns out to be 1. Normally we would not consider arguing from this particular evidence for a random process, despite this *R* value. It is quite clear that we have returned to the basic inference problem concerning pattern and process. Whether we make a spacing approach or try quadrat sampling, we can only conjecture about process from simple pattern evidence.

Further Reading

The Poisson and negative binomial probability models can be found in the probability textbooks described in readings for Chapter Three. Feller (1957) is a good example. It enables the researcher to consider other contagious process models. Most probability models have tables of their values readily available and these usually include detailed description of their operation. For the Poisson distribution, reference should be made to Kitagawa (1952) or General Electric Company (1962), and for the negative binomial Williamson and

Bretherton (1964) should be consulted. The competitive county seat probability model is from geographical literature, as we have indicated, and is described in Dacey (1964a, 1966a, 1966b). Dacey (1964b) further considered the Poisson distribution in the areal case. The approach we adopt is based on Coleman's (1964) mathematical textbook for sociologists; this excellent discussion provides both complementary and supplementary reading, the latter best developed in Haight (1967).

The techniques of nearest neighbor and quadrat analysis derive from the work of ecologists, introduced and presented by Greig-Smith (1964) and Kershaw (1964). Techniques and ideas on point pattern analysis have been developed over a period of more than twenty-five years in ecology and the valuable experience and insight derived from these researches are very well summarized in both books. In the discussion, we have drawn directly on the scale experiments presented by Greig-Smith (1952). These descriptions of ecological studies have been complemented recently by Shimwell (1971); other work has been rigorously extended in Pielou (1969). Pielou particularly indicates that geographers have not learned all they can from ecology, and the book should be considered the starting point for supplementary, extended reading on this subject.

The basic reference for nearest neighbor analysis is Clark and Evans (1954); this work was introduced into geography by Dacey (1962), who is the source for the six-sector study of settlement pattern in Wisconsin. Getis (1964) used both the simple R scale and an R_3 analysis in his study of grocery stores in Lansing, Michigan, and we have reported on his work. Most geography studies have employed the simple R scale to obtain comparisons of point patterns. This is the strategy of King (1962) in his study of U.S. settlement patterns in various small areas and of Hirst (1971a), who traced the changing pattern of administrative centers in Uganda. This approach was developed technically and extended by Dacey (1963), who considered further distance relations without using sectors. Dacey (1967) also applied this approach in analyzing line patterns.

Getis (1964) study of grocery stores also utilized a quadrat analysis, as we have reported. This study is useful for illustrating the application of both techniques to the same set of data. By far the most extensive study of point patterns using quadrat sampling is that of Dacey and of Harvey. We have already referred to Dacey's development of a county seat model; other settlement studies using contagious models can be found in Dacey (1968a, 1969). Harvey's work (1966) includes the extensive analysis of Hägerstrand's diffusion maps. Harvey (1968) also considered the scale problem from a geographical viewpoint and (1967) discussed inferential problems relating to these techniques. Hudson's work (1969a) provides supplementary reading to illustrate technical developments in pattern recognition and (1969b) in using these approaches for substantive modeling of rural settlement patterns.

Map transformation research strategy was developed and described by Tobler (1963). Actual applications of its use are rare, and the work of Getis (1963) on grocery stores in Tacoma, Washington, has not been followed up in geography, despite the advocacy of Bunge (1966). The limitations of this approach were pointed out by Angel and Hymen (1972), and Rushton (1972) proposed an alternative strategy for studying central places on an anisotropic plane.

Nearest neighbor and quadrat analysis are not the only approaches available for studying point patterns. Russian geographer Medvedkov (1967a, 1967b) introduced entropy in studying point patterns of settlement, which apportions the pattern between randomness and regularity. This approach has gained recent popularity and can be followed up in Medvedkov (1970) and Chapman (1970).

The strategy of pattern analysis and inferring processes was introduced by Blaut (1962) and extensively discussed by Harvey (1969) and King (1969a). King (1969b) presented a broad technical review that provides a general introduction to further reading.

Work Table 4.1 Fitting the County Seat Model

Model

$$p(x) = \frac{q\gamma^x e^{-\gamma}}{x!} + \frac{p\gamma^{x-1} e^{-\gamma}}{x!}$$

We have to estimate two parameters p and γ; $q = 1 - p$. (The estimates are denoted by $\hat{p}$, $\hat{\gamma}$, and $\hat{q}$.)

Data

0	0	1
1	1	2
1	2	1

In this map, $n = 9$.

The frequency distribution is:

$x = 0$ $f = 2$
$= 1$ $= 5$
$= 2$ $= 2$
$= 3$ $= 0$
$\Sigma f = 9$

Estimation of Parameters

$$\bar{x} = \frac{\Sigma(fx)}{\Sigma f} = \frac{9}{9} = 1.$$

$$\text{Var}(x) = \frac{\Sigma(f(x - \bar{x})^2)}{\Sigma f} = \frac{4}{9} = 0.4444$$

$$\hat{p} = \sqrt{\bar{x} - \text{Var}(x)} = \sqrt{1 - 0.4444} = 0.7454$$

$$\hat{\gamma} = \bar{x} - \hat{p} = 1 - 0.7454 = 0.2546$$

$$\hat{q} = 1 - \hat{p} = 0.2546$$

Finding Probabiliites and Expected Frequencies

If we let p_x be the Poisson probabilities, with an expectation of $\hat{\gamma} = 0.255$, then $p(x) = \hat{q}p_x + \hat{p}p_{x-1}$ (when $x = 0$, $p(0) = \hat{q}p_x$), which can be calculated as follows:

x	p_x	$\hat{q}p_x$	$\hat{p}p_{x-1}$	$p(x)$	$p(x)\Sigma f$
0	0.7749	0.1973	—	0.1973	1.78
1	0.1976	0.0503	0.5776	0.6279	5.65
2	0.0252	0.0064	0.1473	0.1537	1.38
3	0.0021	0.0005	0.0188	0.0193	0.17

p_x is from the tables of the Poisson distribution. $p(x)\Sigma f$ is the expected frequency for this model, to compare with observed frequencies in the data above.

Testing the Model

For a χ^2 test, expected frequencies are too small. For a Kolmogorov-Smirnov test, $D = 0.0474$. $D(p = 0.2, n = 9) = 0.339$, indicating a high degree of similarity between model (expected frequencies) and empirical evidence (observed frequencies).

Exercise

Fit the county seat model to the case in which the data map above has 1 point in 7 quadrats, 2 points in one other quadrat, and no points in the final quadrat. How good is the fit that we get in this rather special case?

Work Table 4.2 Fitting the Negative Binomial Model

Model

$$p(x) = \frac{(k + x - 1)!}{x!(k - 1)!} p^k(1 - p)^x$$

We have to estimate two parameters, p and k.

Data

0	4	6
1	0	5
0	0	2

In this map, $n = 18$.

The frequency distribution is:

x	f
$x = 0$	$f = 4$
$= 1$	1
$= 2$	1
$= 3$	0
$= 4$	1
$= 5$	1
$= 6$	1
	$\Sigma f = 9$

Estimation of Parameters

$$\bar{x} = \frac{\Sigma(fx)}{\Sigma f} = 2$$

$$\text{Var}(x) = \frac{\Sigma(f(x - \bar{x})^2)}{\Sigma f} = 5.1111$$

$$\hat{p} = \frac{\bar{x}}{\text{Var}(x)} = 0.3913$$

$$\hat{k} = \frac{\bar{x}\hat{p}}{1 - \hat{p}} = 1.2857$$

Finding Probabilities and Expected Frequencies

If we let $p^k = c$ and $q = 1 - p$, then probabilities for $x = 0, 1, 2, \cdots, n$ are given by the successive terms

$$c,\ ckq,\ \frac{ck(k+1)q^2}{2!},\ \frac{ck(k+1)(k+2)q^3}{3!},\ \frac{ck(k+1)(k+2)(k+3)q^4}{4!} \cdots.$$

$\hat{p} = 0.3913$, and $\hat{k} = 1.2857$, and $\hat{q} = 0.6087$. Thus,

$$c = 0.2993$$

$$ckq = 0.2342$$

$$\frac{ck(k+1)q^2}{2!} = \frac{0.2342 \cdot 2.2857 \cdot 0.6087}{2} = 0.1629$$

$$\frac{ck(k+1)(k+2)q^3}{3!} = 0.1086$$

$$\frac{ck(k+1)(k+2)(k+3)q^4}{4!} = 0.0709$$

$$\frac{ck(k+1)(k+2)(k+3)(k+4)q^5}{5!} = 0.0456$$

$$\frac{ck(k+1)(k+2)(k+3)(k+4)(k+5)q^6}{6!} = 0.0291$$

$$\frac{ck(k+1)(k+2)(k+3)(k+4)(k+5)(k+6)q^7}{7!} = 0.0184$$

and so on. These results can be tabulated as follows:

x	$p(x)$ ($\hat{p} = 0.3913$, $\hat{k} = 1.2857$)	$p(x)\,\Sigma f$	*Observed frequencies*
0	0.2993	2.69	4
1	0.2342	2.11	1
2	0.1629	1.47	1
3	0.1086	0.98	0
4	0.0709	0.64	1
5	0.0456	0.41	1
6	0.0291	0.26	1
7	0.0184	0.17	0

Testing the Model

For a χ^2 test, expected frequencies are too small. For a Kolmogorov-Smirnov test, $D = 0.1383$. $D(\alpha = 0.2, n = 9) = 0.339$, indicating that our D value is a long way from the tail of the statistical distribution of D when $n = 9$. We may conclude that there is some degree of similarity between model (expected frequencies) and empirical evidence (observed frequencies).

Exercise

Fit the negative binomial model to the case in which the data map above has 9 points in one quadrat and no points in any other quadrat. How good is the fit that we get in this rather special case?

Work Table 4.3 A Six-Sector Nearest Neighbor Analysis

Data

The point pattern is in the study region of Figure 4.17.

Notice that although we deal only with these 5 points explicitly, points in the border zone may be defined subsequently as neighbors.

Method

1. Draw a 6-sector diagram, or sextant, on tracing paper, consisting of 6 lines radiating from an origin with an angle of 60° between adjacent pairs of lines.
2. Select the first point to be considered in the study region and place the origin of the sextant on the point.
3. Measure the distance from this first point to its nearest neighbors in each of the 6 sectors.
4. These distances are ranked so that the shortest distance is r_{11}, the next shortest is r_{12}, and so on through r_{16}.
5. This operation is repeated for every point (i) in the study region, so that we have a series of 6 ranked distances: r_{i1} and r_{i2} through r_{i6}.
6. Find the average distance for each of the 6 ranks.

These constitute the basic information used in this method.

For an analysis of Figure 4.17, the following average distances are derived:

$\bar{r}_{(1)}$	$\bar{r}_{(2)}$	$\bar{r}_{(3)}$	$\bar{r}_{(4)}$	$\bar{r}_{(5)}$	$\bar{r}_{(6)}$	$\bar{r}_{a6}$
18.6	19.6	20.8	24.2	27.2	33.2	24.8

$\bar{r}_{a6}$ is the average for all 6 sectors.

Nearest Neighbor Statistics

$$\bar{r}_{e6} = \frac{\sqrt{k}}{2\sqrt{n/A}}$$

$n = 5$, $A = 1444$, and when $k = 6$, then

$$\bar{r}_{e6} = 20.6883$$

$$R_6 = \frac{\bar{r}_{a6}}{\bar{r}_{e6}} = 1.1987$$

Poisson Predictions

From Table 4.13, where $d = \sqrt{n/A} = 0.0592$, expected average distances are:

$$\bar{r}_{(1)} \quad \frac{0.5}{0.0592} = 8.4459$$

$$\bar{r}_{(2)} \quad \frac{0.7863}{0.0592} = 13.2821$$

$$\bar{r}_{(3)} \quad \frac{1.0403}{0.0592} = 17.5726$$

$$\bar{r}_{(4)} \quad \frac{1.3034}{0.0592} = 22.0169$$

$$\bar{r}_{(5)} \quad \frac{1.6175}{0.0592} = 27.3226$$

$$\bar{r}_{(6)} \quad \frac{2.1010}{0.0592} = 35.4899$$

Interpretation

The interpretation of these results is a little difficult and is explained in some detail in the text. We can notice that this very simplified example, with 5 points, involves severe boundary problems. The example is used merely to illustrate the method as simply as possible, although in practice we would not consider analyzing a 5-point pattern with a 6-sector method. All empirical analyses involve large numbers of points. One solution that has been suggested for the boundary problem is to omit all points in the study region that do not have all 6 neighbors also within the study region. This suggestion eliminates all the analysis above and emphasizes the artificiality of our example. The analysis above should be treated as purely a means of illustrating the technical aspects of the method.

Exercise

Repeat the exercise above but for the 4-sector case. Compute R_4 and relate the results to Poisson predictions. Do your interpretations for 4 sectors differ from those for the 6-sector analysis?

V

AREAL ASSOCIATION

Some geographers have interpreted the rise of quantitative methods as a challenge to the map as the primary tool of the geographer. It should be clear by now that a far more realistic interpretation is that these new techniques complement and extend traditional cartographic analysis. In Chapter Three, on spatial autocorrelation, and in Chapter Four, we dealt with one map at a time in a search for patterns and processes. In this chapter, we change the emphasis not from maps but from analyzing single map patterns to comparing two or more maps simultaneously. In many ways the areal association studies we describe here are the most explicit illustration of modern spatial analysis building upon and developing a classical geographical method.

Geographers concerned with spatial distributions have not been content traditionally simply to portray phenomena in map form but have usually made some attempt to compare the spatial distributions of different phenomena. The purpose of such map comparison is to find out which phenomena share common locations. We might observe that areas with high annual rainfall also have high wheat yields. We can conclude that rainfall and wheat yields are areally associated. Such studies form an initial step toward the explanation of areal variations of phenomena.

Areal association can be found in most systematic branches of geography. It has been particularly popular in electoral geography, in which French geographers initiated a tradition of explaining voting patterns by visual comparison with maps of various social and economic variables. Similarly, the social geography of cities has advanced traditionally by means of map comparisons of various social phenomena such as crime, disease, class, race, and ethnic membership. However, it is in economic geography, in the work of Iowa geographer Harold McCarty, where the notion of areal association has been explicitly put forward as the path toward a theoretical geography. McCarty has argued that future economic geography theory will be made up of a large number of interrelated principles, statements of the form "Phenomenon x is areally associated with variables a, b, and c." The purpose of our empirical research should be to identify variables a, b, and c in conjunction with phonomenon x. The first step is to map x to produce the *problem map*. This is the spatial distribution we have to explain; it might be the distribution of wheat yields by county over an area of the Great Plains. Next we hypothesize the factors that help determine the problem map. One such determinant of x is probably average annual rainfall. How can we test this hypothesis? One way is to produce a map of rainfall patterns and compare it visually to the problem map. If the relationship is very clear, this approach may be satisfactory, especially if the second map is drawn so that it can be viewed as an overlay on the problem map. With less obvious areal associations, visual comparison of maps can be very unreliable and quantitative approaches to map comparison become indispensable. These take visual comparison a stage further and tell us how much the two phenomena are areally associated by adapting common statistical techniques to spatial contexts. If the success of this transplantation is measured by the ensuing popularity of the quantitative areal association approach, then success is an understatement. In one survey of quantitative geography, it was found that fully two thirds of the techniques used by American geographers in the mid-1960s were areal association techniques. We are clearly dealing with one of the major thrusts of research in modern geography, and we must consider its technical apparatus and methodological implications very carefully. This is our task in this chapter. We begin with visual map comparison in order to illustrate how quantitative areal association develops from and extends the traditional methods of map comparison.

ELEMENTARY METHODS OF MAP COMPARISON

The most common approach to areal association in recent years has been statistical correlation and regression. Before we consider these, we describe briefly some more elementary approaches that have the same general aims and that may successfully achieve them in specific research situations.

Map Overlay

There is no doubt that visual comparison of maps can be useful in searching for relations between spatial distributions. There is no suggestion here that the long history of this method in geography has been in any sense wrong or futile. But roles traditionally performed by visual comparative map analysis can now be much better undertaken with quantitative approaches.

Visual Map Comparison

Consider the two maps in Figure 5.1, showing high income areas and high Republican voting districts in Flint, Michigan, for 1950. Many political scientists have shown that the higher a person's income is, the more likely that person is to vote as a Republican. Figure 5.1 shows this relationship in its spatial form, the two maps indicating a high degree of areal association between the two variables. This is a very simple example whose conclusions will not surprise anyone familiar with the American political system. Map comparison of this form can be used to bring out useful and interesting spatial patterns of association. In Figure 5.2, the pattern of Democratic voting and black neighborhoods are depicted on the same map for four

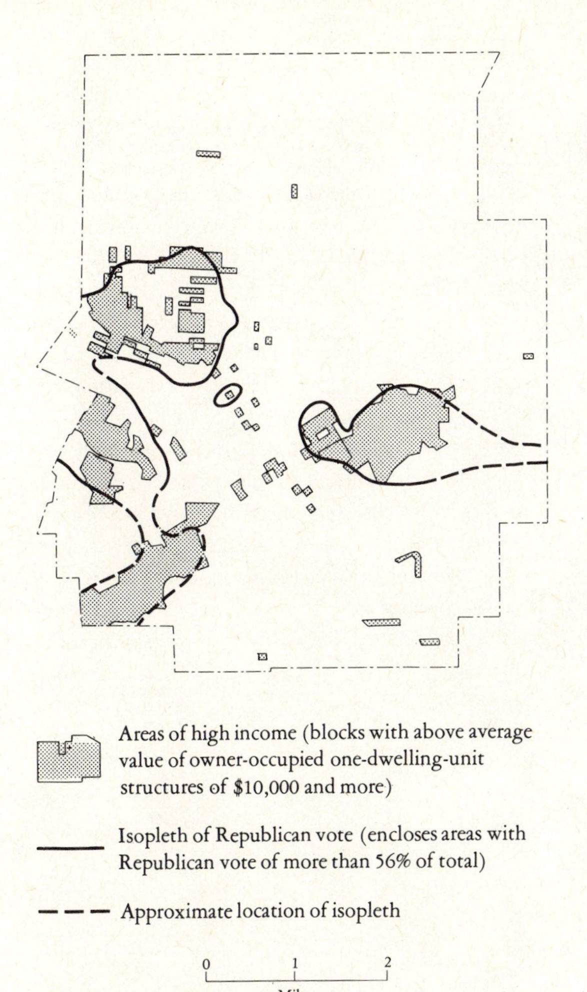

Figure 5.1 High income areas and Republican voting, Flint, Michigan, 1950.
From P.F. Lewis. Impact of Negro migration on electoral geography of Flint. Reproduced by permission from the *Annals* of the Association of American Geographers, vol. 55, 1965.

years—1932, 1942, 1952, and 1962. The black districts are shaded and the Democrat vote is depicted by two critical isolines. This is, therefore, a method of comparison by overlay, the placing of one areal distribution over another. The isolines are chosen so that they define areas that have lower percentages of the Democratic vote than the two black districts. Two isolines are drawn for percentage values of the Democratic vote one percent below the two lowest black neighborhood levels. In 1932, the northern district had a 42 percent Democratic vote and the 41 percent isoline encloses a small area of the city that has a smaller Democratic vote. The southern district had only a 36 percent level of Democratic vote, and the area enclosed by the 35 percent isoline is correspondingly even smaller. In 1932 the black areas were among the most staunch Republican districts in a Republican town. Ten years later, in 1942, we get a very different pattern. The lowest black district in the northern neighborhood had a very high 74 percent Democratic vote and the 73 percent isoline encloses this neighborhood only. This indicates that the lowest Democrat voting level in the northern black neighborhood was higher than most districts in the rest of the city. The same cannot be said about the southern neighborhood. Here the lowest district had a 55 percent level of vote for Democrats and the 54 percent isoline includes most of the typical high-income Republican areas (Figure 5.1) along with the black neighborhood in this now safely Democratic town. The massive reversal of black party preferences had not penetrated this southern neighborhood by 1942 but ten years later both districts conformed to the typical pattern in which black districts form the most solidly Democrat districts—both districts returning more than 77 percent of their vote to the Democrats—a pattern not repeated anywhere else in the city. This phenomenon is extended in the 1960s; the 1962 map shows that the area of solidly Democratic support, this time greater than 78 percent, had grown with the expansion of the black neighborhoods.

This very simple but clear example of map comparison illustrates some interesting spatial contrasts in the changing areal association of Democratic voting and black neighborhoods. The fact that Flint did not experience a comparable black voting reversal in both neighborhoods simultaneously is very clearly brought out. The present-day strength of the Democratic party in black neighborhoods is also forcefully expressed. How do we suggest that quantitative map

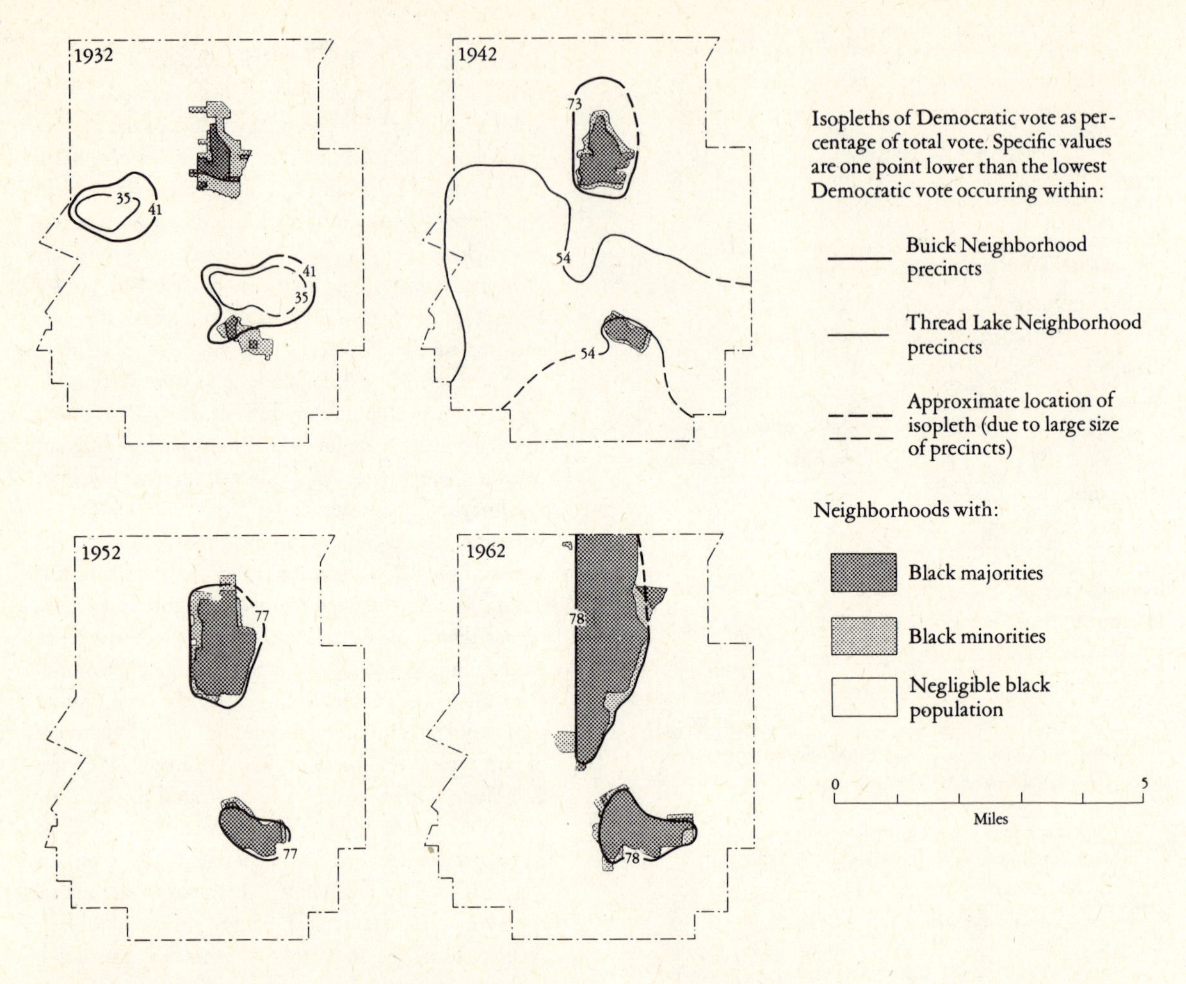

Figure 5.2 Democratic voting and black neighborhoods, Flint, Michigan, 1932–1962.
From P.F. Lewis. Impact of Negro migration on electoral geography of Flint. Reproduced by permission from the *Annals* of the Association of American Geographers, vol. 55, 1965.

comparisons can improve on this very good illustration of traditional visual map comparison?

The basic disadvantage of visual map comparison is that it involves subjective personal assessments that may well vary between researchers. The technique falls down on the basic question of intersubjectivity. An assessment might well be influenced by such extraneous features as map design and the method of portraying variables. In order for this technique to be at all possible, the relation between mapped variables must be very strong. The Flint electoral study fits this category well; two strong electoral relationships are portrayed. We can, therefore, identify this particular example as one in which cartographic methods have successfully portrayed the relations being studied. With weaker relationships, the method begins to be much less effective. Even in the case of the black-Democrat relationship in Flint, we are not able to specify the relative degree of this relationship for any particular year. In fact, enquiring into the degree of relationship—how strong or weak an areal association is—brings us inevitably to quantification.

Numerical assessments of areal association have the basic advantage of being both precise and intersubjective. We can use potentially much more information in the original data than that actually categorized and mapped. Quantification allows for exact specification of the form of relations between variables as we map an areal association problem into the realm of the mathematical function. This very flexible way of describing relations in turn opens the possibility of looking at several variables simultaneously. Such multivariate analysis is largely impossible with simple map overlays, which rapidly degenerate into confusion once more than one overlay has been used. The degree and form of multivariate relations can be accommodated in the quantitative approach to areal association.

We can best describe these advantages of the numerical approach over visual comparisons by illustrations from actual research applications. The use of statistical techniques in the context of areal association studies has not led to the abandonment of maps. The map maintains its important role in the portrayal of spatial variations in the pattern of particular relationships. Statistical approaches have enabled us to extend this role. Let's first consider how comparisons by map overlays can be quantified to assess degrees of association.

The Coefficient of Areal Correspondence

A numerical assessment of the degree of a relation is achieved when we can state precisely, in terms of some relative quantity, by how much two areal distributions correspond. This, in turn, requires that we define exactly what we mean by the phrase two distributions correspond. R.F. Minnick approached this problem by employing a set theory formulation that is basically a numerical extension of traditional map overlay procedures. The result is an explicit measure of areal association known as the coefficient of areal correspondence.

Set theory language is used in this context to formalize what traditional geographers have always attempted to do visually in map comparisons. Let's assume that we have a map portraying the distribution of one phenomenon, such as black residential districts in Flint. We call the area covered by this particular phenomenon set A. We have another phenomenon displayed on an overlay map, such as highly Democratic voting districts; this area is defined as set B. Two areas derived from the two maps are of particular interest in our comparison:

1. The area in which both phenomena are located together, which is simply the intersection of A and B ($A \cap B$).
2. The area covered by either phenomenon, which is simply the union of A and B ($A \cup B$).

Clearly, area $A \cap B$ defines the absolute level of correspondence between the two maps that can be standardized by relating to $A \cup B$ the total area covered by the two phenomena. Thus, a relative measure of areal correspondence (c_A) is given by the formula

$$c_A = \frac{A \cap B}{A \cup B}$$

which relates the area of direct areal correspondence to that of possible correspondence. This simple numerical assessment of areal association is known as the *coefficient of areal correspondence*. If the two distributions are completely separate, then $A \cap B = 0$ and c_A is 0. On the other hand, with two distributions corresponding exactly, $A \cap B = A \cup B$, and c_A is 1. Thus, c_A ranges from 0 for no areal correspondence to 1 for complete areal correspondence. In practice, comparisons between map distributions produce values lying on a scale of areal correspondence between these two extremes.

In Figure 5.3a, the distribution of greater than average population densities in the eastern north-central United States are shown for 1950 and 1960. We do not expect normally to find large changes in population pattern in a single decade, and our visual assessment of Figure 5.3a confirms this expectation. We can use this clearly strong relationship to illustrate c_A. Measuring the areas covered on the map by the intersection and union of the two distributions enables us to compute a very high coefficient of areal correspondence of 0.93. This is interpreted simply as meaning

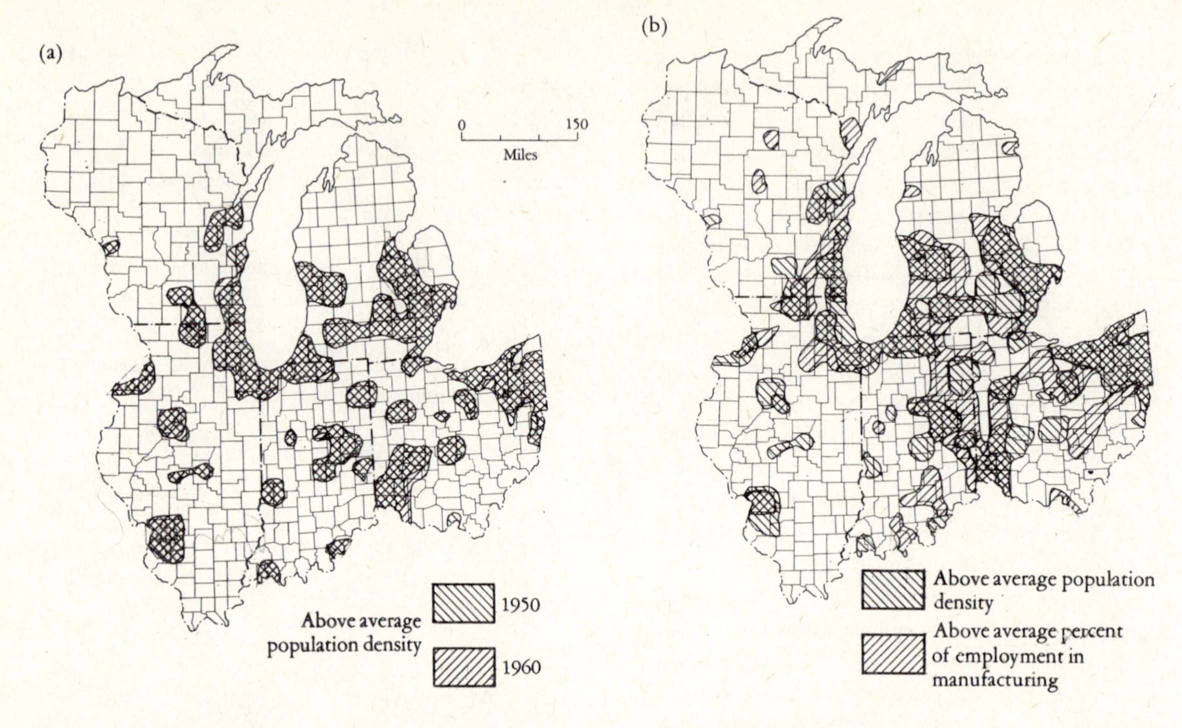

Figure 5.3 Population and manufacturing in eastern north-central United States: (a) population density, 1950 and 1960, and (b) population density and manufacturing employment, 1960.
From R.F. Minnick. *A Method for the Measurement of Areal Correspondence*. Ann Arbor: Michigan Academy of Science, Arts and Letters, 1964, pp. 337 and 342.

that 93% of the area covered by either of the distributions contains both distributions together.

This is our numerical assessment of the areal association depicted in Figure 5.3a. A lower level of correspondence is suggested by Figure 5.3b, in which greater than average manufacturing employment and population density are portrayed for 1960. In this case, $c_A = 0.43$ indicating that more than half of the total shaded area on the map consists of separate above-average density or manufacturing categories.

The coefficient of areal correspondence is a measure of areal association that uses simple nominal data. For each variable, an area is either shaded or not shaded. For some concepts, this gives a reasonable portrayal; we might think that a rural-urban distinction is a useful way of classifying settlements on a nominal scale, for instance.

In our eastern north-central United States example, the variables have been dichotomized about their mean value. In this way, much information is lost. How much more greater than average population density is concentracted in some of the lakeshore areas than in areas similarly categorized further inland? The U.S. census upon which this analysis is based provides the answer to this question but the technique cannot accommodate such information. This approach is limited to situations in which we deal with nominal data or in which we choose to treat data at a nominal level. This need not always be as arbitrary as dichotomizing about the mean value, but it can include the geographer's traditional nominal concept, the region. We can compute c_A to assess the degree of areal association of any two regional patterns. If we wish to compute areal associations using information at more than a simple nominal scale, alternative techniques are required. A very popular example of such a technique is the Lorenz curve.

The Lorenz Curve

The Lorenz curve is a method of plotting two variables on a graph to illustrate the similarities in their areal distribution. Consider Figure 5.4a, in which two variables are displayed on a map of six areal units. Let's begin by assuming that the variables represent number of middle-class voters (x) and numbers of Republican votes (y). The numbers on Figure 5.4a suggest a high degree of areal association, which we can illustrate by constructing a Lorenz curve. The steps are as follows and appear in Figure 5.4b:

1. Calculate the ratio y/x for each areal unit.
2. Rank areal units in terms of these ratios from the smallest (ranked 1) to the largest.
3. Maintaining these ranks, convert variables into percentages of their own total for each areal unit.
4. Cumulate these percentages from rank 1 onward.
5. Plot the cumulative percentages for x and y on graph paper and join the observations to produce the *Lorenz curve* (Figure 5.4 c).

What are some of the more obvious properties of this curve? If both distributions are proportionally identical in each areal unit, the resulting curve corresponds to the diagonal. Deviations away from the diagonal represent differences in the two distributions. If the two distributions are completely separate (where one phenomenon is found the other is always absent) the curve follows the x axis and the vertical boundary of the graph at $x = 100\%$. This is the limiting case in terms of maximum deviation from the diagonal, maximum spatial separation, and minimum areal association. We can use these properties to define an inverse measure of areal association.

The Index of Dissimilarity

The usual way of measuring areal association from a Lorenz curve is to compute the index of dissimilarity (D_A), which is graphically defined as the maximum vertical distance between the diagonal and the curve. It ranges from 0% to 100% for the two limiting cases

(a)

A $x = 30, y = 30$	C $x = 10$ $y = 8$	D $x = 10$ $y = 5$	E $x = 20, y = 19$
B $x = 20, y = 15$			F $x = 30, y = 23$

(b)

Rank by ratio		Convert to percentages		Cumulate	
y/x	Areas	x%	y%	Cx	Cy
0.50	D	8.3	5	8.3	5
0.75	B	16.7	15	25.0	20
0.77	F	25.0	23	50.0	43
0.80	C	8.3	8	58.3	51
0.95	E	16.7	19	75.0	70
1.00	A	25.0	30	100.0	100

(c)

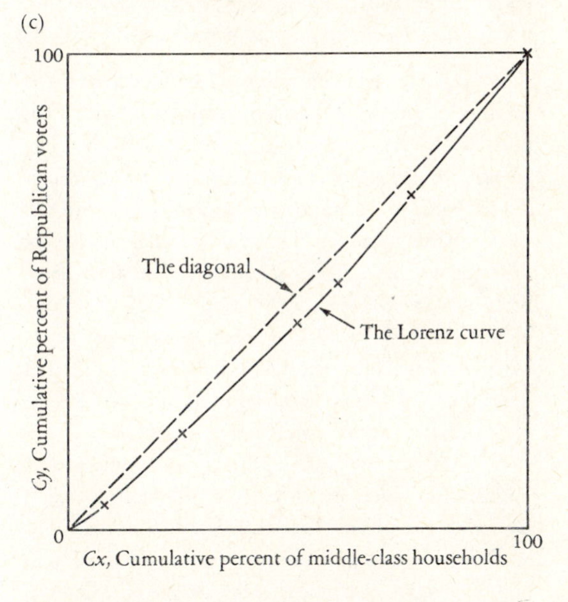

Figure 5.4 Constructing a Lorenz curve. (a) Data: middle-class voters (x) and Republican voting (y) by areas. (b) Data conversion. (c) The Lorenz curve.

discussed above. There are three ways of obtaining D_A:

1. By measuring it from the Lorenz curve on graph paper.
2. By computation from the formula

$$D_A = \max(cx_i - cy_i)$$

using the *cumulative* values already used in constructing the curve (in the example in Figure 5.4b this gives 7.3).
3. By computation from the formula

$$D_A = \frac{\Sigma(x_i - y_i)}{2}$$

in which x_i and y_i are the *uncumulated* percentages for each variable. Using this formula avoids cumulating and we can produce D_A values without drawing the Lorenz curve. Occasionally D_A is calculated by using proportions instead of percentages, in which case it ranges from 0 to 1. In the following discussion, we hold to the convention of using percentages.

We call D_A the index of dissimilarity because it is a measure of differences between areal distributions and, thus, an inverse measure of areal association. The greater the degree of areal association is, the smaller the D_A value is. It is important to identify D_A with this specific name because it is known to spatial analysts by different names in different contexts. The great flexibility and widespread application of D_A has been hidden generally by various aliases it has been given. We illustrate its full potentialities by giving six different interpretations to our variables in Figure 5.4a, beginning with the original voting interpretation.

1. With any two variables such as the numbers of middle-class voters and numbers of Republican votes, D_A gives us a general measure of the dissimilarity in areal distributions. In this interpretation of x and y in Figure 5.4a, we conclude that the degree of difference between the pattern of middle-class voters and Republican votes is only 7.3 percent. They exhibit a high level of areal association.

2. A more specific use of this index is in the context of employment data. If y is the numbers employed in the machine tool industry and x is the numbers employed in the car industry, then D_A is a measure of locational dissimilarity in these industries. In this context, D_A has been called the *coefficient of geographical association*. From our example in Figure 5.4a, we would conclude that employment in the machine tool and car industries is geographically associated because their differences amount to only 7.3%.

3. If we keep an employment figure for an industry as y but change x to population, we produce a measure of the degree of *market orientation* of the industry. In our example, the pattern of machine tool employment is dissimilar from that of total population by only 7.3 percent. This represents its degree of market orientation.

4. If we move population over to make it become y and denote the land area of each areal unit as x, then D_A defines a measure of population concentration. A perfectly even population distribution is reflected in a D_A value of 0, whereas total concentration in one areal unit produces a D_A value approaching 100%. In our example, we are much closer to the former case, indicating that the population distribution is quite evenly spread over the map.

5. If we keep x as total land area of each areal unit but change y to land area under different uses, we produce what has been termed the *coefficient of areal localization*. If the land devoted to a crop is regularly distributed, then D_A is 0, whereas if it is concentrated D_A becomes large. If y is the acreage planted in cereals and x is the total land area, then we would conclude that because D_A is only 7.3, cereals are not very localized in their distribution.

6. Finally, let's define y as the number of blacks and x as the number of whites living in an area. Then D_A is a measure of *residential segregation*. We have a low level of segregation at only 7.3 percent, which indicates a highly integrated region.

Thus D_A is a highly flexible and widely used index. The six examples do not exhaust its utility, because it can be employed on nonareal data outside spatial analysis. We illustrate two contrasting research applications of D_A that relate to areal localization and residential segregation. Table 5.1 shows the degree of areal localization for six horticultural products among the counties of England and Wales in 1956. D_A ranges from 61% to 27%. The high coefficients are

Table 5.1 Localization of Horticultural Production in England and Wales (1956)

Crop	D_A
Celery	61
Bulb flowers (not under glass)	59
Rhubarb	58
Turnips and Swedes	32
Onions	29
Beetroot	27

Source: Michael Chisholm. *Rural Settlement and Land Use.* London: Hutchinson Publishing Group Ltd., 1962.

for crops that require particular conditions for successful farming. Celery requires deep, well-drained soil, and the crop is concentrated in the fens and parts of Norfolk. Bulb flowers grown in the open are concentrated in Cornwall (obtaining the advantage of the early season) and the Holland division of Lincolnshire. The human factor of traditional expertise seems to be the only explanation for the concentration of rhubarb in the West Riding of Yorkshire. The three crops with low levels of localization all tolerate a wide range of conditions and frequently form part of arable rotations. They are found widely distributed in arable country. Hence D_A can give useful descriptive statistics for summarizing the level of areal localization of any type of land use.

Because it is a direct measure of spatial separation of populations, D_A has been employed widely in the social geography of cities as an *index of residential segregation.* The suitability of D_A in this context is due in part to a property of the index that gives it a particularly relevant interpretation in the context of segregation. Consider Figure 5.5a, which depicts a hypothetical city with four census tracts, each having two ethnic groups distributed within them. D_A is computed as $(30 + 30 + 10 + 10)/2 = 40\%$. Let's assume that we must relocate group x in such a way as to make its distribution identical to group y. We can do this by moving 30% of x from A to B and 10% of x from C to D. This results in the distributions depicted in Figure 5.5b. Notice that to even out the distributions—that is, to integrate the two groups—we have had to move 30% + 10% of the population, which equals 40%—the value of the index of dissimilarity. In general, we can interpret the index of dissimilarity as the percentage of one ethnic group that has to be relocated in order to integrate it with a second group. This interpretation makes this index the most widely used statistic for describing residential segregation numerically.

In Table 5.2, the indexes of dissimilarity for Australian-born and overseas-born residents of Brisbane, Australia, are shown. Separate indexes have been computed for different groups of overseas-born residents, showing quite distinctive differences. The residents who were born in the United Kingdom have a residential distribution in Brisbane most like the Australian-born, followed by the New Zealand-born. The three North European groups are less similar in their distribution, but the feature that stands out is the relative segregation of the South European immigrants. For these three groups, well over half their population would have to be relocated to produce a residential pattern mirroring that of the Australian-born residents of Brisbane. This clearly suggests a spatial patterning of groups of immigrants that reflects their respective cultural and ethnic similarities with native Australians.

Similar studies have been carried out for nineteenth-

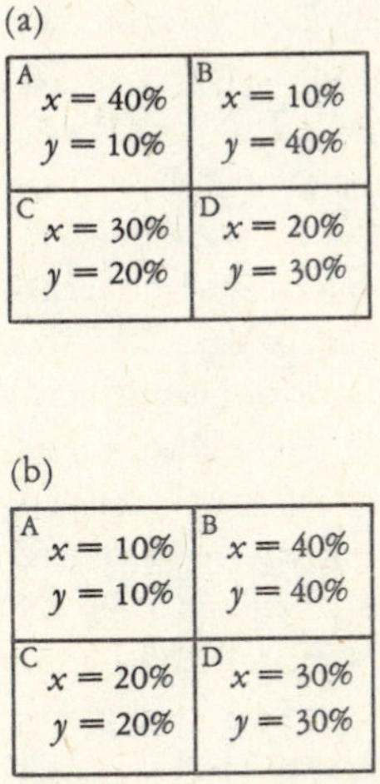

(a)

A: $x = 40\%$, $y = 10\%$	B: $x = 10\%$, $y = 40\%$
C: $x = 30\%$, $y = 20\%$	D: $x = 20\%$, $y = 30\%$

(b)

A: $x = 10\%$, $y = 10\%$	B: $x = 40\%$, $y = 40\%$
C: $x = 20\%$, $y = 20\%$	D: $x = 30\%$, $y = 30\%$

Figure 5.5 (a) Partial segregation, and (b) total integration.

Table 5.2 Dissimilarity Between Australian-born and Overseas-born Residents of Brisbane (1961)

Immigrant birthplace	*Dissimilarity index* (%)
United Kingdom	10
New Zealand	32
Germany	38
Netherlands	43
Poland	48
Italy	60
Greece	68
Yugoslavia	77

Source: Adapted from D.W.G. Timms. *The Urban Mosaic.* London: Cambridge University Press, 1969, Table 3.3, p. 103.

century immigrants to American cities that suggest an equivalent process of assimilation based on cultural difference. Table 5.3 shows a matrix of dissimilarity indexes between various groups making up Boston's population in 1880. Each residential distribution is compared to every other, and because the dissimilarities between two groups is the same whichever is considered first, we require only half a matrix to portray this information. Column 1 shows dissimilarities with the native population and is largely the equivalent of Table 5.2. Here we find the English, Scottish, and Irish immigrants distributed very much like the natives. Other Northwest European immigrants are the next most similar group, followed by the two East European countries, Poland and Russia, with the only South European group at this time, the Italians, being the most dissimilar from the natives. In fact, the Italians are more segregated from the natives than black residents of Boston! With this matrix of dissimilarities, we can take the discussion a little bit further. It should not be thought that because the Italian and the black populations are both segregated from the native whites they occupy the same residential areas. In fact, the situation is quite the opposite. The second D_A value on the bottom row shows that blacks and the Italian-born were the two groups most segregated from one another.

Limitations of the Lorenz Curve

Our examples illustrate that the D_A measure is a useful, flexible tool for spatial analysis. It has notable limitations and disadvantages as a general measure of areal association, however. First among these are its quite strict data requirements. Data for both x and y have to be frequencies or some other direct measure of an area. The variable cannot include negative numbers and must be allocable to a discrete area. These conditions are necessary because of the percentaging and additive cumulative operations. Area and population totals both qualify for analysis with a Lorenz curve. But the same cannot be said for other common variables such as population density (an indirect measure that is nonadditive) and population change (an indirect measure that can take negative values). The most important class of variable excluded from Lorenz curve study are the continuous variables that are particularly numerous in physical geography—altitude, rainfall, temperature, and so on. We cannot conceive of the notion of total altitude or temperature over a region that we can divide among areal units. We can imagine total rainfall over a region and its being apportioned among the areas where it falls, but in fact we collect rainfall data at points so that we can describe it as an areally continuous variable rather than an areally discrete variable. In none of these cases can we measure the areal association of the variable with another variable using a Lorenz curve.

The fact that the Lorenz curve deals with values for discrete areas within a region has several further implications. We illustrate these by continuing our discussion of residential segregation. Table 5.4 shows native white–black D_A values for a selection of northern cities in the first half of the twentieth century. The overall impression given by this table is one of rapidly increasing racial segregation until 1930 followed by stabilization of segregation at very high levels. Segregation seems to have increased at about the same time as the great influx of black migrants into northern cities between 1910 and 1930. This seems at first sight a quite reasonable conclusion. We should notice, however, that the table indicates that the 1910 and

Table 5.3 Residential Segregation Between Selected Populations of Boston (1880)

	1	2	3	4	5	6	7	8	9	10	11
1 Native white											
2 Black	51										
3 England	13	50									
4 Scotland	12	51	7								
5 Ireland	15	54	16	17							
6 Germany	31	54	28	31	33						
7 Norway	40	61	35	36	38	45					
8 Poland	62	54	62	61	56	64	57				
9 Russia	54	58	54	54	50	54	49	39			
10 Sweden	27	50	23	25	29	32	27	55	44		
11 Italy	74	79	74	74	69	78	66	48	34	65	

Source: S. Lieberson. *Ethnic Patterns in American Cities,* Table 19, p. 79. Copyright © 1963 by The Free Press, a division of Macmillan Publishing Co., Inc.

1920 figures are computed from ward data whereas figures for 1930 and 1950 are from census tract data. These indices are based on different areal frameworks. Let's consider what effect this has on the D_A values.

The possible effects of change in areal unit boundaries are illustrated in Figure 5.6a. The rectangle represents a hypothetical city divided into four census tracts with evenly distributed population, 25% of whom are black and totally segregated in the shaded area. For the base units in the first diagram in Figure 5.6a, this total segregation is reflected in a D_A value of 100% between the blacks and the rest of the population. In Figure 5.6b, however, the equally segregated blacks overlap the four census tracts and the resulting D_A value is 0, representing total integration. The value of D_A obviously depends on the relation between the distribution of the black population and the location of the base area boundaries. We can thus arrange the base area boundaries to produce any value of D_A in the simple hypothetical city. This clearly makes the Lorenz curve a fairly inadequate tool for spatial analysis. In practice, however, this problem is not as serious as it seems at first. The example in Figure 5.6a most closely represents the real situation of black segregation because its base areas are ethnically homogeneous. In several metropolitan areas, social scientists have helped define such socially homogeneous census tracts in an effort to overcome the general problem of archival data and their arbitrary areal frameworks (discussed in Chapter Two). In these cases, D_A values are particularly sensitive measures of

Table 5.4 Residential Segregation Between Blacks and Native Whites for Ten Northern Cities (1910–1950)

City	*Wards*		*Census tracts*	
	1910	*1920*	*1930*	*1950*
Boston	64	65	78	80
Buffalo	63	72	81	83
Chicago	67	76	85	80
Cincinnati	47	57	73	81
Cleveland	61	70	83	87
Columbus	32	44	63	70
Philadelphia	46	48	63	74
Pittsburgh	44	43	61	69
St. Louis	54	62	82	85
Syracuse	64	65	87	86

Source: S. Lieberson. *Ethnic Patterns in American Cities,* Table 38, p. 122. Copyright © 1963 by The Free Press, a division of Macmillan Publishing Co., Inc.

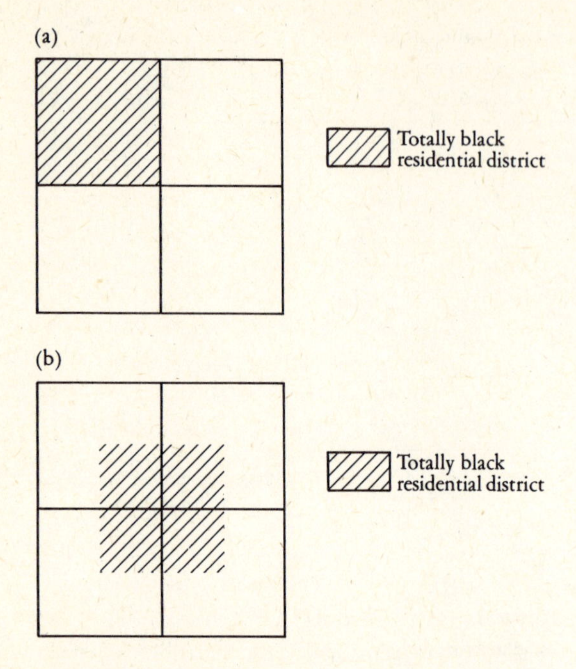

Figure 5.6 Segregation measurement and the location of boundaries.

segregation. Even with arbitrary base units, however, the contrasting results depicted in Figure 5.6 do not occur where there is a large number of base units and the ethnic group is not a very small minority. In such cases, changes in the boundaries of the arbitrary areas do not affect greatly the relative sizes of the D_A values, although direct comparisons between two D_A values derived from different areal frameworks are clearly not permissible.

A much more important effect on results of a Lorenz curve occurs with a change in the scale of analysis. When a relocation of base area boundaries results in a different number of base areas, we are forced to change our scale of analysis. When this occurs, there are easily predictable *systematic* influences on the values of D_A. Consider Figure 5.7, in which the same percentage data are arranged for base areas at two different scales. For four base-units, the D_A between x and y is 50%; for two base-units, D_A is only 15 percent. We can expect generally lower D_A values with fewer base units. This is because, with larger base units, there are less likely to be examples of areal units that have no or few members of one or the other group in them. Any attempt to monitor changes in levels of segregation when changes in scale of analysis are involved produces results as likely to reflect the changes in base areas as the settlement pattern of the ethnic groups.

We have shown clearly that the Lorenz curve is highly dependent on the base units we employ. We can state that the method is generally sensitive to relations within base areas so that a change in base areas producing a change in *intra-area* relations produces a new set of results. It follows from this conclusion that the method is insensitive to the arrangement of distributions *among* base areas. D_A will not distinguish between the two contrasting spatial patterns in Figure 5.8, because only distributions within each areal unit are considered. The researcher who is interested in such patterns must employ techniques of spatial autocorrelation such as those described in Chapter Four. This is, therefore, another limitation of the Lorenz curve.

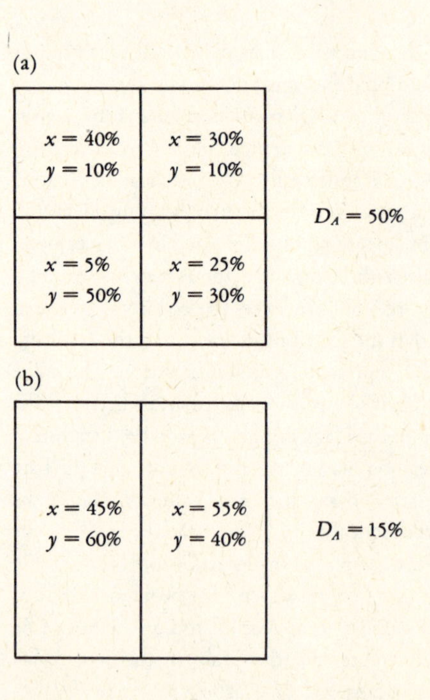

Figure 5.7 Segregation measurement and scale.

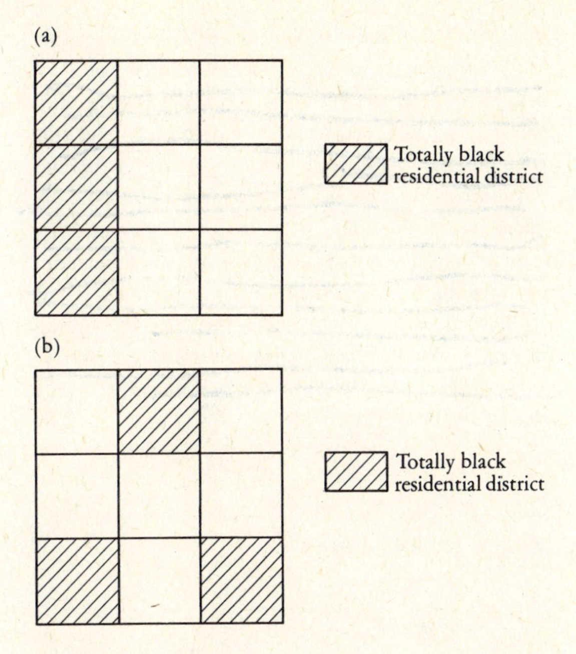

Figure 5.8 Segregation measurement and spatial patterns.

We should not be hasty to dismiss the method, however. The problems and limitations associated with areal units are common to several popular techniques of spatial analysis. They must be considered and borne in mind every time we use a Lorenz curve, but they should not automatically prevent its use. Given frequency data over a set of areal units, the Lorenz curve remains a very useful tool for describing the intra-area relations between various distributions. In many research problems, these intra-area relations are the relevant ones. This is certainly very often the case in analysis of spatial relations of ethnic group residential patterns in cities, as we have illustrated.

We can return now to Table 5.4 and consider the significance of the changing base areas to our interpretations of results. We recognize a change in areal units, but clearly this does not mean that there were no changes in black residential patterns in the cities between the two census periods. The stabilization of the level of segregation between 1930 and 1950 can be interpreted from the table, although the change of areal units means that we have no evidence of the rapid rise in segregation levels suggested previously. In order to show this, we require a consistent set of data over the same areal units for this period. Such data are not readily available for most northern cities, although fortunately we have one such set of data for Cleveland as a result of the census tract program. This gives the following D_A values:

1910	69.2
1920	71.6
1930	82.8
1940	85.8
1950	82.3

which clearly support our original conclusions as derived from Table 5.4. The lesson is not simply to dismiss the Lorenz curve approach because of its dependence on areal frameworks but to admit this limitation and interpret results accordingly.

SIMPLE CORRELATION AND REGRESSION

The Lorenz curve approach to areal association is useful in particular contexts, but as we have noticed there are many situations in which it is not applicable. At the beginning of this chapter, we introduced the idea of areal association with a suggested relation between wheat yields and average annual rainfall. None of the techniques considered above can tackle this areal problem adequately. We can attempt a measure of areal association between yield and rainfall by using the statistical concept of correlation. Furthermore, this can be generalized beyond simple questions of *degree* of relation to consider the *form* of a relation in regression analysis. In this section, we consider correlation and regression analysis in terms of pairs of variables or maps. Such analyses are usually referred to as *bivariate*. This approach can be extended to let us consider more than two variables or maps simultaneously. Such *multivariate* analyses are our concern in the next section.

The Concept of Correlation

The statistical concept of correlation is widely used in many sciences. It is described in all basic statistical textbooks and in textbooks designed to introduce statistics in different discriplines. This textbook is no exception; we introduce the techniques of correlation as a general tool. Our distinctive contribution is an initial discussion and illustration of the application of this statistical approach to map comparison.

Coefficients of Correlation

There is not one but rather several coefficients of correlation. We concentrate on the most commonly used correlation coefficient, which requires interval level data, although we introduce a rank correlation coefficient for ordinal data. Statistical measures of correlation are based on the concept of covariance. As its name suggests, the concept is related very closely to variance as described in Chapter One. The essential difference is that variance is a statistic for describing the variation of a single variable but covariance is a statistic for describing the correspondence in the variation of two variables. Variance is based on *squares* of deviations from the mean, and covariance is based on *products* of two sets of deviations from the mean. The similarity between the two concepts can best be seen in their formulas:

$$\text{Variance} = \frac{\Sigma(x - \bar{x})(x - \bar{x})}{n}$$

$$\text{Covariance} = \frac{\Sigma(y - \bar{y})(x - \bar{x})}{n}$$

Covariance is the average value of the products of pairs of deviations from their respective means. As such, it is an *absolute* measure of the correspondence of the variation of two variables. If two deviations from their respective means share the same sign, the product is positive; if they differ in sign, the product is negative. Large *positive* covariances reflect a great deal of correspondence between the variations of the two variables whereas large *negative* covariances reflect the complete opposite. The actual *size* of covariance depends on the units of measurement of x and y. In order to make our measure of correspondence dimensionless, we standardize for measurement units by dividing the covariance by the product of the standard deviations of the two variables This results in a *correlation coefficient*, which is a *relative* measure of correspondence between two variables and is usually denoted r. Thus,

$$r = \frac{\text{Covariance}}{\sigma_x \sigma_y}$$

which is normally written

$$r = \frac{\Sigma(y - \bar{y})(x - \bar{x})}{\sqrt{\Sigma(y - \bar{y})^2 \, \Sigma(x - \bar{x})^2}}$$

This is the general formula for a correlation coefficient and it can be used with ordinal and interval data. With interval data, the full name of the coefficient is the *Pearson product moment correlation coefficient*. Because this is the most commonly used coefficient, it is often referred to simply as the correlation coefficient, denoted r. With ordinal data, the formula defines *Spearman's rank correlation coefficient*, which is denoted r_s. For ease of computation, the general formula is modified so that r_s is usually defined by

$$r_s = 1 - \frac{6 \, \Sigma d_i^2}{n^3 - n}$$

where d_i is the difference in rank between the ith pair of ranks. (Both coefficients are illustrated in Work Table 4.1.) Both r and r_s range in value from $+1$ to -1.

Let's consider what this range of possible r values represents. Because standard deviations are always positive, their products are likewise positive, so that the sign of the coefficient is derived directly from the covariance. A negative correlation coefficient indicates a tendency for positive deviations from the mean in one variable to be associated with negative deviations from the mean in the other. A negative coefficient, therefore, describes an *inverse relation* between two

indirect

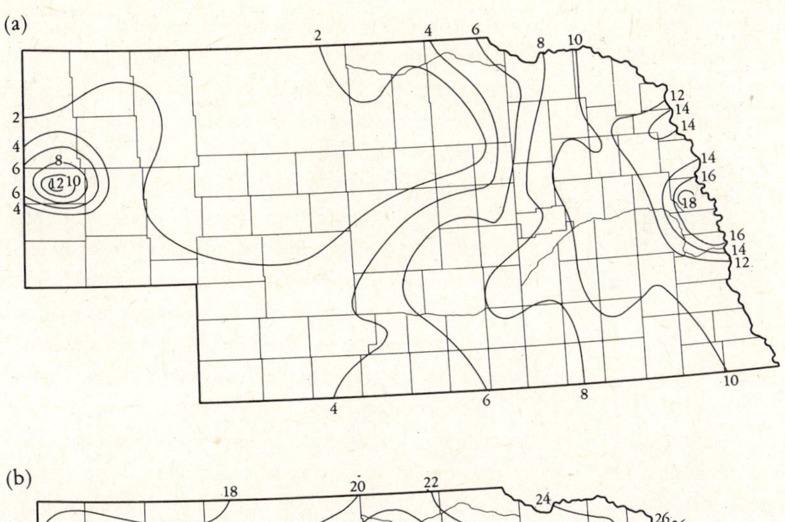

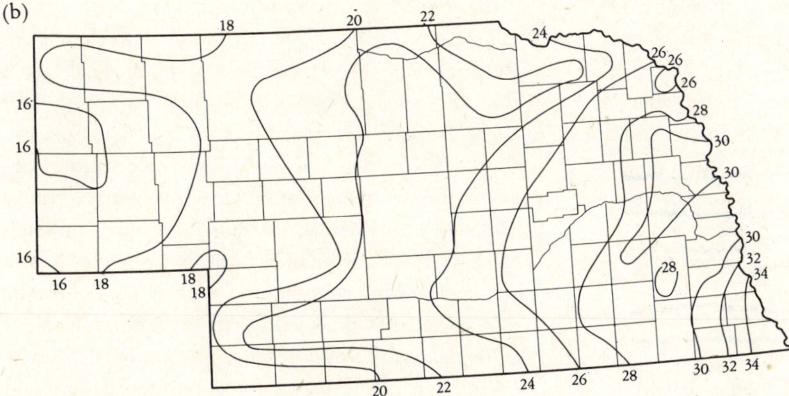

Figure 5.9 (a) Rural farm population (persons per square mile), and (b) average annual precipitation (inches) in Nebraska, 1950.
From A.H. Robinson and R.A. Bryson. A method for describing quantitatively the correspondence of geographic distributions. Reproduced by permission from the *Annals* of the Association of American Geographers, vol. 47, 1957.

variables. A typical example is that of frequency of frost and crop yield. The more frequently that frost occurs the lower the yield is. On the other hand, we might well expect that with higher rainfall the crop yield is higher. This is an example of a *direct relation* that should be reflected in a positive covariance and therefore in an r value of between 0 and $+1$. Little or no relation is indicated by an r value of near 0, which results in computing the covariance from the positive and negative products that largely cancel one another out when summed. This means simply that a positive deviation from the mean in one variable is just as likely to be associated with a negative deviation in the other variable as with another positive deviation.

Let's consider briefly what the two limitating cases $r = +1$ and $r = -1$ actually mean. When $r = \pm 1$, we usually say that we have *perfect* correlation. This is because these limiting cases occur only when the variations in the two variables coincide exactly. This is best illustrated by an example. If one variable has values of 2, 3, 4, and 6 and the other variable has values of 20, 30, 40, and 60, then as the first variable changes from 2 to 3 the second changes from 20 to 30, giving differences of 1 and 10, respectively. The two differences are repeated for the next pair of values: moving from 3 to 4 in the first variable corresponds to a change from 30 to 40 in the second variable. Notice that the change in variable values between objects is *proportionally* identical throughout. This is what is meant by perfect correlation. In our example, the relationship is direct, so that $r = +1$, whereas the variable values of 2, 3, 4, and 6, and 60, 40, 30, and 20 represent an indirect relationship, and $r = -1$. Another interpretation of perfect correlation is that the variation in one variable is wholly accounted for by the variation in the second variable. It is only a very short step between measuring perfect correlation and being able always to predict exactly one variable from the other. In our example for $r = +1$, it will not surprise anyone if a value of 5 for the first variable is found to correspond with a value 50 for the second variable. This idea of accounting for the variation of one variable by the variation of another can be extended to cases other than prefect correlation, and it is given by the *coefficient of determination*, defined simply as r^2. For direct and for inverse perfect correlation, $r^2 = 1$, indicating that all the variation has been accounted for. However, when $r = 0.9$, then $r^2 = 0.81$, which can be interpreted as showing that 81% of the variation is accounted for. This statistical interpretation of r^2 is clearly of interest to spatial analysts as a means of assessing the amount of variation of one spatial distribution that can be accounted for by another. We show why this interesting property obtains when we relate the correlation coefficient to regression analysis below.

Correlation Coefficients as Areal Association

The initial transfer of these statistical concepts and techniques to a spatial context is relatively straightforward in terms both of data and interpretation. The technique requires data in the form of pairs of measurements or ranks on some set of objects. In spatial analysis, the objects are locations defined as areal units such as census tracts, discrete spatial phenomena such as settlements, or sample points from continuous spatial surfaces. Measures from two variables over a set of locations produce data for computing a correlation coefficient. The resulting coefficient indicates the degree and direction of the relation between the two variables over the set of locations.

The application of correlation techniques in this way can best be illustrated with examples from geographical research. We have noticed previously that much of the reasoning of the traditional environmental school of geography can be considered as areal association. We can, of course, take the sort of data employed by this school of geography and answer traditional questions concerning man-environment relations by using correlation coefficients. Figure 5.9 shows the rural farm population density in Nebraska and the average annual precipitation. We can postulate a relation between these two variables by arguing along the lines that rainfall is related to crop yields which in turn relate to the rural population that an area of land can support. We expect a high positive correlation between the two maps. Some sort of relation is apparent visually, both variables declining westward, although the two distributions differ in the extreme west of the state where high population density appears not to be associated with increase in precipitation. By taking a systematic sample of 26 points from locations in Nebraska, we can obtain values for population density and precipitation for each sample location from the two isoline maps. The resulting pairs of values can then be fed into the general equation to produce a correlation coefficient of 0.8, largely confirming our original hypothesis. Notice, however, that we have measured the degree of the relation without having to enter into deterministic assertions about it. We have shown certainly that the distribution of farm population density is related to average annual precipitation, although the relation is not perfect. The coefficient of determination is only 0.64, indicating that more than a third of the variation in population density is still to be accounted for.

Most students of areal association have not been content to study the relation between two variables. Often they are concerned with a particular distribution of phenomena (the problem map) and the purpose of research is to find the determinants of that distribution

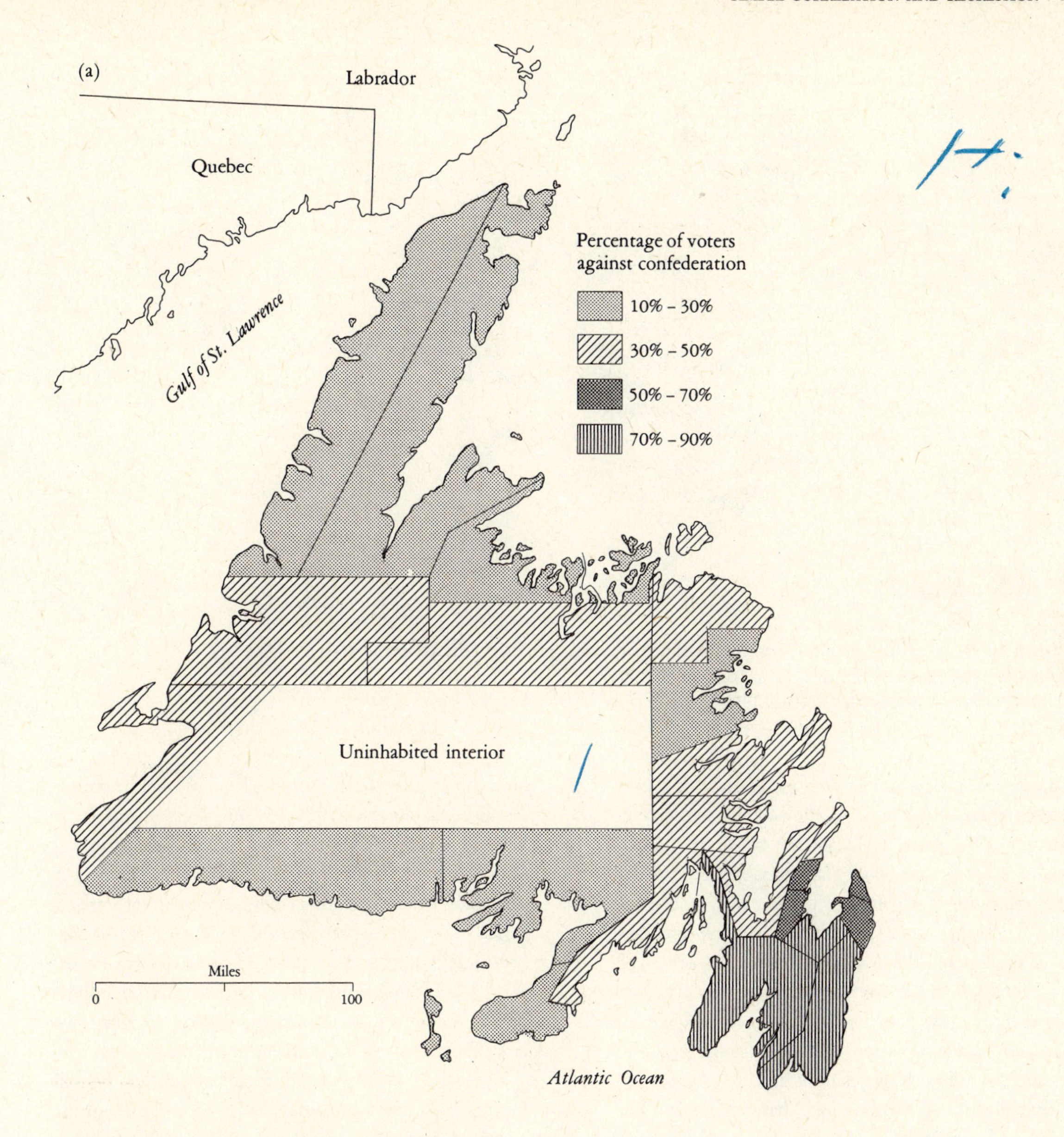

Figure 5.10 (a) Newfoundland Referendum, 1948.
From W. Warntz. A methodological consideration of some geographic aspects of the Newfoundland Referendum. *Canadian Geographer*, vol. I, 1955.

—that is, the variables that are areally associated with its distribution. Three typical examples of such initial problem maps are illustrated in Figure 5.10. Figure 5.10a shows the distribution of votes in favor of Newfoundland becoming an independent dominion in its referendum in 1948, when the majority voted in favor of federation with Canada; 5.10b shows the regional pattern of the percentage of substandard housing in

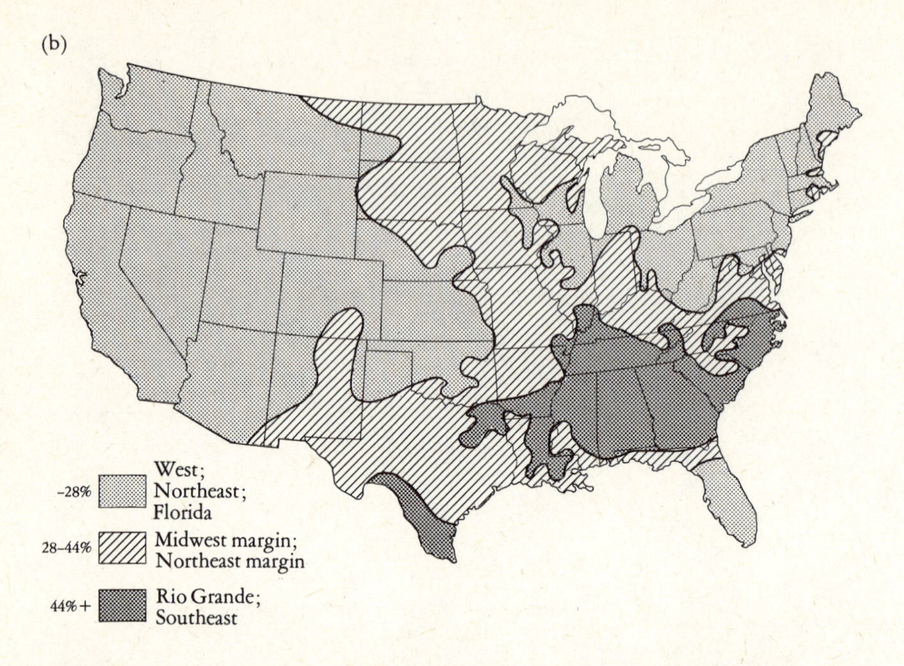

Figure 5.10 (b) Regional pattern of substandard urban housing in the United States, 1950. From George W. Hartman and John Hook. Substandard housing in the United States: A quantitative analysis. *Economic Geography* 32:98, April 1956.

urban places of the United States in 1950; 5.10c shows the percentage of urban population by departments in Peru for 1967. The question the researcher asked initially in each case is, "What variables are areally associated with the pattern shown by the problem map?" In each case, a correlation approach was employed to answer the question. It was hypothesized that the Newfoundland referendum vote is related to such phenomena as per capita income, religion, education, and spatial isolation. This latter concept was measured by using the population potential measure of relative position (described in Chapter Two). The complete list of variables and their Spearman rank correlations with the voting distribution are shown in Table 5.5. This table suggests clearly that the most important determinant of the voting pattern was spatial isolation whereas per capita income was largely unimportant. Table 5.5 also shows the variables chosen to correlate with substandard housing distribution in the United States and their resulting correlations. The high inverse correlations with median income and the high direct correlation with the level of overcrowding are predictable, although the much less important housing age, level of unemployment, and recent population growth are perhaps less predictable. The most surprising result from this table is the relatively strong *negative* correlation between the percentage of foreign-born inhabitants and substandard housing because we usually associate immigrants with poor housing conditions. We return to this particular result when we consider the notion of spurious correlation.

The study of Peruvian urbanization depicted in Figure 5.10c had a purpose slightly different from the previous two. The emphasis was on the regional pattern of development in Peru and the question was whether the level of urbanization within a department is a useful indicator of the level of development as measured by general typical indexes. Table 5.6 lists these indexes and gives Spearman rank and Pearson product moment correlation coefficients. Both coefficients give very high correlations for all indexes, suggesting that urbanization is associated closely with the general

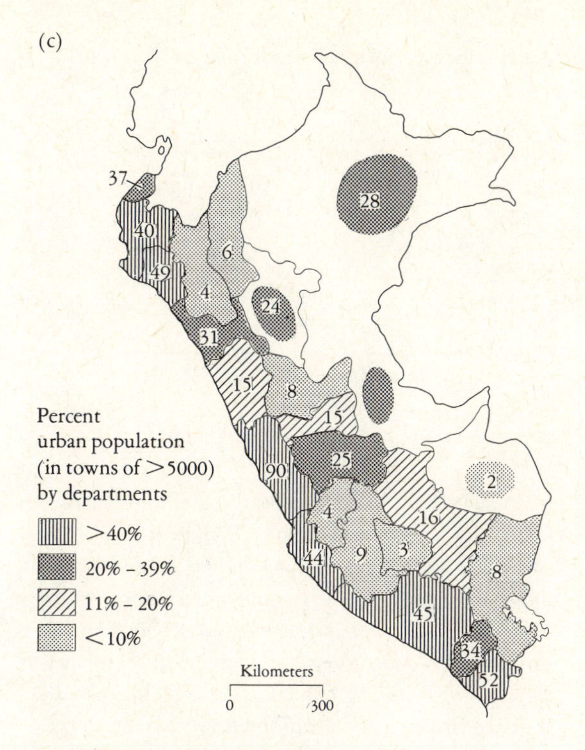

Figure 5.10 (c) Urbanization in Peru.
From C.T. Smith. Problems of regional development in Peru. *Geography* 53:267, 1968.

concept of development in Peru. For our specific purposes, we can use Table 5.6 to consider the relation between the two correlation coefficients. We have already noticed that they are based on the same general formula and that they differ in terms of level of measurement employed. When they are computed for the same data, they usually produce different values because they use different information from the data. Table 5.6 shows different coefficients for each variable. The magnitudes of the coefficient do not differ greatly, and we can generally assume that data for which a high r value has been computed likewise normally produces a similarly high r_s value. The reasons for favoring one coefficient over the other have to do with the assumptions of the two techniques and the ultimate purpose of research. The Pearson coefficient enjoys the advantages of interval data, so the results can be related to mathematical functions derived from the same data. We consider the development of correlation analysis in the context of mathematical functions using regression analysis before we consider the assumptions of the correlation and regression techniques together in the next section.

Most areal association studies have profited from an extension of the technique into the realm of the mathematical function using regression analysis. When the ultimate purpose of research is to produce findings concerned simply with degrees of relationship, this is not the case, and it matters little which coefficient is used as long as we do not compare the coefficient scales directly. One interesting application of areal association has occurred in studies of market orientation, using the potential concept. These studies emphasize the fact

Table 5.5 Determinants of Figures 5.10a and b

Newfoundland referendum voting Variable	r_s	*U.S. Substandard urban housing* Variable	r
Population potential	+ 0.72	Medium income	−0.78
Illiteracy rate	− 0.66	Overcrowding	0.72
Percent Roman Catholic	+ 0.63	Percent foreign-born population	−0.44
Percent employed in trade	+ 0.09	Average age of houses	0.30
Per capita income	+ 0.01	Percent unemployed	0.29
		Population change 1930–1950	0.12

Source: Adapted from W. Warntz. *A Methodological Consideration of Some Geographical Aspects of the Newfoundland Referendum on Confederation with Canada, 1948*, Table II, p. 46. Copyright 1955 by the Canadian Association of Geographers. And adapted from G.W. Hartman and J.C. Hook. Substandard urban housing. *Economic Geography* 32:102, 1956, Table III.

Table 5.6 Areal Associations Between Level of Urbanization and Various Indexes of Development in Peru (1961–1963)

Variable	r_s	r
Percent households with radios	0.916	0.979
Percent households with cooking facilities	0.857	0.957
Percent households with sewing machines	0.752	0.946
Number of industrial establishments per 1,000 population	0.778	0.897
Percent children older than four at school	0.838	0.867
Percent farm units using fertilizer	0.734	0.800

Source: Adapted from C.T. Smith. Problems of regional development in Peru. *Geography*, 53: 263, 1968, Table I.

that a potential surface represents the gross pattern of relative position of any commodity. We can produce potential maps of various commodities simply by interpreting the population element of the potential equation as production figures of the commodity. Figures 5.11a 5.11b, and 5.11c show maps of the gross relative positions of wheat, potato, and ice cream production in the United States. Figure 5.11d is a potential map of population weighted by income for the United States. This may be termed the gross economic potential surface and it represents the pattern of relative position of demand for production, the spatial distribution of the U.S. market. Correlations between production potential surfaces and the economic potential surface provide us with measures of *the degree of market orientation* of a commodity's production. Rank correlations using 48 state centers as sampling points produce the following levels of market orientation: wheat − 0.258, potatoes + 0.767, ice cream + 0.990. Wheat production, with its dependence on climatic factors, is not market oriented; ice cream production most certainly is. These results are clearly of no surprise to anybody familiar with American economic geography. They illustrate the quantitative measurement of the concept market orientation, and further analyses can be extended, in the case of manufacturing industry, to consider the relative pull of raw materials and market. Using 25 random sample points from potential surfaces of income (market), copper mining production (raw material), and copper smelting (production), we find that the correlation between smelting and income potential is − 0.63 whereas with mining it is + 0.98. This illustrates clearly the raw material orientation of copper smelting. Different results are found for petroleum refining in the United States, however, again using 25 sample points from potential surfaces, this time of income, oil production, and refined petroleum. The correlations with refined petroleum are 0.34 for production and 0.60 for income. It seems that the petroleum industry is more market oriented than raw material oriented. These studies are clearly special cases of areal association in which the purpose is to measure some prespecified concept. In this respect, they are similar to the studies of segregation and localization using the Lorenz curve approach. When research consists of a problem map that we are to explain, however, our strategy is to go beyond correlations to the related techniques of regression analysis.

Regression Analysis

All the techniques we have considered so far have attempted to answer simple "How much?" questions. We have been concerned with the amount of correspondence between maps and have been dealing with what are essentially measurement exercises. In Chapter One we identified the question "What are the relationships?" as a more sophisticated application of mathematics. We can answer such questions in the present context by employing regression analysis to complement and extend our correlation analysis. To do this we must first understand the mathematical concept of a function.

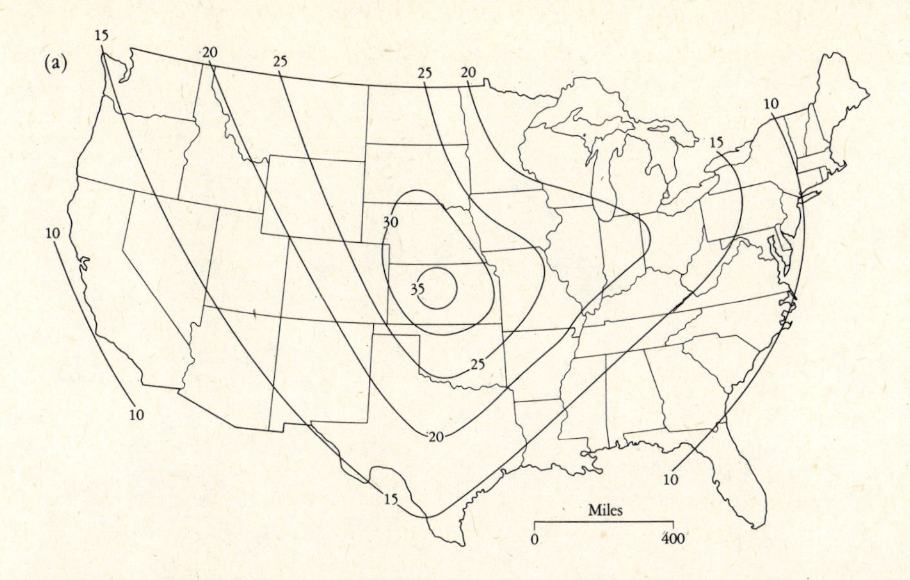

Figure 5.11 (a) Wheat production potential (1940–1949 average) in tens of millions of bushels per hundred miles.
From W. Warntz. *Toward a Geography of Price*. Philadelphia: University of Pennsylvania Press, 1959, pp. 67–69.

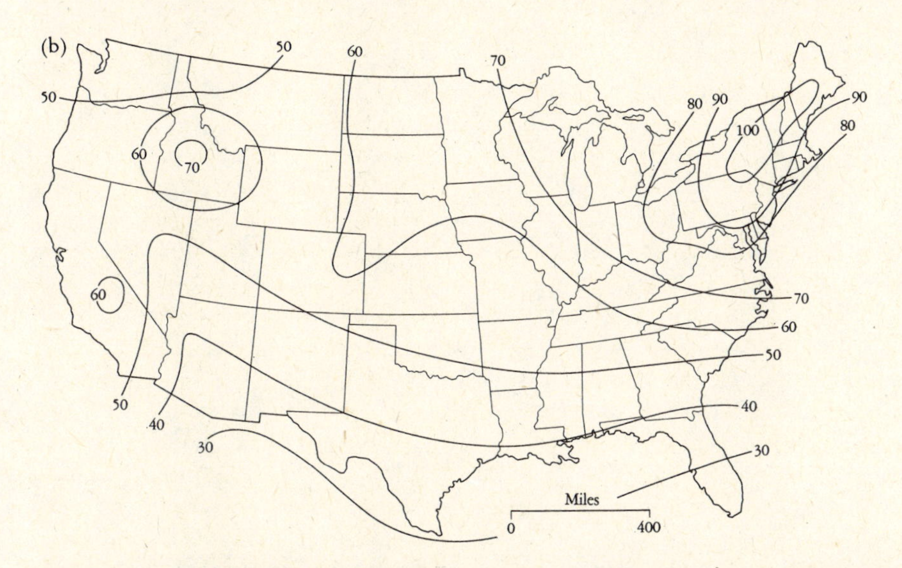

Figure 5.11 (b) Potato production potential (1940–1949 average) in millions of bushels per hundred miles.
From W. Warntz. *Toward a Geography of Price*. Philadelphia: University of Pennsylvania Press, 1959, pp. 67–69.

The Mathematical Concept of a Function

In Chapter One we used as an example of a mathematical model the equation $h = 250d$, in which h represents altitude and d distance from the coast of a small island dome (Figure 1.1). This expression is referred to sometimes as a function. We say in this case that altitude is a function of distance from the coast. Put another way, altitude depends on distance from the coast. We refer to altitude as the *dependent variable* and distance from

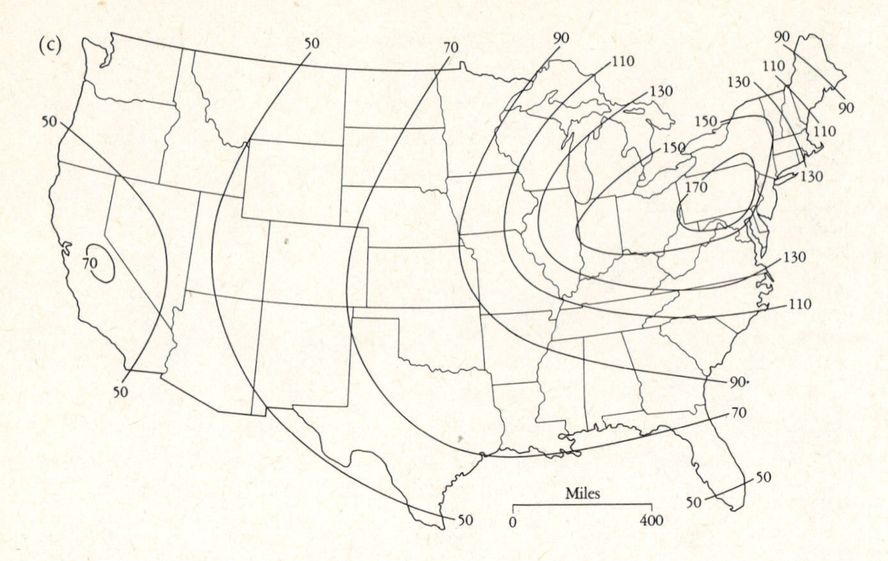

Figure 5.11 (c) Ice cream production potential (1944–1948 average) in millions of gallons per hundred miles.
From W. Warntz. Measuring spatial association. *Journal of the American Statistical Association* 51(276):602, 1956.

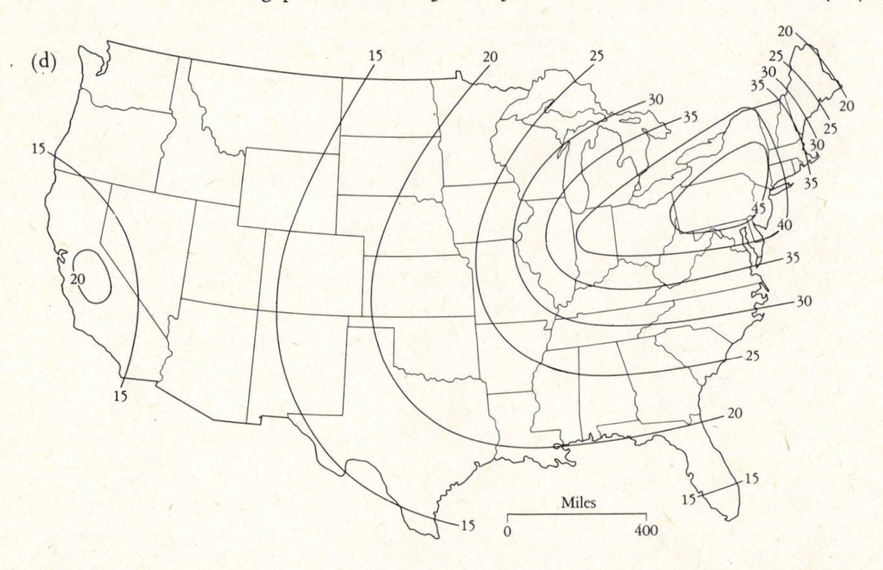

Figure 5.11 (d) Annual gross economic potential (1940–1949 average) in billions of dollars per hundred miles.
From W. Warntz. *Toward a Geography of Price*. Philadelphia: University of Pennsylvania Press, 1959, pp. 67–69.

the coast as the *independent variable*. Let's consider how the mathematical model from Figure 1.1 might have been derived.

Consider the following data from four locations on the island dome: $h = 0$ feet and $d = 0$ miles; $h = 250$ feet and $d = 1$ mile; $h = 500$ feet and $d = 2$ miles; $h = 750$ feet and $d = 3$ miles. These four observations can be plotted on graph paper by calibrating the vertical axis in terms of the dependent variable and the horizontal axis in terms of the independent variable (Figure 5.12a). Notice that the points seem to lie on a straight line. If we draw this line carefully, we produce

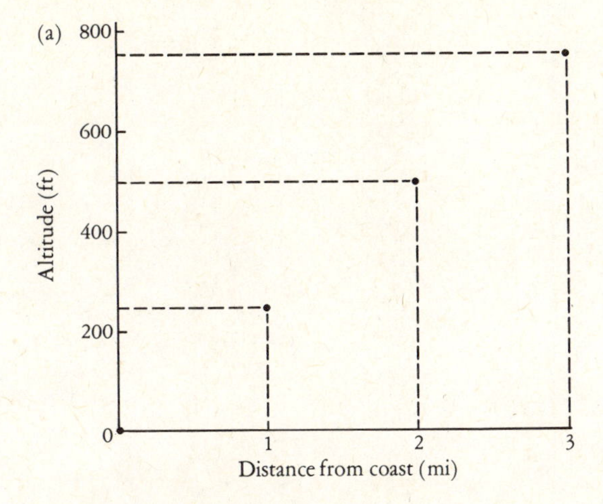

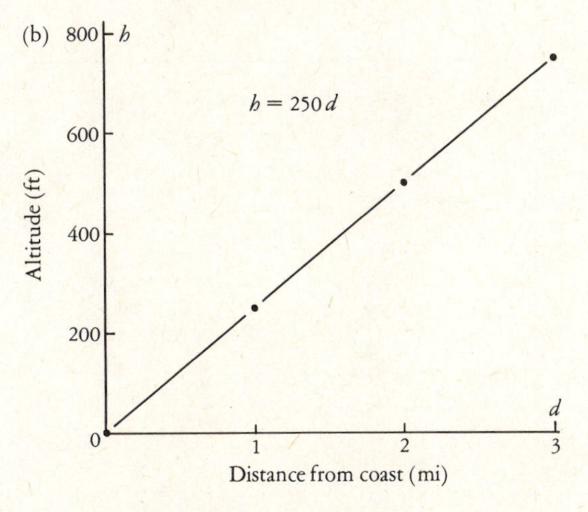

Figure 5.12 The concept of a function.

the situation depicted in Fgure 5.12b. There are two important features of this line. It goes through the origin of the graph, the point where both variables are 0. This means simply that at the coast (where $d = 0$) the altitude is 0($h = 0$). Also, the gradient or slope of the line is 250; the line increases 250 units upward (250 feet) for each 1 unit to the side (1 mile). These two pieces of information specify this line exactly; there is no other possible line that passes through the origin of a graph with a gradient of 250.

Can we predict altitudes at different distances from the coast from these two pieces of information without referring back to the graph? Let's see what altitude we expect at 1 mile from the coast. Because the altitude is 0 at the coast and rises 250 feet for each mile, we predict a height of 250 feet at 1 mile from the coast. Similarly, at 2 miles from the coast the predicted altitude is $(250 \cdot 2) = 500$ feet. The arithmetic operation we carry out in the argument is as follows:

$$h = 0 + 250d$$

which reads: the height at a location is 0 feet (the height at the coast) plus 250 (the gradient) times the distance of the location from the coast. This equation is a mathematical expression of the line in Figure 5.12b. Because it defines a straight line, the equation is termed a *linear function*. Let's look at its general structure.

Any straight line on a graph can be described by a linear function. In the terminology of Chapter One, the straight line is a concrete model of the mathematical function. Linear functions depend on two pieces of information relating to their location on the graph (where they cross the vertical axis) and their gradient. These two properties of a line are usually denoted a and b. In our example, $a = 0$ and $b = 250$. In general, linear functions can be written

$$y = \pm a \pm bx$$

y and x are any two variables replacing the specific h and d variables above, and a and b are the two constants. Several examples of straight lines and their equations are shown in Figure 5.13. Figures 5.13a and 5.13b illustrate different gradients ($b = 0.5$ and $b = 1.5$) and Figures 5.13a and 5.13c illustrate different y intercepts ($a = +1$ and $a = -1$). Figure 5.13d illustrates a negative gradient ($b = -1.5$) in which an increase in the x variable is reflected by a *decrease* in the y variable.

Of what use are these functions to our discussion? We have a potential link between any set of empirical data, which we can plot on a graph, and a simple mathematical model embodied in the linear function. The link is the concrete model of the function—the straight line on the graph. In Figure 5.12a, we have only four observations, although the representation of the function $h = 250d$ ($a = 0$ can, of course, be omitted) in Figure 5.13b fills in the gaps in this data and allows for predictions of height at other distances from the coast. Although we have no empirical observation at half a mile from the coast, we can predict that, when

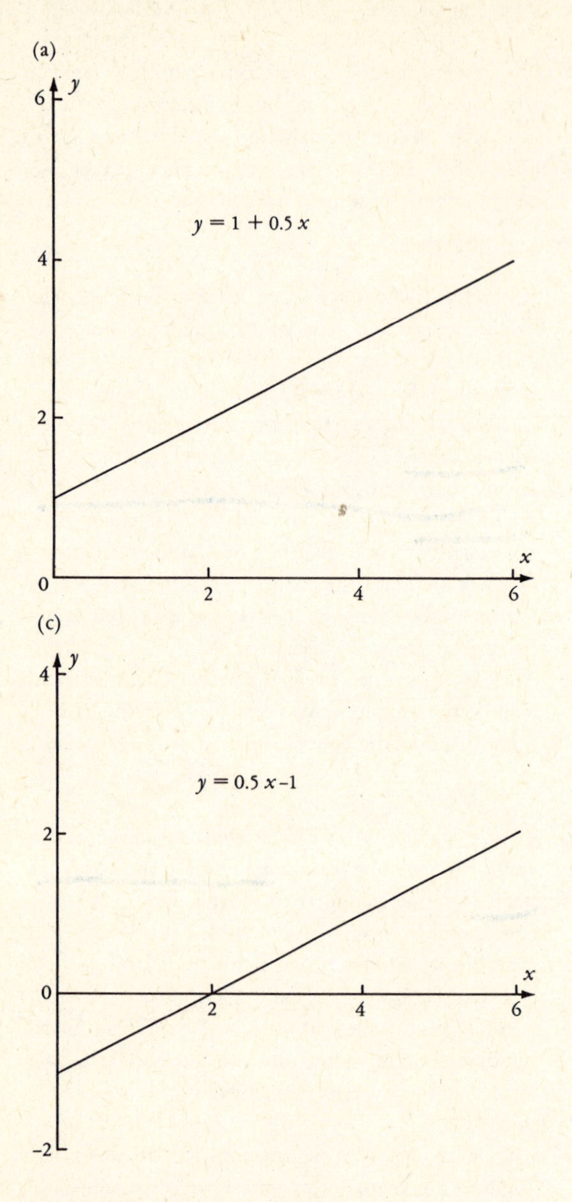

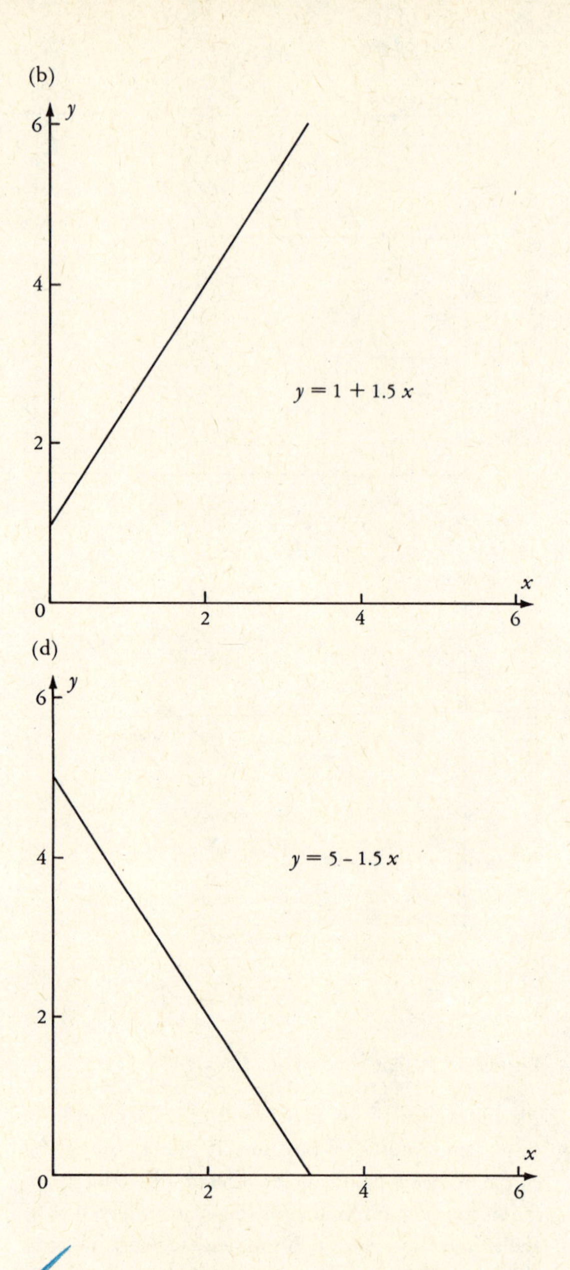

Figure 5.13 Four linear functions.

$d = 0.5$, $h = 250 \cdot 0.5 = 125$ feet. Similar predictions can be made for other distances from the coast. We have a very simple yet very powerful description of the island dome. The purpose of regression analysis is to produce such mathematical models. This is what we mean by going beyond correlation measures of degree of relationship and specifying the relationship *form*.

Fitting a Regression Line

In Figure 5.12a, the four observation points all lie neatly on a straight line. Drawing the appropriate line on the graph is a relatively simple matter. Regression analysis is designed specifically for situations in which observations do not fall conveniently on a straight line.

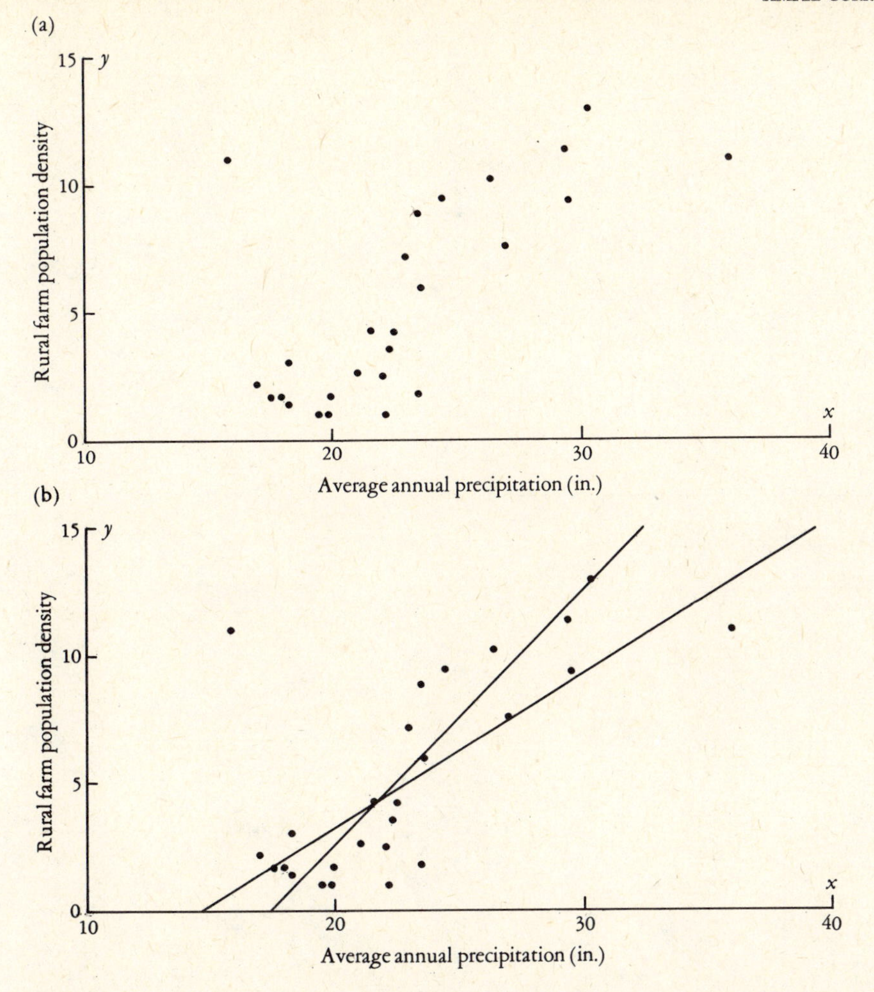

Figure 5.14 The rural farm population—average annual precipitation relationship for Nebraska: (a) data plot, (b) alternative linear trends, (c) measuring the degree of fitness—vertical deviations, and (d) the least squares regression line.

From A.H. Robinson and R.A. Bryson. Reproduced by permission from the *Annals* of the Association of American Geographers, vol. 47, 1957.

Figure 5.14a shows the data points for the Nebraska areal association example (Figure 5.9) when plotted on graph paper. This is a fairly typical set of data of a sort that any social scientist has to deal with. We can see that there is some sort of linear relation between x and y suggested by the data, although the exact nature of the function is obscured by the scatter of data points. This scatter of points is in part the result of the fact that the social scientist cannot control experiments in laboratory conditions as other scientists can do but must observe the behavior of variables in the complex real world. For the geographer, the laboratory is typically some study area such as Nebraska in which variables can be *observed* but not *controlled.* The fact remains that our analysis of areal associations is considerably sharpened when we can specify a mathematical function between two variables. Unfortunately, a freehand attempt to draw a curve through a scatter of data points such as those in Figure 5.14a will be highly subjective and unsatisfactory. Two possible straight line curves have been sketched on our scatter of points in Figure 5.14b. How can we say which is the

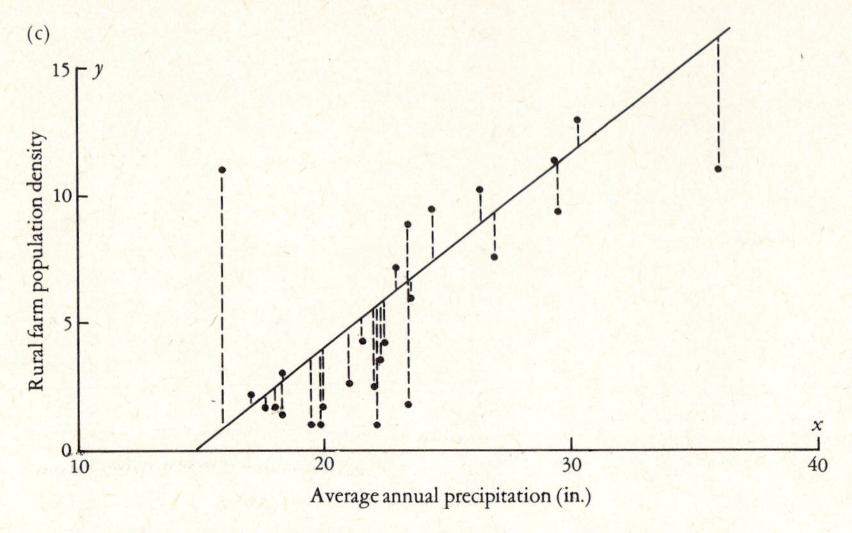

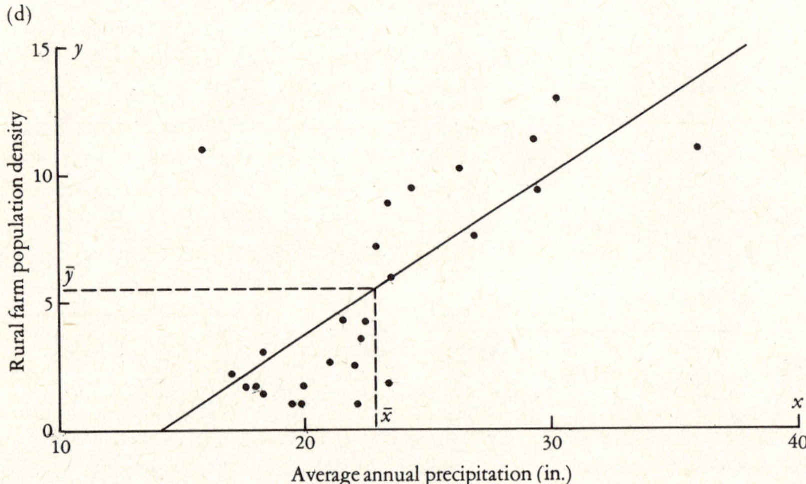

Figure 5.14 *(cont'd)*

correct curve? The freehand approach is clearly not acceptable where there is any appreciable scatter of points. This does not mean that we cannot fit to such data a single line that will be acceptable to other scientists. Regression analysis is a technique for fitting a curve to any scatter of points by providing us with an algebraic solution for estimating the parameters of the equation. We concern ourselves with the simple linear function.

If we are going to produce a single line to describe a scatter of data points, we must first of all decide the criteria we are to use to identify the best-fitting line. The obvious approach seems to be to concentrate on the differences or deviations between a suggested curve and the data points. Such deviations may be measured in terms of x (horizontal deviations) or y (vertical deviations). In Figure 5.14c we illustrate all the vertical deviations from a suggested line. Deviations below the curve are negative and those above the curve are positive. If we square these deviations and sum them, we produce a very simple index for assessing this particular curve. It is obvious that the smaller our index is, the better our curve fits the points. We term this index the *least squares criteria*. Our next step might be to try another line, measure the deviations, recalculate the least squares criteria, and see whether we can produce a better fit. We could continue until our final curve fitted the points as closely as we might think possible.

This procedure is obviously very slow and laborious and it is fortunate that we have an algebraic approach to avoid these geometric experiments. We simply use the so-called normal equations to estimate the parameters of the curve's equation. The resulting curve is sometimes termed the *least squares regression line*, because it minimizes the least squares criteria as described above. In the linear case, with the equation

$$y = \pm a \pm bx$$

in which y is the dependent variable and x is the independent variable, we have to estimate only two parameters, a and b, and these are calculated from the normal equations

$$na + b\Sigma x = \Sigma y$$

$$a\Sigma x + b\Sigma x^2 = \Sigma xy$$

These equations simplify to

$$b = \frac{\Sigma xy - \Sigma x \Sigma y/n}{\Sigma x^2 - (\Sigma x)^2/n}$$

$$a = \bar{y} - b\bar{x}$$

The first is used to calculate gradient b, in this context termed the *regression coefficient*, which is then substituted into the second equation to calculate a, the intercept on the y axis, known sometimes as the *base constant*. We must compute the two mean values, the sum of the cross products (Σxy), and the sum of the squares of the independent variable (Σx^2) in order to calibrate a linear function from a set of data. (A fully worked example is given in Work Table 5.2.)

In the Nebraska example, b is calibrated as 0.5826 and a is then found to be $-$ 7.939. We can, therefore, write out the linear function as

$$y' = 0.5826x - 7.939$$

y' is the estimated rural population density and x is average annual precipitation. We specify the dependent variable as y' rather than y because our equation does not predict rural population densities exactly; we have already decided that this pattern depends on factors other than rainfall. It produces estimates of densities predicted from levels of precipitation. Our equation enables us to estimate that, with 20 inches of average annual precipitation, we can expect a rural population density of

$$y' = 0.5286 \cdot 20 - 7.939 = 3.7$$

We can use this estimate to draw the line of the function on the graph. Because the function defines a straight line, we must find only two points on the line in order to draw it. We already know that the curve crosses the y axis at $y = -7.939$, because this is the value of the base constant. The point $x = 20$, $y = 3.7$ now defines a second point on the curve, which we can use to draw the line (Figure 5.14d). One property of the least squares regression line is that it passes through the point of the two mean values, and this can be used as a visual check on the graph. In the case of the Nebraska data, $\bar{y} = 5.34$ and $\bar{x} = 22.79$, which defines a point exactly on our regression line (Figure 5.14d).

Regression analysis provides us with a calibrated function for any set of data, however widely scattered. Even if visual inspection of a graph of data points suggests no functional relation, we are able to provide a least squares curve to fit the data. Different scatters of data clearly vary in how well they are fitted by their least squares line. In order to assess how well a calibrated function fits a set of data, we compute what is known as the *standard error of estimate*. For every observation, we have an observed value of y and an estimated value of y'. Differences between these values indicate how close the function actually predicts y. These differences are known as *residuals* and are usually denoted z.

$$z_i = y_i - y_i'$$

defines the residual for the ith observation. On the graph, z simply represents the vertical distance between each observation and the regression line. The standard error of estimate is simply a standard deviation for these residual values:

$$SEy' = \sqrt{\frac{\Sigma(y - y')^2}{n - 2}} \quad \text{or} \quad \sqrt{\frac{\Sigma z_i^2}{n - 2}}$$

Notice that we divide by $n - 2$ in these equations. This is the degrees of freedom that we used previously in inferential statistics. It is defined as the number of data points on the graph (n) minus the number of

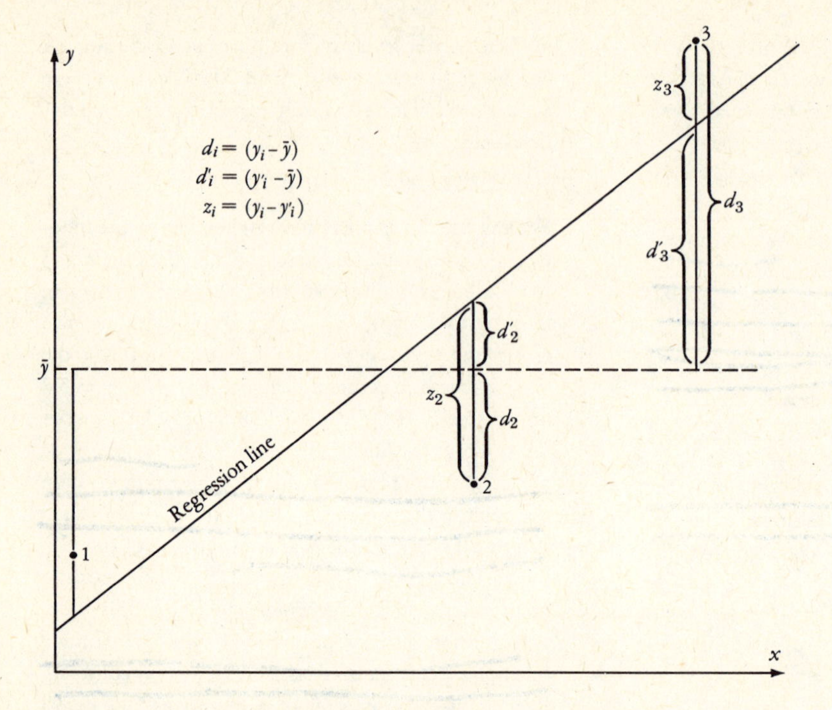

Figure 5.15 Deviations in a regression analysis.
From D.W. Maxfield. Spatial planning of school districts. Reproduced by permission from the *Annals* of the Association of American Geographers, vol. 62, 1972.

parameters used to specify the regression line (2: a and b). Degrees of freedom equals $n - 2$. This is always a simple concept to calibrate but it is usually difficult to understand fully. In this example, the meaning of degrees of freedom is more easily appreciated because of its direct geometrical interpretation. Suppose we have 2 data points on our graph. This is enough to define a straight line, and the regression line lies through them—that is, there are no degrees of freedom ($2 - 2 = 0$). Once we add a point, the regression line depends on 3 points, and there is now $3 - 2 = 1$ degree of freedom. Every time we fit a regression line on a graph, the degrees of freedom of the line is 2 less than the number of data points.

Three Aspects of a Relationship

The correlation coefficient, the regression coefficient, and the standard error of estimate tell us about three separate aspects of a relation between two variables or two maps. We take a new look at the correlation coefficient in the light of our regression analysis, so that we can compare directly these three complementary statistics.

Consider Figure 5.15. It shows a regression line through three points. It also shows the deviations in y that we consider in a correlation and regression analysis. The overall variance of y is measured as deviations of the observations from the mean value of y. These are shown as three lengths, specified d in the diagram. We have introduced the concept of a residual, which is the deviation of the observations from the predictions of the regression equation. These are shown as lengths, specified z in the diagram. We can also consider deviations between predicted values of y and the mean, indicated as lengths specified d' in the diagram. Let's concentrate on the third observation. Length d_3 is given by $y_3 - \bar{y}$, z_3 by $y_3 - y'_3$, and d'_3 by $y'_3 - \bar{y}$. Notice that

$$d_3 = d'_3 + z_3$$

This relation holds for the second observation, although in this case both d_2 and z_2 take negative values (from the above definitions) while d'_3 remains positive. In general, we can notice that

$$d_i = d'_i + z_i$$

Let's build upon this finding. If we square both sides of this equation, we obtain

$$d_i^2 = d_i'^2 + 2d'_i z_i + z_i^2$$

When we sum for all observations, the middle term is canceled out, so we are left with

$$\Sigma\, d_i^2 = \Sigma\, d_i'^2 + \Sigma\, z_i^2$$

The three individual terms in the equation are usually referred to as "sums of squares" terms and they can be viewed as gross measures of variation. Thus we can state that the gross variation about the mean of the dependent variable can be divided into two parts: that gross variation predicted by the regression equation, and that gross variation not predicted by the regression equation. These sums of squares terms are the basis of the three "relative" measures of variation: the standard deviation of the observed values (σ_y), the standard deviation of the predicted values ($\sigma_{y'}$), and the standard error of estimate (SEy'), respectively.

Let's return to the correlation coefficient. In terms of a regression analysis, the correlation coefficient can be defined as

$$r = \frac{\sigma_{y'}}{\sigma_y}$$

that is, the ratio of the standard deviation of the predicted values of y to the actual standard deviation. (The sign of r is the same as for b.) The correlation coefficient is intimately related to a least squares regression line. In fact, we are now in a position to justify our earlier assertion that r^2 measures the proportion of variation of y accounted for by x. From our previous equation, we can define the coefficient of determination as

$$r^2 = \frac{\text{Var}(y')}{\text{Var}(y)}$$

which is the variance predicted by the regression equation (accounted for by x) as a proportion of the total variance of y.

Let's consider the relation between the correlation coefficient and the regression coefficient and standard error of estimate in a regression analysis. Our purpose is to explain the variation of y. The correlation coefficient is simply a ratio of the variation accounted for by one variable and the total variation as measured by standard deviations. It is, thus, a pure ratio that is dimensionless. It measures the *importance* of x as a *determinant* of y. The standard error of estimate, on the other hand, is a direct measure of the actual values of y about their predicted values (that is, the regression line). It expresses the spread of observations about the regression line in units of y. It tells us how accurate an equation is or simply how good x is as a *predictor* of y. Finally, the regression coefficient measures the *amount* of change in y with a change in one unit of x. It is the gradient of the regression line and is measured in terms of units of y per unit of x. These three aspects of a relationship are compared and contrasted in Figure 5.16. In Figure 5.16a rainfall accounts for the highest proportion of variation in yields; in Figure 5.16b, the rate of change in yields per unit of rainfall is greatest; in Figure 5.16c, rainfall is the most accurate predictor of yields.

These three statistical expressions of a relationship extend our bivariate analysis about as far as we can take it in the present context. They are clearly not directly competitive with one another; they are complementary, each emphasizing a different aspect of a relation between two variables. We can best assess them in the light of different research designs. For instance, if the purpose of research is to predict values of y from x, then the important statistic is clearly the standard error of estimate. A high correlation or any regression coefficient is of little interest if there is a wide scatter of points about the regression line. It is of little consolation to people flooded out by an unpredicted high-water level to say that a correlation coefficient is a high 0.8. If the purpose is to explain the dependent variable (the problem map), then it is clear that coefficients of correlation and determination are particularly relevent, because they indicate directly how successful we have been in this task. The widespread use of correlation techniques in areal association

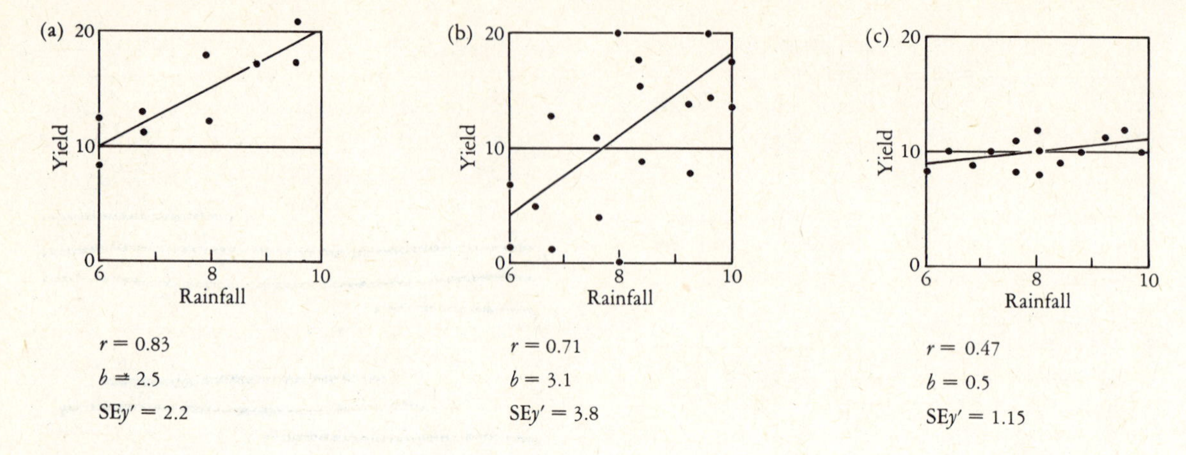

Figure 5.16 Three aspects of linear relationships.
From M. Ezekiel and K.A. Fox. *Methods of Correlation and Regression Analysis*, 3d ed. New York: John Wiley, 1959, Fig. 9–1, p. 149.

studies is justifiable. Many methodologists in the social sciences in recent years have argued that more emphasis be given to the regression coefficient, however. This is because of the view that the functional relation expressed in the regression coefficient is the element of analysis likely to be explicitly incorporated in any theory development.

There is one problem with regression analysis that is not shared by the correlation coefficient. Regression relationships are asymmetric; correlation relationships are symmetric. It does not matter whether x or y is designated dependent or independent; the correlation coefficient is the same. Specification of dependence is important in regression analysis, however. Thus, $r_{xy} = r_{yx}$ but $b_{xy} \neq b_{yx}$. In fact, there are two least squares regression lines for every scatter of data points on a graph—one that minimizes squared deviations of y and another that minimizes squared deviations of x (horizontal deviations versus vertical deviations in Figure 5.14c). Except in the special case of perfect correlation ($r = \pm 1$), where they coincide, one least squares equation cannot be derived from another by simple algebra. If we treat y as dependent and calibrate $y = 10 + 5x$, then $x = 0.2y - 10$ is *not* the least squares fit for deviations in terms of x. This lack of symmetry in regression analysis is not important when we can proclaim definite dependence in our model. In our Nebraska example, it is eminently reasonable to argue that population density depends on rainfall but not that rainfall depends on population density. In this case, there is only one regression line of interest, with population density as the dependent variable, so that no problem exists.

Social science is full of examples of two-way relationships in which dependence is not clearly identified, however. Suppose we find that alienation scores are related to economic status. We could argue that the alienation scores depend on economic status—people with less money are dissatisfied. We could also argue that people willing to work in society obtain their just rewards in terms of economic status and, hence, status depends on a lack of alienation. In this situation, it is not clear which variable is dependent. Our decision affects our final regression equations. The proper solution involves either sophisticated correlation analyses to attempt to determine dependence or a curve-fitting procedure that avoids this particular characteristic of regression analysis. Both approaches are strictly beyond the scope of this book, although reference to them can be found in the suggestions for further reading at the end of the chapter. We continue to assume that we have a given problem map specified as the dependent variable in the research formulation.

FURTHER CORRELATION AND REGRESSION

There are two quite distinct ways in which we extend our discussion of correlation and regression. We noticed in the preceding section that geographers are not usually interested in a single determinant of a problem map but consider several variables. Our technical competence thus far allows us only to consider pairs of variables so that each determinant is evaluated individually. It would seem a useful step forward to put all the determinants into one larger analysis. This is the task of multiple correlation and regression. The next step that we take concerns statistical and mathematical assumptions of the techniques and related problems in applications of this approach. We have concentrated on simply presenting techniques and have paid scant attention to some quite important issues. Because these relate to bivariate and multivariate analyses, we treat them as a whole in the last section of this chapter.

Multiple Regression and Correlation

The first problem that confronts researchers who wish to extend their study from a bivariate to a multivariate situation is simply: "What new variable should be added to the analysis?" Such decisions may be based on several criteria—intuition, underlying theory, or simple guesswork. Another criterion is purely empirical: it is the decision of what extra variable will best add to the variance already accounted for in the bivariate analysis. This approach involves analysis of the pattern of residuals.

Maps of Residuals

The residual values z_i for each observation are an indication of how well the calibrated function predicts the dependent variable for that particular observation. It can be of intrinsic interest to the geographer. As well as using the residuals to derive a summary measure, the standard error of estimate, we can concentrate on the pattern of individual residuals. We have a residual for each areal observation, and we can in fact map residuals to produce a new type of map. Figure 5.17 shows a map of residuals derived from the Nebraska areal association study reported above. The predicted values of rural population density for the 26 sample points are subtracted from the observed population densities. The resulting values are used to draw the isoline map in Figure 5.17. What exactly does this map illustrate? It shows areal variation in the degree of fit of the regression equation. Small residuals are a reflection of very close correspondence between the actual and the predicted values, and these predominate in the south of the state. There is a large area in the center of the state that has large negative residuals, where the actual densities are more than three persons per square mile smaller than the predicted densities. In the extreme west of the state, very high positive residuals are found, including a small area in which the regression equation predicts densities of more than ten persons per square mile smaller than those actually found. This particular observation, which seems independent of the linear trend in the other observations, can be identified readily as the largest residual on all the graphs in Figure 5.14. A residual map shows us clearly where a regression equation overpredicts (negative residuals), where it underpredicts (positive residuals), and where it predicts most accurately (near zero residuals).

Residuals are often standardized before being mapped. The absolute values of residuals are in units of the dependent variable, and they can be made dimensionless in the usual way by dividing by their standard deviation, which in this case is the standard error of estimate. We define a relative residual z_i^* as

$$z_i^* = \frac{z_i}{SEy'}$$

Notice that only the magnitude of the residuals is affected; the relative pattern remains the same. This is, of course, important because it is the pattern that is of interest. Figure 5.18 shows four patterns of standardized residuals for North Island, New Zealand. The initial areal association study concerned the relation

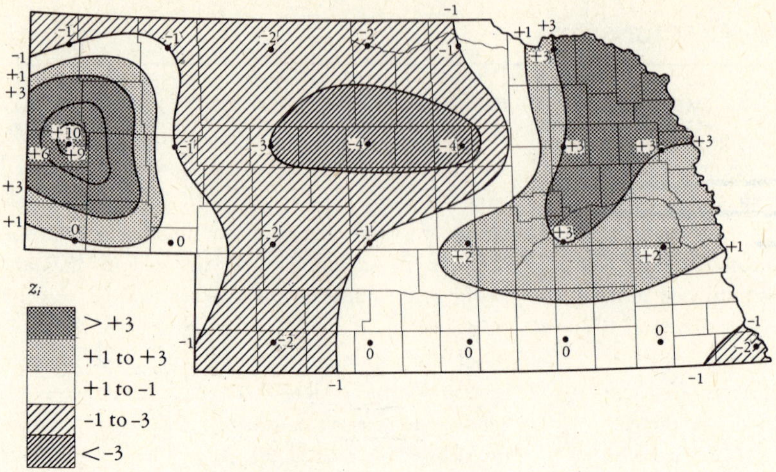

Figure 5.17 Residuals from the rural farm population—average annual precipitation relationship for Nebraska.
From A.H. Robinson and R.A. Bryson. Rainfall relationship for Nebraska. Reproduced by permission from the *Annals* of the Association of American Geographers, vol. 47, 1957.

between dairy cattle and pig farming. The relation is close because skim milk is fed to pigs. The level of areal association has been falling consistently in recent years; $r = 0.92$ in 1950, $r = 0.86$ in 1957, $r = 0.83$ in 1960, and, finally, $r = 0.77$ in 1964. Although the relation remains close, the correlations suggest changes in it. Because it was hypothesized that the distribution of pigs depends on the distribution of dairy cattle, regression equations were specified in this form and residuals from the calibrated equation were standardized. The larger residuals are plotted on the four maps in Figure 5.18. The pattern of fit between the regression equations is remarkably stable over the four time periods. In every case, the southern and eastern parts of the island are very accurately predicted by the equations, whereas the poorer predictions are concentrated largely in the northwest. These include positive residuals, for which there are more pigs than expected, and negative residuals, for which there are less than expected.

It seems that residual maps, absolute or relative, are easy to interpret. But how can they be of use in geographical research beyond merely describing the pattern of accuracy of regression predictions? Analysis of residuals can play an almost classically scientific role in geographical research. If residuals show areas in which regression is a poor predictor, it follows that these are the areas in which factors other than the independent variable particularly influence the dependent variable. The negative residuals in Figure 5.18 suggest areas in which the farms dispose of their dairy produce in such a way as to leave less available for pig rearing. We must investigate this phenomenon by field studies of these areas or consideration of other statistical data that might match the pattern of the residuals. In either case, it is hoped that analysis of the residuals will lead to new hypotheses concerning the distribution of the original dependent variable.

Generation of new hypotheses by analysis of residuals is common strategy in science. In fact, nineteenth-century British philosopher John Stuart Mill identified the method of residuals as one of the four basic scientific methods of investigation. The approach is cyclic. A first hypothesis is tested; the part not explained (the residual) is examined for a second hypothesis; this is tested and the part still not explained is examined, which produces a further hypothesis, and so on. This method can be accomplished within the general framework of regression analysis; in a spatial

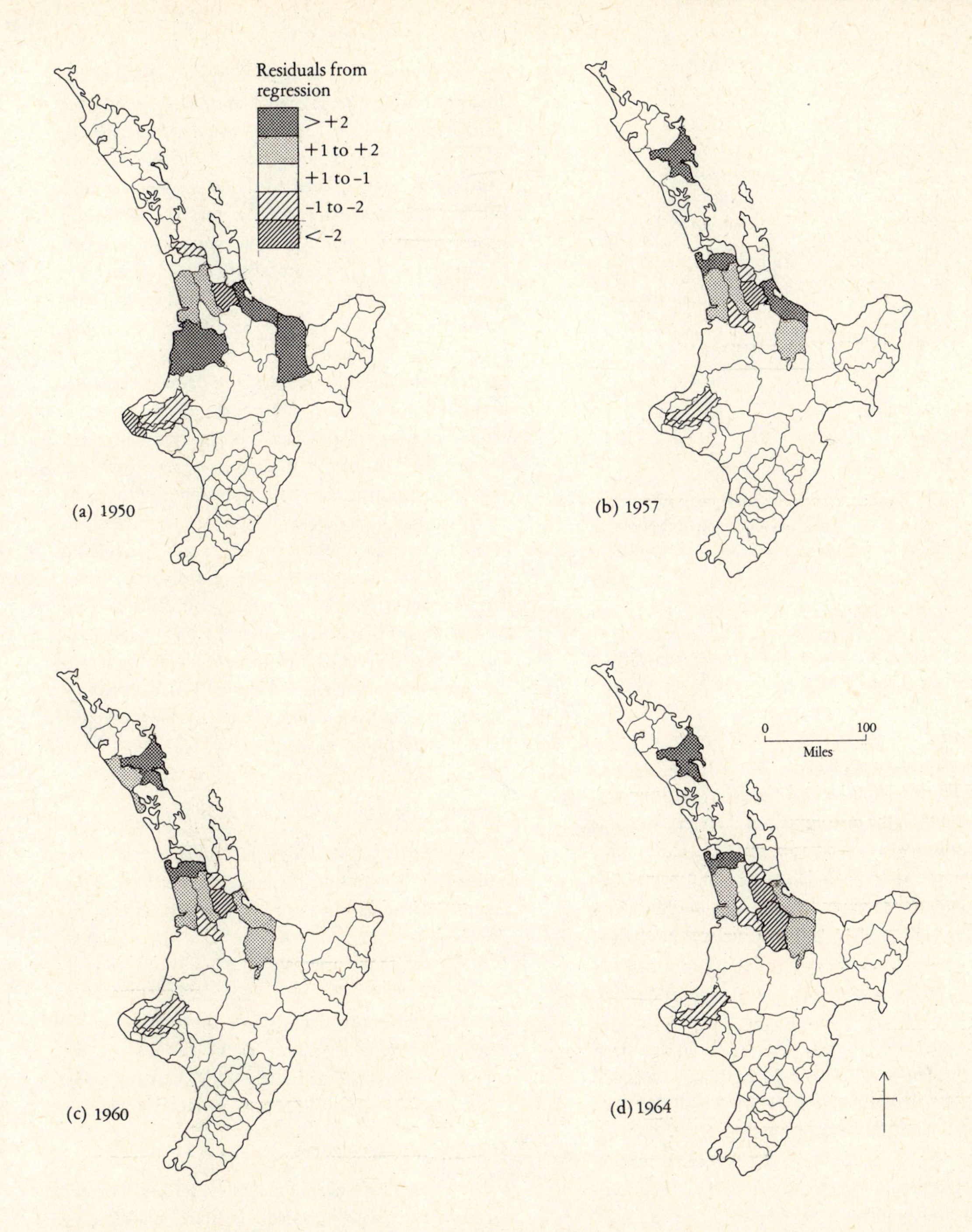

Figure 5.18 Standardized residuals for the dairy cattle–pig rearing relationship, North Island, New Zealand, 1950–1964. From W.A.V. Clark. The use of residuals from regression in geographical research. *New Zealand Geographer* 23:65, 1967.

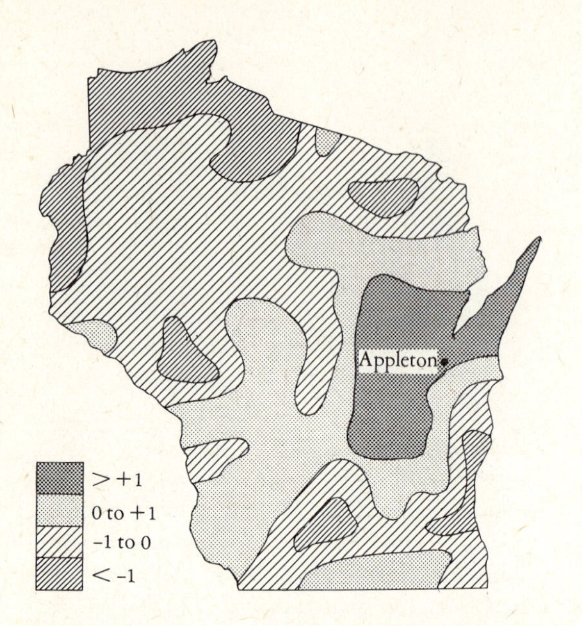

Figure 5.19 Standardized residuals from the McCarthy vote—rural population relationship, Wisconsin, 1952.
From E.N. Thomas. *Maps of Residuals from Regression.* Iowa City: University of Iowa Press, 1960.

context it uses maps of residuals. The classical example of residual strategy in geographical research is an electoral study conducted by Harold McCarty that is whimsically titled "McCarty on McCarthy." This study attempted to explain the areal distribution of percentage votes for Wisconsin Senator Joseph McCarthy in the 1952 election. The initial independent variable was the percentage of the county population that was rural. This produced a correlation of 0.58, indicating that barely more than a third of the variation in McCarthy's vote can be accounted for by rural population percentages. The unexplained variation can be mapped as standarized residuals (Figure 5.19). The location of McCarthy's hometown (Appleton) is also shown on this map. There seems to be a very clear relation between the distance from Appleton and the pattern of residuals; locations near Appleton are positive, indicating underprediction, and locations far away from Appleton are negative, indicating overprediction. This map of residuals has indicated a new hypothesis that the vote for McCarthy is inversely related to the distance from his hometown: the smaller the distance from Appleton is, the larger the vote for McCarthy is. We now have two independent variables that we postulate as being related to the dependent variable. When we correlate the county vote for McCarthy against distance from Appleton, we obtain a disappointingly low r value of 0.18. This means that distance from Appleton is not related particularly strongly to the *original problem map* of McCarthy vote percentages. We have hypothesized only that it is related to the unexplained McCarthy vote percentages after rural population has been considered, however. Thus, the low correlation between McCarthy vote and distance from Appleton is not inconsistent with our analysis of residuals. After all, Appleton is in the most urbanized half of the state. What the map of residuals has suggested is not a second bivariate relation involving a second pair of variables (vote and distance) but a three-variable situation, suggesting that distance from Appleton is important only after the rural population variable has been taken into account. Thus, we hypothesize a multivariate relation in which McCarthy vote depends on rural population *and* distance from Appleton in a single functional relation.

Fitting a Regression Plane and Beyond

Multiple regression analysis is like bivariate regression in that it is based on the mathematical concept of the function. Generally we can write any functional relationship as

$$x_1 = f(x_2, x_3, \cdots, x_n)$$

which reads simply that x_1 is a function of $n - 1$ independent variables numbered x_2 through x_n. We continue to be concerned solely with linear functions, so the general functional relation can be rewritten as

$$x_1 = \pm a \pm b_2x_2 \pm b_3x_3 \cdots \pm b_nx_n$$

What does this linear function represent? In the bivariate case, we illustrated the function's meaning by showing that it defines a straight line on graph paper. With more than two variables, we can no longer represent a linear function so simply.

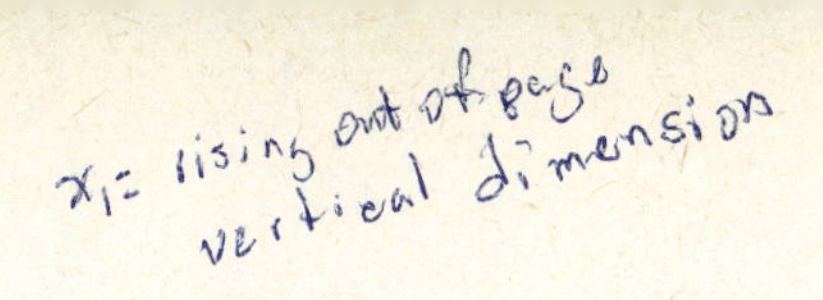

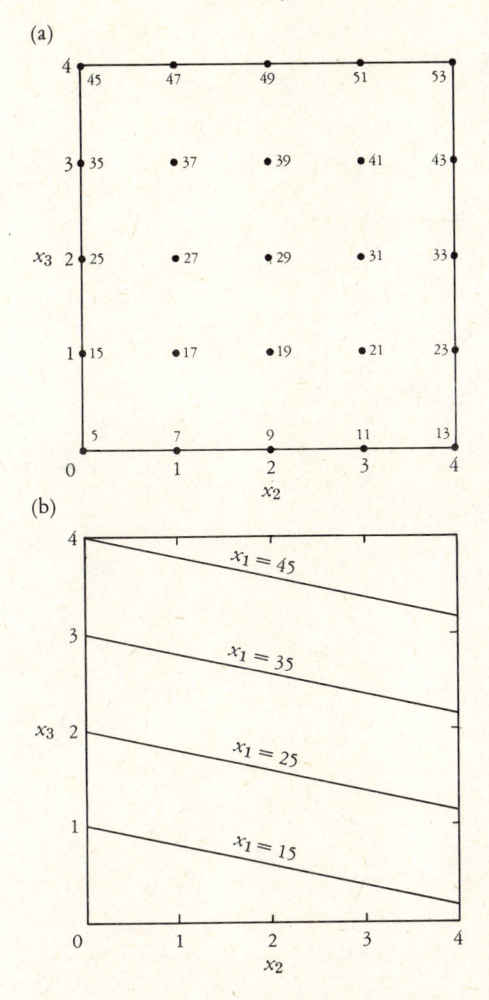

Figure 5.20 Constructing the linear function plane: $x_1 = 5 + 2x_2 + 10x_3$. (a) Values of x_1 for x_2 and x_3 values, (b) contour lines of x_1 values, and (c) the linear function plane.

Let's begin by considering a three-variable case:

$$x_1 = 5 + 2x_2 + 10x_3$$

If we define a space by its x_2 and x_3 axes, we can plot predictions of x_1 on this space. This is illustrated in Figure 5.20a for x_2 and x_3 between 0 and 4. Values of x_1 in this space vary from 5 when $x_2 = x_3 = 0$ to 53 when $x_2 = x_3 = 4$. From such plottings of data, we can construct a model of the surface of values after the manner of contour maps. When we carry out this process, we produce parallel straight-line contours as shown in Figure 5.20b. The pattern of isolines represents a plane in three dimensions (Figure 5.20c). We can say that generally a three-variable linear function defines a plane in a three-dimensional space. The particular plane that is defined depends, of course, on the parameters. The first constant (a) defines the height of the plane it as crosses the vertical x_1 axis where $x_2 = x_3 = 0$. In Figure 5.20c, this occurs at $x_1 = 5$. This constant is comparable directly to the y intercept in bivariate functions. The b constant of the bivariate equation is represented by two parameters. In this case, they define jointly the tilt of the plane. Individually they define gradients, as their bivariate equivalent does. In the case of b_2, this defines the gradient of the line where the function plane cuts the vertical plane at $x_3 = 0$. In Figure 5.20c, this gradient is 2 and relates to the constant multiplying x_2 in the equation. Similarly, b_3 defines the gradient of the line where the function plane cuts the vertical plane at $x_2 = 0$. In Figure 5.20c, $b_3 = 10$.

Three-variable linear functions are clearly very similar to simpler bivariate functions. In fact, the statistical concepts and methods of estimation we used in the last section can be transferred to calibrate multivariate functions. Given a set of data for three variables, we can estimate the parameters of the function

$$x_1 = \pm a \pm b_2 x_2 \pm b_3 x_3$$

from an extended set of normal equations. Because there are three parameters to estimate, we require three equations:

$$\Sigma x_1 = na + b_2 \Sigma x_2 + b_3 \Sigma x_3$$

$$\Sigma x_1 x_2 = a \Sigma x_2 + b_2 \Sigma x_2^2 + b_3 \Sigma x_2 x_3$$

$$\Sigma x_1 x_3 = a \Sigma x_3 + b_2 \Sigma x_2 x_3 + b_3 \Sigma x_3^2$$

which can be simplified to give

$$b_2 = \frac{(\Sigma(x_1 - \bar{x}_1)(x_2 - \bar{x}_2)\,\Sigma(x_3 - \bar{x}_3)^2)}{(\Sigma(x_2 - \bar{x}_2)^2\,\Sigma(x_3 - \bar{x}_3)^2)}$$

$$\frac{-(\Sigma(x_1 - \bar{x}_1)(x_3 - \bar{x}_3)\,\Sigma(x_2 - \bar{x}_2)(x_3 - \bar{x}_3))}{-(\Sigma(x_2 - \bar{x}_2)(x_3 - \bar{x}_3))^2}$$

(c)

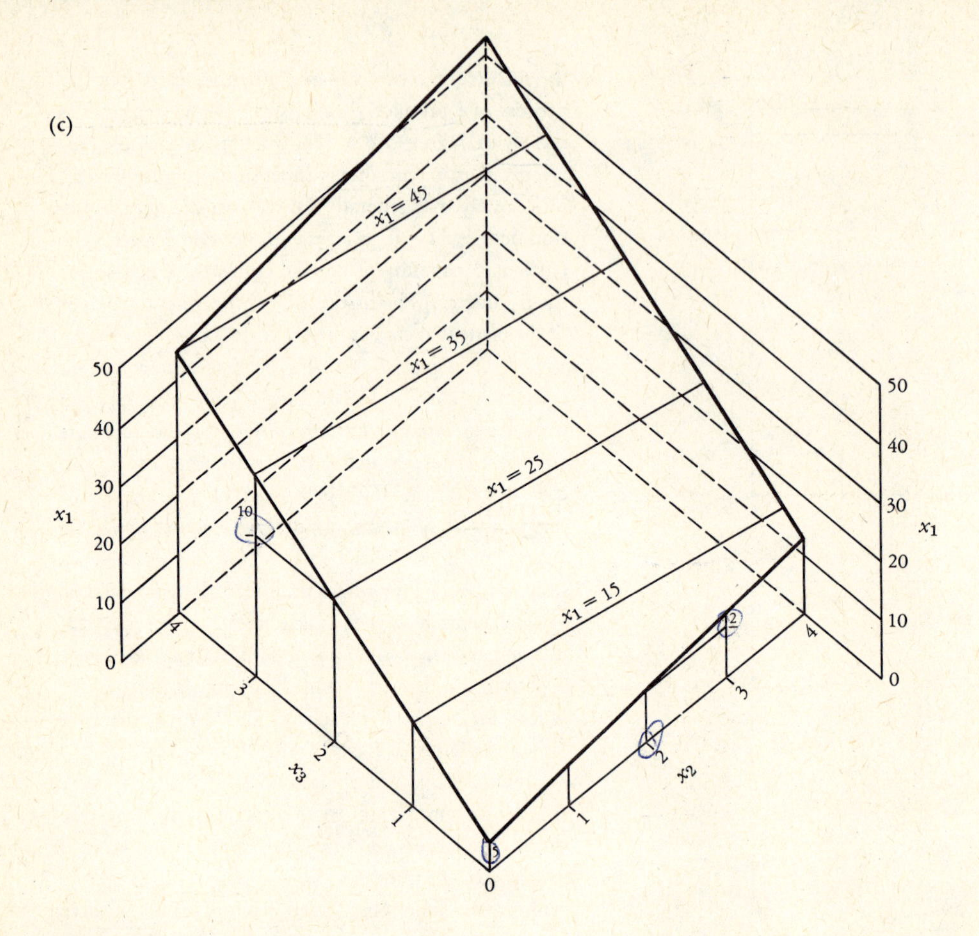

Figure 5.20 *(cont'd)*

$$b_3 = \frac{(\Sigma(x_1 - \bar{x}_1)(x_3 - \bar{x}_3)\Sigma(x_2 - \bar{x}_2)^2)}{\Sigma(x_2 - \bar{x}_2)^2\,\Sigma(x_3 - \bar{x}_3)^2} \quad \frac{-(\Sigma(x_1 - \bar{x}_1)(x_2 - \bar{x}_2)\Sigma(x_2 - \bar{x}_2)(x_3 - \bar{x}_3))}{-(\Sigma((x_2 - \bar{x}_2)(x_3 - \bar{x}_3))^2)}$$

and

$$a = \bar{x}_1 - b_2\bar{x}_2 - b_3\bar{x}_3.$$

These formulas seem complicated but in fact they involve only substituting easily derived quantities. (A fully worked example is given in Work Table 5.3.) This allows us to calibrate the equation

$$x_1' = \pm a \pm b_2x_2 \pm b_3x_3$$

in which x_1' is the value of x_1 predicted by x_2 and x_3. Differences between x_1', and x_1 constitute residuals from the regression equation whose standard deviation constitutes the standard error of estimate of the equation

$$\mathrm{SE}x_1' = \sqrt{\frac{\Sigma(x_1 - x_1')^2}{(n - 3)}}$$

As in the bivariate case, this is a measure of degree of fit, this time of the regression plane to a scatter of points in the three-dimensional space. Notice that, because we estimate three parameters, the degrees of freedom are $n - 3$.

In our presentation of equations for a three-variable function the amount of computation has increased appreciably compared with the bivariate case. As more variables are added, the set of normal equations increases according to the number of parameters to be estimated, and solutions become feasible only with electronic computational aid. This practical problem has been overcome in recent years with the widespread availability of electronic computers and linear regression package programs. We cannot draw concrete models of our functions as we move into greater than three-variable regression, but the algebraic interpretation remains identical. For ease of illustration, we continue to consider the three-variable multivariate case.

The main source of confusion in interpreting multivariate regression parameters occurs in relating the several b constants to the single b constant of the bivariate function. These coefficients in the multivariate case do not simply measure the change in the dependent variable associated with a change of one unit in an independent variable. In the following equations

$$x_1 = a + bx_2$$

and

$$x_1 = a + b_2x_2 + b_3x_3$$

$b \neq b_2$, even though both coefficients relate to x_2.

Coefficients in multivariate regression are usually termed *net regression coefficients*, because they describe change in the dependent variable only after the effects of the other independent variables in the equation have been taken into account—that is, when the other variables are not influencing the relationship or, in other words, when they are constant. In the three-variable case, we can return to Figure 5.20c to see what the net regression coefficients mean when we repeat our description of gradients for three-variable functions in the context of regression. On the vertical plane facing to the right, x_3 is constant because it equals 0 on all parts of the plane. The relation between x_1 and x_2 on this plane is represented by the gradient of 2. In statistical terms, this is the net regression coefficient for x_2 on x_1 with x_3 constant.

Similarly, the regression coefficient of x_3 on x_1 with x_2 at 0 can be seen to be 10. Because the height of the plane at $x_2 = x_3 = 0$ is 5, this gives us the equation

$$x_1 = 5 + 2x_2 + 10x_3$$

It is appropriate at this point to introduce a more complete notation for regression parameters. We use subscripts to indicate what relationship a coefficient relates to and what other variables are being taken into account. In this terminology, $b_{12.3}$ is the net regression coefficient between x_1 and x_2 taking x_3 into account, e.g., $b_{12.3} = 2$ and $b_{13.2} = 10$. We can define the base constant in similar terms: $a_{1.23}$ is the base constant in the equation predicting x_1 from x_2 and x_3. The three-variable equation now becomes

$$x_1 = \pm\, a_{1.23} \pm b_{12.3}\, x_2 \pm b_{13.2}\, x_3$$

and the two-variable case is

$$x_1 = \pm\, a_{1.2} \pm b_{12}x_2$$

This notation distinguishes clearly between b_{12} and $b_{12.3}$.

On addition to the regression coefficients differing in these cases, it is also true that $a_{1.2} \neq a_{1.23}$. As we extend the equation to include further variables, the parameters change again. Thus, in the four-variable case,

$$x_1 = a_{1.234} \pm b_{12.34}\, x_2 \pm b_{13.24}\, x_3 \pm b_{14.23}\, x_4$$

and

$b_{12.34} \neq b_{12.3}$, etc.

Before we continue our discussion of multiple regression analysis, let's consider briefly an example to help us fix ideas and remember the geographical context of this approach. We consider as our problem map the percentages of the urban leftwing vote in the British general election of 1951. Two American-based geographers, Roberts and Rumage, suggested that this voting pattern should be related to several variables having to do with social class, manufacturing, and mining. They employed voting percentages for the Labour party in the 157 largest towns in England and Wales (x_1), which they related to 7 independent variables. The regression equation is

$$x_1 = 40.413 - 0.039x_2 + 0.415x_3 + 0.123x_4 - 0.933x_5 + 0.410x_6 + 0.853x_7 + 0.195x_8$$

in which

x_2 = number of females per 1000 males
x_3 = number of persons living in nonprivate households

x_4 = percent working in manufacturing, mining, and agriculture
x_5 = number of persons in the middle classes
x_6 = number of persons in the working classes
x_7 = percent fulltime students among people 15–24 years old
x_8 = distance to the center of the nearest coalfield

In our notation, we write the first net regression coefficient as $b_{12.345678}$, which may look clumsy but is, in fact, a very succinct way of stating precisely what it defines.

Suppose we want to consider the relative importance of the independent variables with respect to effective changes produced in the dependent variable. Notice that we cannot use the net regression coefficients directly in this role, because they are in units of the variables concerned. We can see from our concrete model in the three-variable case (Figure 5.20c) that net regression coefficients represent gradients and consequently are in units of the dependent variable per unit of the independent variable. If the independent variable is measured in meters and we change the units to kilometers, the net regression coefficient increases a thousandfold; 1 kilometer has exactly 1,000 times the effect of 1 meter in a linear relationship. To compare rates of change in the dependent variable that are directly a consequence of different independent variables, the net regression coefficients are often standardized. In this form, they are known as *beta coefficients* and are defined as

$$\beta_{12.3} = b_{12.3} \frac{\sigma_2}{\sigma_1}$$

Beta coefficients can be computed directly (without computing net regression coefficients) by standardizing the variables into units of their standard deviation about their respective means. Notice that, in this particular case, because the mean of all variables is 0 and the regression line, plane or beyond, always passes through the location of the means, it follows that the base constant vanishes. In the three-variable case, the equation becomes

$$x_1 = \pm \beta_{12.3} x_2 \pm \beta_{13.2} x_3$$

in which x_1, x_2, and x_3 are standardized.

Beta coefficients have been quite common in the calibration of multiple regression functions in geography. In the "McCarty on McCarthy" study, the addition of the variable distance from Appleton (x_3) produced beta coefficients of $\beta_{12.3} = 0.67$ and $\beta_{13.2} = -0.34$. This indicates that x_2, rural population percentage, is relatively more important than distance from Appleton in affecting the percentage vote for McCarthy. These results do not indicate how much of the variation of x_1 can be accounted for by x_2 and x_3 jointly. In the original bivariate study, a correlation of $r_{12} = 0.58$ is reported. We clearly require similar correlation concepts in a multivariate context in order to take our analyses further in this direction.

Multiple and Partial Correlation

In our discussion of bivariate regression analysis, we presented a basic link between correlation and regression by redefining the correlation coefficient as

$$r = \frac{\sigma_{y'}}{\sigma_y}$$

which is simply the standard of deviation of the predicted values of the dependent variable as a proportion of the actual standard deviation of the variable. This logic can be transferred easily into the context of multiple regression, so that we define

$$R_{1.23} = \frac{\sigma_{x'_1}}{\sigma_{x_1}}$$

The capital R indicates a *multiple correlation coefficient* and the subscripts indicate a coefficient relating x_1 to x_2 and x_3. Otherwise, the coefficient is identical to its bivariate equivalent with the predicted dependent variable standard deviation over the actual standard deviation. The only difference is, of course, that the predicted values are from an equation involving several independent variables—in our example, x_2 and x_3. If we square the multiple correlation coefficient, we produce a *multiple coefficient of determination*, which indicates the proportion of the variance of the dependent variable accounted for by the independent variables. This is directly analogous to the bivariate equivalent and its interpretation is based on the same argument as we presented above. Computation of these coefficients is illustrated in Work Table 5.4.

We are now in a position to answer the questions concerning the map of residuals in the McCarty on McCarthy study. The addition of x_3 resulted in a multiple correlation coefficient ($R_{1.23}$) of 0.67, compared with the original bivariate coefficient of 0.58. Thus, whereas originally 34% of the variance of McCarthy's vote was accounted for ($r_{12}^2 = 0.34$), the addition of the distance-from-Appleton variable means that we now account for 45% of this variance ($R_{1.23}^2 = 0.45$). This new model increases the proportion of the dependent variable's variance that is accounted for by from a third to nearly a half, illustrating the benefits derived from the original analysis of the residuals. The eight-variable model of urban leftwing voting in British cities is much more successful. In this case $R_{1.2345678} = 0.898$, indicating that over 80% of the variance of the voting pattern is accounted for by the seven independent variables listed above.

Even though the multiple correlation coefficient gives us an overall picture of the validity of our multiple regression function, it tells us nothing about the individual influences of each independent variable within the model. With such a measure we clearly consider the relative importance of the independent variables in terms of the variance accounted for in the dependent variable and, hence, complement the regression beta coefficients. This is achieved by using the *partial correlation coefficient*, a measure of correlation between the dependent variable and an independent variable after the correlation of the other independent variables has been taken into account. In the three-variable case, the partial correlation coefficient between x_1 and x_2, while taking x_3 into account, is written $r_{12.3}$ and is found by the following procedure. The proportion of variation in x_1 accounted for by x_3 is the simple coefficient of determination r_{13}^2. This leaves $(1 - r_{13}^2)$ of the total variation of x_1 unaccounted for. If we add x_2 to our analysis, the proportion of variation in x_1 that we have accounted for rises to $R_{1.23}^2$, leaving $(1 - R_{1.23}^2)$ unaccounted for. The extra proportion of variance in x_1 accounted for by x_2 is, therefore, $(1 - r_{13}^2) - (1 - R_{1.23}^2)$. The partial correlation coefficient is now based on this extra variance, accounted for as a proportion of the variance originally left unexplained:

$$r_{12.3} = \sqrt{\frac{(1 - r_{13}^2) - (1 - R_{1.23}^2)}{1 - r_{13}^2}}$$

Thus, if x_3 accounts for 80% of the variation in x_1 ($r_{13}^2 = 0.8$), this leaves 20% to be accounted for. If the addition of x_2 leads to an increase in the proportion of variance accounted for of 90% ($R_{1.23}^2 = 0.9$) only 10% remains to be accounted for. Thus, the introduction of x_2 has led us to halving the proportion of variance of x_1 left unaccounted for, so that

$$r_{12.3} = \sqrt{\frac{(1 - 0.8) - (1 - 0.9)}{1 - 0.8}} = \sqrt{\frac{0.2 - 0.1}{0.2}} = 0.71$$

This is, therefore, a second way of measuring the separate effect of independent variables. If we use beta coefficients and partial correlations to compare the relative importance of several independent variables, we get similar results, although they are not identical because of the different way in which the two methods assess the effectiveness of an independent variable. The beta coefficient relates directly to the functional relationship and it concentrates on the changes in x'_1 associated with changes in the independent variable, whereas the partial correlations measure the proportion of variation accounted for separately by the independent variable.

The partial correlation coefficient is a very useful technique in a well-planned research design. In Chapter Four we presented map transformation as a geometric way of holding constant disturbing influences while studying a particular relation. The partial correlation coefficient is a statistical way of achieving the same end. Furthermore, because it is an algebraic rather than a graphic manipulation of data, it is much more flexible and capable of controlling any number of variables in an analysis. Of course, we cannot hold variables constant as we can in the physical laboratory, but we can analyze data in such a way that we treat one or more variables as constant. We illustrate this strategy on our concrete model in Figure 5.20. If we calculate the correlation coefficient between x_1 and x_2 for values that occur only on the left-facing vertical plane, we are in

Table 5.7 Multiple Correlation Analysis in the Great Plains Study

Coefficients of simple correlation	*Coefficients of partial correlation*	*Coefficients of multiple correlation*
$r_{12} = +0.78$	Three variables	$r_{12} = +0.78$
$r_{13} = -0.43$	$r_{12.3} = +0.73$	$R_{1.23} = +0.79$
$r_{14} = +0.58$	$r_{13.2} = +0.12$	$R_{1.234} = +0.90$
$r_{23} = -0.42$		
$r_{24} = +0.26$	Four variables	
$r_{34} = -0.16$	$r_{12.34} = +0.63$	
	$r_{13.24} = -0.10$	
	$r_{14.23} = +0.39$	

Source: A.H. Robinson, J.B. Lindberg, and L.W. Brinkman. A correlation and regression analysis applied to the Great Plains, Table 1. Reproduced by permission from the *Annals* of the Association of American Geographers, vol. 51, 1961.

effect holding x_2 constant. This is because $x_2 = 0$ for every observation. The resulting correlation from this data is identical to the partial correlation coefficient $r_{13.2}$. Where does the simple correlation coefficient r_{13} fit into this scheme? If, instead of drawing our observation from the left-facing plane, we had sampled for any observation in our three-dimensional space, then x_2 would have varied among the observations. In this case, we are not holding x_2 constant, and the correlation coefficient for the x_1 and x_3 values represents simply r_{13}. The partial correlation coefficient $r_{13.2}$ is a much more interesting coefficient than is r_{13}. In fact, r_{13} may be quite misleading in some situations. Let's consider an actual situation in which this is so.

We illustrate multiple and partial correlation analysis with some research that develops from the Nebraska study described previously. We continue to consider rural farm population density for 1950 as the dependent variable (x_1), but this time we extend the study area to include the whole of the American Great Plains from the Canadian border to northern Texas. Three independent variables are chosen for the analysis: x_2 is average rainfall, as before; x_3 is distance from the nearest town of more than 10,000 population; and x_4 is percentage of cropland. Values for these variables are derived for a set of 161 regular, hexagonal base units placed over the study area. For the areally based variables (rural population density and percentage cropland), base unit values are aggregated from county data as described in Chapter Two. Average annual rainfall data is interpolated for the centers of the base areas from isoline maps and, similarly, distances from towns are measured from the base area centers. The complete correlation analysis is given in Table 5.7, showing simple bivariate, partial, and multiple correlation coefficients. From the simple correlations, we can see that the highest areal association is between rural population density and rainfall, with a coefficient very close to that in the Nebraska study. Distance and cropland variables have higher correlations with the dependent variable than with other independent variables. Nonetheless, the partial correlation coefficients are very interesting. With three variables, the degree of association between rural density and rainfall falls only slightly when the distance variable is allowed for. On the other hand, the relation between rural density and distance from towns takes an about turn. There is a change from a moderate, simple negative correlation of -0.43 to a small positive partial correlation coefficient. Thus, by taking into account the effects of rainfall variations, the hypothesized negative association between rural density and distance from towns has disappeared. This can be explained rationally from this pattern of correlations. It relates to the moderately high negative correlation between distance and rainfall (r_{23}) and the smaller negative correlation between distance and cropland (r_{34}). Because distance is related inversely to both these variables, which are in turn related positively to rural density, this alone is enough to produce a moderate, spurious correlation between distance and rural density in a negative direction. When we remove the effects of rainfall, the negative correlation between distance and rural density is lost. With four variables, the direction of this relation is represented by a negative sign for the partial correlation but it remains at a very low level of correlation. The partial correlations between rural density and both rainfall and cropland are smaller than the equivalent simple correlations, which reflects their slight, mutual positive correlation (r_{24}). Finally, the coefficients of multiple correlation confirm these findings concerning the relative importance of the three independent variables. The addition of the distance variable to the model improves only marginally on the original simple correlation between rural density and rainfall, although the addition of the cropland variable appreciably increases the level of correlation to a very high 0.9.

This example of a multiple correlation analysis indicates how careful we must always be in interpreting simple bivariate correlations. We do not live in a bivariate world, and we should not expect too much from analysis of bivariate relations. When spurious correlations are produced, the results may be quite misleading, as the analysis above illustrates. This does not mean that simple correlations are of no use to geographers looking for areal associations. They are clearly of value in any initial analysis of a problem map. We should always bear in mind, however, the relation between hypothesized determinants before drawing conclusions. We may be able to suggest on purely theoretical grounds that the determinants are not related. We usually require empirical evidence of this. Partial correlation coefficients are fortunately available as an extremely useful tool for unraveling multivariate situations, as our example has illustrated.

Before we conclude this section on multiple correlation and regression, we mention a particular approach to multiple regression analysis that uses partial correlations and has been made possible by computers and package programs. This is *stepwise regression*, designed particularly for situations in which there is a large number of independent variables. Instead of calibrating a single regression equation using all independent variables, this approach starts with a bivariate equation and proceeds by adding one variable at a time until the complete equation is finally calibrated. The order in which variables enter the regression sequence is not arbitrary but depends on their contribution to explanation of the remaining variance in the dependent variable. Thus, for the bivariate equation, we start with the independent variable that has the highest correlation with the dependent variable. The second variable to be added is that which has the highest partial correlation coefficient with the dependent variable when the initial independent variable is allowed for. This variable accounts for the highest proportion of variance that is not accounted for in the first equation. This process is repeated by selecting the independent variable with the highest partial correlation coefficient when the independent variables already considered are allowed for. If this approach had been employed with the data for the Great Plains study, the order of variables entering the equation would have been first rainfall, second cropland, and finally distance. Variables are added to the equation in an orderly fashion in accordance with their relative importance.

Technical and Methodological Problems

So far we have introduced correlation and regression techniques in terms of a set of equations whose results have specified desired properties. We have not questioned why the normal equations produce constants that define a least squares fit to a scatter of data points. The position we have reached is as follows. From our discussion of applied mathematics in Chapter One, we can see that we have translated our problem into a mathematical language but have been concerned only with the results of this procedure. From our discussion of the nature of mathematics at the beginning of Chapter One, we know that our conclusions are based on deductions from axioms or assumptions. We have three steps—assumptions, deductions, and conclusions—but we have concentrated solely on conclusions. The reason for this is simply that we are not equipped in this book to master the mathematical language required. There are clearly several dangers in our approach. We are taking conclusions on trust when we have not been told exactly what the bases are. In this section, we do not consider the mathematical deductions themselves but we do enumerate and discuss the assumptions upon which the mathematics is based. This avoids the dangers of misapplication when we consider results and techniques merely as given.

The technical matters are our first concern. Next we ask the basic question, "How useful are these analyses from a geographical viewpoint, anyway?" The answer to this methodological question is ultimately of most importance. Users of areal association ought to be fully aware of both the technical and the methodological problems in statistical and geographical contexts.

Sampling Models

Correlation and regression analysis form an integral part of modern statistics and the concern in statistical theory for samples and populations. Correlation and regression concepts are very closely related, but they derive from two different sampling models. Regression analysis was developed on the assumption that values of the independent variable are selected and then corresponding dependent variable values are sampled. This is known sometimes as the fixed x model. The statistical theory behind correlation coefficients, however, assumes a random sample from a bivariate or multivariate normal distribution for which x and y values are obtained. This may be termed the variable x model. These two sampling models are known sometimes as regression and correlation models, respectively, although these terms are to some extent misleading, because regression and correlation coefficients can be calculated for data derived from either type of sampling model. The sample data produce only estimates of the population correlation coefficient and regression parameters, which in the bivariate case produces the equation

$$y' = \pm a \pm bx$$

which we can rewrite as

$$y = \pm a \pm bx \pm z$$

including the residuals to give exact predictions of y. This form of the regression equation approximates the population regression equation

$$y = \pm A \pm Bx \pm e$$

in which e is an independent error term that may or may not correspond to the residuals derived from the sample regression. Thus, the a and b constants we derive from least squares methods are estimates of the true population parameters A and B. Similarly, the regression line is associated with a sample correlation coefficient r, which is an estimate of the population correlation ρ (Greek rho, pronounced "row").

What relevance has this statistical basis to geographical studies of areal association? Geographers using these techniques ought clearly to be aware of the statistical theory behind correlation and regression so that they can begin to assess its relevance. Let's begin our discussion of this theme by considering how often these sampling models have been employed in map comparison exercises. The fixed x model assumes that a researcher has some specific control over the research design and is particularly associated with laboratory experiments in biological sciences. It is, therefore, of limited direct interest to studies of areal association. The variable x model seems more in keeping with geographical studies of areal association, however. We can cite the work of John Hart and Neil Salisbury, in which they attempted to determine the factors associated with population change from 1940 to 1950 in Midwest American villages. Because the study covered Illinois, Indiana, Iowa, Kentucky, Michigan, Minnesota, Missouri, and Wisconsin, it produced a population of some 3697 villages. This very large population necessitated basing the analysis on a sample of manageable size (400), selected randomly from a listing of all the villages. Values for the dependent variable and independent variables were obtained from a selection process not unlike that in the theoretical variable x model. We consider the normality criteria below. The point we emphasize here is that this type of research design in areal association studies is not common. Some sampling may be involved when variables are represented as continuous variables on isoline maps, as in the Nebraska example we have considered at some length, but it is most common for variables to be available for distinct areal units whose number is not unmanageable. This is true for the other studies reported, ranging from Peruvian provinces and Newfoundland districts to British cities and Midwestern hexagons. In all these examples, the research did not explicitly incorporate a sampling procedure. This does not mean that no sampling was involved. In Chapter Three we illustrated the χ^2 test with an application to Tasmanian settlement types in which we employed a total population of settlements. We concluded nonetheless that the sample and population logic of inferential statistics applied. A similar argument was used to justify z tests for assessing spatial autocorrelation. We can use the same logic for areal association studies. Let's consider a specific example.

The correlations reported for the Newfoundland referendum study were produced with one set of areal units from a vast number of ways the island could be divided into data collection units. The high correlation between voting and Roman Catholic percentages could be a result of this particular set of areal units. If there were no relation between these two variables, the pairs of observation values would still vary between different patterns of areal units, and on rare occasions values as high as 0.63 might occur. The actual set of areal units used to collect data forms only one of many possible patterns of areal units. We can view our results as a particular sample of one set of areal units from a population of the many sets of possible areal units, all for the same area (in this case, Newfoundland). This argument for modifiable areal units can be repeated for all areal association studies that are based on areal units. It does not prove that they are equivalent to the variable x model, because many differences remain, but it shows that results from such studies are not the same as statistical population results, in which parameters cannot vary but are determinate. If we assume that variations in parameter values resulting from modifiable areal units are random and not biased in any sense, then we can treat such discrepancies as the usual sampling error. Hence the approaches developing from the statistician's sampling models may be useful in assessing areal association studies, even when we are dealing seemingly with the total population in an area.

Assumptions and Inferences

Let's proceed in the understanding that we consider areal association studies as sampling exercises. Two questions arise:

1. Are the sample estimates of the relation unbiased representatives of the population parameters?
2. Can we make inferences about the true values of the parameters from the sample estimates?

The answers involve more than simply whether the data can be considered as a sample. Both the fixed x and the variable x models involve a series of assumptions. These have been dealt with explicitly for geographers by Michael Poole and Patrick O'Farrell, and we draw partly on their work. They list seven basic assumptions relating to the sampling models. The first two assumptions relate to basic conditions that must be satisfied before embarking on a regression and correlation analysis; the five other assumptions are specific conditions relating to the mathematical basis of the method.

1. *There is no measurement error in either dependent or independent variables.* This may be relaxed to relate only to independent variables, if we assume that measurement errors in the dependent variable are random and incorporated in the residuals or error term. Measurement error is, of course, very difficult to assess and is never entirely absent, even under laboratory conditions. Outside the laboratory, in geographic and other social science studies, an optimistic view is taken by simply assuming that what unknown measurement error exists does not prejudice the analysis. Our discussion of this topic in Chapter Two is relevant here.

2. *The form of the function that is fitted to the sample data obtains in the population.* This means simply that we must fit the correct function. In our discussion and in most geographical research, simple linear functions are employed. If the data do not show a linear trend, we ought not to fit a linear function to them. Consider Figure 5.21a, which shows selected countries plotted on a graph with the β index of connectivity as the dependent variable and the per capita income as the independent variable. The β values are from Table 2.5 in Chapter Two, where we hypothesized that these values are related to economic development. Figure

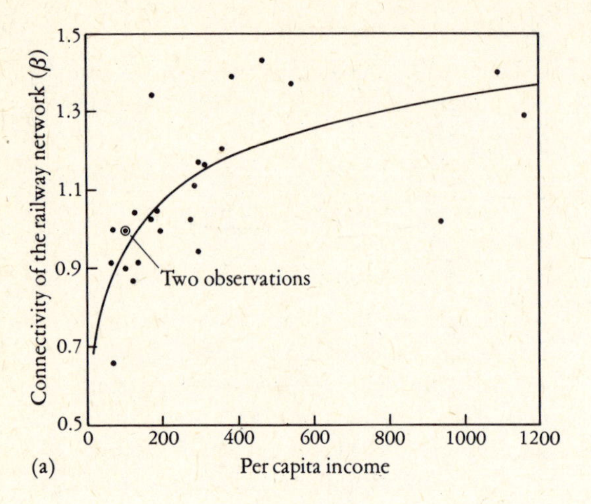

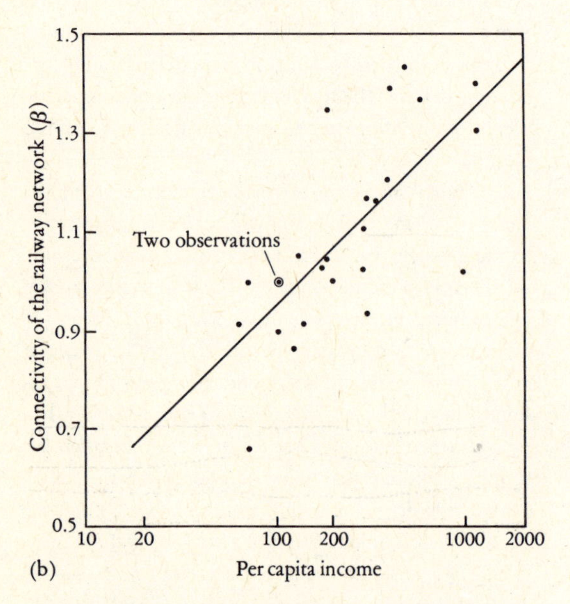

Figure 5.21 The relationship between railway connectivity and per capita income for selected countries (see Table 2.5).

5.21 shows this relation and it is clearly not a simple linear trend. Despite the wide scatter of points, a nonlinear or curved relation is apparent. We cannot use the techniques introduced in this chapter to assess this hypothesis. In such cases, there is a very easy way of making the problem simply linear. We transform the relation into a linear trend. In Figure 5.21b, we plot the same data, but this time the x axis is scaled as logarithms. Notice that this has the effect of straightening out the scatter of points. With data in this form, we can use the procedures discussed above to define a regression line. The only difference is that, instead of using the actual values of x in our equation, we feed into it logarithms of x to calibrate the function

$$y = \pm a \pm b \log x$$

This type of function has been fitted to the data in Figure 5.21b, producing

$$y' = 0.197 + 0.381 \log x$$

in which y' is the estimated β measure of railway connectivity and x is the per capita income. The regression line has also been added to Figure 5.21a; it appears as a curve lying among the nonlinear points. The degree of the relation can be measured as before, and it is computed as $r = 0.715$. This means that 50% of the variance of railway connectivity can be accounted for by the logarithm of per capita income. This procedure of transforming the data into a linear trend so that simple linear regression can be used is very general and can involve a variety of transformations other than the one illustrated above. For highly nonlinear data, for example, both variables may be logged. On some occasions, a linear trend is not easily obtained, indicating that an alternative function is desirable. Nonlinear regression is beyond the scope of this book and can be found in most statistical textbooks that deal with regression analysis in detail.

3. *The mean population values of the dependent variable for different values of the independent variable lie along the regression line.* This is reflected when the residuals of the sample regression line are spread equally on both sides of the regression line at any location on the line.

4. *The population values about the line have equal variances at different locations on the line.* This is reflected when the residuals of the sample regression line are spread evenly about the line over its whole span. This is known as the homoscedacity assumption, and when it is violated the data are said to be heteroscedastic.

5. *The population values about the regression line (the*

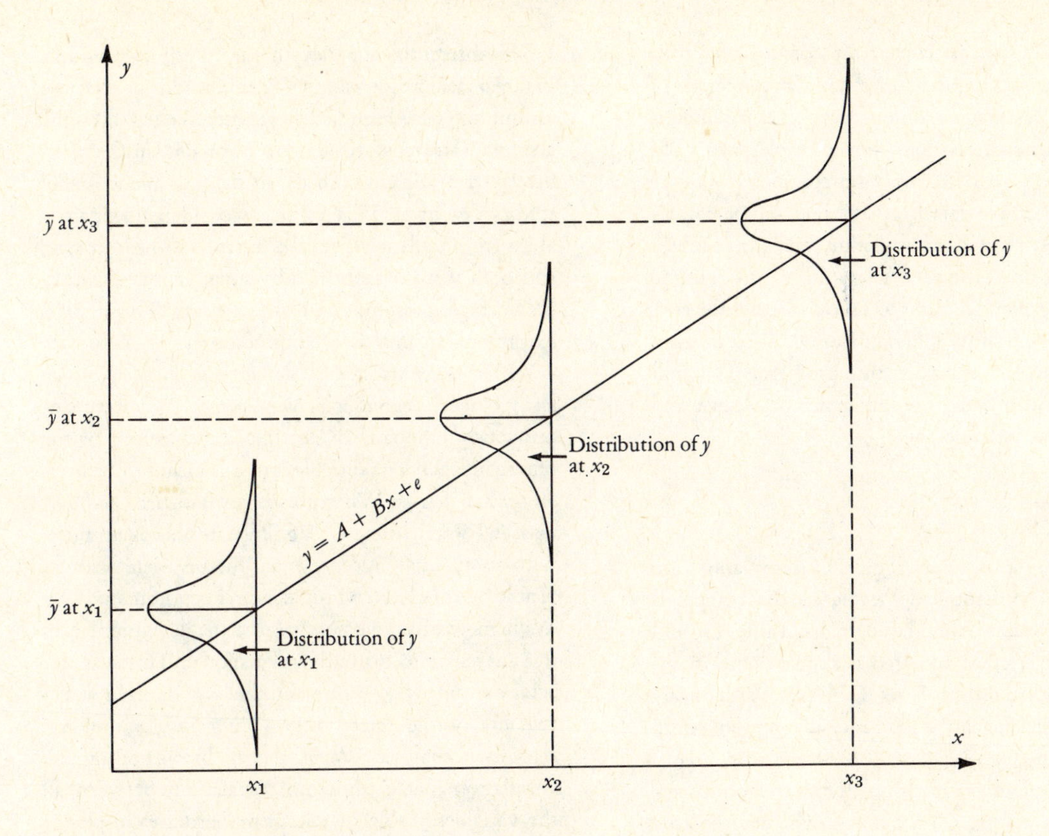

Figure 5.22 Some assumptions of the linear regression model.

error terms) are independent of one another. If this assumption is not satisfied, *autocorrelation* is said to be present. In any sample regression analysis, the error terms are represented by the residuals, and this assumption is clearly violated in the method of analyzing residuals we described above. The whole point of mapping residuals is to find a pattern to them that suggests not only a further variable but also dependence of residuals. Thus, the residuals in the "McCarty on McCarthy" study are clearly spatially autocorrelated (Figure 5.19). This finding casts doubt immediately on the regression equation from which the residuals were computed. Notice that, if this method of residuals continues, with further variables added to the equation until there is no pattern in the residual map, then clearly the final equation does not violate the autocorrelation assumption. Rather than being a hindrance, the spatial autocorrelation of residuals is usually of intrinsic interest to the geographer. The spatial autocorrelation of residuals can be tested (as outlined in Chapter Three) as a two-color map (positive residuals are black; negative residuals are white) or as a *k*-color map that uses the actual residual values.

6. *The population distributions of the variables should be normal; in the variable x model this relates to all variables, and in the fixed x model it relates only to the dependent variable.* When combined with assumptions 3 and 4, the linear regression model can be represented as in Figure 5.22. This normality assumption is necessary when inferential statistics are included in the analysis. In the sampling situation we defined, we can carry out several tests, all of which are parametric. We show how the student's t test introduced in Chapter Three can be applied in regression and correlation analyses.

The questions we ask come from classical inferential statistics; we use sample statistics to infer properties of population parameters. In this case, the parameters concern functional relations. For example, two variables from a population may be completely uncorrelated, but sample data from the population will almost certainly produce a correlation coefficient slightly greater than or less than 0. If we have only sample evidence to go on, we want to know whether the correlation is significantly different from 0. We can use a student's t test to consider this problem. The null hypothesis is that there is no difference between r and 0. The statistic is defined by

$$t = \frac{r\sqrt{n-2}}{\sqrt{1-r^2}}$$

with $df = n - 2$, in which n pairs of observations are correlated. The computed t value can be compared with critical t values from the appendix (Table A.8) and H_0 can be rejected or accepted as appropriate. An example is worked out in Work Table 5.5. With Spearman's rank coefficient, significance can be looked up directly from tables (Table A.10).

With sample data, not only the correlation coefficients need testing. We can test whether the true population gradient is significant given the sample regression coefficient. The standard error of a regression coefficient is

$$\text{SE}b = \frac{\text{SE}y'}{\sigma_x\sqrt{n-2}}$$

We can now test whether the sample b value is different from some hypothesized gradient B by using a student's test:

$$t = \frac{|b - B|}{\text{SE}b}$$

and $df = n - 2$. When we are testing to see whether we have a significant gradient at all, we set $B = 0$, so that

$$t = \frac{b}{\text{SE}b}$$

that is, the number of times b is larger than its standard error. We illustrate this test in Work Table 5.5.

This inferential apparatus has not been used widely in areal association studies. Perhaps this is because assumption 6 has been widely violated. When this is the case, we can transform data to normality in the ways described in Chapter Three. In the variable x model study of Midwest villages, the two independent variables are population size and distance to nearest city, and both were transformed by square roots to satisfy the normality assumption before analysis. This study is exceptional in areal association research.

7. *The independent variables in a multiple regression analysis are independent of one another.* We have not previously emphasized the statistical meaning of the phrase independent variable, but it is meant to be literally what it says. When this is not satisfied, *multicollinearity* is said to exist. We illustrate this assumption with an example We presented the bivariate analyses (Table 5.5) of the Newfoundland referendum study by William Warntz; we now turn to the development of the analysis into a multivariate model. This was not achieved simply by putting all the hypothesized determinants of the referendum voting into a multiple regression analysis. Warntz began by inspecting the correlation matrix showing correlations between all the variables. Table 5.8 shows numerous examples of high correlation between the proposed determinants. The highest correlation in the matrix does not in fact relate to the referendum voting variable; it relates population potential and illiteracy.

Because these are the two variables that correlate highest with the dependent variable, we must ask whether either of these two relations is in fact spurious. A spurious correlation is produced when both variables are independently associated with a third variable and can be detected, as we have shown, by computing partial correlation coefficients. Here we are concerned with this phenomenon in modeling a functional relation in a regression analysis. Let's consider the argument that the high positive correlation between illiteracy and voting for independence is spurious. The more isolated that an area is, the more likely people are to vote for confederation with Canada. Physical isolation reflects the general inaccessibility of the population to social interactions and diffusion of ideas that favor independence and are found in less isolated

Table 5.8 Correlation Matrix for the Newfoundland Referendum Study

	1	2	3	4	5	6
1 Percent vote for independence						
2 Population potential	+0.72					
3 Illiteracy rate	−0.66	−0.80				
4 Percent Roman Catholic	+0.63	+0.41	−0.34			
5 Percent employed in trade	+0.09	+0.58	−0.47	−0.05		
6 Per capita income	+0.01	+0.34	−0.28	+0.31	+0.58	

Source: W. Warntz. *A Methodological Consideration of Some Geographical Aspects of the Newfoundland Referendum on Confederation with Canada, 1948,* Table II, p.46. Copyright 1955 by the Canadian Association of Geographers.

areas. A high level of areal association exists, therefore, between population potential and referendum voting. These isolated areas, with lower levels of social services, have high illiteracy rates. Hence, we find the high areal association between illiteracy rates and population potential. This, in turn, results in a high level of areal association between referendum voting and illiteracy. This argument suggests that the third bivariate relation is spurious and, thus, unimportant. The example illustrates clearly the need to think constantly in terms of functional relations between areal variations and not simply in terms of levels of correlation. In the subsequent development of a multiple regression model, the illiteracy level is omitted from analysis because its contribution as an explanation of the voting pattern is already incorporated in the population potential variable. In fact, three determinants are selected as independent variables: x_2, population potential; x_4, percent Roman Catholic; and x_6, per capita income. This selection of isolation, religion, and economic condition produces a four-variable model

$$x'_1 = 38.8224 + 0.0712x_2 + 0.5179x_4 - 0.0229x_6$$

with

$$R_{1.246} = 0.847.$$

Areal association regression analyses have not always been so careful to avoid multicollinearity. Thus, in the eight-variable multiple regression model predicting urban leftwing voting, many independent variables are almost certainly highly related. For instance, x_5 (percent middle class) is certainly highly negatively correlated with x_6 (percent working class). This voting study is a classic example of multicollinearity in a geographical regression analysis. How important is this indictment of the urban leftwing voting study? How important are each of these assumptions, all of which have almost certainly been violated by numerous geographical analyses? In a bivariate situation, the first five assumptions are required to ensure that the sample coefficients are unbiased estimates of the population parameters. In the multivariate case, assumption 7 must also be satisfied. If we seek to take our analysis beyond the simple knowledge that it is unbiased into the realm of hypothesis testing, we must also satisfy assumption 6, because such testing involves parametric statistics. Let's answer the question: in the eight-variable regression equation relating to leftwing urban voting, it would seem that the least squares method has not necessarily produced unbiased estimates of the net regression coefficients. We return to a consideration of this equation and its use in the final section of this chapter.

Modifiable Areal Units and the Scale Problem

We have used the argument of modifiable areal units as a means of relating areal association studies to the statistician's sampling models. The underlying assumption is that the different estimates of correlation coefficients and regression parameters, resulting from different ways of dividing a study region, represent random sampling errors. This assumption seems reasonable when a study region is divided into the same number of roughly equal areal units. When alternative sets of areal units are at different scales (for instance, aggregates of the original units), then discrepancies in correlation coefficients may be large and systematic.

The *scale problem* in areal association is simply that the correlation coefficient for two variables usually varies systematically when computed *for the same data*, when this data is aggregated into fewer and larger units. All correlation coefficients measure only areal associations between variables relative to the particular scale of the areal units. At first, this problem seems to be a severe limitation on the usefulness of correlation coefficients for finding any generalizations from areal association studies. Early experiments at trying to correct for the size of the areal units failed, and geographers are left with this very explicit scale problem. Statisticians and sociologists have also been aware of the problem, and they have conducted their own variation on a debate of the scale problem, which they usually term *the ecological fallacy*. Sociologists have a long tradition of using data for areal units (especially census tracts) for testing hypotheses. These hypotheses often concern behavior of individual people, and the ecological fallacy then consists of transferring findings about properties of an aggregate of people to statements about the behavior of individual people. A very high correlation between a percentage of Polish-born residents and a crime rate for census tracts does not necessarily mean that the Polish-born people are criminals. All we show with this correlation is that Polish-born people tend to live in areas where the crime rate is high. Similarly, an inverse relation between immigrants who make up the percent of foreign-born and the percent of substandard housing in American cities (Table 5.5) does not necessarily mean that nonnatives live in good housing conditions. This negative correlation says simply that the foreign-born population tends to live in cities with better housing conditions (that is, in northern cities). It says nothing about the housing conditions of individual immigrants.

In an early study explaining this problem, it was shown that the correlation between percentage black and percentage illiterate for the nine U.S. census regions for 1930 is a very high 0.946. When the same data were compiled for states rather than census regions, the correlation dropped to 0.773 and an individual correlation between blacks and illiteracy was found to be only 0.203. Here the scale problem reaches down to the smallest social science scale. Because social sciences are concerned primarily with individual people, one solution that was suggested for this problem is simply to avoid using ecological correlations and concentrate on individual correlations. This sociological solution is clearly unsatisfactory to geographers and, indeed, to other social scientists interested in relations between aggregates of people. For geographers, the area in which a person lives is a very important characteristic and can be described only by using aggregate areal data. Relationships gleaned from such studies do not necessarily attempt to relate to individual behavior. Census tract data might produce a high positive correlation between percentage Jewish in the population and some index of anti-Semitic activity. The obvious interpretation that Jews are anti-Semitic is ridiculous, but the correlation remains interesting from an ecological viewpoint. It suggests that the areas that have the most Jewish families generate the most anti-Semitic feeling. The conclusion is about areas and not individuals, and it is obviously of direct interest to a study of anti-Semitic feeling. This example justifies the use of correlations based on areal data but it does not overcome the scale problem.

It seems that we must accept scale differences in correlations and inquire why. If the question can be answered satisfactorily, then scale considerations can be incorporated constructively in any research design rather than problematically. Let's consider the reasons behind these changes in correlation levels. Suppose that we are interested in explaining the phenomenon of family size. At the level of the individual family, the number of children who are born may be the result of several factors that may be idiosyncratic from the viewpoint of the social scientist. The personality characteristics and particular attitudes of a married couple are important, and even accidents of conception will be reflected in the data. Among a sample of individual families, we can expect a quite wide range of family size—that is to say, the variance of the dependent variable, family size, is relatively large. If we move from an individual level of analysis to data describing the same families aggregated into census tract figures for average family size, we can expect the variation of the dependent variable to be much smaller. The idiosyncratic factors begin to cancel one another out, and we are left with a much smaller range of variance. It may well be that, at this point, the independent variable for proportion of Roman Catholics becomes highly correlated with family size, even though it did not seem very important at the individual level. This follows from our definition of the correlation coefficient as the square root of the coefficient of determination, which measures the proportion of variance of the dependent variable associated with the linear relationship with the independent variable. At the individual level, much of the variance is accounted for by the idiosyncratic factors, so that little is left for religion to explain. When the idiosyncratic factors cancel one another out in data aggregation, religion can begin to play a role as a determinant of family size. This does not mean that the change of scale has caused a change in relationships; it means simply that at one scale the relationship is masked by other factors. Different scales can be expected to produce different degrees of relationship and, hence, of variation in correlation coefficients. It has been found that as areal units become larger, variation in the dependent variable declines and the larger the correlation coefficient is (as our illiteracy-black correlations illustrated). It is important to notice that scale differences do not systematically influence regression coefficients. In fact, if the functional relation between variables always changed with scale, any attempt at scientific generalization would be almost impossible. Because the regression coefficients relate not to the variability of the variables but simply to the change of one variable associated with a change in another, aggregation effects are unimportant. The degree of fit alters radically but the form of the functional relation (the slope of the curve) is affected only slightly by data rearrangements. This feature reinforces our previous statement that regression equations are theoretically more important than simple correlation measures of the degree of relationship.

American sociologist Hubert Blalock has illustrated our argument in his experiments on data aggregation. The correlation coefficient between differences in income for whites and blacks (y) and percentage blacks (x) for 150 southern American counties was found to be 0.54 with a regression coefficient of 0.26. The question Blalock asked is, "What happens to these coefficients when the counties are grouped into larger units?" Let's consider a simple model of this situation.

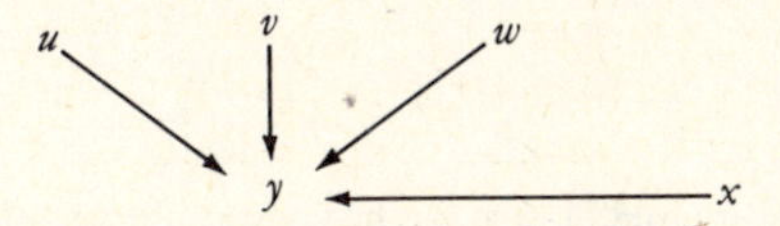

The arrows indicate causal links or, simply, the determinants of y. Thus y depends on x and the *degree* of this dependence is indicated by $r = 0.54$. The *form* of this dependence (that is, the functional relation) is indicated by $b = 0.26$. We have already decided that we do not live in a simple bivariate world, and so we can postulate other variables (u, v, and w) upon which y also depends. We do not have to specify these new variables except to notice that they influence y but not x. Changes in our coefficients with data aggregation depend on what happens to u, v, and w as well as x and y.

The results of three of Blalock's aggregation experiments are shown in Table 5.9. In each case, the 150 counties were grouped into pairs ($n = 75$), fives ($n = 30$), tens ($n = 15$), and fifteens ($n = 10$). Correlation and regression coefficients were computed for the data at each level of aggregation. The methods of grouping are the essence of the experiments. The first two methods are limiting cases, unlikely in real world aggregation but useful here as extreme standards for comparative purposes. The third experiment, then, presents us with the most common form of aggregation, which we expect to fall somewhere between the extremes. Let's consider each experiment in turn.

In the first experiment, the grouping is simply by random assignment. As we aggregate and average variables in their new units, their variation naturally declines. Because the aggregation is random, the observed variance in x declines in roughly the same proportion as the decline in the variance of y. The same is true of the unknown variances of u, v, and w. Hence,

Table 5.9 Aggregation Experiments on Correlation and Regression Coefficients

Method of aggregation	*Coefficient*	*Scale of aggregation*				
		Ungrouped (n = 150)	*Pairs* (n = 75)	*Fives* (n = 30)	*Tens* (n = 15)	*Fifteens* (n = 10)
Random	r_{xy}	0.54	0.67	0.61	0.62	0.26
	b_{xy}	0.26	0.36	0.31	0.27	0.18
Maximizing Var(x)	r_{xy}	0.54	0.67	0.84	0.88	0.95
	b_{xy}	0.26	0.26	0.26	0.26	0.26
By proximity	r_{xy}	0.54	0.63	0.70	0.84	0.81
	b_{xy}	0.26	0.27	0.28	0.28	0.34

Source: Adapted from H.M. Blalock, Jr. *Causal Influences in Nonexperimental Research.* Chapel Hill: University of North Carolina Press, 1964, Table 4, p. 103.

we expect the proportion of the variance in y accounted for by the variance in x to remain roughly constant. The correlation coefficients should not change, apart from small sampling fluctuations. Furthermore, there is no reason to postulate a change in the functional relationship, and the regression coefficient should remain roughly constant. The results in Table 5.9 confirm these expectations, if we remember that with $n = 10$ we can expect quite large random fluctuations. The average values for the coefficients for the grouped data are $r = 0.54$ and $b = 0.28$, compared with the original county values of 0.54 and 0.26. In the first experiment, we find a data aggregation method that does not affect the correlation coefficient.

Consider the second experiment. The counties are grouped to maximize the variance of x. Counties are ranked in terms of their x values and then divided in this form. In the case of $n = 75$, for instance, the two counties with the highest scores on x form the first pair group, and then the counties with third and fourth highest scores on x form the second pair group, and so on. This has the effect of maintaining as much of the variance in x as possible during the various steps in aggregation. But because u, v and w are not related to x, the effect is similar to the effect of random grouping and their respective unknown variances decline more rapidly. Because y is partially dependent on x, its variance declines but not as rapidly as u, v, and w. The relative effect of u, v, and w on y is reduced considerably, leaving x to account for a progressively higher proportion of the variance of y. Hence, we predict a rapidly increasing correlation coefficient, which Table 5.9 confirms. The regression coefficients should not vary, and this too is confirmed.

The third experiment is the one that usually occurs in actual data aggregation. Counties are grouped in terms of proximity, no account being taken of either x or y values. The results in Table 5.9 show that this most common method of aggregation does in fact fall between the preceding experiments in terms of effect on the correlation coefficients. There is the usual rise in value, but this is less marked than in the case in which the variance of x is maximized. The regression coefficients are stable. We can explain the rising correlation coefficients with increasing scale, reported by many researchers, as being the result of the independent variable's spatial autocorrelation whereby grouping by proximity maintains much of the variable's variance. Because most variables we deal with in geography are autocorrelated, this scale effect is very common. Our experiments show that, if the two variables are not autocorrelated, however, the usual proximity aggregation does not maintain either variance, and the correlation coefficient does not increase with scale. These experiments clearly open a whole new way of looking at areal association and its relation to spatial autocorrelation among variables.

In Chapter Two we suggested that systematic biases are a problem only when they are unpredictable. The scale problem begins to seem to have been misnamed.

It is hardly a problem if we can predict its outcome as accurately as Blalock did in his experiments. We must think about what is going on in the relations we study and not rely merely on computation of coefficients to tell us what we want to know. The technical problems associated with such analysis are bound up intimately with methodological considerations. Questions must always return ultimately to the purpose of the research. We have already defined the purpose of much areal association research as the search for models and theories to explain the spatial patterns and processes on the Earth's surface; we must also consider how well actual studies have contributed to this end in practice. Such discussion must be in two parts. First, we can consider the contributions of areal association studies that have been carried out and, second, we can take a general look at the overall method and consider its possible future contributions.

Toward More Theoretical Areal Association Studies

As far as the existing studies of areal associations are concerned, we must conclude that their contributions to developing geographical theory have, to say the least, been disappointing. Despite the widespread use of statistical techniques, the results have been of limited interest and have not been used to draw together generalizations between studies in an attempt to develop theory. Perhaps we can argue that the search for determinants is a part of theory-searching that has not yet borne fruit. Other geographers have been critical of many of these studies because they are devoid of reference to theory. We have a list of variables associated with voting patterns in Newfoundland, substandard housing in American cities, and urbanization in Peru, but what do they contribute to geographic theory? They seem to be nothing more than a series of unconnected empirical statements. Even in the more connected framework of a multiple regression equation, the theoretical implications are by no means clear. We have presented two successful multivariate models, one that accounts for 80 percent of leftwing voting in England and Wales and another that accounts for over 80 percent of the variation in rural population density in the American Great Plains. These are highly successful models from a statistical viewpoint, but how important are they from a geographical perspective? It is not the statistical performance that is the criterion for utility of a model but its geographical relevance. It is not clear how the leftwing voting model adds to our knowledge of patterns and processes relating voting decisions to space. The relevance of rural population density to a theory of rural settlement has not been proven, and our assessment of this study also must be muted. We can appreciate that sophisticated additions to the regression model such as an efficient, comprehensive, stepwise procedure are of minor importance only. The dominant need, if areal association studies are to bear fruit, is that they be part of a research strategy dealing explicitly with the development or testing of theory. The areal association studies we have discussed all tend to be isolated empirical studies waiting for a theory to turn up into which their findings can neatly slip. This criticism is far more fundamental than quibbles concerning the statistical correctness of their data for calibrating their statistics. What does it matter that a significance test showing a correlation significantly greater than zero is not appropriate, if the finding is never subsequently incorporated into any model or theory? It is not statistical significance that is ultimately of importance but, rather, the geographical significance of the results.

Our conclusion is, therefore, that the main problem of areal association is methodological, the lack of theoretical frameworks. This is easier to say than to remedy. Generally acceptable theories have been rare in social science, and in geography in particular. McCarty originally suggested that one source of hypotheses for areal association studies in economic geography should be economic theory. This is indeed one way to relate areal association to theory: we borrow theory from other social sciences. This is, in effect, to take an aspatial model and translate it into a spatial context. The Lorenz curve studies of segregation that show that segregation reflects cultural differences may be considered an elementary example; it translates a loose generalization of sociologists concerning social distances into hypotheses concerning physical distances as reflected in segregation patterns. But because economics has the most developed theory in the social sciences, it follows that it is the most likely source from which to borrow models for translation into space.

The most explicit example of translating an aspatial model into a study of areal association is William Warntz's development of a *geography of price*. When we derived a measure of market concentration of commodities, we used economic population potentials to represent areal variations in the market and commodity production potentials to represent areal variations in production. We can interpret these two variables as direct measures of the spatial pattern of supply (production) and demand (market). In conventional economics the law of supply and demand states that the price of a commodity is related directly to demand and inversely to supply. Instead of relating our two variables to each other, we can use them as two independent variables that determine areal variations in commodity price. The translation of the law of supply and demand into a spatial context produces a geography of price based on the model

$$x_1 = a - b_2x_2 + b_3x_3$$

in which x_1 is areal variation in commodity price, x_2 is areal intensity of commodity supply (production potential), and x_3 is areal intensity of commodity demand (economic population potential). This model has, in fact, been confirmed in several contexts. State price variations for potatoes in the United States can be described by the equation

$$x_1 = 16.32 - 0.99x_2 + 0.131x_3$$

(using the patterns of supply and demand in Figures 5.11b amd 5.11d). This equation accounts for 52 percent of the areal variation in potato price. This encouraging result is repeated when we attempt to explain variations in egg prices for Swedish counties. The egg producer's price (x_1) can be predicted from

$$x_1 = 4025 - 0.0718x_2 + 0.0041x_3$$

in which x_2 is egg production potential and x_3 is population potential. In this case, the coefficient of multiple correlation is 0.71, indicating that about half the variation in egg price is accounted for. The partial correlation coefficients are $r_{12.3} = 0.33$ and $r_{13.2} = 0.63$, indicating that demand is, on its own, associated with much more of the variation in egg price than is supply. We have an interesting and successful transplantation from economic theory to spatial analysis by way of areal association.

The argument calling for a more theoretical basis for areal association studies is clearly supported by history. We can compare our criticism of many of the recent, quantitative areal association studies with early, theoretically oriented research on spatial patterns in the city carried out by the Chicago School of Sociology in the first twenty-five years of the twentieth century. A major research tool of this school was the map and a basic research method was visual map comparison. Despite their technically unsophisticated approach, the theoretical results of their prequantitative map analyses are still reported in geography textbooks half a century later. In recent years these studies have been built upon by geographers using a more sophisticated areal association approach than we have presented. Their approach, called factorial ecology, is the subject of Chapter Six.

Further Reading

Map comparison is discussed as a basic method of study in methodological statements from many schools of geography. The specific areal association arguments we present can be found more fully discussed by McCarty (1954, 1956). Proof of the popularity of this approach in modern geography can be found in Lavalle *et al.* (1967).

The elementary methods of map comparison can all be followed up easily in the literature. Visual comparison with overlays is from Lewis (1965), and its numerical development as a coefficient of areal correspondence is in Minnick (1964). By far the greatest amount of literature is on the Lorenz approach. It was originally introduced into geography by Wright (1937) but much subsequent development has come from outside geography. The technique is discussed generally by sociologists Duncan and Duncan (1955) and by Duncan *et al.* (1961), who consider other measures of spatial separation beyond the simple coefficient of dissimilarity. Applications of this index can be found for the following topics. Isard (1960) and Conkling (1963) discuss application to employment totals. Chisholm (1962) and Johnson (1967) discuss agricultural land use. Application of the Lorenz curve to social patterns is taken up by Duncan and Lieberson (1959), Taueber and Taueber (1965), Lieberson (1963), and Clarke (1971). These references include the sources for some of the discussion in this chapter and give other examples as supplementary material.

Correlation and regression are standard statistical techniques and are described as such in most elementary statistical textbooks and geographic introductions to statistics; Gregory (1968) is one example. Rose (1936) gives an early example of geographical interest in this approach but it was not followed up immediately in geography. The text above describes several more recent but nonetheless relatively early geographical correlation studies by Hartman and Hook (1956) and Warntz (1955). The technical presentation of correlation and regression as they relate to maps was similarly presented about this time by Robinson and Bryson (1957) and Robinson (1962). Other simple examples used in this chapter are in Smith (1968) and Warntz (1956). Other examples in the literature are by Hidore (1963) and Alexander and Lindberg (1961). Robinson (1962) introduces the use of residual mapping, and this is explicitly developed in Thomas (1968) and Clark (1967). We have used examples from all three of these last references, and they can be supplemented by the explicit use of the approach in a research design by Taaffe *et al.* (1963). The multiple correlation study we used is by Robinson *et al.* (1961) and the multiple regression examples are from Roberts and Rumage (1965) and Warntz (1955). Thomas (1960) is a good complementary reference.

The major supplementary reading for this chapter should relate to the final section on the problems of analysis. The most elementary general textbook on these techniques is by Ezekial and Fox (1959) and it is a good starting point. The concepts and assumptions are developed to a more advanced level in Snedecor and Cochran (1966), Draper and Smith (1966), and Guest (1961). These last two references are quite specialized and relate directly to regression analyses. The assumptions of the regression model have been drawn together specifically for geographers by Willis (1975) and Poole and O'Farrell (1971), whose work we have drawn on. The village growth study is by Hart and Salisbury (1965). The scale problem is explicitly dealt with by Robinson (1956), whose weighting solution has been shown to be merely relevant to a special case by Thomas and Anderson (1965). Their inferential solution is far from satisfactory (Curry, 1966), and the approach we have adopted is that of Blalock (1964, ch. 4). The sociological literature on the ecological fallacy can be found in articles by Robinson (1950) and Goodman (1959) and in a more recent review and general clarification by Alker (1969)

Further specific problems in using regression analysis in geographical contexts have been identified and discussed recently by several authors. Testing for spatial autocorrelation among residuals is considered in Cliff and Ord (1972, 1973); the effects of spatial dependence on regression estimates is presented with a solution by Martin (1974) and this is also discussed by Hepple (1974) and Unwin and Hepple (1974); Wrigley (1973, 1975) considers the problem of restricting regression predictions to within a specified range (for example, percentages); Haworth and Vincent (1974) consider prediction errors.

Our discussion of the need for theory follows an early discussion by Burton (1963), and our examples are by Warntz (1957, 1959) and Tegsjo and Oberg (1965). These examples may be supplemented by Knos (1962), which relates a correlation and regression analysis of land values to existing models and theories of urban structure.

Further work on this sort of analysis can lead in several directions. The most common extension is in the direction of factor analysis (the subject matter of our Chapter Six). Correlation analysis of small numbers of variables may be extended in a quite exciting fashion by utilizing causal models. This type of model-building is described in detail by Blalock (1964) and illustrated in geography by Cox (1968) with comments by Taylor (1969). One extension of regression analysis that has been suggested by Till(1973) is the use of major axes lines for specifying linear functions. This enables us to avoid choosing the dependent variables when the variables themselves do not fit this particular type of organization. By far the most common extension of regression is in surface fitting to a set of points over an area. Trend surface analysis is a description of a variable over an area by specifying two locational coordinates as the independent variables. The approach is introduced into geography by Chorley and Haggett (1965); Norcliffe (1969) enumerates some of the problems.

Work Table 5.1 Correlation Coefficients

Purpose

To measure the degree of correspondence between two sets of figures.

Data

Areal unit	y	*rank*	x	*rank*
A	0	1	1	1
B	1	2	3	2
C	5	4	4	3
D	3	3	8	4

Definition

For ranked data, Spearman's rank correlation coefficient is

$$r_s = 1 - \frac{6 \sum d_i^2}{n^3 - n}$$

in which d_i is the difference between ranks.
For interval/ratio data, Pearson's product moment correlation coefficient is

$$r = \frac{\Sigma(x - \bar{x})(y - \bar{y})}{\sqrt{\Sigma(x - \bar{x})^2 \, \Sigma(y - \bar{y})^2}}$$

Arithmetic

For ranked data

y ranks	*x ranks*	*Difference (d)*	d^2
1	1	0	0
2	2	0	0
4	3	1	1
3	4	1	1

For interval/ratio data ($\bar{x} = 4, \bar{y} = 2.25$)

$$r_s = 1 - \frac{6 \cdot 2}{4^3 - 4} = 1 - \frac{12}{60} = 0.8$$

x	$x - \bar{x}$	$(x - \bar{x})^2$	y	$y - \bar{y}$
1	−3	9	0	−2.25
3	−1	1	1	−1.25
4	0	0	5	+2.75
8	+4	16	3	+0.75
		26		

$(y - \bar{y})^2$	$(x - \bar{x})(y - \bar{y})$
5.0625	6.75
1.5625	1.25
7.5625	0
0.5625	3.00
14.75	11

$$r = \frac{11}{\sqrt{26 \cdot 14.75}} = \frac{11}{\sqrt{383.5}} = 0.5617$$

Exercises

Areal unit E has a y value of 2 and an x value of 19.

1. Recompute r_s to include area E. What effect does this extra observation have on the ranked correlation?
2. Recompute r to include area E. What effect does this extra observation have on this correlation coefficient?

Work Table 5.2 Simple Linear Regression

Purpose

To fit a linear function to a set of data points on a graph.

Data

As in Work Table 5.1

Definition

In the equation $y = a + bx$

$$b = \frac{\Sigma xy - ((\Sigma y \, \Sigma x)/n)}{\Sigma x^2 - ((\Sigma x)^2/n)}$$

$$a = \bar{y} - b\bar{x}$$

Arithmetic

	y	x	xy	x^2
	0	1	0	1
	1	3	3	9
	5	4	20	16
	3	8	24	64
Total	9	16	47	90

$$b = \frac{47 - (9 \cdot 16)/4)}{90 - 16^2/4} = \frac{11}{26} = 0.4231$$

$$a = 2.25 - 0.4231 \cdot 4 = 0.5576$$

Regression equation is $y = 0.5576 + 0.4231x$

Exercises

1. Plot the x and y values on a graph and also mark the y intercept (a), the point of the two means ($\bar{x}$, $\bar{y}$), and the value of y when $x = 10$. Draw in the regression line.
2. From Work Table 5.1, we know that area E has values of $y = 2$ and $x = 19$. Plot this new observation plus the original four on a new sheet of graph paper. Find a and b for this larger data set and draw in the new regression line. What effect does this new observation have on your line?

Work Table 5.3 Multiple Linear Regression

Purpose

To fit a function to a set of three or more variables

Data

As for Work Table 5.1, with variables renamed x_1 and x_2 plus a new variable, x_3.

x_1	x_2	x_3
0	1	10
1	3	6
5	4	3
3	8	1

$\bar{x}_1 = 2.25$; $\bar{x}_2 = 4.00$; $\bar{x}_3 = 5.00$

Definition

In equation $x_1 = a + b_2x_2 + b_3x_3$

$$b_2 = \frac{\Sigma(x_1-\bar{x}_1)(x_2-\bar{x}_2)\Sigma(x_3-\bar{x}_3)^2 - \Sigma(x_1-\bar{x}_1)(x_3-\bar{x}_3)\Sigma(x_2-\bar{x}_2)(x_3-\bar{x}_3)}{\Sigma(x_2-\bar{x}_2)^2\Sigma(x_3-\bar{x}_3)^2 - \Sigma((x_2-\bar{x}_2)(x_3-\bar{x}_3))^2}$$

$$b_3 = \frac{\Sigma(x_1-\bar{x}_1)(x_3-\bar{x}_3)\Sigma(x_2-\bar{x}_2)^2 - \Sigma(x_1-\bar{x}_1)(x_2-\bar{x}_2)\Sigma(x_2-\bar{x}_2)(x_3-\bar{x}_3)}{\Sigma(x_2-\bar{x}_2)^2\Sigma(x_3-\bar{x}_3)^2 - \Sigma((x_2-\bar{x}_2)(x_3-\bar{x}_3))^2}$$

$$a = \bar{x}_1 - b_2\bar{x}_2 - b_3\bar{x}_3$$

Arithmetic

$(x_1 - \bar{x}_1)$	$(x_2 - \bar{x}_2)$	$(x_3 - \bar{x}_3)$	$(x_1 - \bar{x}_1)(x_2 - \bar{x}_2)$
−2.25	−3	+5	+6.75
−1.25	−1	+1	+1.25
+2.75	0	−2	0
+0.75	+4	−4	+3.00
			+11.00

$(x_1 - \bar{x}_1)(x_3 - \bar{x}_3)$	$(x_2 - \bar{x}_2)(x_3 - \bar{x}_3)$	$(x_2 - \bar{x}_2)^2$	$(x_3 - \bar{x}_3)^2$
−11.25	−15	9	25
−1.25	− 1	1	1
−5.50	0	0	4
−3.00	−16	16	16
−21.00	−32	26	46

$$b_2 = \frac{(11 \cdot 46) - (-21 \cdot -32)}{(26 \cdot 46) - (-32)^2} = -0.9651$$

$$b_3 = \frac{(-21 \cdot 26) - (11 \cdot -32)}{(26 \cdot 46) - (-32)^2} = -1.1279$$

$$a = 2.25 - (-0.9651 \cdot 4) - (-1.1279 \cdot 5) = 11.7499$$

$$x_1' = 11.7499 - 0.9651x_2 - 1.1279x_3$$

Exercise

Areal unit F has the following values: $x_1 = 6$, $x_2 = 4$, $x_3 = 5$. Recalibrate the multiple regression equation for these three variables with this new observation. What effect does it have on the parameters of the equation?

Work Table 5.4 Multiple and Partial Correlation

Purpose

To find the total and separate correlations of the independent variables on the dependent variable in a multiple correlation situation

Data

As in Work Table 5.3.

Definition

$$R_{1.23} = \sqrt{\frac{\mathrm{Var}(x'_1)}{\mathrm{Var}(x_1)}} = \sqrt{\frac{\Sigma(x_1' - \bar{x}_1)^2}{\Sigma(x_1' - \bar{x}_1)^2}}$$

$$r_{12.3} = \sqrt{\frac{(1 - r_{13}^2) - (1 - R_{1.23}^2)}{1 - r_{13}^2}}$$

$$r_{13.2} = \sqrt{\frac{(1 - r_{12}^2) - (1 - R_{1.23}^2)}{1 - r_{12}^2}}$$

Arithmetic

From Work Table 5.3, $x'_1 = 11.7499 - 0.9651x_2 - 1.1279x_3$

x_2	x_3	b_2x_2	b_3x_3	x_1'	$(x_1' - \bar{x}_1)$	$(x_1' - \bar{x}_1)^2$
1	10	−0.9651	−11.2790	−0.4942	−2.7442	5.4884
3	6	−2.8953	−6.7674	+2.0872	−0.1628	0.0265
4	3	−3.8604	−3.3837	+4.5058	2.2558	5.0886
8	1	−7.7208	−1.1279	+2.9012	0.6512	0.4241
						11.0276

From Work Table 5.1, where $x_1 = y$, $\Sigma(x_1 - \bar{x}_1)^2 = 14.75$, then

$$R_{1.23} = \sqrt{\frac{11.0276}{14.75}} = \sqrt{0.7476} = 0.8646$$

From Work Table 5.1, $r_{12} = 0.5617$ and $r_{12}^2 = 0.3155$.

From Work Tables 5.1 and 5.3,

$$r_{13} = \frac{-21}{\sqrt{46 \cdot 14.75}}$$

$r_{13} = 0.8062$ and $r_{13}^2 = 0.6500$

$$r_{12.3} = \sqrt{\frac{(1-0.6500)-(1-0.7476)}{1-0.6500}} = \sqrt{0.2789} = 0.5281$$

$$r_{13.2} = \sqrt{\frac{(1-0.3155)-(1-0.7476)}{1-0.3155}} = \sqrt{0.6313} = 0.7946$$

Exercise

Use the multiple regression equation from adding areal unit F in the exercise in Work Table 5.3 and calculate revised multiple and partial correlations for this larger data set. What effect does this new observation have on the relative levels of correlation?

Work Table 5.5 Student's *t* Tests for Correlation and Regression Coefficients

Data

Random sample of settlements

Settlement number	1	2	3	4	5	6	7	8	9	10
Population (y) (100s)	1	8	15	20	21	24	28	32	35	36
Distance (x) to nearest larger settlement (mi)	1	3	4	3	6	5	7	9	10	12

H_0

1. r (= 0.929) is not significantly different from 0.

2. r_s (= 0.967) is not significantly different from 0.

3. b_{yx} (= 3.06) is not significantly different from 0.

Statistics

1. $t = \dfrac{r\sqrt{n-2}}{\sqrt{1-r^2}} = 7.1001$ with $df = n - 2 = 8$

2. r_s can be assessed directly from a table of critical r_s values (Appendix Table A) with $n = 10$.

3. $\text{SEb} = \dfrac{\text{SE}\,y'}{\sigma_x\sqrt{n-2}} = 0.3045$

$t = \dfrac{b-0}{0.3045} = 10.049$ with $df = n - 2 = 8$

Probabilities

$t(\alpha = 0.05, df = 8) = 2.306$

$r_s(\alpha = 0.05, n = 10) = 0.564$

Decisions

We can reject H_0 for each case, and we can conclude that the population coefficients differ from 0 at this level of significance.

Exercise

When we add two further settlements, numbers 11 and 12, to our sample, we produce observations of $y = 40$, $x = 2$, and $y = 50$, $x = 5$. These new observations depress all three coefficients so that $r = 0.497$, $r_s = 0.518$, and $b = 2.051$ (SEb $= 1.241$). Are the revised coefficients significantly different from 0 at $\alpha = 0.05$?

VI

FACTORIAL ECOLOGY

When scientists are faced with explaining complicated patterns of phenomena they follow one of two paths. They attempt to simplify the situation either deductively, by making assumptions that dispense with unwanted complications, or inductively by carefully controlling laboratory experiments. The ensuing mathematical analysis usually involves two or three variables. This has been the path in the physical sciences. In Chapter Five we illustrated how aspects of this method can be adapted for spatial analysis in using a map transformation to test central place theory. This experiment was reasonably successful, but it remains a very limited approach for geographers interested in explaining other, more complicated real world situations. The alternative path has been to tackle the complicated pattern directly without excessive simplification. This has long appealed to social scientists in their work outside laboratories and is perhaps best epitomized by the traditional regional geographer's holistic approach. Such research has inevitably lacked rigor. Faced with a complicated situation involving a multitude of variables, researchers have had to make decisions about the ordering of a situation based on personal experience and what little theory was available. This approach has not been intersubjective: one person's regional geography has been another person's poison.

The fact remains that social scientists and geographers, in particular, have remained interested in how things happen in the real world. This inevitably requires analysis of large numbers of variables *in situ*. Two developments in science have made direct numerical analysis of such situations possible. First, there has been the development of multivariate analysis in statistics, which allows for simultaneous consideration of large numbers of variables. Second, the use of such techniques has become feasible with the general availability of electronic computers for researchers. In Chapter Five we met multivariate statistics in a simple form as *multiple regression*. Analyses using this technique usually involve less than ten variables. In this chapter, we are concerned with the techniques of *factor analysis*, which typically involve more than ten variables and sometimes include more than one hundred variables. It has been christened quite aptly a calculus of the social sciences. It is a major tool for structuring complicated real world situations into interpretable patterns. Thus, factor analysis has led a major research thrust in geography in the past decade under the general title of *factorial ecology*.

FACTOR ANALYSIS: THEORY AND CONCEPTS

For the mathematically untrained student, factor analysis has the guise of a very difficult and complicated technique. Textbooks in factor analysis abound with unfamiliar notation from an unfamiliar language—matrix algebra. The use of this powerful calculus enables us to express the assumptions and theorems related to factor analysis in a neat and succinct style. It also has the effect of making the statements unintelligible to people not well-versed in matrix algebra. In our exposition of factor analysis, we assume no knowledge of matrix algebra and consequently sacrifice a succinct general style for a slower, more verbal presentation. We begin historically by considering the research context that stimulated factor analysis as a way of introducing more sophisticated modern applications.

Two-Factor Analysis

Although we have emphasized the large-scale multivariate nature of factor analysis, it evolved from a much simpler measurement problem. It is still useful to consider this original simple problem because it incorporates the basic elements of all factor analyses.

The Initial Theory

Factor analysis has its origins in psychology at the beginning of the present century. Psychologists at that time possessed what seemed the useful concept "general intelligence" but they did not agree on how it should be measured. Their problem was to convert the theoretical concept into an empirical concept that could be measured. Their solution was factor analysis.

Consider a class of ten pupils, each of whom has been tested in four subjects—history, geography, physics, and chemistry. We expect no one pupil to obtain exactly the same mark in every test, but it will probably happen that the pupils who score high on one test (say, history) will tend to score high on the three other tests. One theoretical interpretation of this situation is that every test score reflects two influences—the general intelligence of the pupil and a specific aptitude for the subject that the pupil is being tested on. A simple averaging of pupil scores confuses these two separate influences. Factor analysis allows for their separate identification.

We can summarize the argument in simple algebra:

$$x_1 = f(g) + f(u_1)$$

in which x_1 is the variable made up of test scores on one subject (history), which is a function of the one basic common factor g and a second factor, u_1, that is specific to x_1. Similarly, scores on x_2 (geography) are composed of

$$x_2 = f(g) + f(u_2)$$

Similar equations can be written for the test scores on the other subjects. We can make all four equations more specific if we assume linear functions, so that the equations become:

$$x_1 = w_{1g}g + w_{1u}u_1$$

$$x_2 = w_{2g}g + w_{2u}u_2$$

where w's are weights, relating factors to variables. We assume that the variables have been standardized to zero mean and unit variance, eliminating the base constant from the functions. The weights are the equivalent of the beta coefficient in a regression equation. The weights for the common factor are of particular interest because they relate the observed test scores on the four subjects to the underlying concept of general intelligence. High weightings indicate that test scores on a particular subject reflect closely the underlying intelligence factor; conversely, low weights indicate that test scores in that particular subject do not closely reflect the general factor. We can, therefore, use these weights to derive estimates of this underlying factor.

We have a theory to separate intelligence from test scores on various subjects, which we now have to use to produce a measure of general intelligence. Our theory tells us that it is composed of parts of the test scores of the various subjects. Let's take the results of the first student, whom we call Carl. We have four marks for Carl for history (x_{1C}), geography (x_{2C}), physics (x_{3C}), and chemistry (x_{4C}). We also have weights for each of these subjects: w_{1g}, w_{2g}, w_{3g}, and w_{4g}. We combine these two pieces of information to compute a derived measure of intelligence for Carl:

$$g_C = w_{1g}x_{1C} + w_{2g}x_{2C} + w_{3g}x_{3C} + w_{4g}x_{4C}$$

Because we have four marks for the nine other students, we can compute nine more general intelligence scores for each of them. Thus Helen's general intelligence score is:

$$g_H = w_{1g}x_{1H} + w_{2g}x_{2H} + w_{3g}x_{3H} + w_{4g}x_{4H}$$

We can therefore produce ten scores on general intelligence derived from the original four sets of ten test scores. We create a new composite variable, g.

The basic problem of this approach is easy to identify: How can we put values to these theoretical weights? In the remainder of this section on theory and concepts we attempt to show how factor analysis tackles this problem. The theory behind factor analysis considers the correlations between the original test scores. If the scores for the history test are similar to the scores for geography, then the two sets of scores are highly correlated. This high correlation is interpreted as the result of the subjects' commonly close relationship to the general intelligence factor. If the correlation between two sets of test scores in history and physics is found to be low, this can be interpreted as the result of one or both of them being only mildly related to the general factor. This argument can be interpreted algebraically

$$r_{12} = r_{1g}r_{2g}$$

r_{1g} and r_{2g} are the theoretical correlations between variables x_1 and x_2 and the unknown factor g. This is the basic *cross-products theorem* of factor analysis and it states simply that the correlation between any two variables is equal to the product of the two correlations of the variables with the underlying factor.

To clarify, let's return briefly to the education example. We might find that history has an average correlation with the three other subjects of 0.9. Physics has an average correlation with other tests of only 0.6. This shows that the history test scores simply correspond more closely to the other test scores and, we assume, correspond more closely to general intelligence.

We have looked at the original factor analysis application in some detail because it illustrates many essential characteristics of modern factor analysis. Its logic is based on the cross-products theorem concerning the correlations between the variables and factor and their subsequent use as weightings in the two-factor equation. Modern derivation of the correlations involves the use of matrix algebra and, at this stage, many descriptions of factor analysis would be set into matrix notation. We have already indicated that we will not do this. We are very fortunate in that all the relevant steps in the process of defining factors can be illustrated geometrically. This is true of both the original methods and, as we shall see, the underlying concepts from matrix algebra.

A Geometric Interpretation

Because the cross-products theorem relates to correlation coefficients, it follows that any geometric interpretation of factor analysis must begin with a geometric definition of correlation. We can derive such a definition by considering the location of variables in an object space. We normally plot data on a graph by defining the axes in terms of two variables and locating the objects on the graph by their values on these variables. An alternative method is to define the axes in terms of objects and then locate variables, in this object space, in terms of their measures for the two objects. Two objects might be two towns A and B, as specified in Figure 6.1. We can locate two variables on the graph from their values for the towns. We assume from this point that variables are standardized in terms of unit variance and zero mean. The two variables population size (x_1) and population growth (x_2) have values of $+1$ and -1 for town A and -1 and $+1$ for town B. We plot these on the graph, not just as dots but as lines (or vectors) from the origin. The reason for presenting this data in this way will become apparent.

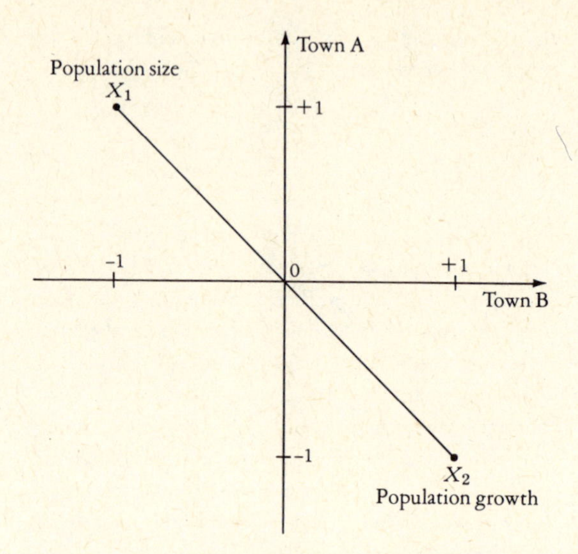

Figure 6.1 An object space.

We met vectors in Chapter Two and found them to be defined in terms of distance and direction. The distance or length of our variable vectors represents the magnitude of their variance and the direction relates them to other variable vectors. Because the variables are standardized, all vectors are of the same length, one unit. The directions are much more interesting. The more alike our two variables are, the smaller the angle between their vectors is. The limiting case is clearly when both have exactly the same values for the two objects and therefore define identical vectors. In this case, they are perfectly correlated ($r = +1.0$) and the angle between the vectors is 0. As they become more dissimilar, the angle between their vectors increases. In the example in Figure 6.1, the angle between the vectors is a maximum 180° because the vectors define a straight line. We can relate this angle to the correlation coefficient by using the trigonometric concept of the cosine. Remembering that for angles ($\theta°$) greater than 90° the cosine is defined as

$$\cos \theta° = -\cos(180 - \theta)°$$

we find that the cosine of 180° is -1. This is the same as the correlation coefficient between these two variables. If the variable vectors had been identical, the angle between them would have been 0° with a cosine of $+1$, the same as the correlation coefficient. Our simple two-object example clearly suggests a definite relation between the correlation coefficient and the cosine of the angle between the variable vectors. Unfortunately, with two objects, the correlation coefficient can take only the two values (-1 and $+1$), as we have discussed. We cannot extend this relation further in two-dimensional object space. It can be shown, however, that the same relationship obtains in multidimensional space in which variables are vectors defined by values for several objects. In general, we say that, with standardized data, the correlation coefficient for two variables is equal to the cosine of the angle between their two vectors in the object space.

We can illustrate a space with three dimensions by using perspective but, beyond this number of axes, illustration is not directly possible. We therefore use simple two-dimensional diagrams, assuming that the

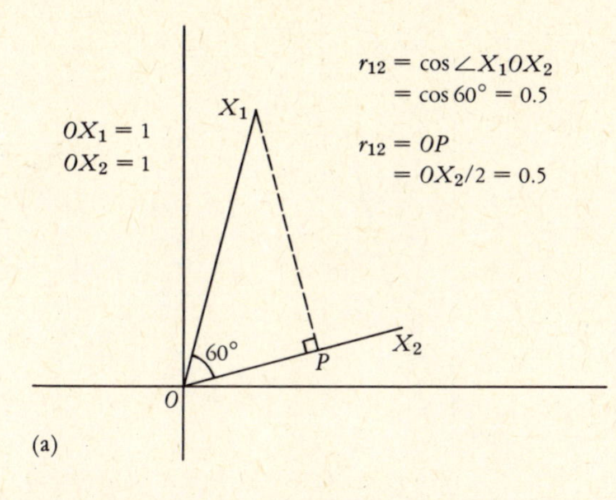

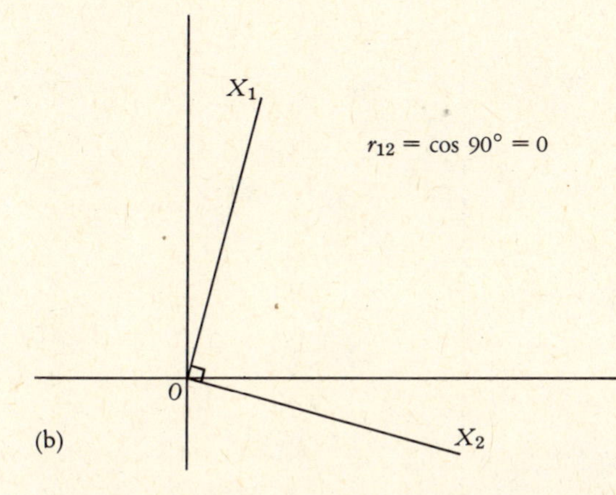

Figure 6.2 Further relationships in an object space: (a) two definitions of correlation, and (b) orthogonal variables.

patterns displayed apply to spaces of higher dimensions. In Figure 6.2, two diagrams of variable vectors are shown, both illustrating further correlations. In Figure 6.2a, the angle between the vectors is 60°, indicating a correlation of + 0.5. If we project one of the vectors onto the other, to define a right angle, we can use the resulting triangle to obtain a second geometric definition of the correlation coefficient. We specify this triangle as x_1 OP in Figure 6.2a and, because we know that

$$r = \cos x_1 \text{ OP} = \frac{\text{OP}}{\text{O}x_1}$$

and $\text{O}x_1 = 1$ (the variance of x_1), it follows that

$$r = \text{OP} = \frac{\text{O}x_2}{2} = 0.5$$

We shall have need of this second geometric definition of the correlation coefficient later.

If we repeat the exercise for Figure 6.2b, we find that the projection produces no distance to represent the correlation, because the vectors are already at 90° to one another. This is, of course, consistent with the relation between the variables, because the cosine of a right angle is 0, indicating no correlation. Such variables are said to be *orthogonal* to one another.

Normally we analyze many more than simple pairs of variables. Let's consider a four-variable example. We might consider the test scores from the initial education example or we might extend the two-variable urban comparison by adding two more variables—say, birth rate (x_3) and death rate (x_4). The correlations between the four variables are shown in the lower lefthand section of Table 6.1. In the top righthand section of this matrix, these correlations are represented by angles derived from considering the correlations as cosines. We can draw Figure 6.3 to illustrate these variables as vectors. We are in a position to interpret geometrically a factor analysis for a common general factor underlying the variables. This common factor is a new vector that best summarizes the pattern of the observed variable vectors. The *resultant* of the original vectors is such a new vector. It

Table 6.1 Relationships Depicted in Figure 6.3 (Cosines = Correlations)

Variable		*Angles*		
	x_1	x_2	x_3	x_4
x_1	0°	10°	60°	90°
	1.0000			
x_2		0°	50°	80°
	0.9848	1.0000		
x_3			0°	30°
	0.5000	0.6428	1.0000	
x_4				0°
	0.0000	0.1736	0.8660	1.0000

is the average direction as defined in Chapter Two and is produced simply by adding together the variable vectors. Such addition can be represented geometrically as a polygon (Figure 6.4a) to which each vector contributes a side; the final vector joining the end point of the additions to the origin is the resultant. It lies through the middle of the original vectors, as illustrated in Figures 6.3 and 6.4a, and as such it can be considered a summary of their interrelationships. Like any other vector, it is defined in terms of direction and length, both of which have important meanings for factor analysis. The overall length of the resultant vector is a measure of how well the factor describes the set of variables. If all the variables are perfectly correlated and define identical vectors, their summation produces a line rather than a polygon, and this line is the resultant (Figure 6.4b). Because each of the variables has unit length, the length of the resultant is n, when there are n variables. Notice that, with n variables each of unit variance, total variance equals n. In this limiting case, therefore, the factor can reflect the variables perfectly, because they are all identical and, hence, it can account for all the total variance. The length of the resultant is n. As the variables become less related to one another, the summation polygon takes on a broader shape, and the resultant does not finish as far from the origin (Figure 6.4c). The actual length of the resultant tells us the amount of variance

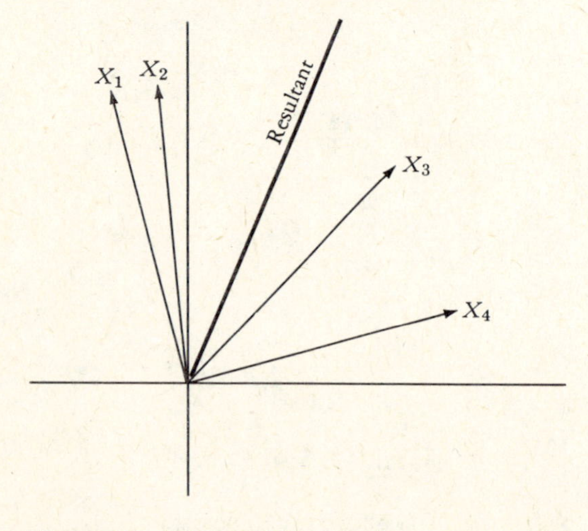

Figure 6.3 A four-variable example.

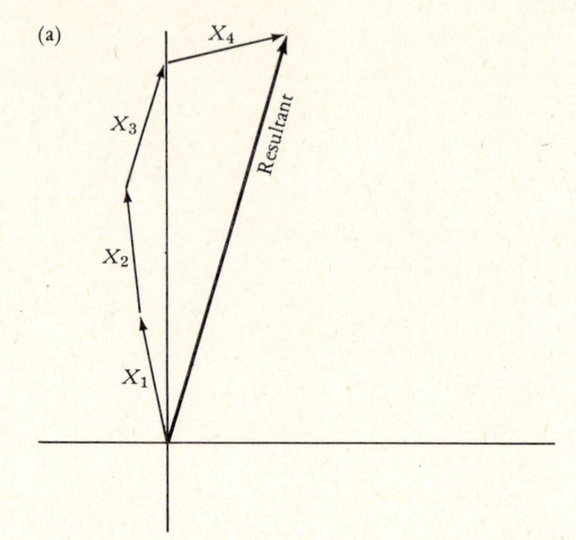

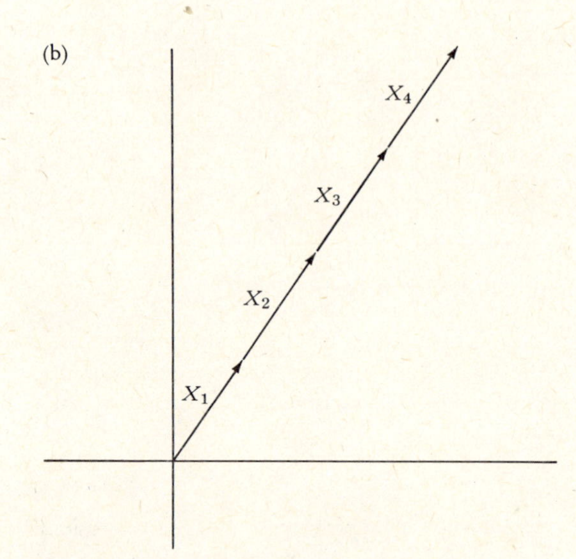

Figure 6.4 Vector polygons.

that the factor accounts for. In the limiting case, in which variables are orthogonal, the vectors cancel one another out when they are summed so that the resultant is zero. There is no common factor between orthogonal variables (Figure 6.4d).

The length of the resultant tells us the overall relation of factor to variables; the direction of the resultant tells us the individual relation between the factor and

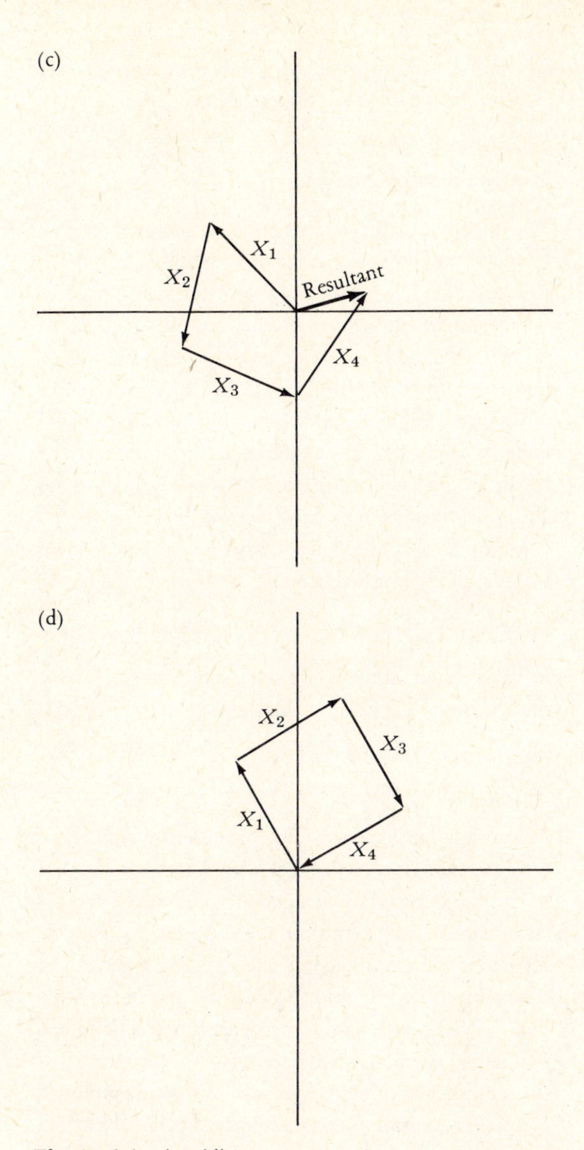

Figure 6.4 *(cont'd)*

each variable. The angle between a variable vector and the resultant indicates how closely the resultant reflects the variable in the space, and the cosine of this angle is the correlation between the variable and the factor. An alternative way of measuring this is to project each variable onto the resultant at right angles (Figure 6.5a), so that the length of the projection represents the correlation, as we have shown. These correlations are often called *loadings* in the context of factor analysis. They have, in fact, more roles than the simple correlation coefficient. As well as being a measure of correlation, they are also the weights in the equations relating the factor to the variables. Hence, factor loadings combine the roles of correlation and regression coefficient. In our simple two-variable example in which $r = 0.5$ (Figure 6.2), the resultant

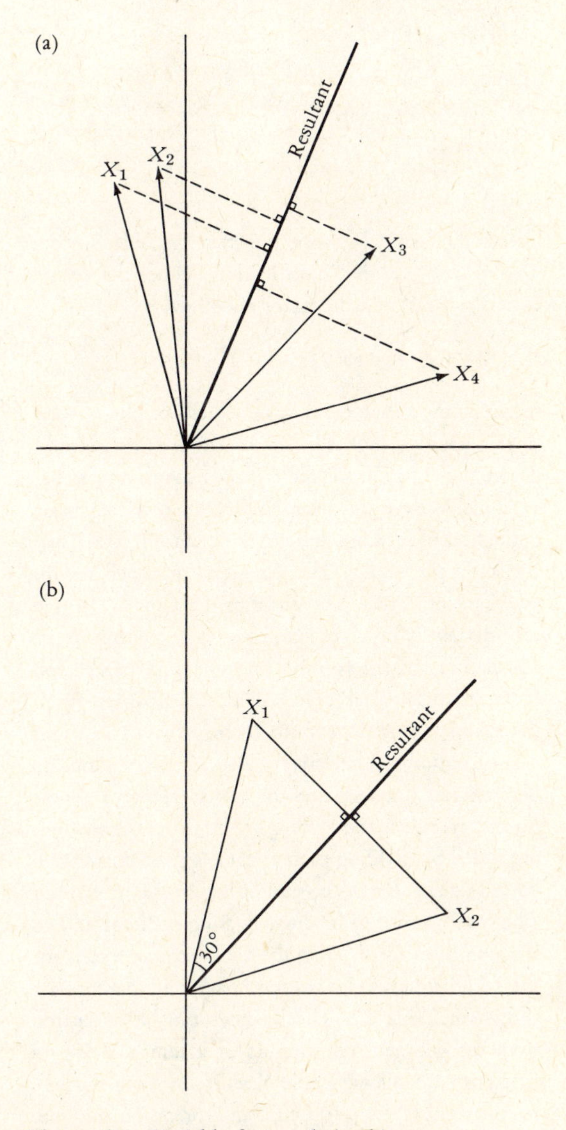

Figure 6.5 Variable-factor relationships.

Table 6.2 Computation of Factor Loadings by Summation Method

	x_1	x_2	x_3	x_4	*Total*
x_1	1.0000	0.9848	0.5000	0.0000	
x_2	0.9848	1.0000	0.6428	0.1736	
x_3	0.5000	0.6428	1.0000	0.8660	
x_4	0.0000	0.1736	0.8660	1.0000	
Total	2.4848	2.8012	3.0088	2.0396	10.3344
$\sqrt{10.3344} = 3.2147$					
Loadings on common factor					
	0.7730	0.8714	0.9360	0.6345	3.215

bisects the angle between the two variable vectors (Figure 6.5b), so that both loadings equal the cosine of 30° = 0.866 and the factor is defined by

$$F = 0.866x_1 + 0.866x_2$$

in which the variables are given equal weight.

We could repeat this approach and graphically derive the equation for the factor in the four-variable example. We use this example instead as an algebraic derivation of loadings and use the diagrams as a visual check on the results. The algebraic approach to this problem of finding loadings is known as simple summation. It involves summing each column in the correlation matrix to find an aggregate correlation between each variable and all other variables. From the cross-products theorem, these gross correlations clearly indicate the relative order of the relationship between the variables and the common factor. These do not represent the loadings. The gross correlations are scaled by dividing each variable's gross correlation by the square root of the sum of the gross correlations. The result is a set of loadings relating each variable to the common factor. For the four-variable example, the correlation matrix, column totals, and loadings are shown in Table 6.2. The loadings can be compared with the geometric illustration of loadings in Figure 6.5. They are, of course, identical. It is not our purpose to prove that the geometric interpretation and the algebraic manipulation are equivalent; proofs are available in factor analysis textbooks.

Multiple Factor Analysis

We introduced this chapter by asserting that we can use factor analysis to structure complicateu real world patterns. We have dealt only with the special case of one common factor in a factor analysis. If all complicated situations we encounter fit the two-factor model, our discussion is adequate. But the world we live in is not that simple, and we must expect that several underlying factors are required to structure a given situation. The original approach to factor analysis has developed into *multiple factor analysis*, whereby several common factors are identified for a group of known variables. Our original equation becomes:

$$x_1 = w_{11}F_1 + w_{12}F_2 + \cdots + w_{1p}F_p + w_{1u}u_1$$

in which there are p common factors plus the factor specific to the variable; in this case it is u_1.

The Principal Axes Method

In multiple analysis, we must produce a whole set of weights to relate each of the p factors to every variable. In practice, our simple algebraic approach becomes rather cumbersome, and this is an appropriate place to introduce what has become the standard approach to extracting factors from a correlation matrix. It is the *principal axes method* and is based on well-established mathematical theorems concerning the properties of the ellipse and matrix algebra. We can employ a geometric interpretation to illustrate essential characteristics and properties.

Let's begin by returning to the interpretation of a set of data in terms of a *variable space*. Each variable forms an axis, and objects are located in the space in terms of their values on the variable axes. The most familiar case is the scatter of dots representing objects on a two-dimensional graph, with the axes defining two variables, although there is no reason why we cannot extend the system to three dimensions and more to include n variables, as we have shown in Chapter Five.

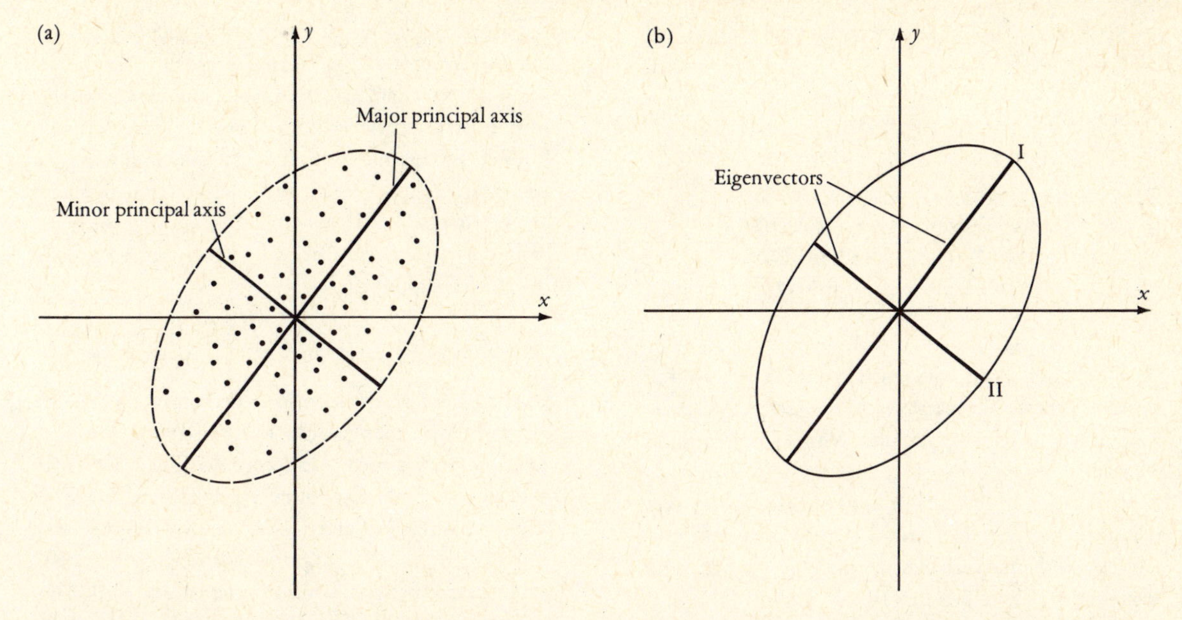

Figure 6.6 Variable relationships as ellipses. a. A swarm of points described by an ellipse. b. Eigenvectors and eigenvalues (eigenvalues are the lengths of the eigenvectors).

For purposes of illustration, we use simple two-dimensional diagrams, remembering that this is a special case in which $n = 2$; in most factor analyses, $n > 10$.

From Chapter Five, we know that the scatter of points in a variable space indicates the relationship between the variables. In the two-dimensional case, if all the points lie on a straight line we have perfect correlation between the variables. Conversely, with no correlation, the points form a circular swarm of points. These two extreme cases of correlation can be viewed geometrically as the two limiting cases of an ellipse. All other swarms of points can be described in terms of true ellipses (Figure 6.6a). Every ellipse is defined by two axes—a major principal axis (the long axis of the shape) and a minor principal axis at right angles to the first axis. If the swarm of points represents the data, these two axes describe two independent sources of variation in the data. The major principal axis is analogous to a regression line and it measures the greatest common variance between the variables. The minor principal axis measures the residual variance not accounted for by the first relationship between the variables. When all points lie on a straight line, the major principal axis measures all the variation, and the minor axis is zero. At the other extreme, with no correlation and a circular swarm of points, the two axes are equal and indeterminate. All data sets lie between these two extremes. With more than two variables, we talk of a hyperellipsoid swarm of data points in the n-dimensional space.

The problem is to define the principal axes of the ellipse. This problem is solved by the principal axis technique. The original variable axes are transformed so that they rotate onto the principal axes of the ellipse (or the hyperellipsoid). In order to do this, we employ matrix algebra. Two sets of basic results are produced—*eigenvectors* and *eigenvalues*. Eigenvectors define the principal axes, and the associated eigenvalues measure the lengths of the principal axes (Figure 6.6b). These two matrix concepts give us our solutions to the geometric problem and the resulting principal axes are the factors we have been searching for.

Table 6.3 Factor Loadings Derived from a Principal Axes Extraction

Variable	*Eigenvector*	
	Factor 1	Factor 2
x_1	+0.81916	−0.57355
x_2	+0.90630	−0.42262
x_3	+0.90631	+0.42261
x_4	+0.57355	+0.81917

A very important question remains. Where are the weights relating the variables to the factors, as specified in our original equation? We can define these weights geometrically if we return to our earlier representation of the data. Instead of making the axes represent the variables, we let them represent the objects, with the variables in this object space. The eigenvectors defining the factors lie in this new geometric arrangement. As in our two-factor example, the relationship between variables and factors is represented by the angles between their respective vectors. Therefore, the weights we are searching for are the cosines of the angles measuring the correlation between variable and factor. Every variable has a correlation or loading with every factor.

We can illustrate the results from a multiple factor analysis using the principal axis method by reconsidering the four-variable example. Four eigenvalues were extracted with values of 2.64275, 1.35721, 0.00004, and −0.00002. Each defines the length of a vector, and in this case there are clearly two relevant vectors. Because there are four variables, total variance is four, and the first eigenvector accounts for $(2.64275/4) \times 100\%$ of this variance = 66.1%. This leaves $(1.35721/4) \times 100 = 33.9\%$ of the total variance accounted for by the second eigenvector. Our four-variable example has, in fact, been chosen carefully so that it can be illustrated in two-dimensions. A four-vector model normally requires a four-dimensional space to contain it. That our first two eigenvalues sum to 4, the total variance, simply confirms what our eyes have seen, that Figure 6.3 is drawn on a two-dimensional plane.

The two eigenvectors are drawn in Figure 6.7, and the resulting loadings are shown in Table 6.3. Notice that the loadings on the first factor are not identical to the loadings calculated by simple summation in the simpler solution in Table 6.2, but the order of the loadings is the same for both cases. We do not expect identical loadings because they derive from different geometric models. Because both models have the same basic purpose, the similarity between the loadings is nonetheless comforting.

Apportioning the Variance

Loadings are equivalent to correlation coefficients and it follows that we can relate them to statistical concepts of variance. Let's consider this. We have noticed previously that every variable is standardized so that its variance is unity. The common factor model with one or more common factors apportions this total variance between common and unique parts. The common variance is the part that a variable shares with other variables so that it is described by the common factors. The unique variance is the part that is unique to that particular variable and it consists of variance accounted for by influences specific to the variable and

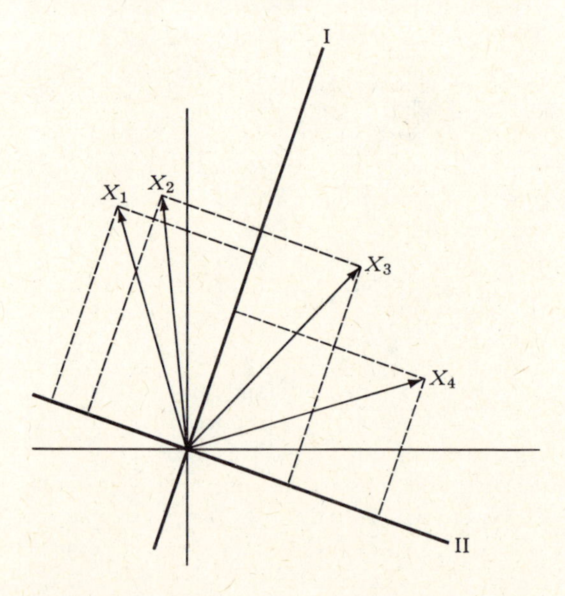

Figure 6.7 Eigenvectors and loadings in the four-variable example.

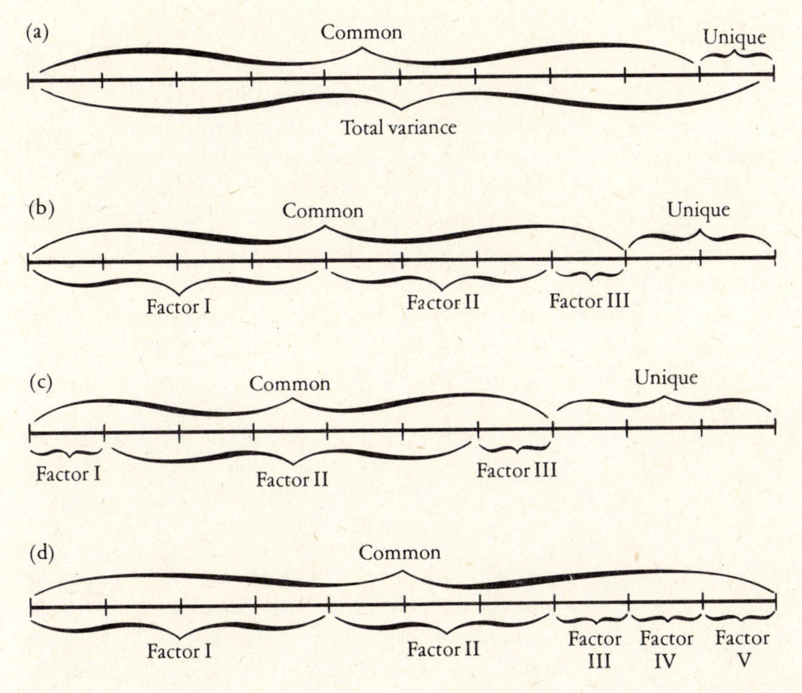

Figure 6.8 Apportioning the variance of a variable. (a) Common and unique variance. (b) Variance accounted for by common factors (1). (c) Variance accounted for by common factors (2). (d) The principal components model.

of other variance resulting from measurement error. Figure 6.8a illustrates this apportioning of the variance. For this variable, 90 percent of the variance is accounted for by the common factors. Every variable has a different breakdown of its variance, depending on its relation to the other variables.

Let's consider this breakdown more closely. If we assume that there are three common factors in our analysis, we might produce two variances such as are displayed in Figures 6.8b and 6.8c. The lengths of variance accounted for by each of the factors are the squares of the loadings, defined as the projections of the variable vectors on the eigenvector. The variable depicted in Figure 6.8b has 40 percent of its variance accounted for by factor 1, 30 percent by factor 2, 10 percent by factor 3, leaving 20 percent uniquely determined. Figure 6.8c shows a different situation, in which factor 2 accounts for 50 percent of the variance and 30 percent is uniquely determined.

In the three examples we have considered, the proportion of variance accounted for by the common factors is 0.9, 0.8, and 0.7. This common variance is termed the *communality* of the variable. Because it is a measure of the importance of the common factors to the variable, it also becomes a measure of the overall relation between the variable and the other variables in the set. This concept of communality is very important in factor analysis theory and leads to circularity of reasoning. The communalities are the values that are fed into the diagonal of the correlation matrix before it is manipulated to produce eigenvectors. Because the communality is the sum of the squared factor loadings, these are not known until the eigenvectors have been determined, but eigenvectors cannot be derived from the correlation matrix until communalities have been included in the matrix diagonal.

This dilemma has stimulated a large technical and methodological literature. The problem has been overcome traditionally by substituting some *estimates* of the communalities in the correlation matrix. This produces eigenvectors that are only estimates of the true common factors. But estimated factors allow new estimates of the communalities to be derived, which can then be used to produce new eigenvectors, and so on. It is assumed that this iterative approach leads to a convergence on the true communalities and factors, although there is no mathematical proof that this does in fact occur.

The question now arises as to what our communality estimate should be. The only explicit guideline is that we know the range within which the true communality lies. The maximum possible communality is obviously one, because the common variance cannot be greater than the total variance of a variable. It is known that the lower limit is equal to the multiple correlation coefficient between that variable and all other variables in the analysis. This value is used often as an initial communality estimate for iterative procedures.

The Principal Component and Common Factor Models

Let's consider the use of communalities of one. It hypothesizes no unique variance for that variable. The equation describing the variable becomes

$$x_1 = w_{11}F_1 + w_{12}F_2 + \cdots + w_{1p}F_p$$

in which there is no specific factor or measurement error. This situation is depicted in Figure 6.8d with five factors. If all variables are hypothesized as having communalities of one—that is, the whole diagonal of the correlation matrix is filled in with unities—the resulting factor analysis is known as a *principal component analysis*. This is a separate factor analysis model that differs technically from the *common factor analysis* model simply in terms of the communalities used. The actual technique of extracting factors from the resulting correlation matrix is the same for both models. The methodological, and indeed philosophical, difference between them is profound.

Principal components analysis assumes a closed model. All the variation in the variables is accounted for by the variables themselves. No outside influences are allowed for. Many factor analysis methodologists have argued that, as a method for modeling real world situations, principal component analysis is totally unrealistic. They argue that, by not allowing for any unique variance in the variables, what unique variance does occur is confused in the solution with the common factors, thus obscuring the true factor structure. Researchers using principal component analysis point to its avoidance of the communality problem, resulting in a mathematically determinate solution. Some mathematicians favor the principal component model on these grounds. The key questions for researchers are:

1. On what occasions do the two models give similar results and when are their results most dissimilar?
2. When the results are dissimilar, how does one choose which model to use?

We can answer the first question simply by referring to the character of the correlation matrix. If all the correlations are large, the communalities are large and the value of one will be a reasonable estimate of the common factor analysis communality. Both models give similar results. Any variable that has low correlations with other variables, however, tends to have a small common variance and large unique variance, so that to insert unity in the communality cell is to use a poor estimate of the true communality. The influence of this variable in the subsequent principal component analysis will be overweighted. If there are many such variables in an analysis, so that the correlation matrix includes a large number of small correlations, the results of a common factor analysis and a principal component analysis will be quite dissimilar.

We can use similar logic concerning the size of the correlation matrix. If there are few variables, the number of correlations is relatively small, so that the communality estimates are relatively more important. Distortions resulting from use of communalities of unity are enhanced with a small data set, and the principal component and common factor analysis solutions can be expected to differ.

When the results from the two models are very similar, the choice of model is not important. When the two solutions are expected to differ, we must face the second question: What model should be used? Almost all such questions in a research context force researchers to trace their steps right back to their research purpose. We can answer the second question by reviewing the pertinent characteristics of each model and relating them to the type of geographical research that employs factor analysis.

We must begin by declaring that the closed model of principal component analysis is unrealistic by definition and thus forms a poor basis for structuring reality. The assumption that all the variance is self-contained in the variables is unlikely ever to occur in reality. The most generous way to view the principal component model is as a simple transformation technique that converts a data matrix with *n* variables, all having unit variance, into a new arrangement of *n* variables in which the first has the largest variance, the second has the second largest variance, and so on, until the final few new variables have insignificant variance. The data may be reasonably described by the first *m* of the *n* new variables. The model is one of data reduction, or simply parsimonious description. Such results may be useful in subsequent analyses or as an initial investigation of a set of data that is largely unknown. As a theoretical tool, the model is severely hampered by its unrealistic assumption.

We seem to have come down in favor of the common factor model, largely because of the principal component model's limitations. What has the common factor model to offer? It is basically a theory about data that, although derived from a particular psychological theory of intelligence, seems to be of general applicability. It seems reasonable that geographers should structure their sets of data describing reality in terms of common and unique variances. The fact is that the principal component model has been used widely in geography. In many cases, it seems that, because of the nature of the correlation matrices used, solutions are not dissimilar to what would have been produced by the common factor model. It is perhaps appropriate to point out that the principal component model can be viewed as simply the limiting case of the common factor analysis model, as the relationship between variables increases and the unique variance declines. In a situation in which correlations are not high and unique variances are large, the common factor model is to be preferred.

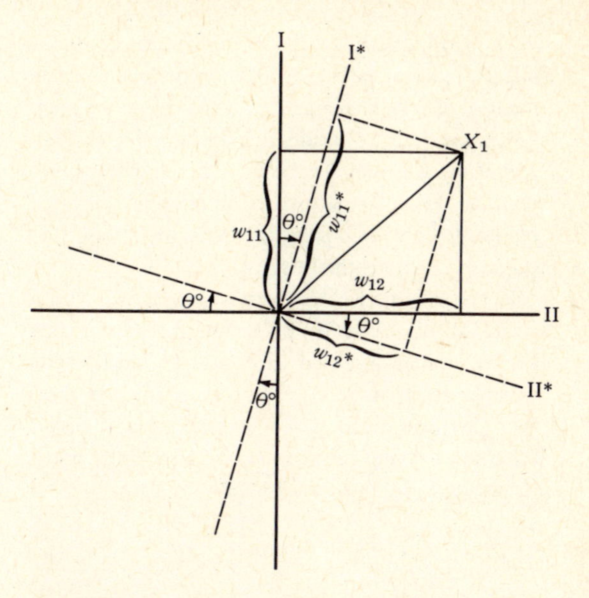

Figure 6.9 Rotation.

Rotation and Factor Comparison

Consider the properties of the factors we have extracted by using the principal axes technique. Each factor is fitted to the data to account for the maximum variance. In practice, this means that the eigenvector is located *between* groups of variable vectors in a sort of compromise position. This is illustrated in Figure 6.7, where there are two distinct pairs of variables. The first factor, however, lies between the two clusters in order to maximize the variance it describes. With this solution, the factors are clearly not good descriptions of the pattern of variables. Factor 1 becomes a general factor with *every* variable loading moderately high on it; factor 2 becomes a bipolar factor with variables from one group of variables loading positively and from the other group loading negatively (Table 6.3). It is clear that an alternative factor solution is often preferable, one in which the two factor axes are located among the two groups of variable vectors rather than between them. This procedure is known as *rotation*.

The axes describing the data that are produced initially in factor analysis are only one set of an infinite number of reference axes that could be determined. Other axes can be considered algebraically or geometrically. Algebraically, they are simple linear functions of the initial factors. Geometrically, they are represented by rotation of the initial axes about the origin; we consider this first.

In Figure 6.9, two eigenvectors are shown with a single variable vector. If we rotate the eigenvector axes $\theta°$, we produce a second set of axes I* and II*. The variable vector now has different loadings on these two new axes, denoted w^*_{11} and w^*_{12}. Algebraically, these new loadings can be found from the linear transformation equations for variable x_1

$$w^*_{11} = \cos w_{11} + \sin w_{12}$$

$$w^*_{12} = (-\sin) w_{11} + \cos w_{12}$$

Using these equations, we can compute a new set of loadings for all variables on any rotation of the original axes. Of course, we are not interested in *any* rotation; we want *particular* rotations that define what we can identify as suitable factors. A hint for this notion of suitable factor is found in Figure 6.7.

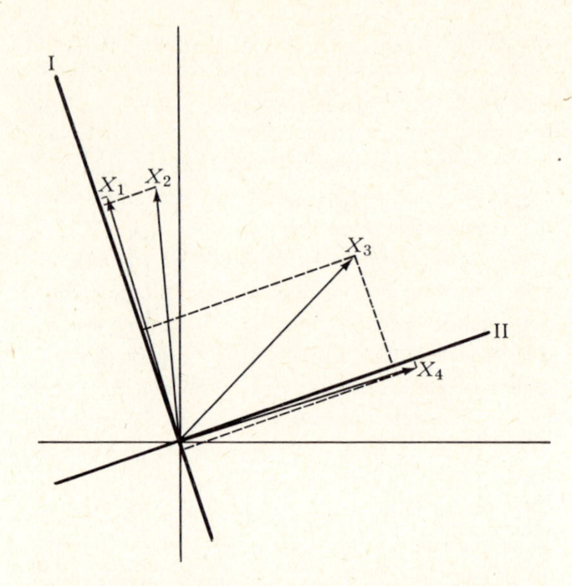

Figure 6.10 The varimax solution to the four-variable example.

Simple Structure and Varimax Rotation

Most factor analysts search for factors that describe groups of closely related variables. In the language of factor analysis, this is to search for *simple structure*. In Figure 6.10, a simple structure solution is indicated, showing that the two rotated factors lie close to the two pairs of variables. The results are that each variable has a high loading on one factor and a low loading on the other. This gives two easily identifiable factors, replacing the general factor and bipolar factor. In the early days of factor analysis, rotation was carried out visually, using two factors at a time. Today we have several computerized rotation procedures that attempt to produce a simple structure automatically. By far the most commonly used is *varimax* rotation, which we describe.

When the concept of simple structure is translated into mathematical form, it must be specified explicitly. For varimax rotation, the criterion focuses on the variance of the squared factor loadings. This variance is at a maximum when all the loadings are zero and one. In this case, every variable vector coincides with a factor and is, therefore, at right angles to all other eigenvectors. The more closely that variables coincide with factors, the higher the loadings are and, hence, the greater the squared loading variance is. If we rotate the factors to the location that *maximizes the squared loading variance*, we have a specific solution for the simple structure concept. This criterion gives the procedure its name. The angle of rotation required to satisfy the criterion is found by an iterative algorithm. The result is a new set of factors with loadings; the rotated axes in Figure 6.10 are the varimax solution to our four-variable example. The new loadings that result from the varimax rotation of the principal axes solution to the four-variable example are shown in Table 6.4. The loadings confirm clearly the geometric evidence that the rotation has produced two factors, each relating to one of the two pairs of related variables.

Table 6.4 Factor Loading from Varimax Rotation

Variable	*Factor 1*	*Factor 2*
x_1	+0.99890	+0.04675
x_2	+0.97561	+0.21948
x_3	+0.45899	+0.88844
x_4	−0.04676	+0.99891

Factor Comparison: The Coefficient of Congruence

The major advantage of rotation for simple structure is clearly in relating results (the factors) to original input (the variables). We consider this procedure of interpreting factors by variables in the next section. An important feature of this improved ease of interpretation of factors concerns us here, the whole issue of factor comparison. We ought not to expect factor analysis studies to be carried out in isolation; researchers, employing the technique in similar contexts, ought to seek to relate their findings. A rotation to simple structure aids factor comparison by its clear definition of the factors. If the members of a set of variables naturally cluster together in various studies, a varimax rotation should identify the cluster in each analysis. If one study uses all the variables in the cluster, the varimax solution should include the resulting factor as one of the main dimensions of the data. If a new study includes some of these related variables, the smaller cluster of variables should still be identified as a factor, if the same relations between the variables exist. The most notable difference should be merely that the factor is less important for the overall description of the second data set. In fact, the same basic factor should always appear in such studies, as long as one or two members of the original cluster of variables is included in the analysis. This very important feature is known as the *invariance property* of simple structure. Loadings obviously vary, but the main difference between the results merely relates to the relative importance of the factor in the various analyses, depending on how many items in the original cluster of variables are included.

The actual loadings between the variables and this recurring factor are not identical in every case. How can we be sure, then, that the factor is basically the same in every analysis? This question is often answered by visual appraisal of the pattern and magnitudes of the loadings. This subjective approach may be adequate in very definite cases, but a measure of similarity between factors from different analysis is clearly helpful. In fact, several measures are available for factor comparison. We describe the *coefficient of congruence*, the most commonly used method. The technique obviously has to concentrate on the q variables common to the two analyses. If the two factors are very similar, they have high loadings on the same variables and low loadings on the same variables. The sum of the products of the pairs of loadings for each common variable is an absolute measure of factor similarity. The actual size of the loadings is taken into account by dividing by an expression that involves these loadings. The full formula for the coefficient of congruence (δ) between two factors e and f becomes

$$\delta_{ef} = \frac{\sum_{j=1}^{q} w_{je}\, w_{jf}}{\sqrt{\sum_{j=1}^{q} w_{je}^2 \sum_{j=1}^{q} w_{jf}^2}}$$

where w's are weights or loadings. The coefficient varies from -1.0 to $+1.0$, depending on the strength and direction of relation between the two factors. Geometrically, it represents the cosine of the angle between the two vectors in the q-dimensional variable space.

Oblique Rotation

Despite the very real improvement in factor solutions brought about by rotation to simple structure, an accurate description of the pattern of variables is not always produced by varimax and similar rotations. Let's return to our four-variable example. It is true that the varimax factors are located near the two pairs of variables, but both factors fail to locate *among* the variable vectors. This is obviously because the rotated solution maintains a right angle between the factors, whereas the two groups of variable vectors are not at right angles. The factors we have considered are known as *orthogonal factors* and have the basic property of being uncorrelated with one another. An alternative to such analysis involves the definition of *oblique factors* that are related and can be interpreted geometrically as factors that do not define orthogonal axes. It is clear from Figure 6.10 that an oblique factor solution gives the best two-factor description of the two groups of variable vectors.

There is something of a debate concerning the relative merits of orthogonal and oblique solutions in factor analysis literature. Proponents of oblique solutions argue on simple epistemological grounds that we cannot assume that the real world is structured in orthogonal dimensions. The orthogonal solution is, after all, only one pattern of relations possible between factors in an infinite number of possibilities, all the remainder being covered by the oblique case. Despite this appealing argument, the fact remains that, outside psychology, almost all factor analyses have used orthogonal solutions. This is certainly true for geography. The reasons seem to be twofold. First, computer programs for oblique rotation procedures have not been readily available for researchers and, second, interpretation of the results is a little more difficult in the oblique case. We consider both problems.

Procedures for oblique rotation of factors are still experimental. There is no widely accepted method of rotation generally used, even in psychology. Several procedures have been developed but, because none has yet been widely accepted, they do not usually find their way into computer program libraries. We do not describe them beyond noticing that they involve iterative algorithms. We ought not to leave discussion of oblique solutions at this point, however. The epistemological arguments for oblique factor analysis are strong and we may reasonably predict greater use of oblique solutions in future. We consider the nature of oblique factor analysis in general terms and indicate why interpretation is more difficult than with the more familiar orthogonal solutions.

The first thing to understand in an oblique factor analysis is that *two* sets of loadings are obtained for relating factors to variables. We can best describe each with reference to a geometric interpretation. Consider Figure 6.11, in which a single variable vector and two oblique factors are illustrated. We can project the variable vector onto the factor axes in two different ways to produce two loadings. If the projection is carried out so that the line from the end of the variable vector to the loading is parallel to the adjacent axis, the result is known as *pattern loading* (Figure 6.11a). If the line from the end of the variable vector is drawn perpendicular to the factor axis, the loading is known as *structure loading* (Figure 6.11b). Both ways of relating factors to variables are needed to interpret the factor-variable relationship. Pattern loading is related directly

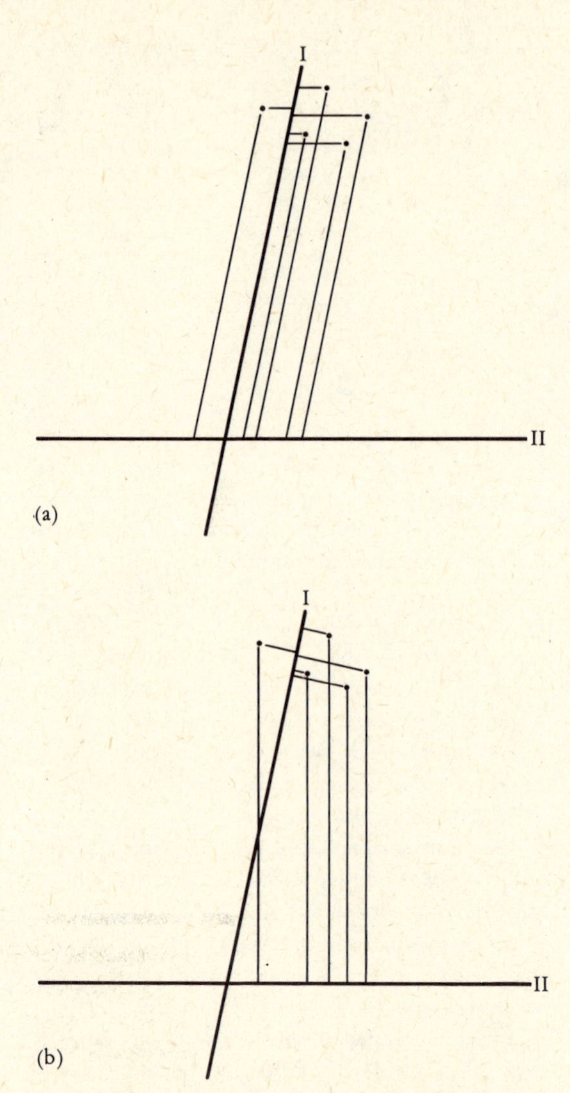

Figure 6.11 (a) Pattern and (b) structure loadings in oblique rotation.

Table 6.5 Factor Loading from Oblique Rotation

Variable	*Factor 1*	*Factor 2*
Pattern loading		
x_1	+1.01726	−0.08386
x_2	+0.97211	+0.09593
x_3	+0.36049	+0.84845
x_4	−0.17111	+1.02814
Structure loading		
x_1	+0.99667	+0.16586
x_2	+0.99566	+0.33456
x_3	+0.56876	+0.93694
x_4	+0.08128	+0.98614

to our original weights and is analogous to a regression coefficient. It can be shown that it is much better for describing the factors as they relate to groups of variables (hence their name). Figure 6.12 and Table 6.5 show how pattern loadings enable us to relate two pairs of variables to one factor each whereas with the structure loadings variables may score high on the other factor (for example, x_3). Of what use are these structure loadings? In many ways they are directly analogous to the *simple* correlation coefficient. The squares of the structure loadings tell us the proportion of the variance of the variable that is accounted for by the factor *and* its interaction with the other factors. As the oblique factors tend toward orthogonality, the two types of loading come closer together so that, with orthogonal factors, the loadings coalesce to produce the single sets of loadings with which we are already familiar and that combine the properties of regression and correlation coefficients.

One further result obtained from oblique factor analysis is a correlation matrix between the factors. This is obviously not applicable with independent orthogonal factors. In our four-variable example, the oblique factors are correlated only at +0.24548 (the angle between them is 76°). Such correlation analysis allows us to explore the relations between the factors we have uncovered. It can even form the basis for further factor analysis defining new higher order factors beyond the original first order factors. Oblique factor analysis clearly opens up a whole new dimension to the analysis.

Before we leave rotation, we should mention one other approach, known as *target rotation*. This may be orthogonal but it is usually oblique. Loadings between variables and factors are hypothesized; we can specify them as simply high or low loadings and denote them with ones and zeros. Factor analysis is used to test the hypothesis that the variables may be described by the factors that are predicted. The initial eigenvectors are rotated until they are transformed into a least squares location with respect to the hypothesized loadings. The criterion is therefore to minimize the sum of the squares of the differences between the loadings and the hypothesized loadings. The solution can be compared to the hypothesized solution, and we can assess whether the variables are adequately described by the hypothesized factor structure.

Target rotation clearly implies a new use of factor analysis in a hypothesis-testing framework. It is particularly useful when we have a theoretical basis upon which to predict the underlying structure of variables. We show how social area analysis predicts such a basis for the urban social geography of a city and can therefore be tested with a target rotation procedure. Before we look at some applications of the concepts and ideas discussed in this section, we must clarify the operational aspects of factor analysis, however.

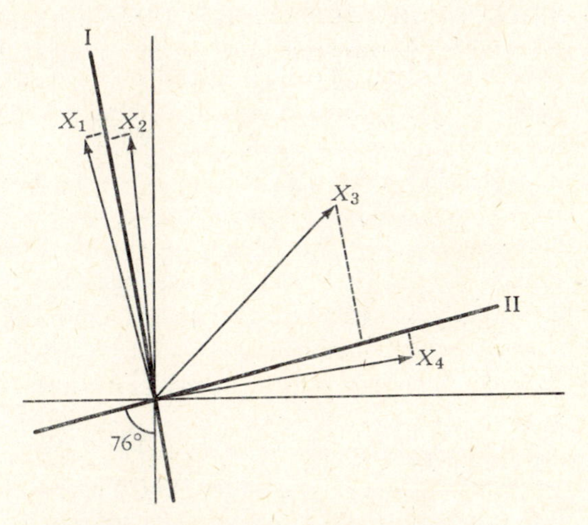

Figure 6.12 An oblique solution to the four-variable example (only pattern loadings are shown).

FACTOR ANALYSIS: OPERATIONALIZATION

We have given the theoretical and conceptual background of factor analysis. To complement this we turn to its operational side. Operationalization means the procedures involved in actually carrying out a factor analysis. We have already noticed that the widespread use of factor analysis is the child of computer technology. It involves huge numbers of calculations and became a common tool only with the general availability of electronic computers.

The facilities available for carrying out a factor analysis vary among computer centers. The general popularity of the technique has led to the widespread availability in recent years of factor analysis programs, in specific computer program libraries and in general package systems. The quality of the service varies between programs; some provide little other than basic routines (the BMD system is one) while others offer a wide range of options extending beyond the material covered in this chapter (the SPSS system is one). Users are not required to have knowledge of computer programming in order to run a factor analysis. All that is required is the typing of systems cards, control cards, and data cards according to instructions in a package manual. The systems cards call the package system into action; the control cards specify the program required and options chosen; the data cards are the data input. The researcher has only to be able to read the package manual and to type cards. The advantages of general availability are obvious. The dangers are equally clear. The technique is available to anybody with access to the system, irrespective of knowledge of factor analysis. With a little technical advice, factor analysis becomes an easy solution for researchers who have amassed a large data set and are not sure what to do with it. Factor analysis, like any other technique, should be incorporated as an integral part of an overall research design.

The factor analysis part of a design can best be viewed as consisting of five matrices. We begin with a data matrix of k cases and n variables, and we convert it into an $n \cdot n$ correlation matrix between variables. The factoring procedure gives an $n \cdot n$ original factor matrix of loadings and an $n \cdot m$ rotated factor matrix with new loadings for variables on m rotated factors (m is usually much smaller than n). We compute a $k \cdot m$ factor score matrix that shows the scores of each rotated factor on the cases. Figure 6.13 shows the sequence of operation. The first two matrices are the input to the factor analysis, and the three other matrices are the output. We consider each in turn.

Input

We have indicated that the input to any factor analysis includes three sets of cards. The system cards are specific to the computer system and need not concern us further. The control and data cards are related directly to the research design. We can view the input to a factor analysis as consisting of two basic parts: control cards, or the coded messages concerning the type of factor model to be employed, and data cards, or the actual data input. Both parts require careful consideration by the researcher.

Model Specification

Two fundamental questions arise every time a factor analysis enters a research design. The particular factor analysis model has to be chosen and the particular rotation procedure must be decided. There are several choices for each. A package program reduces the number of choices, although the more popular ones are normally available. In practice, the two decisions usually reduce to a choice between common factor analysis or principal components analysis and between orthogonal rotation using varimax or oblique rotation.

The principal components and common factor models differ operationally only in that the first uses unities in the correlation matrix diagonal as communality estimates. Program users are normally able to specify the components model or allow the program to estimate the communalities iteratively. Choice of model should be made in terms of the research pur-

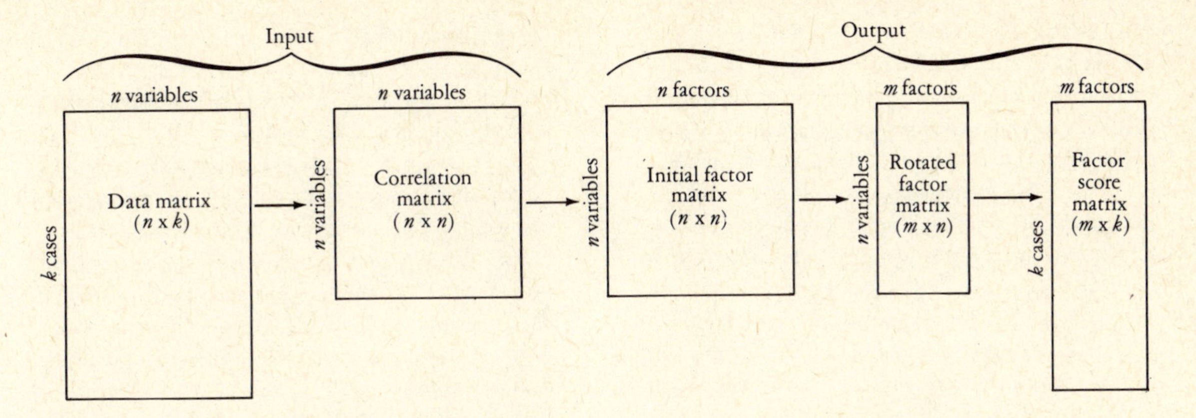

Figure 6.13 The factor analysis procedure.

pose. If the role of the factor analysis is simply to convert a set of data into orthogonal dimensions, principal components analysis is adequate. If the theory of data incorporated in the open model for common factor analysis is accepted for the research, then the common factor model is required. When communalities are high (the variables are highly intercorrelated), the alternative factor solutions should be similar. Often the model decision is of more theoretical than practical importance.

Having chosen the factor model, we usually require a rotation to a simple structure solution. This involves a two-part decision: the number of factors to rotate and the type of rotation to use. The first question may sound trivial: Why not rotate all factors extracted? One of the most useful properties of principal axes extraction is that each successive factor accounts for less and less of the total common variance of the variables. Many of the later factors account for very little variance and may be regarded as being of limited interest. Two types of criteria can be used to decide the number of factors to discard before rotating. The first is theoretical and involves *a priori* hypotheses concerning the number of important factors in the data set. All target rotation procedures involve this type of decision, although other rotation may also be decided on these grounds. The second type of criterion is empirical and relates to the variance accounted for by the factors. This criterion involves various rule-of-thumb approaches. The one most commonly applied is rotation of all factors that have eigenvalues greater than one. This rule derives simply from the fact that eigenvalues indicate the amount of variance accounted for by the factor. The logic is roughly that, because the variables are standardized, each variable in the original data matrix accounts for a variance of one, and by concentrating on factors with eigenvalues greater than one we deal with factors that are empirically more important than the original variables. Factors with eigenvalues less than one are considered unimportant and are not rotated. This rule is obviously neat and simple to operate, but it should not be used without other considerations. Two factors with eigenvalues of 1.1 and 0.9 are similar in importance, and it would be extremely arbitrary to rotate the first and not the second. A look at the sequence of eigenvalues may be instructive of whether discontinuity in values may be observed and used as the cut-off point.

Having decided how many factors are to be rotated, we have to decide on the rotation procedure. If a target rotation is incorporated in the research design, the decision is straightforward, although it should be noticed that this strategy is rarely available in package programs. Otherwise, we normally say that we require a simple structure solution.

Several orthogonal rotations produce different definitions of simple structure, but the varimax method is the most common and is usually recommended as an initial rotation. Next we decide whether to improve our description of the variables by employing an oblique rotation. Sound epistemological arguments favor this approach, although oblique rotation techniques are not particularly well-developed. They normally involve prior selection, by the researcher, of a constant (θ) that actually controls the degree of obliqueness of the solution. In practice, oblique techniques do not match up to the epistemological arguments for obliqueness. The advantage of flexibility in specifying the degree of obliqueness is largely canceled out by the fact that this explicit control seemingly prevents the technique from finding some "natural" oblique simple structure. Oblique rotation should not be avoided, but we ought always to remember that it is still experimental and use it accordingly. It may shed new light on our data patterns but we ought to expect disappointments.

Data Preparation

Having specified the factor model and rotation, we must prepare data input. This involves many problems relating to data collection, measurement, and correlation. The results from a factor analysis are only as good as the data. If the relation between two variables is not linear, the correlation coefficient is not a suitable measure, and the subsequent factors will not reflect this particular relation. Such problems may be overcome by transformation, as we illustrated in Chapter Five. Transformation may also be employed when data normality is required. When the inferential apparatus associated with factor analysis is employed, all the variables in the analysis should be normally distributed. In practice, this causes problems stemming from the size of the data sets employed in factor analysis. The most meticulous researchers consider every variable in turn and select an appropriate transformation to normalize when it is necessary. Others quite commonly employ a blanket logarithmic transformation to cover all the variables. Most researchers do not bother to transform their data at all. This decision is consistent with the fact that the inferential approach has been used minimally in geographical studies. If factor analysis results are to be linked with theory, this is usually accomplished most easily with variables in their original measurement scales. Unless the inferential apparatus of factor analysis forms a part of the research design, data are probably best left in their original and untransformed state.

The problem of variable selection is far more important than transformation. Variable selection may at first seem a strange procedure to consider with a factor analysis. The researcher armed with a factor analysis program and a computer can surely tackle such large data matrixes that variable selection is unnecessary. Is it not the purpose of factor analysis to find the common dimensions of large sets of data? The truth is, of course, that factor analysis is not a crude data-crunching technique. It is based on a sophisticated theory of data. In effect, when we use data we postulate a few common dimensions underly it. This hypothesis demands a theory from us, however poorly it is developed. No variable should be included in a factor analysis unless the researcher has an explicit theoretical or practical reason for including it. If we include all variables available, as we might if we utilize a census report, we are, in effect, developing a theory about the types of variables that census officials ask questions about. If census officials ask a lot of questions about housing conditions, as they do in Britain, we ought to expect housing condition factors. We are neither surprised nor particularly interested in this result.

After we have definite theoretical expectations from our analysis, we must not overload the variables on the expected dimensions. If we postulate a social class dimension and include fifteen variables from twenty on this topic, we must not be surprised if social class evolves as the most important factor in the analysis. The analysis tells us nothing about its general importance beyond the particular range of variables used. Many of the fifteen variables may be redundant and need not be used in analysis to provide us with a social class dimension. The decision whether to include variables can be based on *a priori* theory concerning social class and variable relations plus empirical evaluation of the correlation matrix. The effect of putting in a large number of variables that measure similar concepts is simply to increase the proportion of the variance accounted for by that particular factor.

When variables have been selected, one final check has to be made. It is important that there are more cases than variables, especially when there are few of both. The reason is tied up with the matrix algebra used to extract the factors. There is a limit on the number of factors that can be extracted; one cannot obtain more factors than the smaller of the two sides of the data matrix. If there are twenty variables and five cases, it is possible to produce only five factors from the twenty variables. This limitation can distort the factor pattern and should be avoided. Some psychologists have suggested a ratio of four to one for cases and variables. In our twenty-variable example, we would have eighty cases. This is perhaps erring on the safe side, especially because most geographical applications of factor analysis have been concerned with a relatively small number of final factors. This potential source of distortion must be recognized nonetheless and avoided by ensuring that there are more cases than variables. Having prepared input as we describe, the researcher is ready to carry out a factor analysis. Let's consider now exactly what it is that the computer will give in terms of output.

Output

At the beginning of this section we indicated that output consists of three matrixes—the original factor loading matrix, the rotated factor loading matrix, and the factor score matrix for the rotated factors (Figure 6.13). In fact these matrixes form only a core; any factor analysis program has a rather sizable output with many trimmings. We look at this output piece by piece to give each a substantive meaning and relate it to our discussion of the theory and concepts of factor analysis. Then we consider the interpretation of the output and, finally, we consider the presentation of the results.

What Does the Output Mean?

The output from a factor analysis differs in detail depending on the programs available. There are generally two parts—one relating to the actual factoring process and one that we eventually interpret. The data manipulation parts often include the means and standard deviations of the original variables and the correlation matrix upon which the subsequent analysis is based. There is often information relating to the matrix manipulation such as the determinant and the inverse of the correlation matrix. The matrix algebra does not concern us and is not strictly necessary for interpretation of factor results. We consider only the part of the output that is finally interpreted. We take as our example the actual output from the SPSS factor analysis program for the four-variable case. The layout of the interpreted part of this output is shown in Table 6.6. The options are a principal component analysis followed by a varimax rotation.

The first part of Table 6.6 shows the variables and their estimated communalities. These are all unities because we chose a principal components analysis. This information is followed by the four factors and their associated eigenvalues. The information contained in the eigenvalues is expressed in two ways, as percentages of total variance and as cumulative percentages of total variance. The relative importance of the factors can thus be assessed directly. The irrelevance of factors 3 and 4 is emphasized, for example. The original data is clearly two-dimensional.

The second part of the table shows the factor loadings on the four factors before rotation, the variables in rows and the factors in columns. Hence, x_1 correlates at 0.8 with factor 1. This matrix shows clearly the general and bipolar factors typical of original factor loading matrixes. The third part of the table shows the loadings that result from the varimax rotation. This part of the output is usually of most interest to the factor analyst. Here we have the relations between the factors the analyst has been searching for and the original variables. The matrix is laid out exactly as the preceding matrix, but the values clearly indicate a simpler relation between variables and factors. The fourth and final part of the table shows the factor score matrix. It gives the values for new variables, the factors, on our original observations, the cases. They are derived from the formula that defines factors as a function of the original variables, weighted by their respective loadings from the preceding matrix. In

Table 6.6 Layout of a Factor Analysis Output

Principal component analysis

Variable	*Communality*	*Factor*	*Eigenvalue*	*Percentage of total variance*	*Cumulative Percentage*
x_1	1.0	1	2.64275	66.1	66.1
x_2	1.0	2	1.35721	33.9	100.0
x_3	1.0	3	0.00004	0.0	100.0
x_4	1.0	4	−0.00002	0.0	100.0

Initial factor matrix

Variable	*Factor 1*	*Factor 2*	*Factor 3*	*Factor 4*
x_1	+0.81916	−0.57355	+0.01001	−0.00117
x_2	+0.90630	−0.42262	+0.04112	+0.10169
x_3	+0.90631	+0.42261	−0.10610	+0.09432
x_4	+0.57355	+0.81917	+0.01788	−0.06592

Varimax rotated factor matrix

Variable	*Factor 1*	*Factor 2*
x_1	+0.99890	+0.04675
x_2	+0.97561	+0.21948
x_3	+0.45899	+0.88844
x_4	−0.04676	+0.99891

Factor scores

Case	*Factor 1*	*Factor 2*	*Case*	*Factor 1*	*Factor 2*
1	−2.4	−1.2	6	−0.8	+1.6
2	−2.2	−1.5	7	+1.2	+1.8
3	−1.2	+0.2	8	+1.0	−0.4
4	+0.5	+0.9	9	+1.9	−2.0
5	+0.5	+1.3	10	+1.2	−0.7

geography this factor score matrix has often been of major interest to researchers.

This presents a typical factor analysis output (this particular output was the source for Tables 6.3 and 6.4). We can understand what the matrixes mean to the analyst. We must now be able to interpret the results within the research context. Interpretation is essentially subjective and requires that we give our factors names.

Factor Labeling

The labeling procedure involves going back to our original data. In our hypothetical four-variable example, the cases are ten counties on a small island and the variables are percentages of farmland sown in

x_1 wheat
x_2 barley
x_3 temporary grass
x_4 permanent grass

We label the factors by trying to find a name that expresses the pattern of variable loadings. Obviously, we look for the high loadings on a factor and try to produce a general name that reflects these. We might term factor 1 the arable factor, because it loads high on both crop variables; factor 2 might be named the pasture factor, because both grassland variables load

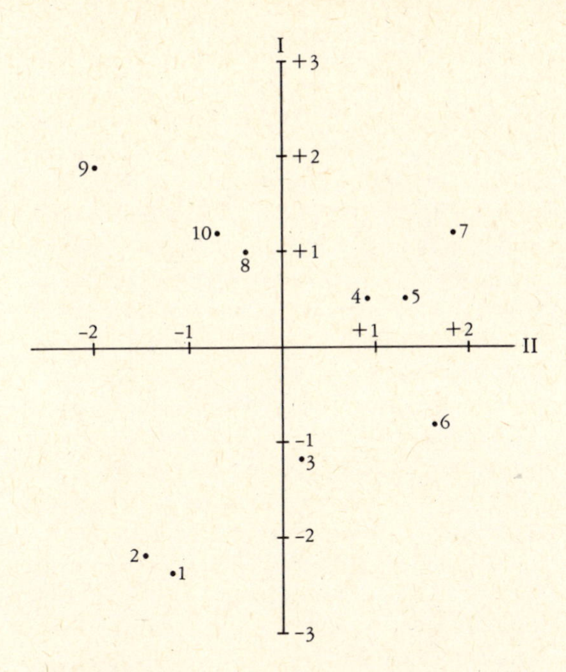

Figure 6.14 The factor space in the four-variable example.

high. We can conclude that the agricultural patterns on this island, as represented by four variables, can be described in terms of arable and pasture dimensions.

The labels are a succinct description of the factors, having been derived from inspection of the loadings. Our hypothetical example is very easy to interpret. In practice, with many more variables and more complicated patterns of loading, labeling is more difficult. If a factor has high loadings with twenty variables, it is difficult to produce a single name that reflects all the variables. Interpretation should subtly try to suggest the common factor among the variables. Loadings of near zero often give clues to the meaning of a factor, because they indicate what the factor does *not* represent. It must be remembered, furthermore, that negative loadings are as important as positive loadings for interpretation. Our label might be in the form of two names to represent a bipolar factor. In a regional study, a factor might load positively on variables reflecting urbanization and negatively on variables reflecting rural phenomena; we might term this an urban-rural dimension.

The purpose of labeling factors is to convey their meaning to researchers and other interested persons. The problem is that terms used to name factors in social science have many extraneous connotations. In communicating results, these extra meanings may cause

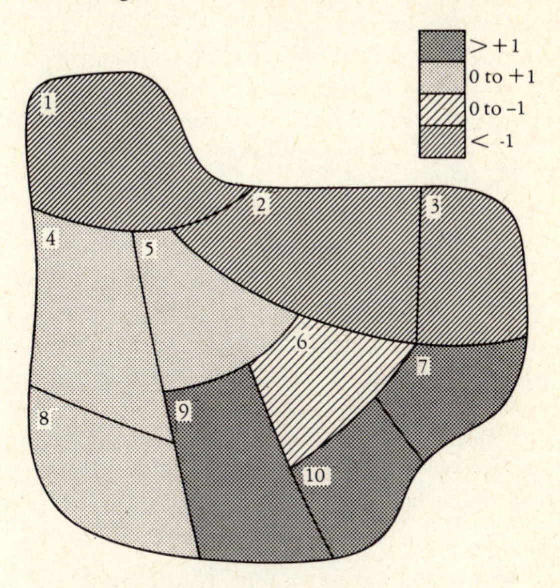

(a) Factor I: Arable dimension

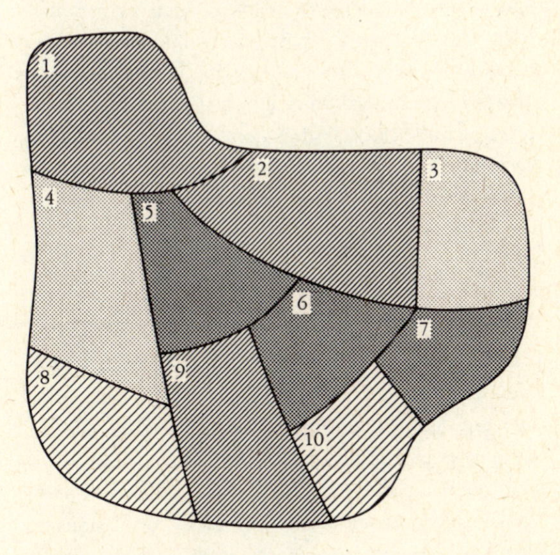

(b) Factor II: Pasture dimension

Figure 6.15 Factor score maps in the four-variable example.

confusion and misunderstanding. This is overcome to some extent when a body of knowledge and language based on factor analyses has been built up. There may be general agreement on terms to describe factors that result from the inclusion of certain sets of variables in the analysis. Because the process is subjective, there is the danger of stretching the evidence to identify a factor commonly accepted as existing. This tendency can be checked and corrected by the use of coefficients of congruence between factors to see how closely the factor from the new analysis corresponds to the comparable factor in previous studies. We illustrate the use of this check in the next section.

The final stage is the actual presentation of the results. This usually involves both loadings and the factor scores. The loadings are best presented simply as tables from the output. When large numbers of variables are used, only those loading high on a factor may be presented. The table is essential in all presentations if readers are to satisfy themselves concerning the factor interpretation.

Factor Spaces and Factor Score Maps

The factor scores lend themselves to graphic presentation. Geographers are particularly interested in this part of the analysis because it relates the results to the observations as areal units and, hence, the spatial patterns of the factors. Factor scores can be presented in two different ways, in factor space or geographical space. Abstract factor space refers to a space defined by two axes representing two of the factors. In our four-variable example, the two factors arable and pasture can be represented by vertical and horizontal axes, as in Figure 6.14. Because we have scores for each case on both factors (Table 6.6), we can locate the cases in this space. This sort of presentation is general for factor analyses in all fields; it enables us to view the similarities between the cases in terms of the two factors. County 1 is most like county 2 and least like county 7 in terms of arable and pastural patterns of land use. Distances between the counties in this space can be used to classify the counties into groups or regional types.

If the observations are areal units, another form of presentation is possible. Because the columns of the factor score matrix show the values for every areal unit on the factors, it follows that each column's values can be mapped to show the spatial pattern of the column factor. This type of presentation is popular in geography. Figure 6.15 shows the maps of the two factors derived from our four-variable example. It shows that the arable pattern has a southerly distribution whereas the pastoral pattern cuts it from northwest to southeast across the island. This is concrete expression of the whole factoring process in terms of maps, which we can discuss and incorporate in our knowledge of agriculture on the island.

Thus we return to maps. In the final section of this chapter, we deal exclusively with actual examples of factor analysis in geographic research. It is only in this way that we can begin to appreciate the potentialities and the limitations of this approach.

FACTORIAL ECOLOGIES

The adoption of factor analysis in spatial analysis is identical to that previously described for correlation and regression. The objects or cases studied become areal units and we proceed as with other objects while remembering the problems inherent in areal individuals. We move from ecological correlations to ecological factor analysis or, as it is more usually termed, factorial ecology.

Why is analysis using areal units called ecological? This certainly seems to be a dilution of the term, which is usually taken to mean the relation between an organism and its environment. Early sociologist Robert Park envisaged the subject human ecology as relating people to their environment, as a sort of basic substratum to a full understanding of human society. Human ecology was for him truly ecological, drawing on concepts such as competition and succession from animal and plant ecology. Park realized that the best evidence for such behavior was most clearly found in early twentieth-century American cities with their many competing ethnic groups. Thus, in practice, most of human ecology became the ecology of urban areas. Furthermore, because empirical evidence was spatial in character, ecological analysis became synonymous with analysis using areal data. Human ecology soon developed into a subject very similar in content to modern urban social geography. This is reflected at the present day in the interpretation of factorial ecology as the application of factor analysis to the areal study of social patterns in cities. Factor analysis has found most most application in geography in this area of research, and so we use this theme to illustrate the use of the technique in a practical research context. We follow this with a more general discussion of factor analysis in geographical research, where we broaden the definition of factorial ecology beyond the confines of urban studies.

Social Patterns Within the City

Before we can appreciate the role of factor analysis in urban social geography, we must assess the position reached in this discipline before the developments of factorial ecology. Early in the 1950s there were two opposing groups of researchers debating the areal structure of cities. On the one hand, the remaining human ecologists were revising and modifying their classical position; on the other hand, a new group of researchers, the social area analysts, were challenging their authority. Let's consider each in turn.

The major legacy from classical human ecology is undoubtedly the *concentric ring model* of the city, first proposed by Ernest Burgess in the mid-1920s. This model proposes that a great number of social patterns decline with distance from the center of a city and that they can be conceptualized as a set of concentric bands about the city center. This is one of two spatial models from this period. The second came from land economics and proposed that cities could be much better conceptualized as a series of sectors running from center to periphery, each characterized by a fairly homogeneous social status. This is Hoyt's *sector model.* A third model was later postulated in geography, incorporating parts of the two other models arranged around several nuclei. The *multiple nuclei model* is hardly a model at all because it predicts a particular pattern that differs for each city and is simply a sort of conceptualized complexity.

The social area analysts, despite their name, were not as explicitly spatial as the human ecologists. They hypothesized that urban populations can be differentiated along three major dimensions. These researchers, led by Eshref Shevky, proposed measures of each of these dimensions by deriving indexes from available census variables. Economic status was represented by an index combining variables that describe occupation, education, and rent levels in a census tract. Similarly, a family status index was produced from variables describing fertility, the number of women in the labor force, and proportions of single-family dwelling units. Ethnic status was represented by a single variable describing proportions of subordinate ethnic groups in the census tracts. With these indexes, every census tract in a city could be given a value along one of the three dimensions. The first two were considered to be the most important, and census tracts were classified in an index space defined by economic and family dimensions. Census tracts occurring together in this space were said to form the basic social areas of the city. Notice that there is no reason to suppose that tracts forming a social area are contiguous in the physical space. In fact, social area analysis says nothing about the spatial structure of the city beyond hypothesizing similar tracts that define social areas.

Target Rotation to the Social Area Dimensions

The initial application of factor analysis to the studying of internal social structures of cities was stimulated by social area analysis. One of the earliest proponents of this approach, Wendell Bell, used factor analysis to test the Shevky proposition that three distinct dimensions of urban social differentiation can be identified. Bell's original application of the technique remains one of the clearest examples of factor analysis in its hypothesis-testing role.

The method of studying the city as developed by social area analysts involves exact specification of the three basic dimensions of economic status, family

Table 6.7 Testing the Social Area Analysis Model in Los Angeles (1940)

Variable	*Initial factor loadings*			*Hypothesized factor matrix*			*Rotated factor loading*		
	1	2	3	1	2	3	1	2	3
Occupation	+0.886	+0.075	−0.233	+	0	0	+0.482	+0.193	−0.094
Education	+0.777	+0.511	+0.088	+	0	0	+0.319	−0.044	+0.282
Rent	+0.963	+0.390	−0.361	+	0	0	+0.653	−0.192	−0.189
Fertility	+0.913	−0.189	−0.089	0	+	0	+0.109	+0.562	+0.176
Women in the labor force	+0.646	−0.560	−0.185	0	+	0	+0.148	+0.617	−0.193
Percent single-family dwellings	+0.485	−0.635	+0.065	0	+	0	−0.147	+0.727	+0.015
Ethnic groups	+0.465	−0.447	+0.444	0	0	+	−0.109	+0.044	+0.576

\+ = Hypothesized high loading
0 = Hypothesized low loading
Source: W. Bell. Economic family and ethnic status: An empirical test. *ASR* 20:47–49, 1955, Tables 2–5. With permission of the American Sociological Association.

status, and ethnic status from the seven key variables. The first two indexes were produced by arbitrarily combining three variables and giving them equal weighting. Producing such an index for a set of census tracts in no way validates the hypothesis that a particular dimension exists as a basis for social differentiation. One could just as easily produce an index by combining rent with fertility and produce an alternative dimension for the census tracts. Bell recognized that factor analysis can have two roles. It can analyze interrelation among the seven variables to see whether the three dimensions exist as a valid description of the variable set. Once the dimensions are recognized as factors it can also combine the variables by weighting them with respect to the variance accounted for by the factors. The dimensions are then much less simply arbitrary indexes.

Bell made his initial test of the social area analysis dimensions by analyzing a data matrix consisting of the seven key variables for 570 census tracts in the Los Angeles area for 1940. He used the common factor model and extracted factors from the correlation matrix by using a simplified variant of the principal axes method known as the centroid method. The resulting factor matrix is shown as Table 6.7, and a typically very general first factor can be observed. To test the social area analysis dimensions, he used target rotation. He predicted that large and small factor loadings are consistent with the social area dimensions (Table 6.7). The factor axes were then rotated obliquely in such a way that this hypothesized factor matrix was reproduced as closely as possible. The resulting loadings are presented along with the predicted pattern and are very similar. The hypothesized high loadings are all much larger than the hypothesized low loadings. In fact, we can state clearly that for Los Angeles in 1940 Skevky's seven key variables can be combined to form three separate dimensions or patterns. Bell repeated the analysis for San Francisco census tracts and reached comparable conclusions. It would seem that factor analysis confirms one of the basic tenets of social area analysis.

Because nearly all subsequent applications of factor analysis in geography have produced orthogonal factors, Bell's study gives us a rare opportunity to illustrate results from an oblique solution. Table 6.8 shows the factor correlation matrix for the Los Angeles analysis. The pattern is very interesting in the light of the many subsequent orthogonal solutions, because the three factors are found to be quite highly correlated. In particular, the ethnic factor correlates in a negative manner with economic status. This eminently reasonable finding, that areas with high ethnic scores tend to have a low economic status, is not allowed to emerge in

Table 6.8 Correlation Between Factors in Los Angeles (1950)

Factors	*Economic status*	*Family status*	*Ethnic status*
Economic status	1.00	−0.50	−0.73
Family status	−0.50	1.00	+0.15
Ethnic status	−0.73	+0.15	1.00

Source: W. Bell. Economic family and ethnic status: An empirical test. *ASR* 20:47–49, 1955, Tables 2–5. With permission of the American Sociological Association.

any orthogonal solution. It is also interesting that family status is correlated moderately in a negative manner with economic status. Subsequent factorial ecologists tend not to take into account these original findings of Bell's analysis.

Social Area Dimensions from Larger Data Sets

In some ways it can be argued that Bell's application of factor analysis has not extended the social area approach beyond a simple testing of the structure of the seven key variables. There is no questioning of the validity of the choice of these seven variables. They are meant to be representative of very general concepts, and it is reasonable to ask whether they really represent a set of basic dimensions that relate to much larger sets of variables. Modern census data give researchers many variables and, with electronic computers, there is no reason why they should limit their analyses to seven variables chosen before computers were available. What we are really asking is: "How invariant are these social area dimensions when analysis covers a wider range of variables describing social patterns within a city?"

The answer is important. If the social area dimensions disappear in a more comprehensive analysis of related variables, the three factors identified by Bell might not be general dimensions but simply a product of the original variables. The validity of the social area model would be thrown into severe doubt. Moreover, the original variables were chosen partly for reasons of availability in American census data. Because census data in other countries do not exactly duplicate these original variables, it is important that the dimensions derived by Bell's study be produced by similar sets of social variables if cross-national comparisons between cities is to be at all possible.

The importance of these arguments was recognized early in factorial ecology and research was designed to test them. In one such study, C.F. Schmid and K. Tagashira analyzed 42 variables from the 1960 census for the 115 census tracts of Seattle, Washington. The variables cover a wide range of phenomena, including population change, age, sex, race, family patterns, socioeconomic status, education, occupation, employment, and housing. The 42 variables clearly represent a comprehensive description of social patterns in Seattle for 1960. The variables were correlated, the first eight factors were extracted using the principal axes method, and axes were rotated on varimax criteria. The three social area dimensions from Bell's study were identifiable as factors 1, 2, and 5. Ethnic status was extracted as the fifth (rather than the third) factor because it is represented by only 3 of the 42 variables. The analysis was then repeated on a reduced data matrix comprising 21 variables, selected in terms of their subjectively assessed importance. Once again, the three social area dimensions appear, this time as factors 1, 2, and 4. The analysis was repeated with ten selected variables, rotating six factors, and this time the social area dimensions emerge as the first three factors. The ten variables are listed in Table 6.9 and their loadings are given for the three factors identified as the social area dimensions for each of the analyses. The consistently high loadings on the same factors for analyses with different numbers of variables shows that, in this particular case at least, the social area dimensions are invariant to the number of variables used and do not depend solely on use of the original key variables.

Having shown how the constructs of social area analysis represent general basic dimensions of social differentiation in an American city, we can now go on to consider what has become the typical factorialecology for the city. We take as our example research by Robert Murdie from the University of Chicago Geo-

Table 6.9 Invariance of Factors in Seattle (1960)

Final 10 variables	*42 variables*	*21 variables*	*10 variables*
Family status			
Percent under 15	−0.993	−0.988	−0.966
Percent married	−0.889	−0.952	−0.903
Mean household size	−0.954	−0.989	−0.934
Percent females in labor force	+0.789	+0.811	+0.827
Economic status			
Median value of owner-occupied units	+0.947	+0.884	−0.830
Percent male professional and technical occupations	+0.922	+0.913	−0.933
Median education grade	+0.917	+0.892	−0.882
Mean monthly rent	+0.852	+0.796	−0.682
Ethnic status			
Percent black	+0.879	−0.881	+0.995
Percent other nonwhite	+0.705	−0.573	+0.560

Source: C.F. Schmid and K. Tagashira. Ecological and demographic indices. *Demography* 1:197–211, 1964, Tables 6, 7, 11. With permission of the Population Association of America.

graphy Department which pioneered in this field. The study is of Toronto for 1951 and 1961, although we concentrate on the 1951 analysis. It is typical of many recent studies in terms of its large data matrix: 235 census tracts and 86 variables. Computer technology and the multivariate nature of the technique are being fully exploited. The variables include a very comprehensive coverage: 22 on employment characteristics, 19 on dwelling characteristics, 8 on earnings, 8 on ethnic origin, 7 on education, 6 on population structure, 6 on religion, 4 on family structure, 3 on household structure, and 1 each on population change, language, and location. Six factors were extracted from the resulting 86 times 86 correlation matrix, accounting together for 72 percent of the total variance. They were rotated on varimax criteria, and the first three factors with their major loadings are shown in Table 6.10. It is immediately noticed that these first three factors seem to correspond to the social area constructs. This gives further support to the generality of these constructs: they are unraveled from the huge 86 times 86 correlation matrix as important orthogonal dimensions. The three other dimensions are the extra dimensions necessary to account for the extra variations relating to the wide range of variables analyzed. The inference is clearly that, with large data sets, the social area constructs are necessary but not always sufficient dimensions for describing the data.

We can assess the degree to which the factors relate to the original social area constructs by looking at their variable loadings. In Table 6.10, the ten variables with the highest loadings are given for each factor along with the actual loading values. All the variables that the social area analysts predict will make up the three dimensions seem to be in the correct columns. Thus, education and occupation variables load highest on the factor interpreted as economic status. Although it does not make the top ten, the variable median rent also loads high here (− 0.658) as predicted. The family status factor has high loadings for types of dwelling and also for the female labor variable. (Outside the top ten, distance from center loads at −0.596.) Finally, the third factor has high loadings for variables describing religion, ethnic groups, and language, as predicted. We can conclude that the labels used in Table 6.10 are reasonable and that we can legitimately link these

Table 6.10 Factor Loadings for the First Three Factors for Toronto (1951)

1. Economics status		*2. Family status*		*3. Ethnic status*	
Variable	Loading	Variable	Loading	Variable	Loading
13+ years Schooling	−0.924	Percent owner-occupied dwellings	−0.889	Percent of population born in British Isles	−0.920
Percent employed in transportation	+0.897	Percent tenant-occupied dwellings	+0.884	Percent Protestant	−0.893
Percent employed professionally	−0.893	Percent dwellings with washing machines	−0.823	1–4 years Schooling	+0.786
Percent employed in manufacturing	+0.885	Percent apartment dwellings	+0.804	Percent Anglican	−0.799
Percent employed in management	−0.884	Percent females in labor force	+0.771	Non-English or French speakers	+0.758
Percent earning $4,000+	−0.836	Percent employed in service	+0.697	Percent Jewish	+0.758
5–8 years Schooling	+0.830	Percent single detached dwellings	−0.696	Percent Ukranian	+0.685
Percent dwellings with refrigerators	−0.807	Percent of population 5+ not in school	+0.664	No schooling	+0.658
Persons per room	+0.795	Percent unemployed	+0.635	Percent Polish	+0.655

Source: Adapted from R.A. Murdie. *Factorial Ecology of Metropolitan Toronto, 1951–1961*. University of Chicago, Department of Geography Research Paper No. 116, 1969.

dimensions directly with the original social area constructs.

We have concentrated on the 1951 factorial results for Toronto. A similar analysis was carried out for 1961 data from which six factors were extracted and rotated. There is no need to describe the results although we use them to illustrate the use of the coefficient of congruence to compare factors. The 1961 analysis yields an economic status factor and a family status factor, although the ethnic status dimension is divided into two factors, Italian and Jewish. Table 6.11 shows the coefficients of congruence between loadings for these four factors and the three factors extracted from the 1951 data. In every case the 1961 factor has the highest congruence with its 1951 counterpart. This is true even of the two separate ethnic dimensions, although the Italian factor has fairly high congruence with all three 1951 factors. This use of the coefficient of congruence shows nonetheless a stability of these basic dimensions over time in Toronto, even while the ethnic pattern divides into two separate factors.

Cross-National Comparisons

Stability of dimensions within one city over a short time period is not a particularly surprising result. The coefficient of congruence is much more useful as part of a research design for comparing factors between cities. This is particularly the case when the cities are on different sides of national boundaries. Our concern has been with North American cities, but the social area dimensions should relate to modern urban societies all over the world. Cross-national studies of cities are required to validate this extension of the model over space, and the coefficient of congruence has a basic role in this comparison.

A pioneer in factorial ecology, Frank Sweetser, employed a cross-national research design in his careful comparison of the ecological structures of Boston, Massachusetts, and Helsinki, Finland. He used the 1960 census for both cities and attempted to match up variables from these two sources. In the end, he was able to match twenty variables to which he could give

Table 6.11 Coefficients of Congruence Between Factor Loadings in Toronto (1951 and 1961)

1951	1961			
	Economic status	*Family status*	*Italian ethnic status*	*Jewish ethnic status*
Economic status	+0.92	+0.08	+0.61	+0.09
Family status	+0.29	+0.73	+0.42	+0.37
Ethnic status	+0.35	+0.19	+0.82	+0.82

Source: Adapted from R.A. Murdie. *Factorial Ecology of Metropolitan Toronto, 1951–1961.* University of Chicago, Department of Geography Research Paper No. 116, 1969.

general descriptive labels, but in no case were the American and Finnish census variables identical. The two sets of variables were closely parallel. For instance, room crowding was measured for Helsinki by the variable number of persons per 100 rooms but for Boston by the percent of housing units with more than one person per room. The biggest difference is probably in the ethnic variables; for Helsinki this was percent speaking Swedish as the principal language and for Boston it was percent foreign-born. We have a nonetheless carefully selected set of 20 comparable variables for 70 statistical areas in Helsinki and 441 comparable census tracts in Boston. Two 20 times 20 correlation matrixes yield 3 factors each, which were rotated on varimax criteria. Because only one ethnic variable was included in each study, no ethnic status dimension appears for either city, and we concentrate on the first two factors in each analysis; these can be interpreted readily as economic status and family status factors.

Table 6.12 shows the major loadings on each of these two factors for both cities. The loadings are divided into those with similar values for both cities and those with differences. Both types of loading are of interest in this cross-national comparison. There are more loadings exhibiting similar values in each case, and these also include the variables most readily identified with the two labels used. Thus, educational variables consistently load high on economic status in both cities, and fertility and nonfamily population load high on family status in both cities. The variables with contrasting loadings indicate some basic differences between the organization of American and Finnish urban societies. The high status of clerical occupations in Helsinki as compared to Boston is notable, as is the negative loading of detached dwellings on the economic status variable for Helsinki. This reflects basic differences in housing situations in the two cities, which is further illustrated by the variable home ownership and its loading on family status. The differences in ethnicity are not particularly surprising given the two definitions of the variables. Although all these differences are interesting and point to specific contrasts between Helsinki and Boston, the overall impression is one of similarity between ecological structures. This conclusion is supported by coefficients of congruence of +0.744 between the two economic status factors and +0.792 for the two family status factors. Despite specific differences, there is a high degree of congruence between the ecological structures of these two cities, which produces dimensions consistent with two of Shevky's original constructs.

The continuing emergence of factors that reflect the original social area indexes is indeed impressive. It should be emphasized that the social area constructs do not automatically appear whenever a large number of variables is analyzed for urban subareas. They seem to appear in some form or other when some of the variables entering the analysis correspond to, or are similar to, the original Shevky variables. When this is not the case, we must not expect factor structures resembling those we have described. In a factor analysis, as with any other technique, the results of analysis depend on what is put in. No ethnic status factor emerged for Boston, for example, because only one

Table 6.12 Economic and Family Status Factors in Helsinki and Boston (1960)

Variables	*Helsinki*	*Boston*
Similar economic status loadings:		
Low educational status	−0.99	−0.92
Professional and manual occupations	+0.98	+0.94
Blue-collar occupations	−0.98	−0.90
High school status	+0.98	+0.96
College status	+0.97	+0.87
Room crowding	−0.85	−0.79
Housing defects	−0.72	−0.68
Different economic status loadings:		
Ethnicity	+0.73	−0.38
Clerical occupations	+0.73	−0.04
Detached dwellings	−0.50	+0.56
Proportion male	−0.55	−0.18
Residential stability	+0.39	−0.06
Similar family status loadings:		
Fertility	+0.83	+0.82
Elementary school age	+0.75	+0.87
Nonfamily population	−0.73	−0.56
Retirement age	−0.67	−0.82
New housing	+0.52	+0.76
Different family status loadings:		
Working women	−0.16	−0.61
High school age	+0.03	+0.49
Home ownership	+0.12	+0.64

Source: F.L. Sweetser. Factor structure as ecological structure. *Acta Sociologica* 8:213, 1965, Table 3.

such variable was included in the analysis. This clearly does not mean that there is no largely black residential area in Boston.

We illustrate this clearly by using an early factorial ecology of Liverpool, England, by Elizabeth Gittus. This analysis used variables from the 1961 English census for Liverpool enumeration districts. The English census differs from that in many other nations in its very wide range of questions concerning details of housing conditions. This is a result of the census reorganization of 1911 following on public exposure of poor housing conditions by late-nineteenth-century philanthropists such as Charles Booth. For the Liverpool study, 31 variables were chosen, relating mostly to various population and housing characteristics. The housing bias of the census is reflected in the fact that 21 of the variables relate in some way to housing conditions. Not surprisingly, the factors produced from a principal components analysis of this data differ markedly from the Toronto study, and there is no sign of social area constructs. The first two factors are by far the most important; they account for 48% of the total variance. The first five variables in terms of loading are shown for these two factors in Table 6.13. We have clearly two new dimensions relating to housing conditions, one reflecting the proportion of the population that is overcrowded and the other the proportion of the population that is sharing dwellings. Given the input of variables, these two factors are not unexpected.

It should be clear that a factor analysis is only as good as the input variables. This truism is sometimes

Table 6.13 Factor Loadings for the First Two Factors for Liverpool 1961

Factor 1		*Factor 2*	
Variable	*Loading*	*Variable*	*Loading*
Persons per room	+0.97	Households sharing dwelling	+0.91
Households < $\frac{1}{2}$ person per room	−0.87	Households sharing dwelling without WC	+0.91
Population in households 6+	+0.85	Households renting private furnished accommodations	+0.87
Households > $1\frac{1}{2}$ persons per room	+0.82	Dwellings containing 2+ households	+0.86
Households 6+	+0.77	Dwellings containing 3+ households	+0.85

Source: Adapted from E. Gittus. *An Experiment in the Identification of Urban Subareas.* Transactions of the Bartlett Society, 1964, Table 2, p. 121.

stated as a humorous alternative between DIDO and GIGO—that is, data in, dimensions out, or garbage in, garbage out. The first alternative is our aim; the second is by no means unknown. Factor analysis has suffered from its general convenience for producing apparent results from large sets of disorganized data. This put-everything-in syndrome produces garbage in the paper output. A basic rule for any exercise in factor analysis is that *no variable should be incorporated in the analysis unless it can be justified in terms of the research purpose.*

Research purpose is usually guided by theoretical considerations, and variables are chosen accordingly. An analysis may have alternatively an applied purpose, in which case choice of variables may be unrelated to theory but related to some specified problem. Most factorial ecologies of the city have consistently used sets of variables related to the basic social area model. The result has been a consistent set of dimensions that seem to have general validity. The study of Liverpool is an exception that seems to be of little theoretical interest, at least at the present time. The factors can be interpreted as representing nineteenth-century philanthropic influence on the British census rather than basic criteria for differentiation in the city. This does not necessarily mean that the analysis is garbage. Gittus did relate the work to planning problems in Liverpool and it seems wholly reasonable that patterns of housing conditions should be of interest to planners to some theorists of urban structure. Every study using factor analysis must be judged in terms of its purpose.

Analysis of Variance and the Spatial Models

Geographers are interested in the findings of factorial ecology in terms of the basic dimensions of urban areas, but they remain concerned primarily with spatial expression of these dimensions. We have considered only the relations of factor to variable in the form of factor loadings. Now we concentrate on the factor scores that relate factors to the original objects. We consider the factors of the city census areas to have been interpreted and consider the spatial pattern of named factors—economic status, family status, and ethnic status as presented for Toronto in 1951. These are obviously of particular theoretical interest, and it will become more apparent why we have chosen them.

The scores for the three factors are mapped in Figure 6.16. The resulting distributions do not present striking patterns on visual inspection. Economic status seems to produce largely positive scores to the north and negative scores to the northeast but the distinction elsewhere is by no means clear. Family status has its highest positive scores located centrally with negative scores found largely away from the center. The pattern for ethnic status seems even more amorphous, suggesting some high level of concentration. These maps and these personal impressions are not particularly helpful. To be of any use at all they must be supplemented by a methodology that enables us to judge objectively whether the distributions in fact correspond to some

(a) Factor I: Economic status

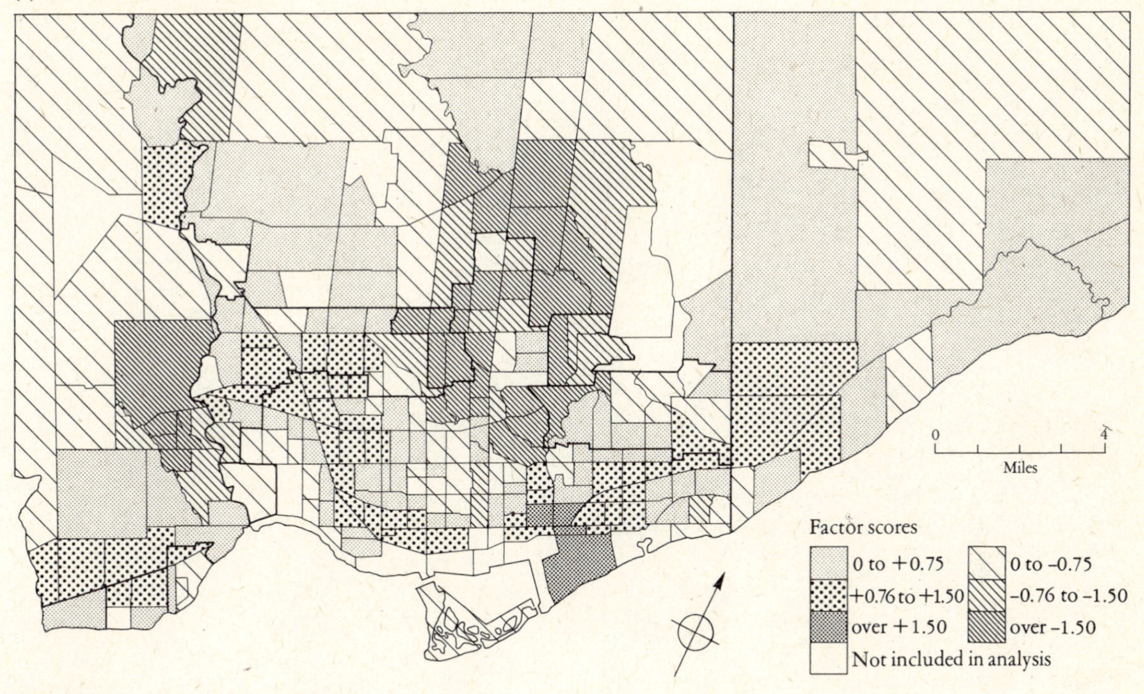

Figure 6.16 Factor score maps for Toronto, 1951.
From R.A. Murdie. *Factorial Ecology of Metropolitan Toronto, 1951–1961.* University of Chicago, Department of Geography Research Paper No. 116, 1969.

specific spatial model. This has been done by Murdie, using *analysis of variance*.

Analysis of variance involves a simple statistical test of significance based on a statistic termed *Snedecor's F ratio*. The objects for which we have measures on a particular variable or factor are arranged in a set of *a priori* classes. For each class, we can calculate separate mean and variance values. The variations among the different class mean values can be used to find *interclass variance*, and the new class variances define *intraclass variance*. The total variance displayed by a variable can be allocated to either variance. This is what is meant by analysis of variance. With the Snedecor *F* test, we ask: "Are the observations from the different classes in fact significantly different from one another?" The null hypothesis is that there are no significant differences between the various class means. An absolute measure of difference between class means is given by the interclass variance. The logic of the test argues that, for the class means to be significantly different, interclass variance must be of much greater magnitude than intraclass variance. If the variations within classes are as large or nearly as large as the variations between class means, obviously we could not confidently assert that the classes reflect anything more than a simple arbitrary division of a reasonably homogeneous set of observations. On the other hand, if the variations between classes are much larger than the variations within classes, it would seem that our *a priori* classes in fact divide the observations into distinctive groups. The statistic *F*, based on this reasoning, is simply the ratio between the two components of variance:

$$F = \frac{\textit{interclass variance}}{\textit{intraclass variance}}$$

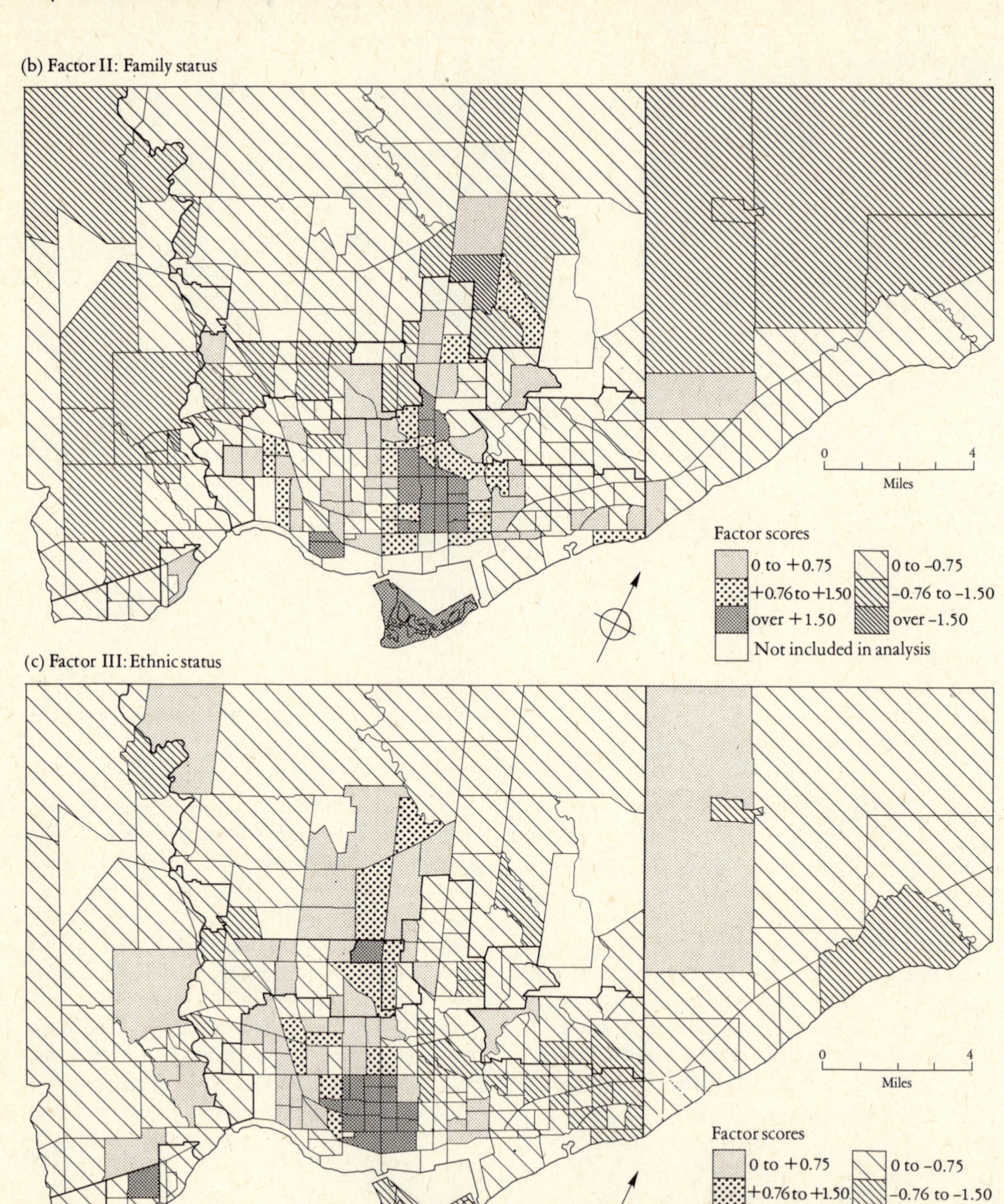

Figure 6.16 *(cont'd)*

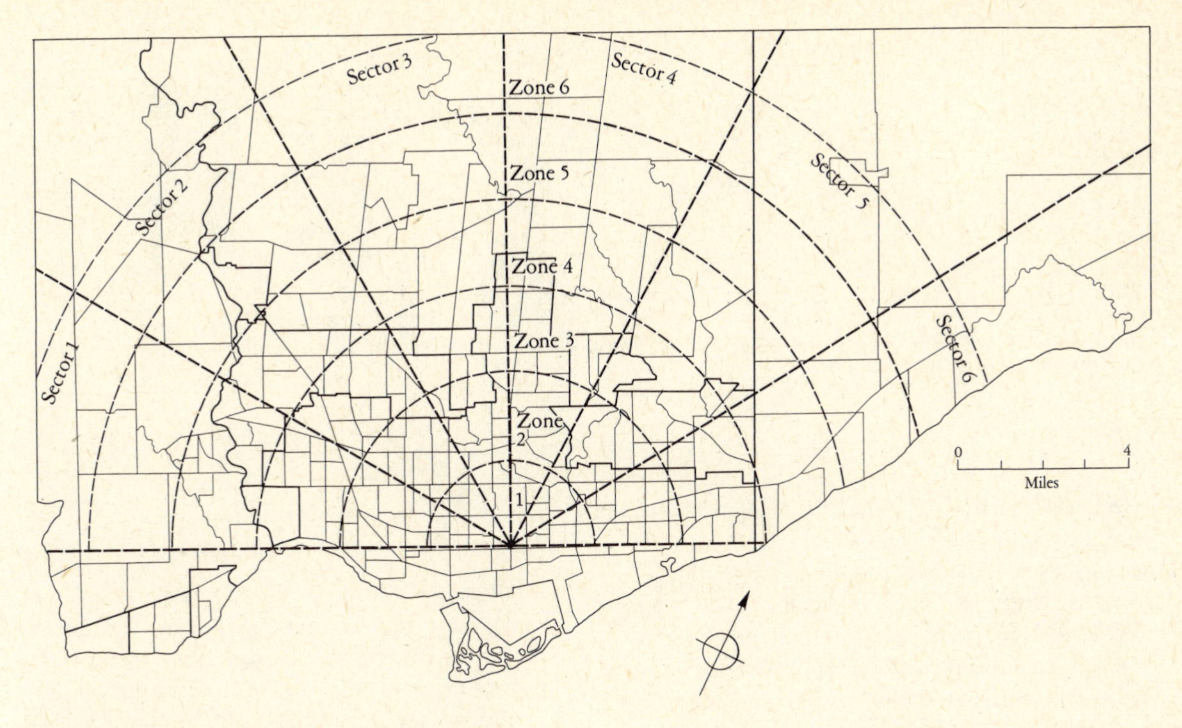

Figure 6.17 Zones and sectors for Toronto, 1951.
From R.A. Murdie. *Factorial Ecology of Metropolitan Toronto, 1951–1961.* University of Chicago, Department of Geography Research Paper No. 116, 1969.

The larger the ratio is, the less likely it is that the classification is arbitrary. Tables of the F ratio are available for finding critical values of the statistic at the usual significant levels (Table A.11). An example is presented in Work Table 6.1.

In factorical ecology, the variables are the factor scores and the classes that have been hypothesized are the sectors and zones of the classical ecological models. We might allocate every census tract in a city to one of five concentric zones and carry out an analysis in terms of variance of family status scores within zones and variance of scores between zones. The consequent Snedecor F test would help us ascertain whether the spatial allocation of tracts produces significantly different levels of family status. If the zones are significantly different from one another at some acceptable level of probability, we can use this evidence to support the hypothesis that the family status factor has a spatial pattern consistent with Burgess's concentric ring model.

This is essentially the logic employed in relating factorial ecology dimensions to the classical spatial models of the city. A slightly more sophisticated form of analysis is actually employed. The technique we have described is more correctly termed *one-way* analysis of variance. "One-way" refers to the fact that only a single classification of the objects is employed. In *two-way* analysis of variance, two classifications are used and objects are allocated to a cross-tabulation of the classifications. For urban ecology, this allows for the simultaneous assessment of the two basic models (concentric and sectoral). This involves setting up a table with sector classes that define columns and concentric ring classes that define rows. Every census track can then be allocated in the cells of this table. With the data in this form, average scores for a factor can be

Table 6.14 Analysis of Variance for Economic Status in Toronto (1951)

Source of variation	*Sum of squares*	*Degrees of freedom*	*Estimate of variance*	F	*Critical value of* F *at* $\alpha = 0.05$
Total	1,477,602	210	7,036		
Sectors	292,923	5	58,584	11.9	2.27
Zones	20,028	5	4,006	0.8	2.27
Interaction	300,528	25	12,021	2.4	1.58
Within cell	864,115	175	4,938		

Source: Adapted from R.A. Murdie. *Factorial Ecology of Metropolitan Toronto, 1951–1961.* University of Chicago, Department of Geography Research Paper No. 116, 1969.

calculated for each cell. Column means indicate differences between sectors for the factor scores, and row means indicate differences between concentric rings. We can identify two types of variance in the data, *between-sector* variance measured from the column means as well as *between-zone* variance measured from the row means. Two further sources of variation can be recognized—*within cell* variance and variation between individual cells, irrespective of their summation into sectors or rings. This last source of variation is known as the variance resulting from the *interaction* of the two classifications. In two-way analysis, the variation in the variable is allocated to four sources between each of the two classes and between and within the individual cells (Work Table 6.2).

We have avoided discussion of the actual procedures for calculating the various types of variance. They are calculated like any variance. For instance, the between-sector variance is found by treating each of the column means as observations, finding their deviations from the overall mean, which are each squared before being summed. This gives us the between-sector sum of squared deviations (or simply sum of squares), which is then divided by the degrees of freedom ($n-1$, where there are n sectors) to give the variance. The variance is simply the sum of the squares divided by degrees of freedom. When we view it in this way, we can present the two equalities in the analysis:

1. The sum of squares of the various types of variance sum to the total sum of the squares.

2. The respective degrees of freedom sum to the degrees of freedom for the total variance.

It follows that the actual components of the variance do not sum to the total variance. These two elements of the variance, the variances themselves, and the resulting *F* ratios are usually presented in a table that effectively summarizes any single analysis. We are now in a position to discuss such a table resulting from the Toronto factorial ecology example.

The zones and sectors defined for Toronto in 1951 are shown in Figure 6.17. The sectors are defined in terms of six 30° angles about the point of maximum land value, and the rings are defined by six concentric two-mile zones about this same point. Both classifications are spatially arbitrary and are not based on *a priori* knowledge of Toronto's social patterns. When tabulated against one another, these two spatial classifications give 36 cells into which all census tracts are allocated. Factor scores for each tract for the three basic dimensions can now be analyzed.

Let's consider economic status first. The analysis-of-variance table is presented as Table 6.14. Notice that the sum of squared deviations for the sectors is far greater than for the zones. When the various sums of squares are adjusted by degrees of freedom to produce variances, the source of variation resulting from sectors is found to be overwhelmingly the most important. In fact, the between-sector variance is almost twelve times larger than the unaccounted variance within the cells. The *F* ratio is extremely highly significant when compared to the critical value of F at $\alpha = 0.05$. We can easily reject the null hypotheses that there are no differences between sectors in terms of economic status and conclude that economic status is clearly sectoral in pattern. With the concentric zones, we get a quite different story. The between-zone variance is actually less than that remaining in the cells, and F is less than one and far below any critically significant level. We cannot reject a null hypothesis in this case: economic status does not seem to be concentric in pattern. The variation from interaction seems important, producing an F ratio much smaller than for sectors but nonetheless significant at the level of probability that was used. We can, therefore, conclude that the pattern is not simply sectoral but that there is some other pattern of concentration of scores in particular sector-zone cells.

We have presented this analysis-of-variance table to illustrate the actual make-up of the Snedecor F test. In practice, our interest focuses almost solely on the size of F. We can summarize the analysis of all three factors—family, ethnic, and economic status—in a single table of F ratios (Table 6.15). This shows that, whereas economic status is primarily sectoral in pattern, family status is overwhelmingly concentric, with an extremely large F ratio. The interaction F ratio is significant and the sectoral F ratio is significant at $\alpha = 0.05$. Thus, there is some evidence for a sectoral aspect in the family status pattern and there are definite cells of concentrated similar scores. Ethnic status shows very significant F ratios for all three types of variation. It seems that this pattern is both markedly sectoral and zonal, with scores concentrated in particular cells.

We can summarize as follows. Economic status is largely sectoral in pattern and thus bears remarkable similarity to the pattern proposed by the original sector model. Family status is largely concentric and, hence, is consistent with the pattern of the classical Burgess model. Ethnic status has a pattern that combines elements of sectors and rings and their interaction—a sort of mixed spatial model similar, at least superficially, to the so-called multiple-nuclei model. It seems that a methodology combining factor analysis and analysis of variance has allowed students of urban social structure to bring together the two opposing approaches to studying the city that we identified at the beginning of this discussion: human ecology and social area analysis. In fact, it is almost a classic example of an empirical synthesis of competitive theories. This is clearly no mean feat.

Table 6.15 Snedecor F Ratios for Three Factors in Toronto (1951)

	Between sectors	*Between zones*	*Interaction*
Economic status	11.9*	0.8	2.4*
Family status	2.8†	40.0*	2.1*
Ethnic status	22.4*	17.8*	9.5*

*Significant at $\alpha = 0.01$
†Significant at $\alpha = 0.05$
Source: Adapted from R.A. Murdie. *Factorial Ecology of Metropolitan Toronto, 1951–1961.* University of Chicago, Department of Geography Research Paper No. 116, 1969.

Trials and Tribulations of Factorial Ecology

The study of social patterns within cities was chosen to illustrate the application of factor analysis because it furnishes a neat example of the technique as it developed from previous studies and helped synthesize them into a unified body of knowledge. Perhaps the example, and our presentation in particular, is *too* neat. There are certainly many loose ends to be tied together and quite basic extensions to be added to our study of urban social patterns. These are being achieved to some extent in cross-cultural factorial ecological studies of cities. It is not our purpose to follow up this theme, although it is necessary to correct the impression that factor analysis is being received unconditionally as a basic tool for future geographical research. Criticism of factorial ecology studies is basically of two different types. There are problems relating to the technique of factor analysis and there are problems resulting from geographical applications of the technique. We consider these in turn.

We must first extend our use of the term "factorial ecology" beyond the explicitly narrow confines of the

Table 6.16 Factorial Ecology of Toronto: Change Variables (1951–1961)

Factor 1: suburbanization		*Factor 2: ethnic change*	
Flush toilet (%)	+0.895	British-born (%)	−0.766
Bath (%)	+0.880	Roman Catholic (%)	+0.710
Commercial occupations (%)	+0.856	Anglican (%)	−0.686
Furnace heat (%)	+0.817	Italian-born (%)	+0.677
Population density	+0.774	United Presbyterian (%)	−0.625
Median earnings	+0.775	Language not French or English (%)	+0.588
Population potential	+0.756		
Mortgages (%)	+0.747		
Managerial occupations (%)	+0.746		
Wage-earner heads (%)	+0.715		
Professional occupations (%)	+0.706		
Population change (%)	+0.701		
Distance from center	+0.701		

Source: Adapted from R.A. Murdie. *Factorial Ecology of Metropolitan Toronto, 1951–1961*. University of Chicago, Department of Geography Research Paper No. 116, 1969.

city, however. Factorial ecology was originally limited in scale, but it has been extended recently to cover any application of factor analysis to areal (that is, ecological) data. We use the term in this broader meaning. Analyses at regional and international scales qualify as factorial ecology, and we meet such studies below.

The Factor Analysis Debate

Factor analysis has probably generated more controversy than any other quantitative technique in geography and the social sciences. Debates are sometimes characterized as arguments between two opposing camps—the pro-factor analysts, drawn heavily from the ranks of the psychologists, and the anti-factor analysts, mostly statisticians and mathematicians. Both extreme views, that factor analysis is a misbegotten fraud and that it is the panacea of all social science, are clearly overstatements. The urban ecology example furnishes proof of the utility of the technique. It is also true that other applications of the technique could have been chosen that would have shown factor analysis producing irrelevant or obvious conclusions. We balance our presentation by taking a more critical look at some examples.

We used Robert Murdie's study of Toronto; let's consider another part of his analysis. As well as describing the structure of Toronto at particular points in time, he also attempted to look at the pattern of change in Toronto from 1951 to 1961. His method was basically the same as his cross-sectional analyses, except that the data matrix consisted of change variables. These are simply measures of the differences between variables for census tracts at two points in time. A principal components analysis was carried out on this 56 variable times 277 census tract matrix, and he rotated to simple structure. Six factors were extracted, accounting for 57 percent of the total variance. In fact, only the first four were clearly interpretable. We consider the first two.

Factor 1 was labeled suburbanization and accounted for 19.6 percent of the variance. Table 6.16 shows the major loadings and Figure 6.18a shows the pattern of factor scores. The interpretation of this factor is obvious from loadings and scores. Some critics would say that the factor is a little *too* obvious. Toronto, like many other cities, experienced peripheral growth in the 1950s. This is well known and we do not need a principal components analysis to confirm it. What factor analysis techniques can achieve is specification of

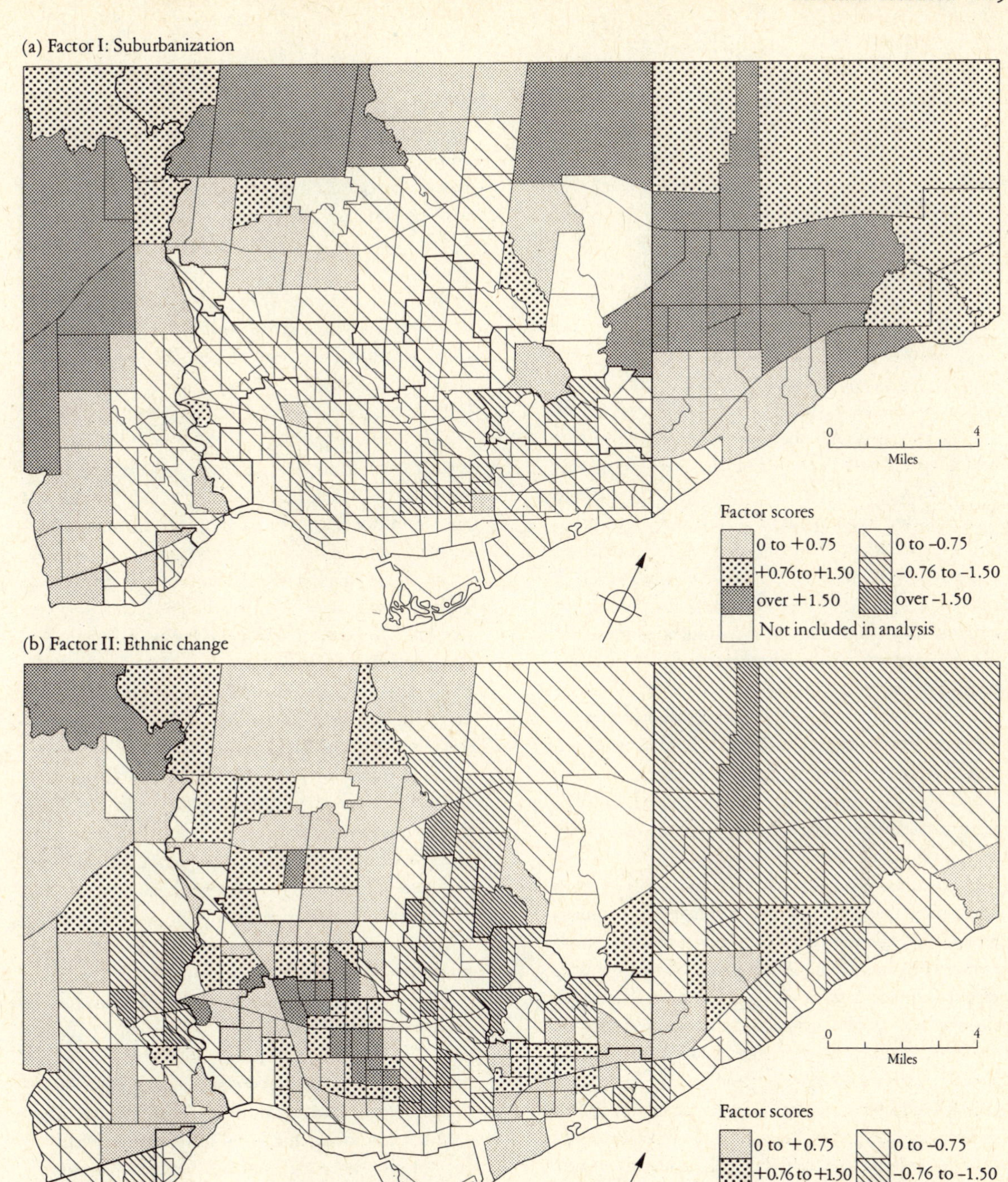

Figure 6.18 Factor score maps for change variables, Toronto, 1951–1961.
From R.A. Murdie. *Factorial Ecology of Metropolitan Toronto, 1951–1961.* University of Chicago, Department of Geography Research Paper No. 116, 1969.

a measure of the particular phenomenon under analysis. This involves careful selection of variables based on *a priori* knowledge. The emergence of the suburbanization factor from a larger analysis of change seems to produce only a description that is not particularly novel and not necessarily useful. A similar argument can be made with respect to factor 2, labeled ethnic change and accounting for 9.4 percent of the variance. The table of loadings (Table 6.16) is again easily interpretable by anyone familiar with Toronto, as is the map of factor scores. The study of ethnic change can be traced back to early use at the University of Chicago of the ecological concept of succession. The essence of the empirical research that has developed from this beginning is the identification of particular groups and their relations with other groups. This cross-ethnic change factor seems to have little to offer beyond a simple description of where changes in population composition occur (Figure 6.18b). In both cases, it seems fair to suggest that the spatial distribution of some of the individual variables is much more useful and informative in understanding change than is the composite factor.

The contrast between these findings and the results from the cross-sectional analysis, which as we have shown relate to previous ecological studies, is quite spectacular. It is hard for the most fervent member of the pro-factor analysis camp not to suspect that factor analysis might be redundant in some way.

If the criticisms of factor analysis involved only suggestions that the technique is sometimes inappropriate, the debate we are discussing would be mild. Factor analysis presents us with more problems than simply inappropriate use. Perhaps the most distinctive characteristic of the technique is its great degree of flexibility. This can be viewed as paradoxically both a blessing and a problem. The flexibility has methodological and technical aspects. First there is the flexibility in purpose, as we have shown. Factor analysis can be used in research contexts ranging all the way from data description through hypothesis testing to the structuring of theory into matrix algebra language. Factor analysis incorporates a whole family of related techniques. For the most part, geographers have used principal components and common factor analysis. Many other forms of factor analysis exist (sixteen have been identified recently by Brian Berry in a factorial ecology review) and some are becoming available in computer package programs. We can expect a wide use of factor analytic techniques in future studies. It is unfortunately not clear which forms of factor analysis are most suitable for which research problems. We do know that different techniques inevitably produce different results. A factorial ecology of a city based on common factor analysis will be different from a factorial ecology of the city based on, say, alpha factor analysis.

This situation is clearly disturbing and will remain with us until geographers can sort out exactly what it is they want from factor analysis and can match this with available techniques. It is fair to say that the past predominance of common factor analysis, and of principal components analysis in particular, has been the consequence of their availability in handy package programs; it has not been based on evaluation of their suitability for problems at hand. It cannot be expected that research in factorial ecology will come to a halt while these basic problems are being solved. In fact, part of the criteria for assessing the suitability of any technique must be based on results and their interpretation in the light of current knowledge in the field. From this viewpoint, many studies using available package programs have been reasonably successful, as our section on urban ecology has shown. This discussion highlights the need that results be presented and accepted in a modest manner, avoiding any suggestion, however large the data matrix, that the researcher is producing absolute truth.

We must consider the use of factor analysis in geography as experimental. Many of the problems faced by users of factor analysis are not specific to this particular technique but are shared, perhaps less explicitly, by many other quantitative techniques. Consider the common criticism of factor analysis that you get out only what you put in. We have consistently argued that this assertion is true. But it is in no way a unique property of factor analysis. It relates to *all* empirical analysis. No analysis can create more empirical evidence; it can merely re-order available evidence into a form that is more easily incorporated into the existing body of knowledge. Correlation coefficients start with two variables and present us with a measure of their correspondence; point pattern analysis expresses a dot map in terms of pattern type; factor analysis transforms a data matrix into a set of general dimensions. Another criticism is that, whatever the form of data matrix is, factor analysis will *always* give an answer. It will even find dimensions in a table of random numbers, if we program it to do so. *All* techniques give an answer. A regression analysis on two columns of random numbers will discern and calibrate a mathematical function. Related statistics, the standard error of estimate and correlation coefficient, will reveal the validity of the analysis. Similarly, in factor analysis the small range in eigenvalues will indicate that a table of random numbers has very little underlying structure. We conclude that many of the criticisms that have been made of factor analysis are not unique to this set of techniques, although they may be valid criticisms of specific applications because the approach is comparatively new.

Alternative Data Inputs

We have always used the same form of data matrix for input, with n variables represented by columns and k areal units represented by rows. Correlations between columns give us the $n \cdot n$ correlation matrix from which factors are extracted. These analyses are generally termed *R-mode* analyses. This is not the only form of data input that has interested geographers. We present variations in this section.

The simplest modification we can make to the initial data matrix is simply to transpose it. This makes

Table 6.17 Q-Mode Analysis of United Nations Voting Behavior (1963–1964)

Country	*Factor 1*	*Factor 2*	*Factor 3*	*Factor 4*	*Factor 5*	*Factor 6*
Denmark	0.90	0.12	−0.02	−0.27	−0.01	−0.17
United States	0.81	0.07	0.23	−0.27	0.09	0.23
France	0.59	0.01	−0.48	−0.02	−0.23	0.27
Chad	0.12	0.87	0.17	0.01	−0.03	0.06
Ivory Coast	0.08	0.73	0.35	−0.04	0.27	−0.04
Togo	0.09	0.62	0.49	−0.02	0.23	−0.01
Ghana	−0.09	0.14	0.88	0.17	−0.11	−0.04
Iraq	−0.24	0.15	0.73	0.30	0.25	−0.04
Sudan	0.00	0.24	0.60	0.24	0.05	−0.09
Czechoslovakia	−0.42	−0.04	0.28	0.85	−0.02	−0.02
Cuba	−0.36	0.00	0.28	0.85	−0.07	−0.02
Albania	−0.27	0.01	0.49	0.59	−0.05	−0.07
Lebanon	0.09	0.16	0.46	0.08	0.66	0.10
Libya	0.21	0.44	0.45	0.01	0.54	−0.05
Portugal	0.23	−0.25	−0.06	−0.44	−0.08	0.68
Spain	0.52	0.13	−0.11	−0.26	0.09	0.66
Venezuela	0.70	0.52	−0.01	−0.07	0.13	−0.02
Jamaica	0.59	0.51	0.03	0.06	0.32	−0.19
Israel	0.43	0.35	−0.04	−0.18	0.04	−0.31

Source: Adapted from B.M. Russett. *International Regions and the International System.* Chicago: Rand-McNally, 1967, Table 4.3.

the columns represent the areal units and the rows represent variables. Correlations between these columns produces an $k \cdot k$ matrix of relations between areal units from which factors may be extracted. This is known as *Q-mode* analysis. Q-mode analysis is easy to understand, but what does it mean? To answer this, we must go back to the original correlations. Correlations in this context measure the degree of correspondence between two *objects* in terms of their respective values on the variables. Two areal units that have similar characteristics in terms of the variables record a high correlation. Therefore, a Q-mode factor analysis produces dimensions from the pattern of object vectors in variable space. All that happens is that we turn around the analysis to produce clusters of objects in a variable space instead of groups of variables in an object space.

The interpretation of a Q-mode factor analysis is the same as in R-mode analysis except for this reversal of roles. Loadings are for objects, indicating how strongly they correlate with a factor. Because the factors represent clusters of objects, with areal units they can be interpreted as regions or regional types. It is, therefore, surprising that Q-mode factor analysis has not found more favor in geographical studies. We illustrate this approach by using the work of political scientists in their attempts to ascertain the dimensions of nations. These are global studies using nation states as units to produce a political ecology.

Bruce Russett has attempted to develop a theory of international integration. He argued that it is important to specify any *sufficient* conditions for successful integration but it is also possible to hypothesize situations that may be considered as *necessary* conditions. He identified one such condition as that nation states should have similar political attitudes toward international affairs. Political attitudes are usually measured by using the voting behavior of countries in the United Nations. The data matrix, therefore, consists of some 108 countries with 66 roll calls for the 18th Session of

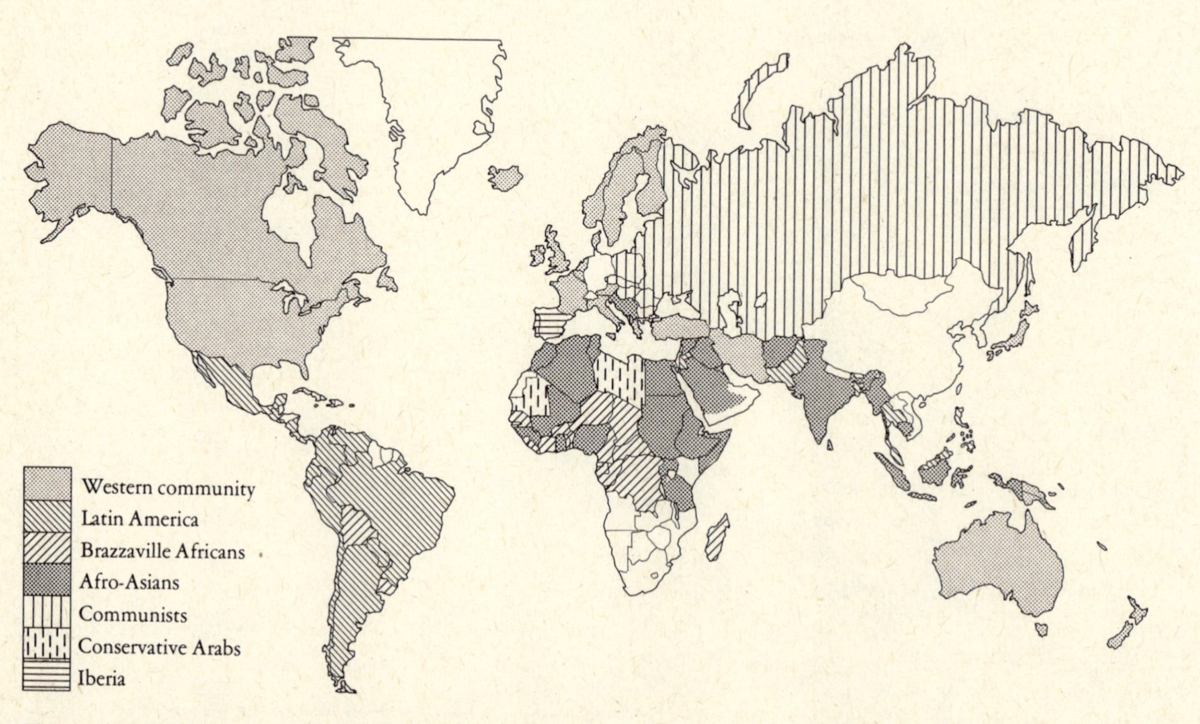

Figure 6.19 World regions of United Nations' voting behavior.
From B.M. Russett. *International Regions and the International System*. Chicago: Rand-McNally, 1967, Fig. 4–1, pp. 72–73.

the General Assembly (1963–1964). On each vote each state was coded 2 (affirmative), 1 (abstain), or 0 (negative). Thus, we have a data matrix consisting of 0s 1s, and 2s, each row indicating the pattern of votes among countries and each column showing the individual voting behavior of each country. This is set up for a Q-mode analysis producing a 108 times 108 correlation matrix indicating the degree of similarity in the voting behavior of pairs of countries. Extracting factors from this matrix by using the principal components model with a rotation to simple structure produces the results shown in Table 6.17. Here we see a selection of 19 of the 108 countries. The loadings are interpreted as correlations between country and factor. This analysis is very successful in identifying the major groups of countries in terms of their voting behavior in the United Nations. All the western European states and European-settled states load high on factor 1, which Russett confidently termed the Western community cluster. The allies of these European-peopled states (Japan, Taiwan, Iran) are included in this cluster. The next three factors identify a spectrum of voting behavior from what Russett termed the Brazzaville Africans (factor 2) through the Afro-Asians (factor 3) to the Communists (factor 4). Notice that this latter group loads negatively on factor 1 while, conversely, the western community states load negatively on factor 4. Two small factors, 5 and 6, identify small distinctive clusters labeled conservative Arabs and Iberia, which consists of only Spain and Portugal with

Table 6.18 Factor Loadings and Scores for Indian Commodity Flows (1959–1960)

Loadings for destinations		*Scores for sources*	
Factor 1			
M. Pradesh	0.67	Bombay state	1.63
Punjab	0.51		
Hyderabad	0.58		
M. Bharat	0.60		
Rajasthan	0.78		
Saurashtra	0.89		
Kutch	0.79		
V. Pradesh	0.79		
Bombay port	0.81		
Saur. Ports	0.90		
Factor 2			
Andhra	0.50	Madras state	2.87
T-Cochin	0.89	M. Pradesh	1.18
Cochin port	0.91		
Madras port	0.59		
Other Madras ports	0.79		
Pondicherry	0.79		
Factor 3			
Assam	−0.88	Calcutta	−3.46
Bihar	−0.58	U. Pradesh	−1.64
Orissa	−0.66		
West Bengal	−0.67		
Tripura	−0.90		

Source: Adapted from B.J.L. Berry. *Essays on Commodity Flows and the Spatial Structure of the Indian Economy*. Geography Research Paper No. 111, 1966. Reprinted with permission of the author and the University of Chicago Department of Geography Research Papers.

their peculiar voting record. One final cluster of states can be identified in the countries that load medium to high on the first two factors. These, largely Latin American, states seem to vote in a pattern midway between the Western community and Brazzaville African positions. Detailed inspection of these results shows the technique to be quite sensitive to different political positions of countries within the clusters. France loads as the lowest of the NATO allies on factor 1 and Albania loads lowest on factor 4. This analysis achieves an impressive ordering of the original data matrix into clear clusters of states.

Geographers can employ Q-mode factorial ecologies to define their regions. The pattern of world regions in terms of U.N. voting behavior is illustrated in Figure 6.19. The analysis remains static in the sense that areal patterns are compared but no reference is made to the flows between areas that produce these patterns. Factorial ecology is essentially an extension of the areal association approach. Thus, whether we use the R or the Q mode, it usually defines patterns of phenomena in space rather than identifying the underlying processes producing the patterns. This criticism can be partially answered in two different ways. When factorial descriptions are integrated with a theory that incorporates processes, then the factorial ecology has a legitimate place in the theory by complementing studies of actual processes. This is true, to a limited extent, of social area analysis. Second, processes can be incorporated into a factorial ecology by analyzing the pattern of flows.

A flow matrix is an array of data showing interactions between areal units. It is a square matrix in which every row represents an interaction origin and every column represents a destination. If the data consist of flows of coal, the value in the cell at row i and column j denotes the amount of coal flowing from area i to area j. If the columns are now correlated to produce a correlation matrix, the correlations show the degree of similarity between the sources of flow for the two destinations under consideration. Two areas that obtain the commodity from a similar set of areas have a high correlation. A factor analysis of the matrix, therefore, produces clusters of areas that are similar in terms of their supply areas. We have a technique for defining *functional regions*. The factor

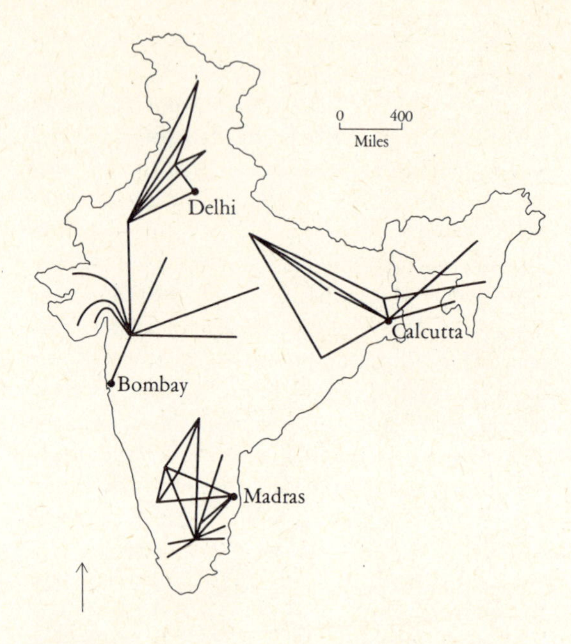

Figure 6.20 A factor analytic definition of functional regions in India.
From B.J.L. Berry. *Essays on Commodity Flows and the Spatial Structure of the Indian Economy*. University of Chicago, Department of Geography Research Paper No. 111, 1966.

loadings indicate which areas are correlated with a factor, while the factor scores indicate the most prominent sources of shipment to areas clustering around the dimension.

An example clarifies this. The procedure has been applied by Brian Berry to commodity flows between 36 Indian states. We look at the results from a principal axes factor analysis with varimax rotation for the flow matrix describing total flows of 63 commodities. Ten factors with eigenvalues greater than one were rotated and we consider the first four, which account for 22.5 percent, 13.5 percent, 10.9 percent, and 11.6 percent of the total variance. Table 6.18 shows the loadings over 0.5, indicating the main consuming areas, and the factor scores over 1.0, indicating the main supply areas relating to three factors. Factor 1 describes the pattern of flow from Bombay state to the surrounding areas. The four factors in fact define functional regions around Bombay, Madras, Calcutta, and Delhi. The results can be mapped by drawing the flows as lines between origins and destinations to define the four functional regions (Figure 6.20).

The production of functional regions from flow matrixes represents a classic example of factor analysis in the role of simplifying and clarifying a complicated data set. The original 36 times 36 matrix includes information on 1,296 individual flows. Factor analysis has decomposed this myriad of flows into some basic independent patterns that we find correspond to what geographers had defined as functional regions.

Location Variables in Factorial Ecology

Factorial ecology qualifies as spatial analysis because it uses areal units as its objects. Some geographers have argued that, because we are interested in the locations of the areal units in space, we should include location variables in factor analyses. This particularly geographical argument seems a highly suitable subject with which to conclude our discussion. Location variables have been included in data matrices for factor analysis in four distinctive ways. We consider each.

The simplest form in which location can enter a data matrix is as two coordinate axes defining the map that contains the areal units. In the Indian study, a factor analysis was carried out on a data matrix containing 98 variables describing Indian states. Two were latitude and longitude. What influence does inclusion of these two location variables have? If we turn to the results, we find that they load separately on two different factors (Table 6.19). The third factor picks out west-east differences in India's agriculture and we find longitude loading at +0.66 on the factor. The production of jute similarly loads positively because of its eastern location pattern, while jowar loads high in a negative position because of its western location pattern. In contrast, factor 5 picks out northern specialization in grains. Thus latitude loads negatively at −0.69 whereas production of maize and wheat load positively. The factor scores maps (Figures 6.21a and 6.21b) for these two dimensions clearly indicate the east–west and north–south contrasts suggested by our interpretation of the factors.

This analysis seems perfectly valid and the inclusion of the location variables seems geographically satisfying. Satisfaction does not withstand closer scrutiny. At a philosophical level, there is circularity of method when we map the factor scores. Because longitude and

Table 6.19 Loadings for Selected Factors in India (1959–1960)

*Factor 3 = 7.6% of common variance**		*Factor 4 = 6.4% of common variance†*		*Factor 5 = 5.8% of common variance**	
Jowar	−0.79	Total	−0.88	Maize	+0.60
Bajra	−0.53	Urban	−0.67	Wheat	+0.78
Groundnut seed	−0.64	Rural	−0.82	Grain	+0.66
Sesame seed	−0.49	Literate	−0.72	Latitude	−0.69
Cotton fiber	−0.70	Manufacturing	−0.59		
jute fiber	+0.44	Urban manufacturing	−0.60		
Edible oils	−0.47	Grain production	−0.45		
Cotton gins	−0.74	Gur production	−0.59		
Longitude	+0.66				
Proportion of population urban	−0.42				

*Variables relate to production unless otherwise noted.
†Variables relate to potential unless otherwise noted.
Source: Adapted from B.J.L. Berry. *Essays on Commodity Flows and the Spatial Structure of the Indian Economy*. Geography Research Paper No. 111, 1966. Reprinted with permission of the author and the University of Chicago Department of Geography Research Papers.

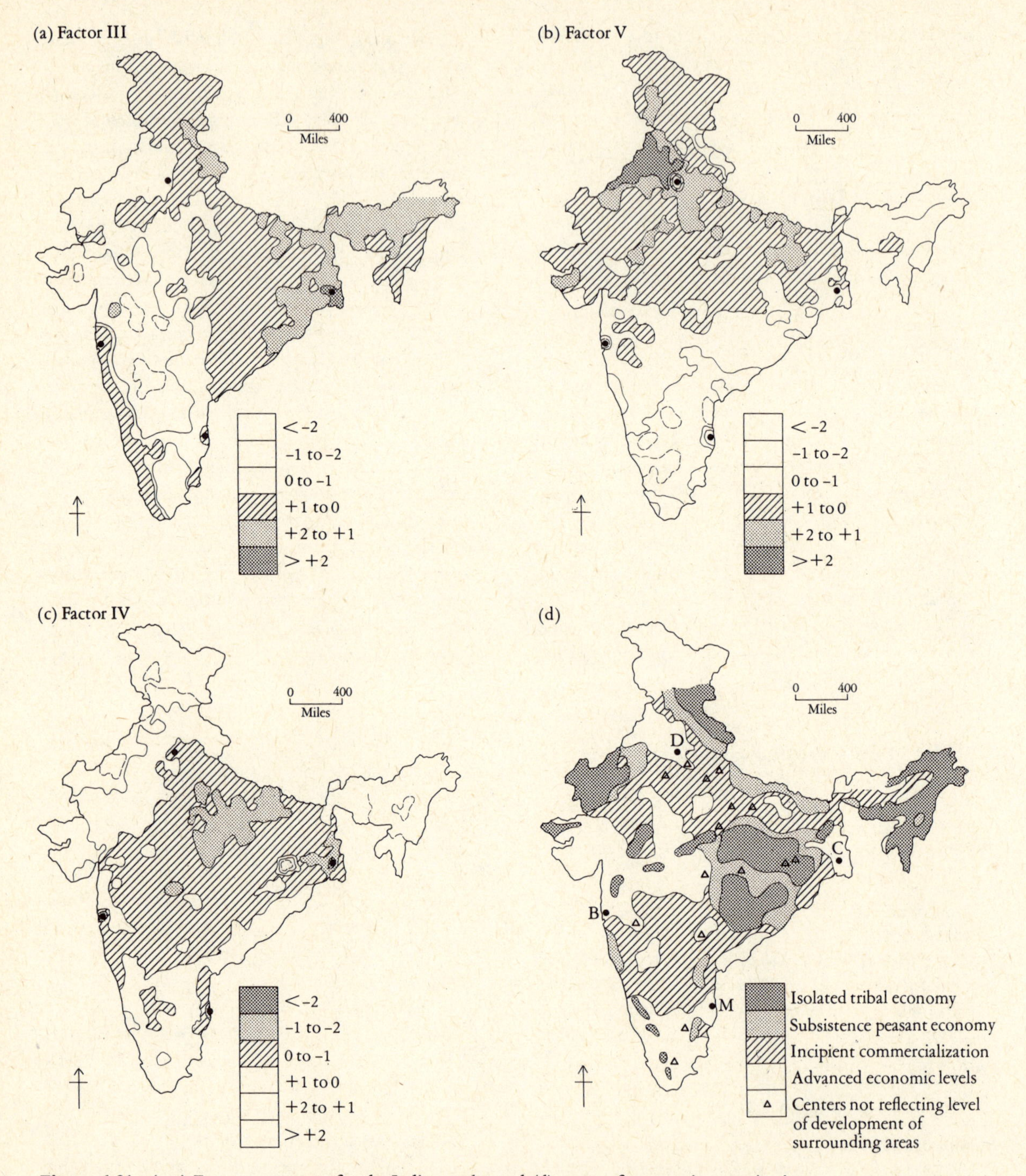

Figure 6.21 (a–c) Factor score maps for the Indian study, and (d) types of economic organization.

latitude contribute to the calculation of the scores, when we map the scores we are, to some extent, locating location. We may differ on how important we think this circularity may be, but the question becomes less important when we realize that the factors extracted from a geographical matrix that includes coordinate axes variables are *not* invariant to the choice of axes made. Because there is an infinite number of orthogonal axes that can describe a space, it follows that there is an infinite number of factorial solutions, once we admit coordinate location variables into the analysis. In the Indian example, longitude and latitude are quite arbitrary axes, despite their conventional use. Their inclusion in the analysis helps us formulate factors from variables that happen to have north–south or east–west variations in their distributions. This results in factors 3 and 5. If axes describing northeastwardness and southeastwardness had been used, these would have helped create new factors relating to other variables. All we can say for latitude and longitude is that, because the analysis is at a subcontinental scale, they become reasonable surrogates for rainfall and temperature patterns, respectively, and help produce two agricultural factors that probably would have emerged in any case, although possibly lower down in the rank order of factors.

Instead of including location as two arbitrary coordinates, we may have some hypothesis or theory that enables us to consider location as a single variable in the form of distance from some selected point. In studies of the city, for example, it has been shown that many variables vary in terms of distance from the central business district. In fact the classic ecological generalization of concentric rings is associated with many studies identifying gradients of intensity of variables from the center of the city. Murdie used such evidence as justification for including the variable that he called distance from the intersection of peak land value in his factorial study of Toronto. We have noticed how this particular variable loads high on the family status factor (page 258). We noticed how the map of the factor scores (Figures 6.16) also suggests a concentric pattern. This is to some extent the consequence of the circularity of method, because we are again locating location on this map. Circularity is even more important when we use analysis of variance to test the hypothesis that the factor scores are concentric in pattern. Notice that the alternative classical model, the sector model, is not incorporated in the factor analysis by including direction as a variable (we noticed the problem of its simple measurement in Chapter Two). In fact, the inclusion of distance from the center because it is totally independent of sectors, helps produce a nonsectoral pattern in factor scores when it loads high on the factor. This variable therefore contributes to the fact that family status is found to be only very moderately sectoral.

Our conclusion must surely be that inclusion of location variables adds nothing to the analysis except confusion in mapping and circularity in argument if spatial hypotheses are subsequently tested. We must not overestimate the importance of location variables in the analyses we have described, however. The contribution of location to total variance in analysis is small in both cases—in the Indian study two variables in 98; in the Toronto study, one variable in 126. They tend to distort the alignment of one or two factors and help predetermine the spatial pattern of the scores.

We can carry out a factor analysis with only location variables in the analysis. This constitutes the third way in which location has been incorporated in factorial ecology studies.

Theoretical geographer William Bunge has argued that location can be considered a single variable by considering it in terms of nearness. Iowa is locationally more similar to Illinois than to California. In fact, a simple distance matrix of places in which cells indicate distances between places can be viewed as indicating the pattern of locational similarities. It can be used as input in a factorial ecology.

In the global political ecology we referred to, we indicated that ecological patterns relating to political integration were being investigated. One condition for integration that Russett considered was simple geographical proximity. States near one another are more likely to integrate than are states far apart. This feature was therefore incorporated into the analysis by carrying out a special factorial ecology. A distance matrix was constructed, containing the great circle distances between the political capitals of the 126 countries being considered. The result is a square 126 times 126 symmetrical matrix. All the values were then standardized by dividing them by the single largest distance in the matrix (New Zealand to Spain, 12,400 miles) and subtracting the result from 1. Short distances are represented by values approaching 1 and long distances tend toward 0. Because the distance from a country's capital city to itself is 0 miles, unities are inserted in the diagonal and the two Spain–New Zealand cells become 0. This new matrix of proximities ranging from 0 to 1 represents the locations of the world's countries.

Because this data matrix is square and symmetric, we can bypass the traditional computation of a correlation matrix and extract factors directly. This is known as *direct factor analysis* and uses the principal components model. This model is based on mathematics that applies to any matrix and does not relate to correlation matrices only. Some researchers think that the dependence of factor analysis on correlation matrixes is somewhat unfortunate, and there has been talk of liberating the technique from its statistical origins. We can best view this as another example of the extreme flexibility in factor analysis. With direct factor analysis, researchers remain closer to their original data. There is, in fact, no need for distances to be transformed in the manner we described except that it aids interpretation. When we use a matrix with

Table 6.20 Direct Factor Analysis Based on Proximities

Country	*Factor 1*	*Factor 2*	*Factor 3*	*Factor 4*
Denmark	0.85	0.31	0.27	0.25
German Democratic Republic	0.85	0.31	0.28	0.26
Czechoslovakia	0.85	0.29	0.26	0.29
Ireland	0.79	0.39	0.21	0.28
U.S.S.R.	0.79	0.24	0.40	0.24
Egypt	0.70	0.17	0.34	0.50
El Salvador	0.22	0.94	0.11	0.07
Honduras	0.22	0.94	0.10	0.09
Panama	0.22	0.94	0.05	0.15
United States	0.42	0.80	0.16	0.10
Canada	0.48	0.76	0.20	0.07
Brazil	0.20	0.72	0.02	0.47
Philippines	0.25	0.05	0.91	0.10
Taiwan	0.30	0.09	0.90	0.07
South Vietnam	0.27	0.04	0.90	0.24
People's Republic of China	0.40	0.15	0.83	0.07
India	0.49	0.04	0.71	0.33
New Zealand	0.34	0.47	0.70	0.09
Cameroun	0.43	0.28	0.21	0.80
Zaire	0.37	0.26	0.25	0.80
Gabon	0.40	0.30	0.20	0.80
Chad	0.51	0.25	0.24	0.73
Kenya	0.39	0.13	0.40	0.72
Malagasy Republic	0.25	0.09	0.47	0.69

Source: Adapted from B.M. Russett. *International Regions and the International System.* Chicago: Rand-McNally, 1967, Table 10.1.

values ranging from 0 to 1, the magnitudes of the loadings are similar to what we are familiar with in analyses of correlation matrices. By inverting distances, similarities in location are measured by the higher values in the same way that similarities in variable patterns are represented by higher correlations.

Let's turn to the results from this direct factor analysis of proximities. Eighty-nine percent of the variation was accounted for by four factors, which were then rotated. The four factors are sufficient to classify all the 126 countries, with every country loading above 0.5 on at least one factor. A selection of the results is laid out in Table 6.20. Russett named these factors Europe, western hemisphere, Asia, and Africa, respectively. The classes do not coincide exactly with the usual use of these terms, but they are reasonable geographical labels. The Middle East states load highest on the Europe factor and are classified with Europe although they also have sizable loadings on the Asia and Africa factors. The western hemisphere factor coincides exactly with countries usually classified under that head.

What does a factor analysis of areal units using proximities produce? Why are there four factors in this case and why are there these particular four factors? The answer is simply that the factor analysis is picking up clusters of small countries. The countries are small, distances between their capitals are small, proximities are large, and thus a cluster of small countries produces a set of large proximities to be extracted. In Table 6.20, we find that it is these small countries that load highest on each factor. At the global scale, there are four areas containing groups of small countries—northwestern Europe, Central America, Southeast Asia, and west central Africa. Countries in these four areas head the loadings on each factor. With a different pattern of areal units, a totally different set of results is likely to occur. This is the ultimate example of areal units influencing a spatial analysis; the results reflect simply the pattern of areal units. If each of the fifty United States had been included separately in the analysis, we could have expected confidently a fifth North American factor. These American states are already integrated politically, however, partly because of their proximities. In fact, the political integration being investigated can be viewed as the eradication of various factors. In this particular study, the results were deemed useful because they allowed the proximity condition to be compared directly with other conditions for integration, such as the political attitudes that were also factor analyzed. Factorial ecologies of the location variable alone would seem generally to have little utility, particularly when areal units are arbitrary to any degree.

Our conclusions about location variables in factorial ecology have been quite pessimistic. One reason may be that we have insisted on using simple absolute location; it has been argued that aggregate relative position is the way in which location should be viewed in theoretical geography. From our discussions in previous chapters, we might fairly conclude that we can represent relative position quite adequately by using the potential concept. This line of argument was accepted in Berry's India study; he included six measures of potential in the analysis relating to total, urban, rural, literate, manufacturing, and urban manufacturing populations. It is perhaps disappointing that they all load high on the same orthogonal dimension as factor 4, which includes only two other variables with high loadings (the production of grain and gur; see Table 6.19). Rather than contributing to different aspects of the Indian social and economic structure, these different relative location measures are numerous enough, and intercorrelated enough, that they produce their own factor, which may be labeled inaccessibility because the potential measures load negatively on it. Only the three urban and manufacturing potentials load high elsewhere, all on the general first factor.

This result seems at variance with expectations because relative position concepts have been included only on the assumption that they relate to the other variables. Why, therefore, do we have an independent potential factor in this analysis? The answer lies partly in the nature of potential variables and partly in the spatial structure of India. Potential variables, by their very nature, produce smooth spatial distributions; other variables often include quite irregular spatial patterns. In India, large population concentrations in four widely separate zones around Calcutta, Bombay, Madras, and Delhi produce level potential surfaces.

Many isolated areas in central India that still have a tribal economy do not have very low potentials because their locations are between the four population concentrations. These isolated tribal areas are not isolated in terms of potential variables. The inaccessibility factor is, therefore, misleading; we illustrate its pattern in Figure 6.21c along with a map of areas of different economic organization.

We can conclude that it is not necessarily the idea of including a relative location variable in the factorial ecology that is at fault here but, rather, it is the particular measure that was chosen. In less developed parts of the world, the decline of a population's influence with distance can be expected to be greater than for modern Western countries. Thus potential variables that have been found useful in the United States are almost certainly less useful in India. This seems to be an example for which potential experiments, such as we described in Chapter Two, are called for—the distance exponents should be allowed to vary. In this way, the influence of the major concentrations of population can be made to decline more rapidly with distance and areas that have remained truly isolated can be measured as isolated by the potential variables.

The inclusion of location variables in their four guises is consistent in one respect—their disappointing contribution to analysis. A geographer's natural inclination to add location variables to a data matrix seems to require curbing. Factorial ecology studies are essentially spatial, in that their objects are areal units. The addition of location variables seems only to confuse. Spatial patterns are certainly of interest but these can be considered subsequent to the factor analysis in studying factor scores. The analysis of variance strategy is particularly suitable in this context. The use of relative location variables may be an exception to this generalization, although their inclusion in a study must be carefully thought out. This applies, of course, to every variable in a factor analysis. We have stated that no variable should be included in a factorial ecology unless a sound reason can be given for it. The fact that the researcher is a geographer is not sufficient reason to include location variables in a factor analysis.

Further Reading

Factorial ecology is an interesting example of time lag in the application of a technique in a discipline. Statisticians (Kendall, 1939) and sociologists (Hagood *et al.*, 1941) were carrying out factorial ecologies a generation before interest was aroused in geography. One advantage of this late arrival of geographers is that, from a teaching viewpoint, some very good elementary introductions to what has traditionally been considered a difficult subject are now available. Much of our discussion is drawn from Rummel (1970), which is concerned explicitly with *applied* factor analysis. This work is particularly easy to read because the examples are political and can be more easily related to geographical research than traditional psychological examples can. Two excellent introductory books by psychologists can also be recommended—Child (1970) and Comrey (1973). Furthermore, Cattell, a pioneer in this field, has produced two very clear and highly recommended articles on factor analysis and its role in research (1965a and 1965b). For students who prefer articles, reference may also be made to Rummel (1969).

Further understanding of factor analysis requires knowledge of matrix algebra. This is given in Rummel (1970), although a social scientist's introduction to this mathematics is available by Horst (1963). It is also introduced in a geographical context by Gould (1967) and developed further by Tinkler (1972). The best technical statement on factor analysis, given this knowledge, is probably that by Harman (1967). For inferential topics relating to factor analysis, reference should be made particularly to Lawley and Maxwell (1963). The flavor of the so-called factor analysis debate can be obtained by comparing Cattell (1965a and 1965b) with Ehrenberg (1962).

In our illustration of the utility of factor analysis, we have been forced to over-emphasize some of the literature at the expense of the rest. Our examples have concentrated largely on cities and can be found in Bell (1955), Schmid and Tagashira (1965), Sweetser (1965), Murdie (1969), and Gittus (1964). This literature is very well summarized in Timms (1969). The general extension of factorial ecology is presented by Berry (1971a) and Rees (1971) in geography and in Dogan and Rokkan (1969) more generally. Alternative references for specific factor analysis research designs are as follows.

Principal components analysis is used by Carey (1966), Henshall and King (1966), and Spence (1968). Common factor analysis is used in Thompson *et al.* (1962), Smith (1968), and King (1969). Oblique rotation is used in Hughes and Carey (1972), Davies and Lewis (1973), and Giggs (1973). Direct factor analysis is used in Russett (1967) and Berry, Barnum, and Tennant (1962). The earliest example of factor analysis of interaction matrixes is by Garrison and Marble (1964); there is a more complete discussion in the Indian study (Berry, 1966), and recent examples are by Goddard (1970) and Clark (1973a). For use of analysis of variance with factor scores, reference should be made to Johnston (1970b), who comments on the application reported above (Murdie, 1969), and Ray (1969), who applies it at a national scale. As a general method, analysis of variance is used by Zobler (1958) for regional construction, and good descriptions of the method are to be found in Blalock (1960) and Snedecor and Cochran (1966).

In recent years, geographers have become aware of several problems in using factor analysis and a small literature has developed. Problems are dealt with in general by Janson (1969) and specific topics are covered as follows: factor labeling in Palm and Caruso (1972); factor score computation in Joshi (1972); transformation in Johnston (1970a) and Clark (1973b); DIDO or GIGO in Berry (1971b); rotation in Davies (1971a, 1971b, 1972) and Mather (1971, 1972). This critical vein is most well developed in Williams (1971), and the problems are summarized and discussed in Clark, Davies, and Johnston (1974).

Factor analysis is closely related to other research techniques and approaches in geography. Techniques of nonmetric scaling perform measurement duties similar to factor analysis and may be considered a sort of nonparametric factor analysis. The starting point for consideration of this topic is Golledge and Rushton (1972). Factor analysis is also closely related to classification procedures. In fact, it often forms part of a classification or regionalization operation. This theme can be followed up by reference to Spence and Taylor (1970).

Work Table 6.1 One-Way Analysis of Variance

Data

Census tract factor scores for an economic status dimension classified into three sectors. Overall mean is

$$\bar{x} = \frac{(4.0 + 0.1 - 4.1)}{16} = 0$$

	North	*Southeast*	*Southwest*
	+2.1	+0.9	−2.2
	+0.2	−1.2	−1.3
	+0.1	+0.8	+0.5
	+0.9	+0.1	−1.0
	+0.8	−0.5	−0.1
	−0.1	—	—
Total	+4.0	+0.1	−4.1
Number of tracts	$n_N = 6$	$n_{SE} = 5$	$n_{SW} = 5$
Means	$\bar{x}_N = 0.6667$	$\bar{x}_{SE} = +0.0200$	$\bar{x}_{SW} = -0.8200$

H_0

There are no differences between the three sectors in terms of economic status.

Analysis of Variance

Let SS refer to a sum of squared deviations from a mean.

$$\begin{aligned} \text{total SS} &= \Sigma(x_i - \bar{x})^2 \\ &= (2.1 - 0)^2 + (0.2 - 0)^2 + \cdots \\ &\quad + (-0.1 - 0)^2 \\ &= 16.8600 \end{aligned}$$

$$\begin{aligned} \text{between-class SS} &= \Sigma(n_c(\bar{x}_c - \bar{x})^2) \\ &= 6\cdot(0.6667 - 0)^2 \\ &\quad + 5\cdot(0.02 - 0)^2 \\ &\quad + 5\cdot(0.82 - 0)^2 \\ &= 2.6670 + 0.002 + 3.3620 \\ &= 6.0310 \end{aligned}$$

Total SS = between-class SS + within-class SS

$$\begin{aligned} \text{Within-class SS} &= \text{total SS} - \text{between-class SS} \\ &= 16.8600 - 6.0310 \\ &= 10.8290 \end{aligned}$$

We can now construct an analysis of variance table (for n scores and k classes).

Source of variation	*SS*	*df*	*Variance*
Total	16.8600	$n - 1 = 15$	1.1240
Between class	6.0310	$k - 1 = 2$	3.0155
Within class	10.8290	$n - k = 13$	0.8330

Statistic

$$F = \frac{\text{between-class variance}}{\text{within-class variance}} = \frac{3.0155}{0.8330} = 3.6200.$$

Probability

The critical variance ratio from Table A.11 is $F(df_1 = 2, df_2 = 13, \alpha = 0.05) = 3.81$.

Decision

From this analysis, we can see that we just fail to reject H_0 at $\alpha = 0.05$. This means that we have not been able to show that our sectoral classification of cities is not significantly different from any arbitrary classification of census tracts at this significance level.

Exercise

The analysis barely fails to find a significant sector pattern of economic status at $\alpha = 0.05$. Although we reject H_0, we might suspect that this is the consequence of the small sample of tracts. Two new tracts are added to the data, one for a northeast sector (score = 1.5) and one for a northwest sector (score = −1.5). Do these additions lead us to different research conclusions concerning the pattern of economic status in this city?

Census tracts are rearranged as follows:

Northeast	*Northwest*	*South*
+2.1	+0.9	−1.2
+0.2	+0.8	+0.5
+0.1	−0.1	−0.5
+0.9	+0.1	−2.2
+0.8	−0.1	−1.3
+1.5	−1.5	−1.0

Is the F statistic significant at $\alpha = 0.05$ with this new sectoral arrangement?

Work Table 6.2 Two-Way Analysis of Variance

Data

Census tract factor scores for an economic status dimension classified by sectors and zones

Zones	Sectors			Total	Mean
	Northeast	*Northwest*	*South*		
1	+1.2, +1.6	+0.5, +0.1	−1.1, −2.1	+0.2	+0.0333
	$\bar{x}$ = +1.4	$\bar{x}$ = +0.3	$\bar{x}$ = −1.6		
2	+0.9, +1.1	−0.2, +0.4	−2.0, −1.6	−1.4	−0.2333
	$\bar{x}$ = +1.0	$\bar{x}$ = +0.1	$\bar{x}$ = −1.8		
3	+1.3, +1.3	−0.6, −0.4	−1.0, +0.6	−1.2	+0.2000
	$\bar{x}$ = +1.3	$\bar{x}$ = −0.5	$\bar{x}$ = −0.2		
Total	+7.4	−0.2	−7.2	0	
Mean	1.2333	−0.0333	−1.2000		0

H_0

1. There are no differences between the sectors in terms of economic status.
2. There are no differences between the zones in terms of economic status.
3. There is no significant pattern of combinations of zones and sectors in terms of economic status.

Analysis of Variance

total SS = between-cell SS + within-cell SS
Within-cell SS is often referred to as unexplained SS because it is not accounted for by the classification.

between-cell SS = between-sector SS + between-zone SS + interaction SS

The latter term relates to combinations of columns and row variation that remains after column and row SS have been accounted for. For our analysis, we consider

total SS = between-sector SS + between-zone SS + interaction SS + unexplained SS

These can be calculated as

$$\text{total SS} = (1.2 - 0)^2 + (1.6 - 0)^2 + \cdots + (+0.6 - 0)^2 = 23.9200$$

$$\text{between-cell SS} = 2(1.4 - 0)^2 + 2(1.0 - 0)^2 + \cdots + 2(-0.2 - 0)^2 = 21.6800$$

$$\text{between-sector SS} = 6(1.2333 - 0)^2 + 6(-0.333 - 0)^2 + 6(-1.2000 - 0)^2 = 17.7726$$

$$\text{between-zone SS} = 6(0.0333 - 0)^2 + 6(-0.2333 - 0)^2 + 6(0.2000 - 0)^2 = 0.5730$$

$$\begin{aligned}\text{interaction SS} &= \text{between-cell SS} - \text{between-sector SS} - \text{between-zone SS} \\ &= 21.6800 - 17.7726 - 0.5730 \\ &= 3.3344\end{aligned}$$

$$\begin{aligned}\text{unexplained SS} &= \text{total SS} - \text{between-cell SS} \\ &= 23.9200 - 21.6800 = 2.2400\end{aligned}$$

We can now construct an analysis of variance table (with n scores, k sectors, and r zones).

Source of variation	*SS*	*df*	*Variance*
Total	23.9200	$n - 1 = 17$	1.4071
Between cell	21.6800	$kr - 1 = 8$	2.7100
Between sector	17.7726	$k - 1 = 2$	8.8863
Between zone	0.5730	$r - 1 = 2$	0.2865
Interaction	3.3344	$(k - 1) \cdot (r - 1) = 4$	0.8336
Unexplained	2.2400	$n - kr = 9$	0.2489

Statistics

$$F = \frac{\text{between-sector variance}}{\text{unexplained variance}} = \frac{8.8863}{0.2489} = 35.7023$$

$$F = \frac{\text{between-zone variance}}{\text{unexplained variance}} = \frac{0.2865}{0.2489} = 1.1511$$

$$F = \frac{\text{interaction variance}}{\text{unexplained variance}} = \frac{0.8336}{0.2489} = 3.3491$$

Probabilities

From Table A.11

$F(df_1 = 2, df_2 = 9, \alpha = 0.05) = 4.26$
$F(df_1 = 2, df_2 = 9, \alpha = 0.05) = 4.26$
$F(df_1 = 4, df_2 = 9, \alpha = 0.05) = 3.63$

Decisions

1. We reject H_0 and we can very confidently assert that economic status has a sectoral pattern.
2. We cannot reject H_0 and we can be equally confident that economic status does not have a concentric pattern.
3. We cannot reject H_0 although we barely fail to reach the 0.05 threshold level. There may well be some characteristic sector–zone interaction patterns of economic status, although our evidence does not quite support this view.

Exercise

Repeat the analysis with the following data for a different city. These represent economic status factor scores. Is this dimension arranged sectorally or zonally or is there interaction between these classes?

Zones	*Sectors*			
	North	*East*	*West*	*South*
1	+1.6+1.8	+0.6+0.8	+0.0−0.2	−1.0−1.2
2	+1.0+1.2	+0.2+0.0	−0.6−0.8	−1.6−1.8

VII

APPLIED SPATIAL ANALYSIS

One very clearly discernible trend in modern geography is the search for more relevance. The idea of relevance has been interpreted in different ways by different geographers. In its most broad sense it is synonymous with applied geography, the application of pure geography to practical problems in the real world. The impact of professional geographers on policy-making has been, with the exception of a few notable examples, unfortunately sporadic. The question is whether the "new" geographers with their quantitative tools will have more success than their predecessors. Evidence is beginning to accumulate that suggests that the answer may well be yes. Part of this evidence is the emergence of a set of specialized tools called applied spatial analysis.

We can identify five roles that have been claimed by applied geographers. The most ambitious is the generalist who overviews specialists. In this role, geography is viewed as a broad, holistic discipline; geographers contribute to an applied project by synthesizing the work of specialists in a project team composed of economists, sociologists, botanists, and the like. This notion of applied geography has declined as the idea of geography as a holistic discipline has gone out of favor. Other geographers argue that their colleagues should enter the world of business and government and sell their spatial specialism. This involves geographers in research on store location, for example, and is part of an area of the discipline identified as marketing geography. In the public sector, the geographer's spatial viewpoint has been exploited in physical planning, which is institutionalized in most Western countries. The two other roles of the applied geographer are as much related to an individual's perception of her or his role in society as they are to education as a geographer. One group of geographers has become concerned with what is loosely termed social justice; a radical geography is emerging in which geographers consider how disadvantages suffered by certain sections of society relate to location. Other geographers specialize in the problems of the so-called underdeveloped world. They are often concerned with such things as man-land relations in agricultural projects. All but the first of these roles concerns the geographer as a specialist. The latter four use the techniques of applied spatial analysis.

There are at least three basic strategies in applied spatial analysis. The most simple can be termed description. It follows in a long tradition in geography that provides information for groups of decision-makers who may or may not include geographers. The classic example is Dudley Stamp's land utilization survey in Britain in the 1930s; it provided land use maps that helped formulate agricultural policy for the country. Most techniques from areal sampling to factor analysis can provide descriptions of phenomena that are of interest to decision-makers. In this chapter, we concentrate on the other basic strategies of prediction and optimization.

All decision-making is concerned ultimately with the future, and for this geographers employ a method of spatial prediction. The simplest prediction involves merely the extrapolation of current trends. Of much greater interest are questions of the "What if?" variety. We might ask: "What if we locate a new shopping center five miles to the north of a particular central business district? What effect will this have on sales in the central business district? Where will the shoppers at the new center come from?" Such questions relate to retail patterns and cannot be answered by extrapolating trends because the new shopping center's current sales are zero—it has not yet been built. In such situations, prediction must be based on some understanding of the system we are trying to manipulate. For this we may use social gravity models.

Furthermore, instead of asking "What's there?" or "What might be there in the future?" we sometimes ask "What ought to be there?" In this case, the strategy is to specify the best spatial pattern given a set of criteria and constraints. If we want to serve a population with a new hospital so that total distance from patients to hospital is minimized, we ought to recognize immediately that what we require is the areal median location or point of minimum aggregate travel (discussed in Chapter One). By locating a new hospital at the areal median, we use the best, or optimum, location given the stated criterion. Finding the optimum pattern of one location can be extended to multiple location problems, and in the final section of this chapter we consider how to do this when we discuss location-allocation problems in general.

The three strategies of description, prediction, and optimization do not coincide with the categories of applied geography identified above. Optimizing techniques are useful to private and public planners as well as to those concerned with social justice in the city and in underdeveloped countries. The techniques we describe under these strategies can have roles in pure as well as applied research. Some researchers have attempted to develop the gravity model, for example, into a general theory of spatial interaction in geography. Nevertheless, all the approaches we describe were developed in applied contexts and have their greatest use in applied research. They certainly do not exhaust the content of applied spatial analysis, but they represent a well tested sample.

GRAVITY MODELS

In this section, we look at spatial analysis techniques developed in relation to problems in urbanized societies. Patterns of new development are of widespread concern and different countries produce different strategies of control and planning. Britons talk of town and country planning while Americans generally speak of urban and regional planning. We use the terms town and urban planning interchangeably and are interested largely in the approaches urban planners have adopted for predicting spatial patterns.

Much early town planning was very disappointing in that it was rather static, following its architectural heritage. British town planning in the 1950s was centered on the development plan, which was basically a normative land use map. A major breakthrough occurred early in the 1950s with the linking of traffic flows to land use data in the United States. The argument was simply that different types of land use generate different and variable traffic flows. This thesis is very elementary and somewhat obvious but it is the basis for understanding, and therefore being able reasonably to predict, in urban planning. Its acceptance brought fundamental changes in the planning of land uses and traffic, including new and improved tools of applied spatial analysis. These changes are reflected in the large regional *land use transportation studies* carried out for several American metropolitan areas in the 1950s—the Chicago Area Transportation Study and the Tri-State New York Metropolitan Transportation Study are two examples. After 1962 studies such as these became requisite to urban qualification for federal aid for road construction. In Britain, such studies were not undertaken until the middle of the 1960s, the earliest being the West Midlands Transportation Study.

The basic model for all these studies is the gravity model. Derived by analogy from Newtonian physics, it has been greatly reformulated to suit social science. The model originated independently in several disciplines; it has been suggested that in fact we have a new discipline of "social physics." The population potential measure of relative position we discussed in Chapter Two comes from it. We are concerned with the gravity model only as a predictive tool. Its application can be traced back to the work of W.J. Reilly, whose "law of retail gravitation" is concerned purely with marketing and is remarkable since it predates the land use transportation studies by more than two decades. The application of Reilly's ideas in transportation led to the breakthrough we mentioned. Reilly's original work and subsequent developments in retail studies present us with a convenient way to begin. Let's look at the law of retail gravitation before we consider the variety of gravity and other models that have developed from it.

The Law of Retail Gravitation

In 1931 W.J. Reilly brought together his earlier studies on retail trade in a book entitled *The Law of Retail Gravitation*. He identified two simple rules concerning the flow of retail trade. One is that the larger a town is the more trade it attracts, and the other is that towns attract more trade from near than from far locations. These two rules are brought together in the law of retail gravitation, which may be expressed as follows: "Two cities attract retail trade from any intermediate city or town in direct proportion to the population of the two cities and in inverse proportion to the square of the distances from these two cities to the intermediate town." This may be written algebraically:

$$\frac{b_{ix}}{b_{iy}} = \frac{p_x}{p_y}\left(\frac{d_{iy}}{d_{ix}}\right)^2$$

b_{ix} and b_{iy} are the amount of trade drawn to towns x and y from location i; p_x and p_y are the populations of these two towns; d_{ix} and d_{iy} are their respective distances to i. The ratio of the trade flows is the product of the population ratio and the square of the reciprocal distance ratio. (The inclusion of the exponent 2 for the distance ratio is based on empirical evidence from field surveys.)

This formula is very simple and logically plausible. Two towns x and y with equal populations and equal distances from i receive the same amount of trade from i—that is, $b_{ix}/b_{iy} = 1$. When p_x is twice as large as p_y but the distances remain equal, x receives twice as much trade as y. When the towns have identical populations but x is half as far as y is from i, then x receives four times more trade from i than y.

We recognize Reilly's rules as basic gravity hypotheses, but his final formulation of law differs from what we customarily see as a gravity model. We can show easily that his formula is derived directly from the normal gravity model, however:

$$b_{ix} = k\,\frac{p_x p_i}{d_{ix}^{\,2}}$$

and

$$b_{iy} = k\frac{p_y p_i}{d_{iy}^2}$$

are both in the usual gravity formulation. The ratio between the two interactions can be written

$$\frac{b_{ix}}{b_{iy}} = \frac{kp_x p_i d_{ix}^{-2}}{kp_y p_i d_{iy}^{-2}}$$

$$= \frac{p_x}{p_y}\left(\frac{d_{iy}}{d_{ix}}\right)^2$$

Reilly's law is a modified form of the usual gravity model for assessing the relative attraction of cities. Its application in the marketing field has been quite distinctive within the broad realm of social physics. This is particularly true when the model is used to define the limits of trade areas around towns. We describe this early role of the gravity model in applied spatial analysis before we return to spatial flows in the retail and other contexts.

Hinterland Delimitation

How can we use Reilly's formula to find the limit of a town's trading area? This question is surprisingly easy to answer. If the boundary line between the trade areas of two towns is defined as the point at which the two towns draw the same amount of trade, then we have the ratio

$$\frac{b_{ix}}{b_{iy}} = 1$$

and

$$\frac{p_x}{p_y}\left(\frac{d_{iy}}{d_{ix}}\right)^2 = 1$$

Let's concentrate on one of the breakpoints: d_{iy}. We can define d_{ix} as $d_{xy} - d_{iy}$ when d_{xy} is the distance between the two cities x and y so that

$$\frac{p_x}{p_y}\left(\frac{d_{iy}}{d_{xy} - d_{iy}}\right)^2 = 1$$

Therefore,

$$\frac{d_{iy}}{(d_{xy} - d_{iy})} = 1 + \sqrt{\frac{p_y}{p_x}}$$

which can be reorganized so that

$$d_{iy} = \frac{d_{xy}}{1 + \sqrt{p_x/p_y}}$$

Distance d_{iy} is the distance along the route from y to x to the point i of equal attraction—that is, at which they attract equal trade. It is a limit of the trade area of town y as we have defined it.

This rearrangement of Reilly's formula gives only the limit of the trade area of y in the direction of town x. The whole trade area of y can be delimited by carrying out this same exercise in the direction of adjacent rival centers (w, z, and so on). Each time we obtain another limit in a particular direction, and the limits are joined to delimit the whole hinterland of y. This is the area in which our formula predicts that town y will be dominant in that it will attract more trade to it than to any rival town.

Let's consider a simple example. Suppose that town x has 50,000 people and town y has 200,000 people and they are 20 miles apart. Then

$$d_{iy} = \frac{20}{1 + \sqrt{50{,}000/200{,}000}} = \frac{20}{1.5} = 13.33$$

This means that we reach the trade area of town x 13.33 miles from town y. Town y is shown as y_1 in Figure 7.1 where this breaking point between the two towns is shown. Figure 7.1 also shows four further rival centers which have populations and distances from town x as follows:

Town	*Population*	*Distance from x*
y_2	50,000	18
y_3	100,000	10
y_4	200,000	40
y_5	30,000	10

If the method above is applied to this new data, breaking points are found at the following distances: 9, 5.8, 26.7, and 4.4 miles, respectively, from y_2, y_3, y_4, and y_5. We are now in a position to draw up a predicted trade area for town x. The lines joining x to each rival center are each divided by perpendicular lines at the breaking points and these perpendiculars join together to enclose the trade area of x as shown in

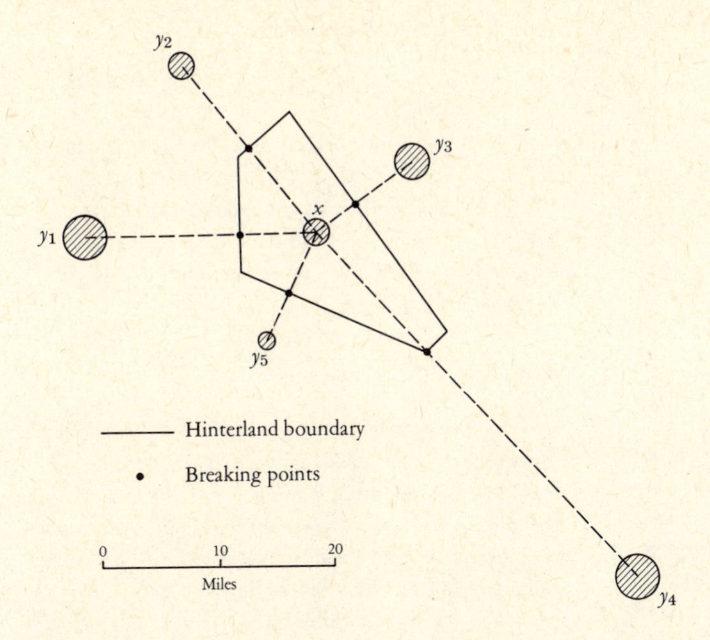

Figure 7.1 Hinterland delimitation (circles are proportional to populations).

Figure 7.1. Notice how the hinterland of x spreads out toward the smaller town y_5, is balanced equally with town y_2, but is severely curtailed by its larger rivals y_1, y_3, and y_4. This is the gravity principle at work in hinterland delimitation.

The obvious question is: How accurate is this method for delimiting hinterlands? Reilly's law is based on a large amount of research on retail areas. He tested his formulation by determining by field survey the breaking-points between towns and then comparing the result with his law's prediction. He avoided the strict geometric nature of our solution by measuring intertown distances along the roads between the towns. Using field survey, Reilly found that the breaking-point between St. Louis and greater Kansas City for high grade goods was at Tipton, Missouri. This is 121 miles by road from Kansas City. When he used the 1930 census population figures for St. Louis and Kansas City, his law predicted that the breaking-point would occur 123 miles along the road from Kansas City Other early comparisons between gravitational predictions and field survey results were found to be similarly impressive.

This sort of very favorable evidence accounts for the long history of application of Reilly's method. It can be used in marketing research when present-day market area patterns are required. Reilly suggested that the law enables large retail firms to coordinate advertising with markets. It is simple, reasonably accurate, and avoids costly field investigation. Thus, the retail trade territory of any town or city can be approximately determined using only census population totals and Reilly's breaking-point formula. For future predictions, field surveys are, in any case, impossible, and a reliable predictive equation is necessary. Reilly's law serves this purpose; if future populations can be predicted, then the future trade area pattern can be drawn accordingly. It can therefore be used as a tool for planning future retail provision.

Consumer Probabilities and Potential Market

The major criticism of Reilly's hinterland delimitation method as a practical tool is that it is unnecessarily restrictive in defining the market. It can define areas of dominance adequately, but this is not same as defining

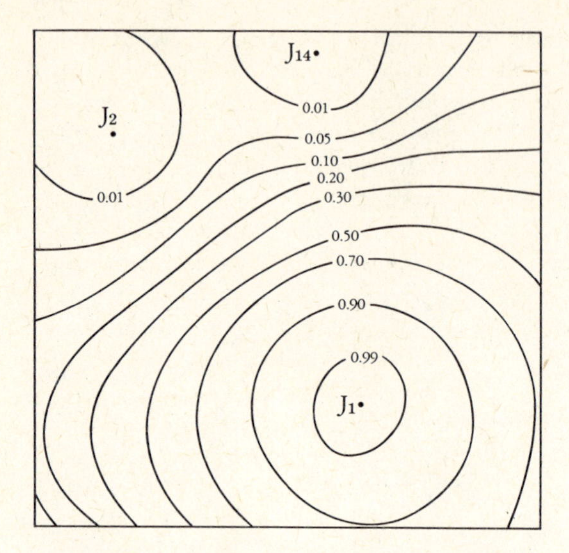

Figure 7.2 Probability map of consumer patronage. Isolines indicate the probability of a consumer-patronizing shopping center, J_1.

the market. All empirical evidence suggests that market areas around towns overlap as, in fact, Reilly's original formula predicts. The trade area boundaries we describe are simply the points at which half the trade goes to each town. Beyond this point, one town dominates but Reilly's law still predicts some trade (below one half) for the other town. Thus, we must take this into account when we attempt to measure the total market of a given town.

The major theoretical criticism of Reilly's law is that it merely describes shopping behavior and does not tell us why observed regularities occur. David Huff has shown how both the practical and the theoretical objections may be overcome by casting the law into a probabilistic framework. He was able to develop a behavioral model from the viewpoint of the consumer. Huff postulated that the probability that a consumer at point i will travel to a shopping center j (p_{ij}) is a ratio of utility of the particular shopping center to the consumer (u_{ij}) and the total utility of all shopping centers that may be considered by the consumer. This may be written:

$$p_{ij} = \frac{u_{ij}}{\sum_{j=1}^{k} u_{ij}}$$

for a set of k competing centers. We can introduce gravity concepts as they relate to utility. We can postulate that the utility of a shopping center increases with its size but decreases the farther a consumer has to travel to reach it. We can measure the utility of a shopping center as:

$$u_{ij} = \frac{s_j}{t_{ij}^{\lambda}}$$

s_j is the total number of square feet of selling space in shopping center j and t_{ij} is the time required to get from point i to the shopping center. λ is an exponent to be calibrated. Notice that there are refinements to every part of our gravity formulations. The mass element relates directly to retailing (s_j) rather than a general population variable. The effect of distance is modified in two ways. Time is substituted for simple distance measures, and the exponent is allowed to take values other than two. These improvements are the result of attempting to make the model more theoretically realistic and empirically sound.

Given our definition of utility, we can rewrite the probability expression as:

$$p_{ij} = \frac{s_j/t_{ij}^{\lambda}}{\sum_{j=1}^{k} (s_j/t_{ij}^{\lambda})}$$

We can construct a map of the probabilities of consumers patronizing a given center. We assume that the study area is divided into a set of small areal units (i). Two sets of data are required—the selling space footages for each center (s_j) and the travel times from each i to each j. A value for λ is required also. A value of 2 may be assumed, although a value is usually estimated from other empirical data for shopping trips. Given this data, the probable consumers in every areal unit who will travel to a shopping center can be calculated. Figure 7.2 shows the probability map for consumers with respect to center J_1 in part of Huff's study area. This map relates to clothing purchases for which a value of $\lambda = 3.191$ was used. The isoline for a probability of 0.5 is the equivalent of the breaking-point in the hinterland delimitation approach.

This probability map is an advance on Reilly's method, but it is not enough for a thorough analysis of a center's potential market. It is clear that we must know the distribution of population or, better still, income for retail expenditure in the study area. There might be a 0.99 probability that a consumer from an area will patronize a center, but this might be fewer shoppers in absolute terms than in an area with only a 0.4 probability. This occurs when the first area has a much smaller population than the second area. It is obvious that we must consider this spatial variation in income for retail expenditure in assessing each center's potential market. This can be done by incorporating the consumer expenditures (c_i) in each area into our formula, so that

$$c_{ij} = c_i \frac{s_j/t_{ij}^{\lambda}}{\sum_{j=1}^{k} (s_j/t_{ij}^{\lambda})}$$

c_{ij} is the amount of expenditure from consumers in area i that is spent in shopping center j. The potential market for the shopping center (m_j) is

$$m_j = \sum_{i=1}^{n} c_{ij}$$

in which there are n consumer areas.

The formulas we have developed from Huff's work are essentially the same as those proposed in the mid-1960s by Indian geographer T.R. Lakshmanan and American urban planner W.G. Hansen. They termed their approach a market potential model, and the final summation equation is very similar to the potential models we introduced in Chapter Two. In Chapter Two we actually use the phrase market potential in relation to Harris's work. To avoid confusion over terms, we use the phrase potential market model to describe the approach adopted here.

The model of Lakshmanan and Hansen differs from Huff's equations only in terms of their use of a distance friction factor d_{ij}, derived from the Baltimore Metropolitan Area Transportation Study. Their study was part of an investigation of the Metrotown Plan for Baltimore. This form of regional development envisages a series of suburban towns with full-scale retail centers situated around Baltimore. The purpose of the study was to see whether the large commercial cores envisaged for the Metrotown concept could be realized. Their strategy was first of all to develop a model that could adequately describe the pattern of retail sales in the mid-1960s. The result was the model we have described, which was found to give a good fit to known retail sales in six large shopping centers in the Baltimore area. The second step involved projecting the shopping goods expenditure forward to 1970 and 1980. This was done by using predicted future populations and per capita expenditures available from a study of the Baltimore Regional Planning Council. Future distances were based on the planned highway network, leaving only the estimates of floor space for retail centers to be calculated. This value was varied to allow the evaluation of different patterns of shopping centers. Because the Metrotown centers are envisaged as full-sized shopping centers, two sizes were predicted: 500,000 square feet of shopping floor space in the areas of denser development and 250,000 square feet in the other districts. With these projections and future centers, we can run the model and compute potential markets for 1970 and 1980.

How do we assess the viability of the proposed centers and the patterns they make up? The simplest criterion is computation of sales per square foot. Sales are predicted by the gravity model and the footage is part of the input. It was assumed that viable shopping would require sales per square foot of between \$50 and \$55. This criterion was based on patterns of shopping-center viability in the mid-1960s. If a center has lower sales per footage than this figure, it is probably too large for its potential market. High sales per footage indicate that its potential market would probably lead to its increasing in size or finding that a rival shopping center would cut into its market.

Using these criteria, twenty-five alternative future retail patterns were set up and evaluated within six areas of metropolitan Baltimore. Different spatial patterns were hypothesized and their potential markets were measured. The evaluation of three of these

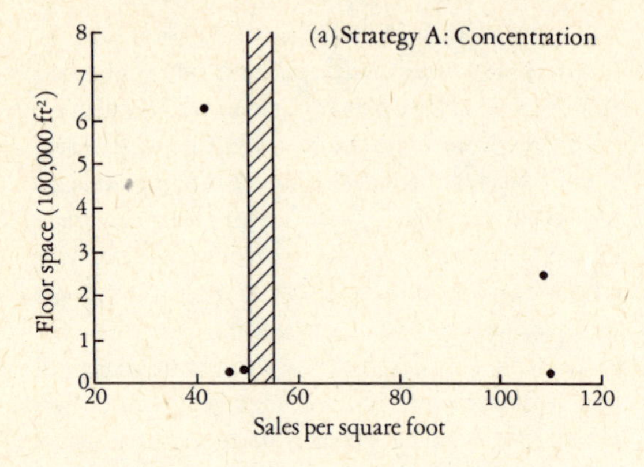

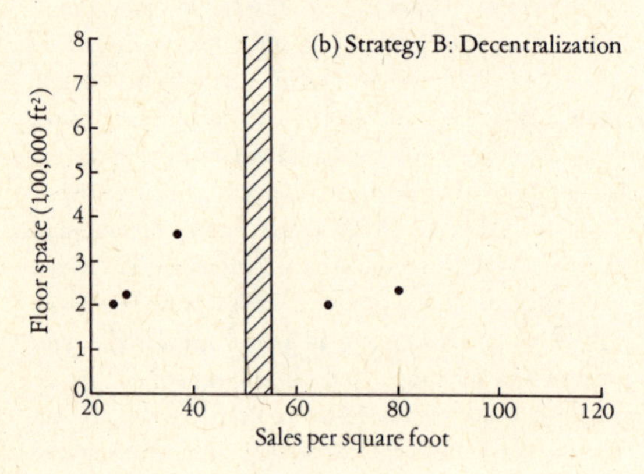

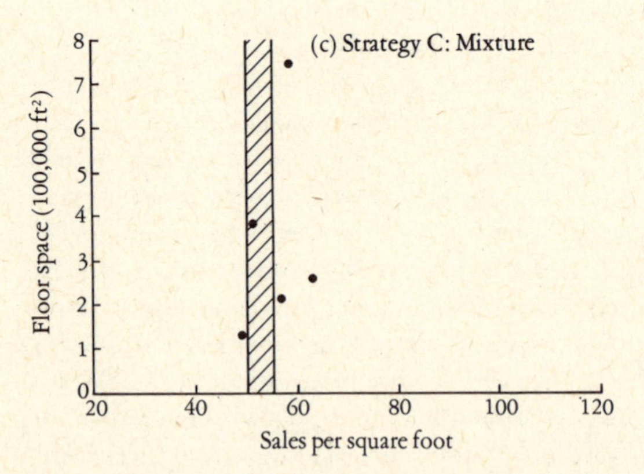

Figure 7.3 Evaluation of future shopping patterns, Baltimore, 1970. Shaded column represents "ideal" sales per square foot.

From T.R. Lakshmanan and Walter G. Hansen. A retail market potential model. Reprinted by permission of the *Journal of the American Institute of Planners*, vol. 31, no. 2, May 1965.

strategies is illustrated in Figure 7.3. This diagram shows graphs of size (floor space) against sales per square foot. The five shopping centers of Baltimore's southern region are plotted on these graphs for 1970 as predicted from 1962. There is one graph for each strategy. Strategy A was one of concentration, with future shopping space allocated to existing central centers. Strategy B was one of decentralization, with future space allocated to new and existing centers in outer areas. Strategy C is a mixture of these two. Figure 7.3a shows that strategy A leads to two relatively small centers with extremely high sales per square foot of over $100. This reflects the uneven retail coverage in the area, these two centers having a market influence much greater than their size would warrant. Such a pattern is unrealistic under normal market conditions because both centers would in all probability have grown naturally in size. The decentralization strategy has a similar effect. One shopping center has very high sales per square foot ratio but much more relevant are the three centers with low ratios below 40. This again reflects an uneven retail coverage, this time producing unviable outer centers. Strategy C produces, as we might expect, a much more realistic pattern, with a much smaller range of sales per square foot values about the ideal range.

Strategies A and B were simple exploratory alternatives to illustrate the evaluating criteria. Alternative C was designed to provide a much better level of retail service. Various patterns were evaluated and a preferred system of centers was identified. This enabled the prediction of the most likely future retail growth patterns, which are summarized in Table 7.1. The Baltimore area is divided into three concentric zones. In the prediction for the first period, 1962–1970, most retail growth is in the inner belt; between 1970 and 1980 the area of most growth is in the outer belt. The satellite communities retain constantly low retail growth. Within this pattern, however, different levels of growth can be identified and they provide important guides to commercial policy. Three existing shopping centers were discovered to have considerable potential for future growth; in addition, three new locations had great possibilities for growth.

Table 7.1 Projected Growth of Shopping Floor Space in Baltimore (1962–1980)

Zone	*1962–1970*		*1970–1980*	
	Million sq. ft.	*Percent*	*Million sq. ft.*	*Percent*
Inner belt	1.90	50	1.04	26
Outer belt	1.36	36	2.44	60
Satellites	0.54	14	0.56	14

Source: T.R. Lakshmanan and Walter G. Hansen. A retail market potential model. Reprinted by permission of the *Journal of the American Institute of Planners*, vol. 31, no. 2, May 1965.

The conclusion of this study with respect to the Metrotown policy was that nine potential candidates as Metrotown centers were judged viable by 1970 and six more by 1980. The Metrotown policy appears to be consistent with the operation of current urban growth processes.

Let's conclude this section by emphasizing the differences between the two gravity model approaches we have described. The hinterland delimitation method concentrates on the shopping centers. Data are required, in the first instance, only for centers that are considered in pairs. The potential market method, on the other hand, is more directly concerned with consumers, and these are allocated to every center from all parts of the study region. In contrast to the precomputer-age Reilly model, this second model requires much more computation, inevitably involving electronic equipment. Part of the computation includes the calibration of the model's exponents. We have not fully considered this problem but we do so in our description of a practical application in the Northwest Regional Shopping Center study.

Impact: A new Regional Shopping Center for Northwest England

In June 1963 the first planning application for an out-of-town regional shopping center in Britain was submitted as the culmination of a Ministerial enquiry. The sponsors of the shopping center commissioned the Town and Country Planning Department at the University of Manchester to undertake an impact study on the effects of the proposed shopping center. The result was in the form of two reports, one employing the hinterland delimitation approach and the other the potential market model. This gives us a unique opportunity to compare these two approaches within the context of a single applied research example.

The question the research team attempted to answer was of the classical "What if?" variety. They asked specifically, "What if a large regional shopping center were built at Haydock in south central Lancashire?" This site was at the intersection of the major north–south highway through Lancashire (the M6) and the main east–west route (the East Lancs Road). Two sizes of shopping center development were considered, although we describe only the results from the more realistic proposal for a smaller Haydock center. The area potentially affected by this new center was defined as northwest England, stretching from the Pennines to the coast and from Stoke to Lancaster, including part of North Wales. Within this area a hierarchy of existing shopping centers was identified, but we consider only the two highest grades: Manchester and Liverpool in the first and, beneath that, twenty-one existing centers plus Haydock.

The first approach for assessing the impact of Haydock was a modification of the original Reilly breaking-point method. Travel distances, weighted by type of road, were used instead of simple road distance, and average total sales were used instead of population to indicate the attraction of a center. Incorporating these modifications, breaking-points were found between each pair of adjacent towns by using 1961 data so that hinterlands could be constructed (Figure 7.4a). The next step was to construct a similar map for 1971 from predictions of population, incomes, sales, and roads for this date. Two hinterland maps were constructed for 1971, one assuming the current system of centers and the other assuming a new center of secondary grade at Haydock. Comparison between these two predictions forms the basis for the assessment of impact. The two maps are shown in Figures 7.4b

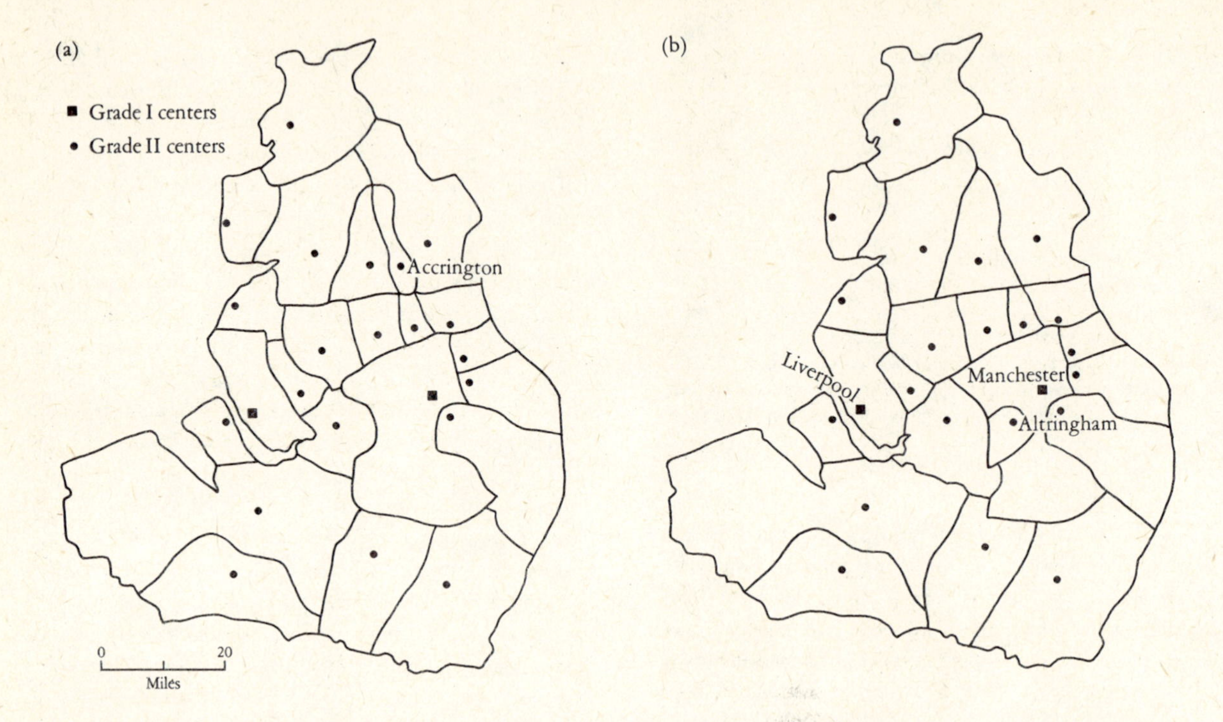

Figure 7.4 The Haydock shopping center study. a. Hinterlands in 1961 for grade 1 and 2 centers. b. Predicted Hinterlands in 1971 for grade 1 and 2 centers. c. Hinterland predictions in 1971 including Haydock. d. Predicted sales loss to Haydock in 1971.
From the Department of Town and Country Planning, University of Manchester. *Regional Shopping Centres in North West England, 1964.*

and 7.4c. Changes in hinterland boundaries between Figures 7.4a and 7.4b represent simple variations in regional growth, with Warrington gaining in importance and Altringham entering as it grows from the third grade to a second-grade center. Accrington is missing from the map because it was expected to decline from second to third grade. The addition of Haydock to the map in Figure 7.4c is not overtly dramatic. It secures a small hinterland about it, as we expect, although it is unusual in that it includes two detached sections. These outliers result from using travel times as the distance measures and they identify two areas around junctions for the M6 motorway. The most important result, however, relates to the predicted sales that Haydock captures that would otherwise go to Wigan, St. Helens, Warrington, Preston, Southport, and Manchester. In fact, Haydock is predicted as having a little first-grade shopping facilities and so it also captures some of the market from Liverpool and Bolton. Table 7.2 represents the essence of the final conclusions in the research project. The new center has only a major impact on three towns Wigan, St. Helens, and Warrington. In the latter case, the population growth in the area is enough to sustain continued growth in the existing shopping center. In marked contrast, the development of Haydock would mean an actual decline in the total sales at St. Helens. Using this and other evidence, the Ministry of Housing and Local Government turned down the request from the developers to build a new shopping center, and northwest England today has essentially the same pattern of shopping centers that it had in 1961.

Because this practical decision had already been made, the second analysis of the impact of Haydock is largely academic. The research team began to have reservations about the results they had obtained from the

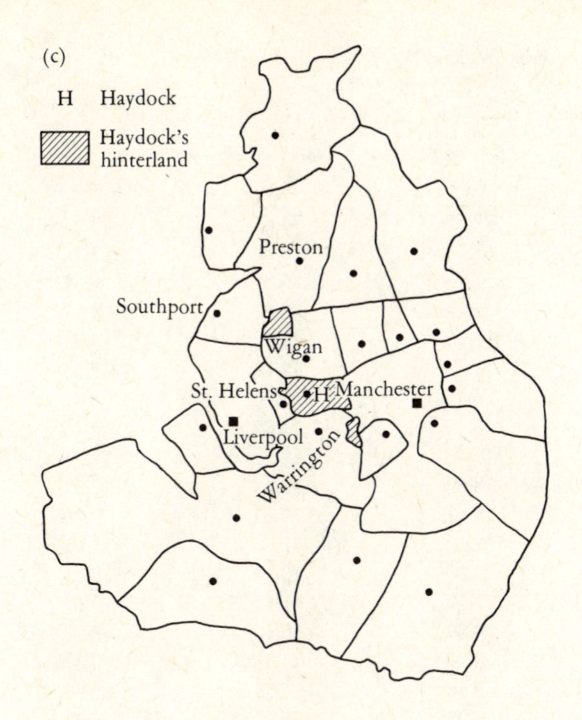

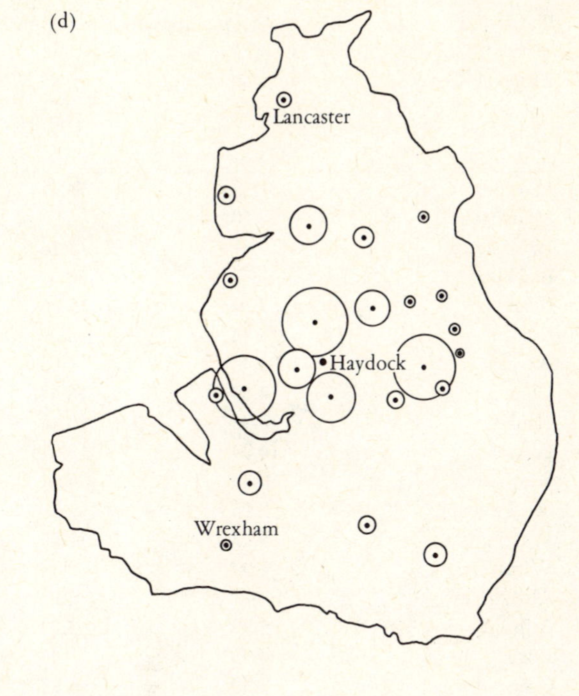

Figure 7.4 *(cont'd)*

hinterland approach. The method of Lakshmanan and Hansen was then still new but incorporating the idea of overlapping markets seemed to be a much more realistic way of describing potential sales. They used the modified formula

$$s_j = \sum_{i=1}^{n} c_i \cdot \frac{a_j^{\beta}/d_{ij}^{\lambda}}{\sum_{j=1}^{k} (a_j^{\beta}/d_{ij}^{\lambda})}$$

s_j is the predicted sales in shopping center j. Because data on sales floor footage were not available, a special measure of attraction (a_j) was devised, based on the number of retail facilities available at a center. The major difference occurs with the use of two exponents, λ for time-distance and also β for attraction. Let's consider how these were calibrated with 1961 data.

The data consisted of 244 origins (i) and 23 centers (j) for 1961. The attraction exponent β was set at 2.0 and the distance exponent was λ at 0.50. Using these values, the formula was used to predict sales in each of the 23 centers. These sales figures were then compared with the known sales figures for 1961 by using a χ^2 goodness-of-fit test. Keeping $\beta = 2.0$, this procedure was repeated for λ in steps of 0.05 to $\lambda = 5.00$. In each case, a χ^2 goodness-of-fit test indicates how well the particular specification of the equation predicts the 1961 sales figures. When this sequence of values was completed, the whole exercise was repeated for values of β in steps of 0.1 to $\beta = 3.5$. The result is a matrix of χ^2 measures of goodness of fit for 1456 predictive equations, with β ranging from 2.00 to 3.50 and λ ranging from 0.50 to 5.00. The particular combination of β and λ with the lowest χ^2 identifies the equation giving the closest degree of fit. These were $\beta = 3.0$ and $\lambda = 2.6$, so that the equation is

$$s_j = \sum_{i=1}^{n} c_i \cdot \frac{a_j^{3}/d_{ij}^{2.6}}{\sum_{j=1}^{k} a_j^{3}/d_{ij}^{2.6}}$$

This equation was used to predict shopping center sales in 1971.

Table 7.2 Predicted Sales in Major Shopping Centers in Northwest England (1971)

Shopping Center	*1961 total sales*	*Hinterland 1971*		*Potential market 1971*	
		No Haydock	*Haydock Grade 2*	*No Haydock*	*Haydock Grade 2*
Liverpool	61.93	74.27	72.25	77.63	75.24
Manchester	57.10	64.70	62.50	70.00	67.72
Birkenhead	9.61	11.16	11.16	9.88	9.77
Blackburn	7.83	10.20	10.20	8.83	8.58
Blackpool	17.50	18.86	18.86	18.99	18.83
Bolton	15.42	15.09	14.47	18.18	17.43
Burnley	5.43	6.95	6.95	5.98	5.91
Chester	14.69	16.44	16.44	14.82	14.48
Hanley	10.96	13.40	13.40	12.52	12.23
Preston	14.29	18.91	18.69	18.41	17.51
Southport	10.58	9.61	9.61	9.63	9.46
Stockport	9.15	11.10	11.10	11.76	11.61
Warrington	8.43	17.59	15.99	10.74	9.23
Wigan	8.99	10.49	9.50	11.13	8.52
Altringham	—	4.89	4.89	5.40	5.18
Ashton	5.99	5.27	5.27	4.45	4.42
Bury	4.93	5.46	5.46	5.58	5.52
Crewe	5.50	5.95	5.95	5.42	5.27
Lancaster	5.78	7.26	7.26	5.75	5.65
Oldham	8.61	7.78	7.78	8.56	8.48
Rochdale	6.76	6.34	6.34	6.10	6.04
St. Helens	8.89	6.66	6.19	7.82	6.97
Wrexham	7.71	8.22	8.22	7.06	7.01
Haydock	—	—	8.12	—	15.21

All figures are in millions of pounds sterling at 1961 prices.
Source: Department of Town and Country Planning, University of Manchester. *Regional Shopping Centres, Part Two: A Retail Shopping Model,* Tables 1 and 2, pp. 31 and 33, December 1966.

The results from this prediction are shown in the two righthand columns of Table 7.2. If we first compare the straightforward predictions (without a Haydock center) with the hinterland predictions, we find relatively small but interesting differences. The two largest shopping centers, Liverpool and Manchester, gain appreciably because their sales are no longer restricted by the hinterland boundary prediction. Most other, smaller centers are predicted to have smaller sales by the second model, with the exception of towns that were tightly hemmed in by hinterland boundaries—St. Helens and Altringham, for example. The main difference, when we consider the predictions including the Haydock center, is that in the second model Haydock captures sales from every center in the study area, however small the amount (Figure 7.4d). Even Lancaster and Wrexham lose some sales to Haydock, although the principal towns that suffer in relative terms (Table 7.2) are Warrington, St. Helens, and Wigan. The overall result for Haydock is interesting. When we use the hinterland method, Haydock is hemmed in by the surrounding trade areas, but the market potential approach predicts appreciable freedom from reliance on the local area so that predicted sales are almost doubled.

The authors of the second study were particularly

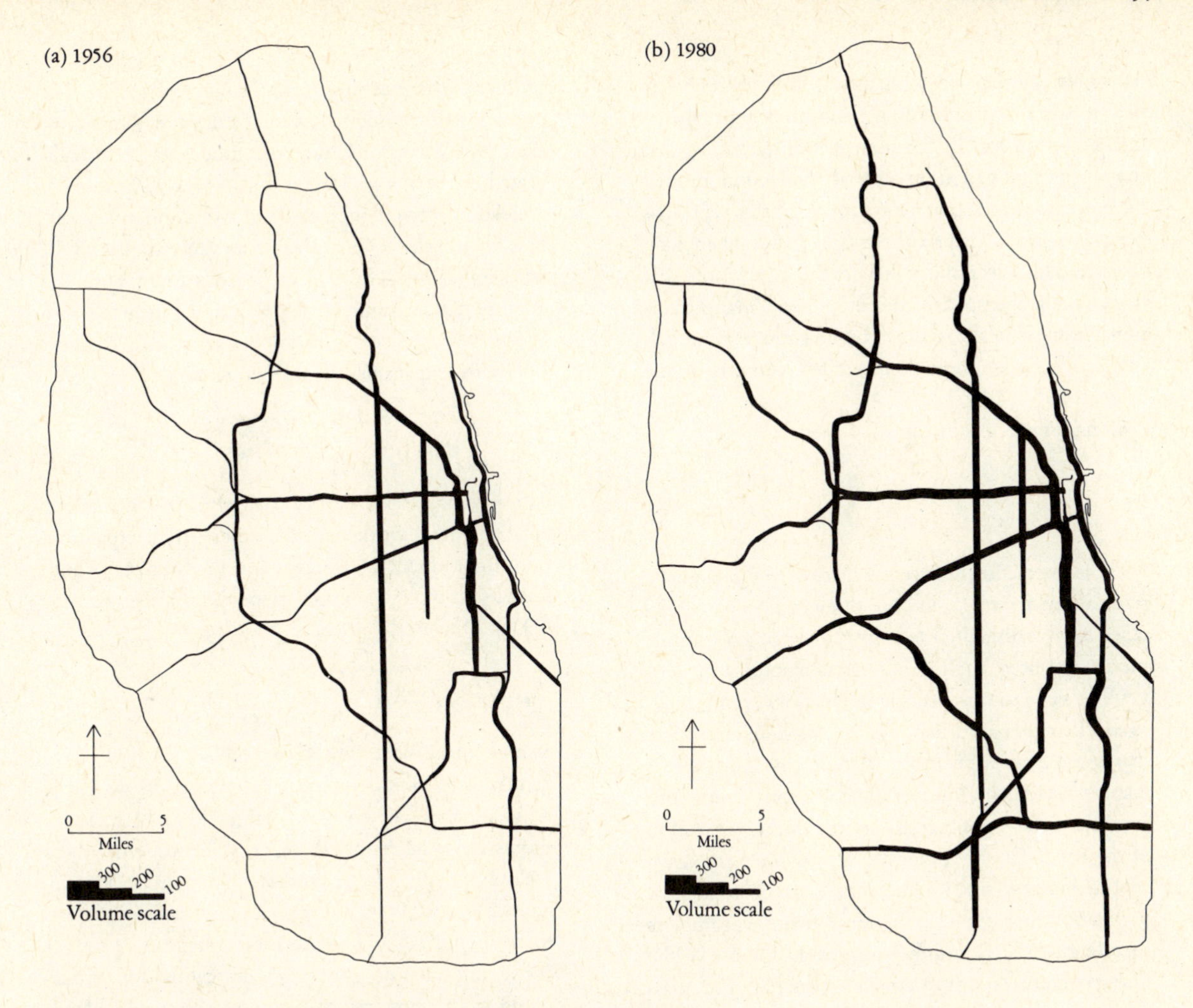

Figure 7.5 Travel volumes on Chicago expressways, 1956 and 1980.
From the State of Illinois. *Chicago Area Transportation Study*, vol. I, 1959, p. 89.

pleased with the way their new approach gave similar results leading to the same conclusions as in the first study. We must not forget the differences that we discussed above, however. These can be related directly to the contrasting ways in which the two approaches allocate consumer expenditure. This difference is brought out well in Figures 7.5a and 7.5b. The Reilly method requires little computation relative to the market potential model. Ease of computation seems to be the main advantage of this approach. With the general availability of electronic computers, the potential market approach becomes widely available for any large-scale studies of the "What if?" variety.

The Family of Gravity Models

We have concentrated on one type of movement (shopping) although we have recognized that the gravity model is a framework for studying spatial interaction in general. The urban planner is concerned not merely with shopping centers but also with housing, factories, recreational facilities, and the roads connecting them.

Movements associated with these activities, for journeys to work or recreation, are also modeled as spatial interaction. In fact, it has been in transportation that the major innovations in spatial interaction models have occurred in recent years. Much of this work has been associated with mathematical geographer Alan Wilson. His research has led to the conclusion that there is not one spatial interaction model, the traditional gravity model; there is a whole range of models that are basically more sophisticated versions of the original.

Consider the original gravity model:

$$I_{ij} = k\frac{p_i p_j}{d_{ij}^{\,b}}$$

I_{ij} is the interaction between two areas i and j that have populations p_i and p_j and are distance d_{ij} apart. Distance is typically raised to some power b. This simple model has been applied in various contexts in geography. Let's assume that the interactions are journeys to work within an urban region. We soon appreciate that the use of populations to represent "masses" is inappropriate. The number of people commuting into an urban zone is not directly related to the population of the zone. This is obvious if we consider the usual central business district that has a small residential population but typically attracts many commuters because of its large number of jobs. In the use of the gravity model in most applied research, the mass variables are specified in terms of the interaction being studied. Therefore, in journey-to-work models, the masses employed are the total number of journey-to-work origins (O_i) in a zone and the total number of destinations in a zone (D_j). O_i represents the number of workers leaving a zone—the trips produced by the zone—and D_j represents the number of jobs in a zone estimated from the number of trips attracted to it. The new transport gravity model becomes:

$$T_{ij} = k\frac{O_i D_j}{d_{ij}^{\,b}}$$

T_{ij} is the number of trips from i to j.

This modification of the original gravity model has many implications. The most obvious is the degree of dependence between the interaction and the masses, because they are defined from the same set of data. The problems that ensue may be illustrated with a simple example. We assume that the model has been calibrated so that $k = 1/1000$ and $b = 2$. The town is divided into two zones, east and west, one mile apart. Eastside has all the jobs (1000) and Westside has all the workers (1000). This is a prosperous town with no unemployment, and so population is attracted into Westside while new jobs are attracted into Eastside. Before the expansion, the total flows are:

$$T_{we} = \left(\frac{1}{1000}\right)\left(\frac{1000 \cdot 1000}{1^2}\right) = 1{,}000$$

This is to be expected, of course, with 1000 jobs in one zone and 1000 workers in the other. After expansion, both the number of workers and the number of jobs double, so that we expect twice the number of trips (2000). Our model predicts

$$T_{we} = \left(\frac{1}{1000}\right)\left(\frac{2000 \cdot 2000}{1^2}\right) = 4{,}000$$

While workers and jobs have doubled, the flows have quadrupled! This is clearly nonsense. Casting the gravity model in terms of origins and destinations shows how the model is logically inconsistent: there are not enough jobs or workers to require 4000 journeys to work. We know in fact how many journeys there should be. We use this information to produce constraints on our gravity model calibration procedures, so that nonsense results cannot occur. It is from the specification of these constraints that different types of gravity model can be identified.

Gravity Model Constraints

We term the gravity model we used above the *completely unconstrained gravity model.* Our argument has suggested that the first modification of the model is to ensure that it predicts the correct total number of flows. This can be written

$$\sum_{i=1}^{n}\sum_{j=1}^{n} T_{ij} = \sum_{i=1}^{n} O_i = \sum_{j=1}^{n} D_j$$

in which there are n zones. All this constraint equation tells us is that the total number of trips should be the same as the total number of origins and the same as the total number of destinations. This very simple constraint can be incorporated in our definition of k by making it into a scaling factor. We compute the total trips without k (T^*_{ij}) and then scale this total down to the correct number by multiplying it by

$$k = \frac{\sum_{i=1}^{n} O_i}{\sum_{i=1}^{n} \sum_{j=1}^{n} T^*_{ij}}$$

or

$$k = \frac{\sum_{j=1}^{n} D_j}{\sum_{i=1}^{n} \sum_{j=1}^{n} T^*_{ij}}$$

In our initial example, $k = \frac{1}{1000}$ was the appropriate scaling factor before growth. With the addition of workers and jobs, we require a new value for k. Flows predicted without k are

$$\frac{2000 \cdot 2000}{1^2} = 4{,}000{,}000$$

whereas O_i and D_j are only 2000; therefore,

$$k = \frac{2000}{4{,}000{,}000} = \frac{1}{2000}$$

so that the model, constrained for total flows, is

$$T_{we} = \left(\frac{1}{2000}\right)\left(\frac{2000 \cdot 2000}{1^2}\right) = 2000 = O_i = D_j$$

This total flow constraint is the simplest of all. In fact, it is often considered so obviously necessary, and therefore trivial, that this second model is sometimes called the unconstrained model. We term it the *total-flow constrained model* to distinguish it from the original, unconstrained one. It can be written as

$$T_{ij} = k\,\frac{O_i D_j}{d_{ij}^{\,b}}$$

where

$$k = \frac{\Sigma O_i}{\Sigma\Sigma T^*_{ij}} = \frac{\Sigma D_j}{\Sigma\Sigma T^*_{ij}}$$

T^*_{ij} is unscaled predicted flow.

Why do we require more constraint on our model? Let's return to the two-zone city and make it a little more realistic by specifying some jobs in Westside and some workers in Eastside. Eastside has 900 of the total 1000 jobs and Westside has the remaining 100; Westside has 800 workers, leaving 200 for Eastside. Now, journeys to work can be calculated as

$$T_{we} = k\left(\frac{800 \cdot 900}{1^2}\right) = 720{,}000\,k$$

and

$$T_{ew} = k\left(\frac{200 \cdot 100}{1^2}\right) = 20{,}000\,k$$

Therefore,

$$k = \frac{1{,}000}{740{,}000} = \frac{1}{740}$$

so that

$$T_{we} = \frac{720{,}000}{740} = 973$$

and

$$T_{ew} = \frac{20{,}000}{740} = 27$$

This tells us that, of the 1000 trips, 973 start in Westside and finish in Eastside while only 27 go the other way. The total number of trips is correct ($973 + 27 = 1000$), but the model is still at variance with our original information. According to this, there are 100 trip origins (that is, 100 workers living) in Eastside, but only 27 of the predicted trips start there. On the other hand, Westside produced 973 trips whereas we know that only 900 origins are located there. Thus, we must have another constraint for the production of trips. This *production-constrained gravity model* is written

$$T_{ij} = A_i\,\frac{O_i D_j}{d_{ij}^{\,b}}$$

where A_i is a set of scaling factors for all O_i. There is now a special constant for every origin zone. In our case, there are two zones with origins, and so we need two A constants. These are defined as

$$A_i = \frac{1}{\sum_{j=1}^{n} (D_j / d_{ij}^{\,b})}$$

in order to satisfy the constraint:

$$\sum_{j=1}^{n} T_{ij} = O_i$$

This says simply that the number of trips leaving zone i must be the same as the number of origins in zone i.

In our two-zone example, we now meet a problem we have not previously considered: self-distances. When we start summing in the expression above, one j between 1 and n will in fact be the origin i under consideration. Thus, a distance d_{ii} is required. If we define d_{ii} as 0, we find ourselves in the realm of infinities as soon as we divide. We take up the problem again later, but at the moment we simply define d_{ii} as 0.5. Thus,

$$A_e = \frac{1}{(900/0.5^2) + (100/1^2)} = \frac{1}{3{,}700} = 0.000270$$

and

$$A_w = \frac{1}{(900/1^2) + (100/0.5^2)} = \frac{1}{1300} = 0.000769$$

The equation now reads

$$T_{ew} = A_e \frac{O_i D_j}{d_{ij}{}^b} = \frac{0.00027 \cdot 200 \cdot 100}{1^2} = 5.4$$

The number of workers living in e and working in e is

$$T_{ee} = \frac{0.00027 \cdot 200 \cdot 900}{0.5^2} = 194.4$$

Hence, we have accounted for all 200 trips originating in Eastside: 5.4 go to Westside while 194.4 remain within industrial Eastside (0.2 commuters are lost because of rounding errors, with constants defined to six decimal places).

Trips originating in Westside can be similarly apportioned:

$$T_{we} = \frac{0.000769 \cdot 800 \cdot 900}{1^2} = 553.68$$

and

$$T_{ww} = \frac{0.000769 \cdot 800 \cdot 100}{0.5^2} = 246.08$$

so that

$$T_{we} + T_{ww} \simeq 800 = O_w$$

Notice that, as we begin to add constraints, the equation becomes a bit more complicated, because we have a different constant A_i for every zone in order to calibrate the production-constrained model. We can repeat these arguments in terms of destination. In the total-flow constrained model, which ignores within-zone movement, 973 trips are predicted to Eastside whereas only 900 jobs are there for the taking. There is clearly also a place for an *attraction-constrained gravity model*, specified as:

$$T_{ij} = B_j \frac{O_i D_j}{d_{ij}{}^b}$$

B_j is a set of scaling factors for all D_j. These are defined by

$$B_j = \frac{1}{\sum_{i=1}^{n} (O_i/d_{ij})}$$

to satisfy the constraint

$$\sum_{i=1}^{n} T_{ij} = D_j$$

which simply says that the number of flows into zone j must equal the number of destinations in zone j.

In our simple two-zone example, $B_e = 0.00117647$ and $B_w = 0.0025$, so that both T_{ew} and T_{ww} equal 50, which together sum to the D_w value of 100. Similarly, we can apportion the 900 jobs at Eastside between $T_{we} = 847$ and $T_{ee} = 53$.

We have identified three gravity model constraints and produced three different models. We can include all three constraints in a single comprehensive model which is termed the *production-attraction constrained gravity model*. This is the most complicated model of the family and may be written as:

$$T_{ij} = A_i B_j \frac{O_i D_j}{d_{ij}{}^b}$$

where

$$A_i = \frac{1}{\sum_{j=1}^{n} (B_j(D_j/d_{ij}{}^b))}$$

and

$$B_j = \frac{1}{\sum_{i=1}^{n} (A_i(O_i/d_{ij}{}^b))}$$

which satisfies the constraints

$$\sum_{i=1}^{n} T_{ij} = D_j$$

for every D_j, and

$$\sum_{j=1}^{n} T_{ij} = O_i$$

for every O_i, and each in turn ensures that

$$\sum_{i=1}^{n} \sum_{j=1}^{n} T_{ij} = \sum_{j=1}^{n} D_j = \sum_{i=1}^{n} O_i$$

The distinctive feature of this new model is that it includes *two* sets of scaling factors for *both* origins and destinations. From the equations, we can see that in order to calculate A_i we have to know B_j but we have to know A_i before we can determine B_j. There is no direct mathematical solution to this problem, but the sets of constants can be found experimentally by using an iterative process. This method is a very common means for producing results from this type of problem.

The procedure is as follows. We begin by giving either A_i or B_j a set of arbitrary values. For instance, we might start with every A_i as unity. The values of unity are then fed into the equation to solve for B_j. These are our *first estimates* of B_j. The first estimates are then fed into the equation to solve for A_i giving *revised estimates* for A_i. These new values for A_i are used to find revised estimates of B_j which in turn produce better estimates of A_i, and so on. The iterative procedure continues until it is decided that improvements from further iteration are so small as not to warrant any more effort. At this point, we have our *final estimates* for A_i and B_j which we can incorporate into our model to produce the production-attraction constraint.

Clearly the iterative procedure is very simple in concept but it entails large numbers of calculations in practice. Hence, A_i and B_j are normaly calculated with the help of an electronic computer. Even in our simple two-zone example, we cannot illustrate the computations with a few calculations. (Iterations for this example are laid out in Work Table 7.1.) The algorithm is a converging algorithm as described in Chapter One. In this case, solutions converge after 8 and 9 steps, to produce

$A_e = 0.419762 \qquad A_w = 1.485443$

$B_e = 0.000656 \qquad B_w = 0.000207.$

Trip predictions are also calculated in Work Table 7.1 to act as a check on the solutions. For this particular case:

$$T_{we} + T_{ww} = O_w = 800$$

$$T_{ee} + T_{we} = D_e = 900$$

$$T_{ew} + T_{ww} = D_w = 100$$

$$T_{ew} + T_{ee} = O_e = 200$$

which satisfy all constraints simultaneously.

We have produced a family of gravity models identified in terms of constraints on equations. The obvious question is: Why do we need these different versions of the gravity model? The way to answer is to show how the constrained models are used in applied spatial analysis. The beauty of this typology of gravity models is that it brings together several equations derived for different purposes into a single framework. We can show that the last three constrained models correspond directly to planning applications of gravity formulations.

We illustrate this argument by returning to the retail gravity model and show how it fits into the family of gravity models. Let's recall the potential market model:

$$c_{ij} = c_i \frac{s_j/t_{ij}^{\lambda}}{\sum_{j=1}^{k} (s_j/t_{ij}^{\lambda})}$$

Each term in this equation can be related directly to parts of the production-constrained model. To demonstrate this, we rewrite the equation:

$$c_{ij} = \left(\frac{1}{\sum_{i=1}^{k} (s_j/t_{ij}^{\lambda})} \right) \left(\frac{c_i s_j}{t_{ij}^{\lambda}} \right)$$

c_{ij} is the consumer expenditure in j from i—that is, it is the equivalent of T_{ij}, our interaction term. c_i is consumer expenditure in i, the equivalent of O_i, the origin of flow. s_j is sales in j, the equivalent of D_j, the destination of flow. The distance factor is represented by travel time t_{ij}^{λ}, which is the equivalent of the simpler $d_{ij}^{\,b}$ factor. Given these equivalents, we can see that the retail gravity model is essentially identical to the production-constrained model, with the A_i constants included directly in the equation.

How can we interpret the retail gravity model in the light of this equivalence? This is clearly a production-constrained model because our interest is focused on the attraction end of the interaction. Consumer expenditures are given and we wish to assess the potential of a set of shopping centers. It is obvious that if we constrain the attraction we are answering the question before analysis. In the context of studies such as the northwest England example, we estimate consumer expenditure patterns initially so that we can predict the resulting sales pattern. Hence, we use a production-constrained (but open attraction) gravity model.

Let's look at another typical use of the gravity model in planning. Suppose that planners have decided the location of a new industrial park that will provide a given number of jobs. The question we ask is: Where will the labor force be drawn from? This is to ask for an attraction-constrained model. The pattern of destinations (the industrial park) is given and we wish to predict where the journey-to-work origins will be. We must leave the pattern of origins unconstrained if we are to predict the residences of the potential work force. Such *residential location models* correspond directly to the attraction-constrained gravity model.

What of our production-attraction constrained model? It suggests that our interest for prediction is on neither end of the interaction but is concentrated on the flows themselves. This model is used when origins and destinations are given and we wish to predict the pattern of flows. It is used therefore to predict traffic flows to facilitate estimation of future transport needs. At the urban scale, it is used as a journey-to-work model; at the larger, regional scale, it becomes a model for predicting interregional commodity flows. As a transport model, it has been rivaled in application in recent years by another type of spatial interaction model called the *intervening opportunities model.*

The Intervening Opportunities Model

The intervening opportunities model is not strictly a gravity model because it does not depend on a distance factor. In theory and practice, however, it is so similar to the gravity formulations that we consider it here. It was proposed originally in 1940 by sociologist Samuel Stouffer as a method of explaining intraurban migration patterns in Cleveland, Ohio. It was developed independently as part of the Chicago Area Transportation Study by Morton Schneider late in the 1950s, and we present his version of the model. Schneider was dissatisfied with the basic gravity model approach because he considered that it lacked any real explanation of urban travel patterns. The result was a new theory of urban travel.

The basic idea behind the intervening opportunities model is that the probability of a trip $p(T_{ij})$ beginning at i and terminating at j is directly proportional to the number of opportunities at j, D_j, and inversely proportional to the number of opportunities nearer to i than j $({}_iD_{j-1})$. Notice that a distance factor does not enter the model directly but defining ${}_iD_{j-1}$ requires knowledge of a distance factor (usually travel time) so that opportunities can be ranked in terms of distance from the origin. Consider the Chicago Area Transportation Study. Suppose that a particular traveler requires a carton of milk. Milk is a common commodity and we assume that 1 in every 200 possible destinations can provide this good. If the traveler selects destinations randomly, he can expect to find a milk retailer 1 time in 200. The traveler will not wish to travel any farther than necessary, however, and he will begin at the nearest destination. The probability that his trip will terminate there (that the trip purpose is fulfilled) is $\frac{1}{200}$. The probability that he will not terminate his trip at the first destination is, therefore, $(1 - \frac{1}{200})$. If this is the case, he goes on to the second destination. The chance of finding this second destination suitable is again $\frac{1}{200}$. The probability that this destination will be the terminating point, however, is less than $\frac{1}{200}$ because there is a chance that he may not have reached this far, if the first destination was the terminal. In fact, $(1 - \frac{1}{200})$ multiplied by the probability of being satisfied at the second destination $\frac{1}{200}$ is the probability of arriving at the second destination. We can see that the first probability incorporates the notion of intervening opportunities; the second is simply the opportunities available at a destination. The probability of terminating the trip at the third nearest destination is the chance of passing the first destination $(1 - \frac{1}{200})$ multiplied by the chance of passing the second destination $(1 - \frac{1}{200})$, multiplied by the probability of satisfaction at the third destination $\frac{1}{200}$. We can generalize to say that the probability of terminating the trip at destination n is the probability of getting past the first $n - 1$ destinations: $(1 - \frac{1}{200})^{n-1}$ multiplied by the chance of stopping $\frac{1}{200}$.

To operationalize this model we must make it applicable to aggregate data in zones rather than at the individual decision-making scale we used above. We denote the probability that a trip will be satisfied at any random destination as l. This is a reflection of the purpose of the trip: trips for common needs have relatively high probabilities (they are widely available, as in the milk example) but more specialized trips have lower probabilities. The probability of a trip reaching zone j from i is $(1 - l)^{{}_iD_{j-1}}$, in which ${}_iD_{j-1}$ represents all the destinations (or opportunities) from i to $j - 1$ (intervening opportunities between i and j). The probability that a trip will be satisfied after zone j has been considered is $(1 - l)^{{}_iD_j}$, in which ${}_iD_j = {}_iD_{j-1} + D_j$. It follows that the probability of stopping at zone j is therefore the difference between the two expressions

$$p(T_{ij}) = (1 - l)^{{}_iD_{j-1}} - (1 - l)^{{}_iD_j}$$

Because D_j is included in ${}_iD_j$, the larger that the number of destinations is, the greater is the chance that the trip will terminate in zone j. The total number of trips from i to j is the expression above multiplied by the specific number of trips beginning in zone i—that is, O_i, so that

$$T_{ij} = O_i\{(1 - l)^{{}_iD_{j-1}} - (1 - l)^{{}_iD_j}\}$$

This formula applies to the group of trips associated with probability l. For computing all trips between i and j, we expect to use the formula above for several groups of trips, each with different values for l.

This model was calibrated, tested, and used as a predictive tool by the Chicago Area Transportation Study team. From other parts of their study, they were able to obtain travel times between zones and thus were able to rank opportunities. The principal problem is obtaining values for l. The solution in the Chicago study was to divide all trips into two groups—local trips with high l values and special trips with smaller l values. Both l values were then derived empirically. The higher probability was specified to produce typical proportions of intrazonal trips. The lower probability was calibrated to produce the approximate total network travel mileage over the study area. The actual figures (0.000021 and 0.0000023) were found by experimental trial and error. With the calibration completed, the model was tested against survey data collected for 1956 and was found to reproduce a travel pattern substantially the same as that reported by previous survey.

Let's turn to the basic purpose of developing the model to predict future traffic patterns; the future specified in the Chicago Area Transportation Study is 1980. Three sets of data are required on opportunities (destinations), on relative travel times, and on appropriate l values. Another part of the study predicted 1980 destinations, and the future road network was expected to be very similar to the 1950s one in 1980. Known plans for road improvements were included and, because only relative travel times are required for ranking zones in terms of time from an origin, this was adequate for defining intervening opportunities. Future l values are more difficult to predict. The solution was to calibrate the higher l value so that similar proportions of trips would be intrazonal as in 1956. An l value of 0.000018 was derived for local trips. The lower probability was calibrated after a trial run to ensure that total mileage would be approximately the same as predicted for 1980 in a previous part of the study. This produced an l value of 0.00000175. Using these predicted data, zone-to-zone movements were calculated for 1980 and allocated to travel routes. The predicted traffic pattern for the expressway system is shown in Figure 7.5b and may be compared with the 1956 pattern in Figure 7.5a. An even growth of traffic over the whole region is indicated, although the proportion of traffic on expressways is predicted to fall from 48 percent to 38 percent of the total. This is because the wider spacing of the expressways in the outer areas makes them less likely to be used by the many new commuters in the suburbs. The total vehicle rates of travel in both years are shown for seven rings around the Chicago Loop in Table 7.3, and this confirms this changing pattern of travel. It was predicted that most suburban traffic will utilize arterial streets in 1980.

This great shift in travel pattern at first seems sur-

Table 7.3 Vehicle Miles of Travel as Estimated for 1980 over the 1956 Chicago Road System

Ring	*Vehicle travel (thousands of miles)*		*Percent change*
	1956	*1980*	
0, 1	2,590	2,934	+13
2	3,447	3,947	+15
3	4,316	5,075	+18
4	5,348	6,921	+29
5	5,308	8,595	+62
6	7,007	16,690	+138
7	6,330	22,892	+262
Total	34,346	67,054	+95

Source: State of Illinois. *Chicago Area Transportation Study*, vol. II, Table 23, p. 91, 1959.

prising. It can be shown that this prediction is consistent with observed trends in travel volume densities, however. It is directly a consequence of increasing suburbanization and car ownership. The intervening opportunities model indicates some dimensions of the highway transport problem that such social trends precipitate.

Problems in the Use of Predictive Interaction Models

We have presented a sample of the predictive tools of spatial analysis that the planner and the applied geographer use. Planning decisions based on such analyses typically involve very large sums of money. New regional shopping centers and urban highways involve present-day society in huge monetary expenditures for the future. The validity of these methods is not simply academic; it is also of public interest that the accuracy of the predictions upon which plans are based be carefully checked.

The problems and limitations of the gravity and associated models can be viewed at three levels of generality. First, there are the largely empirical problems that the models share with many other quantitative methods. Second, there are the more philosophical issues surrounding the role of prediction in human activities, which the gravity models share with other predictive tools. Third, there are some basic controversies associated with the whole macro approach incorporated in spatial interaction models. We consider each.

We have discussed most problems that the spatial interaction models share with other quantitative methods, and we need only enumerate them here. The models are only as good as the input they receive. Data collection and measurement errors are important. The models include three basic variables (production, attraction, and distance) but these have been specified and measured in different ways in different applications. Distance, for example, has been measured in estimated travel costs, estimated travel times, and road lengths. All these measures have some unknown error because the values are calculated between zone centroids. Such errors are accepted as part of the price of working at aggregate scale. The specification of the areal framework of an analysis affects the results, too. Different sizes of zone produce different amounts of intrazone flow. The measurement of intrazone distance (d_{ii}) is somewhat arbitrary. The outer boundary of a study area must be considered carefully because no region is totally isolated from the outside world. In the Northwest England study, intraregional shopping flows underestimated the importance of Hanley in the southeast of the study region, which received much trade from beyond the region boundary.

Having solved or accepted such problems, there is next the problem of calibrating the model. Most attention is focused on the value of the distance exponent (b) in the equations. In our hypothetical example for two zones, we assumed an inverse square function of distance in the tradition of the classical gravity model. In practice, distance exponents are estimated empirically from the data used in the particular study. One way of doing this is to employ least squares regression procedures, using a transformation strategy as described in Chapter Five. It might well be that different forms of calibration and function produce substantially different predictions. If this is the case, it would seem that planning studies ought to provide us with several predictions of future patterns rather than a single prediction based on the particular model. The two northwest England retail studies are instructive in this context, but a wider variety of models and parameters might be even more informative as we try to ascertain what sort of patterns we can expect.

Issues of prediction as a procedure in gravity model studies are more subtle than the empirical problems. There are two threads to criticism of it. First, the gravity model is essentially a static model doing a dynamic job. It is static in that it describes a system of interaction at one point in time. In its original form it ignores changes in behavioral processes that may be going on. Much of this behavior is summarized in the single distance exponent but there are many reasons for not expecting this value to be constant over time. Future trends in exponents have been employed as part of the calibration for predictive gravity models.

The second issue over prediction is much more basic and throws doubt on the whole procedure. It has been argued that gravity models produce self-fulfilling prophecies. If a large flow of traffic is predicted and a large highway is built to cope with future traffic, almost inevitably the road will carry the full capacity of traffic predicted for it. This need not be the consequence entirely of correct prediction from initial conditions. New roads generate traffic. People willing previously to travel on public transport may well be attracted to journey by car with the opening of a new freeway from their place of work to their suburb, and this is particularly important in predicting traffic levels. In other uses of gravity models, this seems to be much less of a problem. A new shopping center does not automatically fulfill its sales prediction. Overpredictions might be reflected in empty stores. This whole question of circular causation must be considered carefully in any application of gravity model predictions.

We come to the criticism of gravity models that some people have considered to be the most basic. It is sometimes argued that the gravity model provides us with a gross description of behavior but does not begin to explain this behavior. This is often coupled with criticism that the gravity model is a crude physical analogy with no theoretical basis in social science. The best way to consider this argument is to divide social science research into two basic types of approach —microanalysis and macroanalysis. Microanalysis deals with behavior of individuals and draws on social psychology and its decision-making models. It is clearly at this level of analysis that we can begin to attempt explicit explanations of behavior, but the scale is often unsuitable in contexts that involve many hundreds or even millions of decisions. In the Chicago traffic studies, for instance, it was out of the question to attempt to ascertain the particular reasons for each individual trip and incorporate them into a final model; in Chicago there are simply too many trips and varieties of reasons for them. This does not mean that we abandon modeling of large numbers of trips. It indicates simply the need for a different approach. Macroanalysis studies behavior in the aggregate. The gravity model can be used in this category. This does not mean that individual factors such as employee loyalty are assumed not to exist. Such factors largely cancel themselves out, and aggregate patterns of behavior may be described by a few structural variables. We admit that the gravity model is very simple and does not include many factors that people consider when planning to shop or take any other trip. However, it does seem to specify the main structural determinants of travel patterns and is, therefore, usually a reasonable predictor of aggregate behavior.

LOCATION-ALLOCATION PROBLEMS

The techniques in this section attempt to define best or optimum solutions to clearly stated problems. They do not all inevitably succeed but all share this aim. These techniques are part of the field of *operations research*; in many ways, this has become the private sector's equivalent of the public sector's urban planning. Many problems dealt with in operations research have to do with economic and business topics. A sizable proportion include a spatial dimension and are of increasing interest to geographers. This is most explicitly true for location-allocation problems.

A location-allocation problem involves two basic

elements—a set of consumers distributed spatially over an area and a set of facilities to serve them. A question can then be stated: "How can we allocate facilities to best serve the consumers?" This verbal articulation of the problem suggests further questions. How do we decide which consumers are to be served by which facility? Will all the facilities have the same capacity or will some serve more consumers than others? The answers depend on the exact specification of the particular location-allocation problem, and we consider this below. The most important characteristic of the problem to notice is its extreme generality. We have used the terms "consumers" and "facilities" as general nouns but they may be replaced by many specific terms with no significant effect on the type of problem. Consumers may be farmers, school children, infertile men, factories, and voters. They may be served by facilities such as agricultural advisory services, schools, clinics, raw material sources, and electoral polling stations. Each of these correspondences defines a location-allocation problem.

A little thought about our definition of a location-allocation problem reveals that it consists of origins (consumers), and destinations (facilities), and that it presumably flows between them just like gravity model problems. The gravity model, however, tries to describe how a spatial pattern of flows actually does or will exist. The location-allocation approach, on the other hand, tells us how a spatial pattern of flows should exist, if it were to conform to some specified criteria. There is no explicit attempt to describe actual patterns. The optimum pattern is defined as either a standard against which actual patterns may be compared or a pattern to be implemented by some centralized planning authority.

Solving location-allocation problems is the essentially geographical procedure of searching for spatial efficiency. Economic geographers have long dealt with such problems, although they may not have identified them explicitly as location-allocation problems. Some early solutions include such traditional geographical tools as map overlays. We describe solutions to simple allocation problems in the next section. We follow this by the strictly mathematical solutions of linear programming. Finally, we consider complicated problems in which exact mathematical procedures are not feasible, so that computer techniques for deriving approximate solutions are necessary. It should be emphasized that all three types of approach, from map overlays to what we call heuristic algorithms, have been useful in solving practical problems. It is fitting, therefore, that we conclude this book on spatial analysis with the aspect of quantitative geography that many geographers think can contribute the most, at least in the short run, to human welfare.

Simple Allocation Problems

Simple means small in size. The procedures we describe deal with problems that are relatively small. We have already encountered one simple location problem in Chapter One. This is the well-known Weber problem, which attempts to find the point of minimum aggregate travel among a set of locations. If locations are consumers, then this median center is the optimum location for a single facility—say, for minimizing patient travel to a hospital. It is a simple problem involving only one facility but it is difficult to define and usually requires an iterative procedure. The problems we describe below can be solved manually; they consist of a simple geometric procedure for defining regions around known centers and a map overlay for adding facilities to an existing set.

Proximal Solutions for Allocating Consumers

Consider the pattern of six centers in Figure 7.6a. The question is: "How can we apportion the area on this map into regions about each center so that every location in a region is nearer to that region's center than to any other center?" The set of regions that satisfies this criterion is known as the *proximal solution*. A simple geometric method of producing this solution was proposed at the beginning of the twentieth century by climatologist A.H. Thiessen. His problem was to define regions around rainfall stations so that rainfall totals for several stations in one area could be weighted by the area they represented in computing an average rainfall for the whole area. It was clear that several

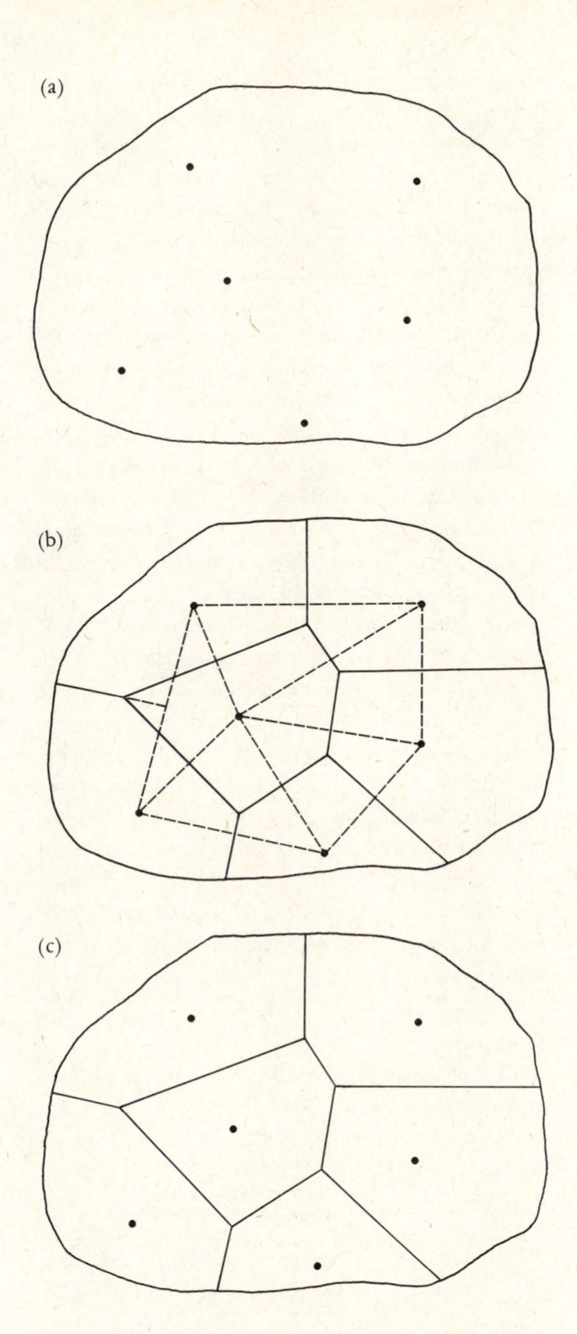

Figure 7.6 Constructing proximal regions.

similar records of rainfall for stations close together should be weighted less than more remote stations. The method he devised was proximal: each rainfall station was said to be representative of all the locations that were nearer to it than to any other station. The regions he derived have come to be known as *Thiessen polygons.*

Thiessen polygons are constructed as follows (Figure 7.6b). Each center is joined to its neighboring centers; these lines are then bisected vertically; the bisectors meet in threes or terminate at the border of the area and, hence, define a set of regions (Figure 7.6c). The result is a proximal solution. Because every location is nearer to its center than to any other center, it follows that we have an allocation of consumers to a fixed set of facilities that minimizes travel distance. If the facilities are taken as given, and our purpose is to minimize travel distance, Thiessen polygons provide an optimum set of service or catchment areas. It is of theoretical interest to notice that Thiessen polygons derived from a regular triangular pattern of centers are Christaller's hexagons, beloved of central place theorists.

This technique is clearly very simple in both method and result. It assumes, for instance, that the facilities can cope with any allocation of consumers that Thiessen polygons produce. In practice, of course, different facilities are of different sizes, and this may require consideration in drawing their catchment areas. Such considerations take us beyond Thiessen polygons.

Let's consider the services provided by the Public Health Department on the island of Anglesey in 1972. We consider the locations of dental clinics (Figure 7.7a), ambulance services (Figure 7.7b), and general clinics (Figure 7.7c). These have been chosen because the Anglesey County Council decided that each of these services can be provided for the inhabitants of the island from merely four locations. Furthermore, different sets of four locations have been chosen for each service, although of course the potential consumers remain the same. The southeastern part of the island is served from a different center for each service. If we assume that people will want to be served by the nearest clinics and ambulance stations, then the proximal solution is appropriate. The requisite Thiessen polygons are included in the maps in Figure 7.7. We can use these sets of regions to estimate the spatial efficiency of each service provision. The island is divided into some 600 kilometer squares for which

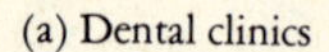

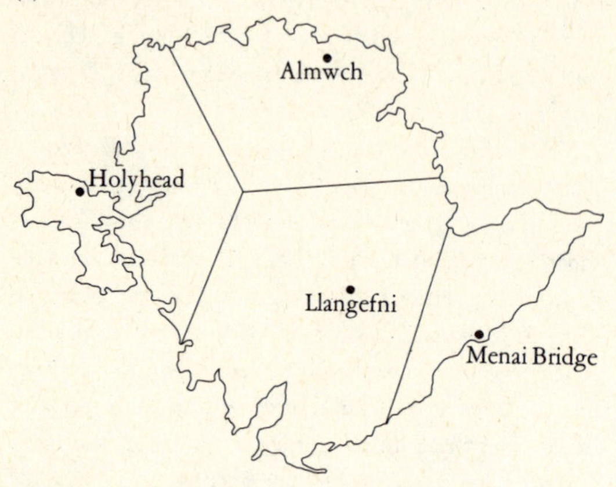

(b) Ambulance services

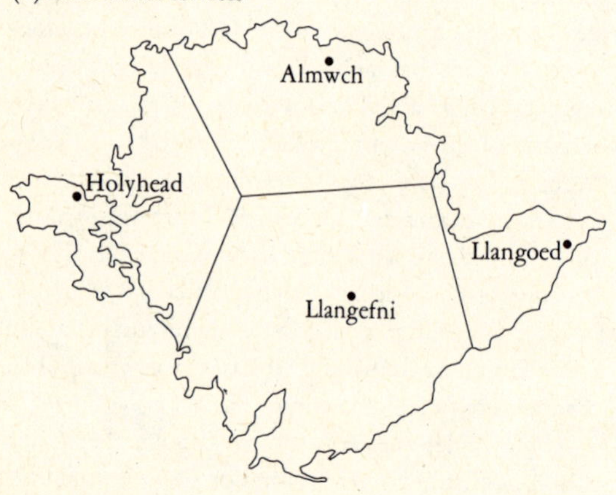

(c) General clinics

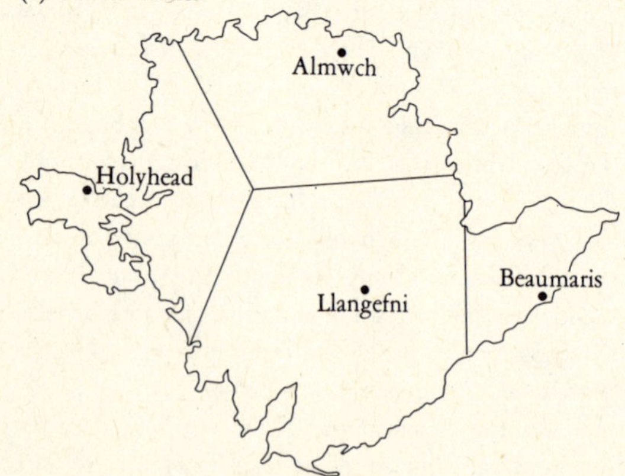

Figure 7.7 Provision of public health facilities on Anglesey.

populations are known. Using this grid, the total distance from each center to its consumers is found by multiplying each grid population by the distance from its allotted center and summing products for all grid cells in the Thiessen polygon. This is done for each region for all three sets of facilities. The total population kilometers for a set of facilities indicates the spatial efficiency of their locations with respect to their consumers—it quite literally measures how close the services are to the population served. The results are shown in Table 7.4. Holyhead and Amlwich are the same in each case, but they differ in defining Llangefni's catchment area on its eastern border. We can see from the table that spatially the most efficient fourth center is Menai Bridge, as used by the provision of dental clinics (Figure 7.7a). This is good news for sufferers of toothache on Anglesey.

What general conclusions can we draw? The method illustrates a way of evaluating spatial efficiency, but we should notice practical limitations on Thiessen polygons. In this case, the geometric nature of the solution, defining nearness solely in terms of straight distance, may lead to real distortions. Moreover, we must not think that the four centers used by the dental clinics are themselves the optimum four locations for services on Anglesey. There are almost certain to be four other locations better placed to serve this population. Remember that the Thiessen polygons define only optimum solutions for catchment areas with

Table 7.4 Spatial Efficiency of Public Health Facilities on Anglesey (1972)

Center	*Population Kilometers (thousands kilometers)*		
	Dental clinics	*Ambulance services*	*General clinics*
Amlwch	38	38	38
Holyhead	70	70	70
Llangefni	113	142	127
Menai Bridge	45		
Llangoed		33	
Beaumaris			39
Total	266	283	274

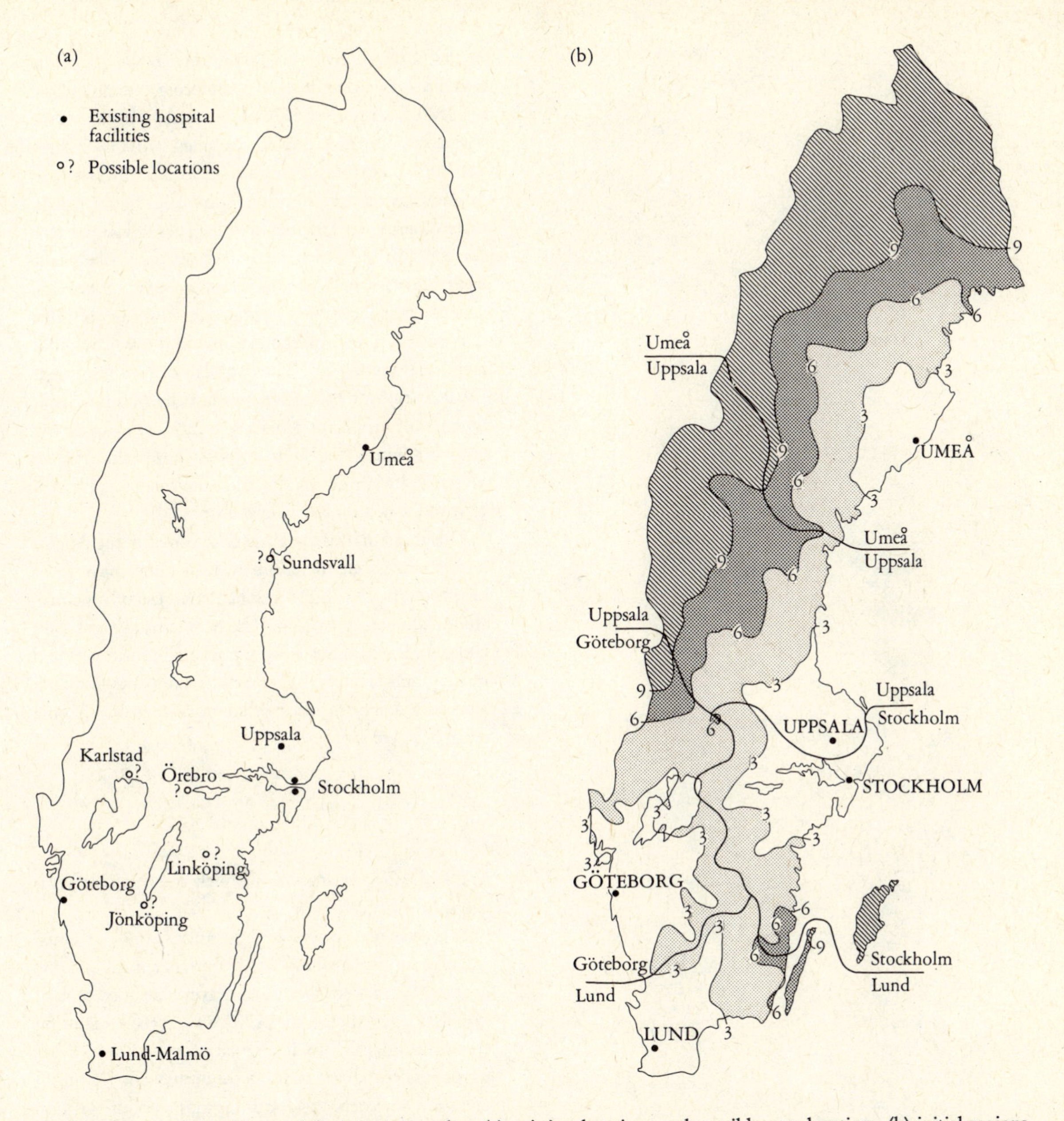

Figure 7.8 Specialized hospital facilities in Sweden: (a) existing locations and possible new locations, (b) initial regions and isochrones, and (c) optimum solution for two additions and revised isochrones.
From Sven Godlund. Population, regional hospitals, transport facilities and regions. *Lund Studies in Geography*, Series B, No. 21, 1961.

centers already given. But let's consider the possibility that the Anglesey County Council has decided to increase its provision of dental clinics to six. What would be the best locations for the two extra clinics? Such problems have been solved using map overlays for much larger situations than Anglesey clinics.

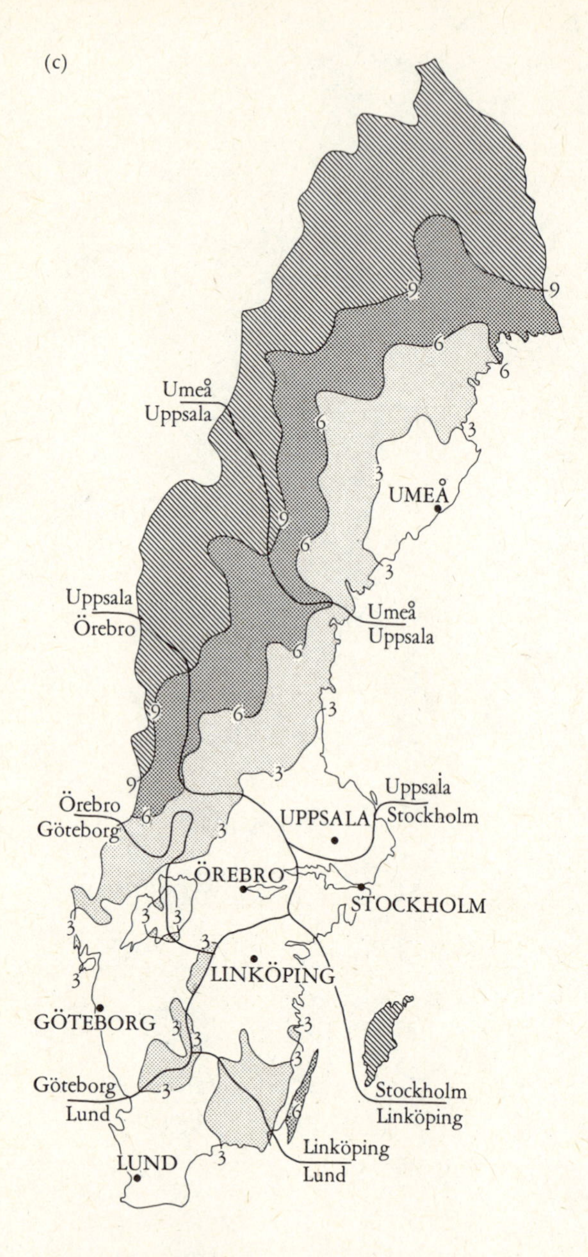

Figure 7.8 *(cont'd)*

Allocating Extra Facilities with a Map Overlay Approach

The classic example of the use of traditional map analysis by a geographer to solve a policy problem is in the work of Sven Godlund. Late in the 1950s the Swedish government was considering extending the supply of certain specialized hospital facilities to the Swedish population. These included artificial kidney machines that were already available in six of Sweden's hospitals at the five locations shown in Figure 7.8. In the expansion of facilities the planners envisaged two further sites on which these services would be made available. Because such facilities are part of hospital service, their location is restricted; in this case, possible sites were limited to five medium-sized towns without facilities (Figure 7.8a). The problem can be expressed: "Which two of these five sites are the optimum pair when added to the original five locations?" The criterion for defining the best sites was simply the final set of seven locations that would minimize total travel time to the nearest specialized hospital.

This is essentially the same criterion as in the Anglesey example, but the more realistic time measure of nearness replaces simple distance. This use of time precludes Thiessen polygons. The problem involves only choosing 2 extra sites from a possible 5, however, and it is quite tractable by complete comparison of all alternatives. The number of combinations of 2 from a set of 5 is given by

$$\frac{n!}{(n-r)!} = \frac{5!}{(5-2)!} = 20$$

We make 20 assessments and pick the one that gives the shortest travel time.

For each of the sets of 7 locations, lines of equal travel time (isochrones) were constructed around the centers. These were then placed over a detailed map of population predicted for 1975. By careful counting from this material, total population travel times were computed for all 20 possible solutions. The optimum solution of isochrones and regions is illustrated in Figure 7.8c. If this exercise is carried out for the 5 initial locations, the isochrones and regions in Figure 7.8b are produced. Comparison of the two maps illustrates the impact of this location decision in terms of consumer travel time. The 5 original locations have a total travel time value of 3.93 million hours, and this is reducible to 3.39 million hours with the optimum

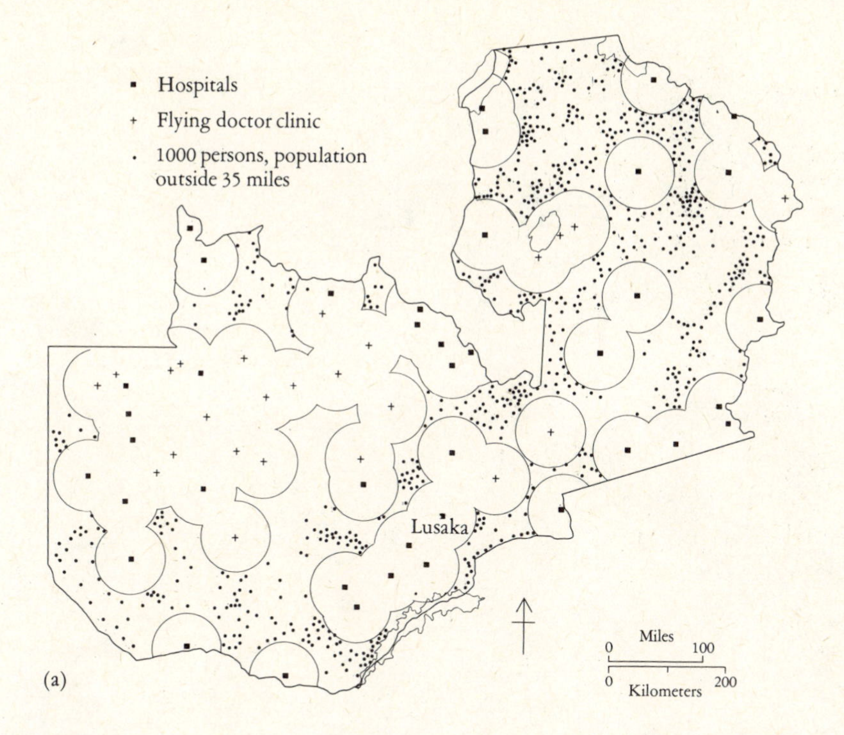

Figure 7.9 Flying-doctor facilities in Zambia: (a) distribution of doctor services, July 1969, and (b) proposed sites of future services.
From N.D. McGlashan, ed. *Medical Geography*. London: Methuen & Co. Ltd., 1972, Figs. 7–1 and 7–2, p. 100.

solution of 7 locations. Notice that the best extra locations are from middle Sweden, not northern Sweden where there seems to be a dearth of specialized facilities (Figure 7.8a). This is simply because the solution places facilities as near to people as possible, and most people live in the southern half of the country.

Geographers seem to have a better reputation in Sweden than in many other countries. Godlund's very clear and practical application of traditional geographical methods is one of the reasons that this is so. The Swedish government has adopted his proposals (Figure 7.8c) for their planning of advanced hospital facilities.

A modification of Godlund's approach for extending facilities was used in Zambia to present proposals for the future development of flying doctor services. Modifications were necessary because of the different circumstances, but the problem is easy to specify: "At what locations should the Flying Doctor Service place its clinics so that every person in Zambia lives within thirty-five miles of medical service?" The distance of thirty-five miles was defined by government policy. Mary Jackman attempted to answer this question as an exercise in applied medical geography.

The first step was to map the existing pattern of medical facilities, including ground-based clinics and hospitals. These are shown for 1969 in Figure 7.9a. A circle with a thirty-five-mile radius is drawn around each location to indicate areas served adequately by the government criterion. The western parts of the country, with their strong missionary activities, are much more densely served than the northern sections. This map was then compared with a population map of Zambia on which population is indicated by dots, each representing a thousand persons. In this case, there are no obvious sites for new development, and any of the dots can be considered a potential future clinic. In fact, in order completely to satisfy govern-

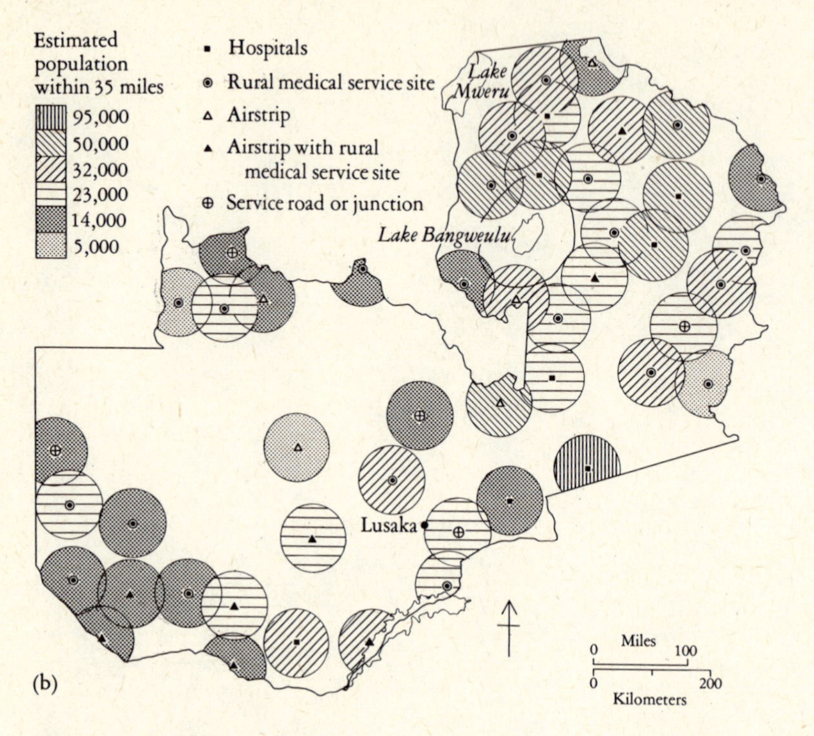

Figure 7.9 *(cont'd)*

ment policy that all the population should be within thirty-five miles of medical facilities, forty-eight potential sites were chosen (Figure 7.9b). A huge development such as is suggested by this analysis is clearly not immediately feasible. Priorities must be listed, and this is done by identifying the proposed sites that have the most population within the thirty-five-mile sphere of influence. Four levels of priority are identified. The highest goes to the proposed center in the southeastern part of the country with a service population of more than 50,000. Next comes the six centers with service populations greater than 32,000, five of which are in the north. Most of the remainder can be classified as of medium priority, although we may separate the seventeen centers that have a potential catchment of less than 14,000, the majority of which are in the southwest. From this sort of analysis it is clearly possible to recommend ordered expansion of flying doctor and associated medical facilities toward the government goal while ensuring maximum benefit at each stage of expansion.

Linear Programming

The methods we have discussed are simple practical procedures designed to solve specific types of problems. Since World War II, a formal mathematical approach to optimizing problems has been developed under the general title of *linear programming*. The basic mathematics of this approach is centuries old, but the method was not actually derived until 1947. The originator is American mathematician George B. Dantzig from a project for planning diversified activities of the U.S. Air Force. In any period, the U.S. Air Force has certain goals to achieve and a variety of activities to help achieve them. Dantzig's method allows us to identify the best combination of activities from among alternatives. This specific problem and its solution was very quickly generalized into the linear program we know today. Problems that had been previously identified but not solved now seemed soluble with this approach. One that has remained a favorite of linear programmers is the diet problem. We outline its

Table 7.5 The Diet Problem

Nutritional elements	*Food 1*	*Food 2*	*Food n*	*Minimum standard*
1	a_{11}	a_{12} · · ·	a_{1n}	c_1
2	a_{21}	a_{22} · · ·	a_{2n}	c_2
·	·	·		
·	·	·		
·	·	·		
m	a_{m1}	a_{m2} · · ·	a_{mn}	c_m
Prices	p_1	p_2 · · ·	p_n	

structure to illustrate situations amenable to linear programming.

The diet problem is famous because it is the first economic problem solved by linear programming; Dantzig gave the solution in 1947. If a diet is to be acceptable, it must contain levels of nutritional elements greater than specified minimum standards. But a diet does not consist of nutritional elements individually; it is made up of a variety of foods, each containing different amount of each element. The problem is: "What combination of foods will provide an acceptable diet at the lowest cost?" The data required to answer this question are set out in Table 7.5. The cells of the table indicate the amount of each element in one unit of food. Thus a_{11} is the amount of element 1 in food 1. The bottom row gives the different prices for one unit of each food.

We can state the diet problem for mathematical analysis. Let x_i be the quantity of food i that is included in the diet. We minimize the objective function (ϕ) in which $\phi = p_1x_1 + p_2x_2 + \cdots + p_nx_n$ subject to

$$\begin{aligned} a_{11}x_1 + a_{12}x_2 + \cdots + a_{1n}x_n &\geqslant c_1 \\ a_{21}x_1 + a_{22}x_2 + \cdots + a_{2n}x_n &\geqslant c_2 \\ \vdots \qquad\qquad & \quad \vdots \\ a_{m1}x_1 + a_{m2}x_2 + \cdots + a_{mn}x_n &\geqslant c_m \end{aligned}$$

This statement is a *program*. Because all the relations contained in it are linear, it is a linear program. This particular linear program says simply that we are to minimize total cost by a combination of food items that reaches the minimum standard for each nutritional element.

Two characteristics of the program should be noticed in particular. First, it involves large numbers of constraints that have to be satisfied. Second, these constraints are not necessarily equalities; all that is required is that the standards be equaled but they may also be exceeded. These two features are the reasons that linear programming was so late in being developed. Despite its subsequently widespread application, no simple direct method has been evolved for solving the problem. Iterative solutions have been devised requiring repetitive steps that bring us continually closer to the final result. Dantzig's original approach, the *Simplex method*, has been supplemented by numerous procedures. We are concerned with one particular type of linear program known as the *transportation problem*. It has certain simplifying features that make it easy to compute.

The Transportation Problem

The transportation problem is an explicitly spatial linear program. It is like the diet problem in that it was formulated and solved before the development of linear programming, although it is now incorporated as one of the most familiar and widely used types of linear programming. It deals with the spatial allocation of scarce resources to specified markets. We divide a study area into deficit and surplus regions in terms of some specified commodity or resource. We can consider coalfields as surplus coal regions and the remainder of an area as deficit coal regions. The transportation problem can be expressed as: "What pattern of commodity flow between regions will satisfy the deficits while minimizing total transport costs?"

Table 7.6 illustrates the basic elements of any transportation problem. There are deficits (d_j) and surpluses

Table 7.6 The Transportation Problem

Surplus regions	*Deficit regions*				*Surplus*
	1	2	3	m	
1	c_{11}	c_{12}	c_{13} . . .	c_{1m}	s_1
2	c_{21}				s_2
3					
.	.	.	.	.	.
.	.	.	.	.	.
.	.	.	.	.	.
n				c_{nm}	s_n
Deficit	d_1	d_2	. . .	d_m	

c_{ij} = cost from region i to region j.

(s_i) in the final row and column, respectively. We can write the program as follows. Let x_{ij} be the flow of the commodity from i to j. Minimize objective function ϕ

$$\phi = \sum_{i=1}^{m} \sum_{j=1}^{m} c_{ij}x_{ij}$$

in which c_{ij} is transport cost from i to j. This is subject to

$$\sum_{j=1}^{m} x_{ij} \leqslant s_i$$

$$\sum_{i=1}^{n} x_{ij} = d_j$$

This program has basically the same form as the diet problem but we have simplified its statement by using sigma notation. Notice also the equality for d_j. In transportation problems, it is usually assumed that the deficit regions are satisfied exactly. It is also very common for the first constraint to be an equality, so that all a surplus is exported, and $\Sigma s_i = \Sigma d_j$. We normally add one further constraint:

$$x_{ij} \geqslant 0$$

which means simply that we ensure no negative flows (flows back to surplus regions from deficit regions). In linear programming, as in the real world (until recently), we do not send coals to Newcastle.

How do we compute a pattern of flows that satisfies these constraints while minimizing ϕ? The procedure is quite simple, although it is tedious in application. Normally the tedium is left to the computer, but for illustration we work out an optimum solution for a very simple transportation problem involving three surplus regions (or origins) and three deficit regions (or destinations) as shown in Figure 7.10a. Table 7.7 shows the essential data for the problem, with $\Sigma s_i = \Sigma d_j = 100$, indicating that 100 units of the commodity are to be moved. The first step is to derive an *initial feasible solution*. By feasible we mean a solution that satisfies the constraints but does not necessarily, and usually does not, minimize ϕ. This initial flow pattern may be derived in several ways; we employ a method that takes the cost (c_{ij}) into account. The stages are shown in Table 7.8. First, rank the costs by giving the least cost a rank of 1, the next least expensive 2, and so on. Now concentrate on the cell that is ranked 1 (this is in the last column in the first row, under C).

Table 7.7 The Simple Transportation Problem (Costs Between Regions)

From surplus region	*To deficit region*			*Surplus*
	A	B	C	
1	16	10	2	60
2	12	4	6	20
3	9	7	5	20
Deficit	50	30	20	100

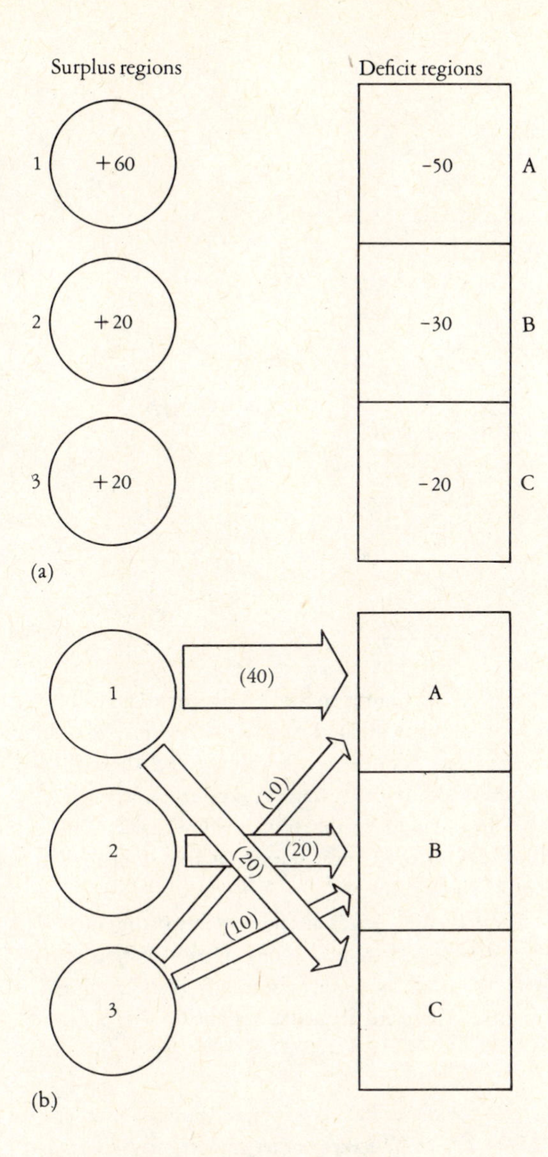

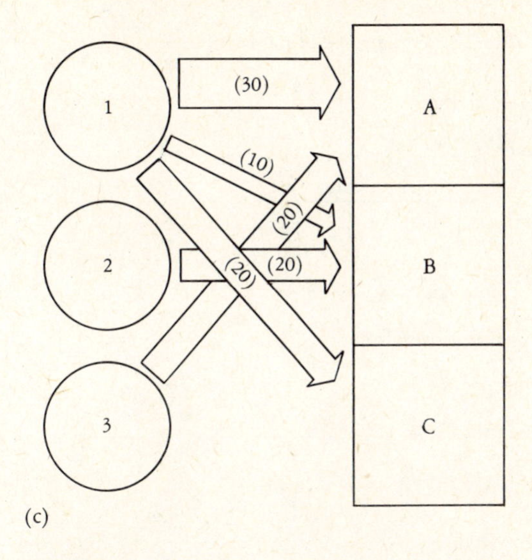

Figure 7.10 Solutions to the problem in Table 7.7 (figures refer to surpluses and deficits). (a) Surpluses and deficit regions, (b) initial feasible solution flow pattern, and (c) optimum solution flow pattern.

Allocate as much of the surplus for region 1 to region C as possible. In this case, region C requires a total of 20 units and all this can be provided by region 1. Therefore 20 is allocated to region C, as we see under "initial feasible solution," and the totals around the table are adjusted accordingly. Region C now requires no further commodity and region 1 has only 40 units for allocation.

This procedure is now repeated for the cell that is ranked 2 (it is in the second row under B). Deficit region B requires 30 units in all, but region 2 has a surplus of only 20, and 20 is all that can be allocated to this cell. Commodity totals available around the margins are adjusted again. This is repeated for all cells in turn.

Notice that the cell that is ranked 3 (in the bottom row under C) has zero flow allocated to it, because all C's needs have already been met by region 1. The final solution has in fact only $n + m - 1$ nonzero cells, indicating flows where there are n surpluses and m deficits. This is a basic theorem of linear programming. In this case, there should be no more than $3 + 3 - 1 = 5$ flows. The part of Table 7.8 called "initial feasible solution" shows that we do in fact produce exactly five flows, and this solution is illustrated in Figure 7.10b.

The method for producing this solution ensures that the constraints are satisfied, but we have said nothing about the objective function. We can now calculate ϕ as

$$(40 \cdot 16) + (10 \cdot 9) + (20 \cdot 4) + (10 \cdot 7)$$

$$+ (20 \cdot 2) = 920$$

Table 7.8 Finding the Initial Feasible Solution

Ranked costs

From		*To*	
	A	B	C
1	9	7	1
2	8	2	4
3	6	5	3

Initial feasible solution

From		*To*			*Outstanding totals*	
	A	B	C			
1	40	0	20	60	40	0
2	0	20	0	20	0	
3	10	10	0	20	10	0
Outstanding	50	30	20			
totals	40	10	0			
	0	0				

Thus, the total transport costs involved in the flows in Table 7.7 are 920. We must now ask whether this is the lowest cost possible for a flow pattern that satisfies our constraints. If it is the lowest, then it is the optimum solution. We can check this matter by computing *shadow prices* and *opportunity costs*. The steps in calculating these are shown in Table 7.9

Shadow prices are computed for every surplus and deficit region—that is, for every row and column in the table. We call the shadow prices for the surplus regions u_i and for the deficit regions v_j. We have to find three u_is and three v_js. We can compute their values from

$$v_j = c_{ij} + u_i$$

and

$$u_i = v_j - c_{ij}$$

once we have the value of one of them. Therefore, we arbitrarily set one of the shadow prices at 0. We make $u_1 = 0$. Other shadow prices are now computed by using only c_{ij} values (Table 7.7) for cells that have flows—nonzero cells from our existing solution. Thus, v_A can be computed from u_1 because the cell in column A of the first row is nonzero. Therefore, $v_A = 16$. However, we cannot compute v_B because the cell for it (in column B of the first row) is a 0 cell. We can compute u_3 because we know v_A and the cell in the third row under column A is nonzero. Therefore, $u_3 = 16 - 9 = 7$. In this way, shadow prices can be found for all rows and columns (as shown in the top part of Table 7.9).

These shadow prices are now used to compute the opportunity costs for the 0 cells in the existing solution. We must find four opportunity costs ($\bar{c}_{ij}$) in our example. These are defined as simply

$$\bar{c}_{ij} = v_j - u_i$$

so that for the cell in row 2 under column A, $c_{2A} = 16 - 10 = 6$. The three other opportunity costs are found similarly (as shown in the bottom part of Table 7.9).

We can now decide whether our existing solution is in fact the solution with the lowest cost. This is done by comparing $\bar{c}_{ij}$ with actual costs (c_{ij}). If all c_{ij} is larger than all corresponding $\bar{c}_{ij}$, then the existing solution is optimum and no improvements can be made. On the other hand, if one or more $\bar{c}_{ij}$ is larger than the corresponding c_{ij}, then there will be a better solution involving less cost. If we compare Table 7.9 with Table

Table 7.9 Finding Shadow Prices and Opportunity Costs

Shadow prices (cell values are c_{ij} for nonzero cells from Table 7.8)

			A	B	C
$u_i =$	0	1	16		2
	10	2		4	
	7	3	9	7	
$v_j =$			16	14	2

$u_1 = 0$
$v_j = c_{ij} + u_i$
$u_i = v_j - c_{ij}$

Opportunity costs ($\bar{c}_{ij} = v_j - u_i$)

i, j	A	B	C
1		14	
2	6		−8
3			−5

7.7, we can see that, in fact, $\bar{c}_{1B}$ is 14 (Table 7.9) while c_{1B} is only 10 (Table 7.7). This indicates that our initial feasible solution is not the optimum solution, and we must improve it in order to minimize ϕ.

Now we use our initial feasible solution as a starting point for an iterative process that will conclude with the optimum solution. There may be several instances of $c_{ij} < \bar{c}_{ij}$, so we pick the cell with the largest difference. In our example, there is only one instance of $c_{ij} < \bar{c}_{ij}$, and therefore we choose it—the cell at row 1 and column B. We use this cell as a pivot about which we improve our existing solution. The idea is to allocate flows to this cell and make the necessary adjustments to other cells. The process is illustrated in Table 7.10.

We indicate that we are to add flows to the cell at row 1, column B, by placing a plus sign in the cell. This means region 1 will not be able to supply either A or C in the manner of the existing solution because it has only 60 units of the commodity to export, some of which will go to B. Therefore, we must take away some of the flows in one of these two cells in order to start the flow from 1 to B. We consider taking part of the flow from the cell at 1,A, so that we put a minus sign in this cell. If, however, take we part of the flow from 1 away from A, its deficit must be made up from another source. In this case, the only other surplus region that supplies A is region 3, so that this flow will have to be increased. Therefore, we put a plus sign in cell 3,A. Once again, the move has other implications. By diverting some of the surplus in region 3 to A, it follows that region 3 will not be able to meet its previous commitments elsewhere. In particular, we have to decrease the flow to region B. Thus, we put a

Table 7.10 Adjusting the Existing Solution

Change cycle

	A	B	C
1	−	+	
2			
3	+	−	

Revised solution ($\delta = 10$)

	A	B	C
1	30	10	20
2	0	20	0
3	20	0	0

Table 7.11 New Shadow Prices and Opportunity Costs

Shadow prices (cell values are c_{ij} for nonzero cells from Table 7.10)

			A	B	C
$u_i =$	0	1	16	10	2
	6	2		4	
	7	3	9		
$v_j =$			16	10	2

Opportunity costs

	A	B	C
1			
2	10		−4
3		3	−5

minus sign in cell 3,B. But region B will now have to be supplied from another source to make up for the loss of trade from region 3. Notice, however, that we have already made arrangements for this by putting a plus sign in cell 1,B. This means that we need make no further suggested adjustments. We can see now that the purpose of the exercise is to find a circular route of complementary + and −. This is not always possible. For example, if we had put the first minus sign in cell 1,C, instead of 1,A, we would have found that we could not cope with this loss unless we were to initiate a new nonzero flow pattern.

We now have the pattern of adjustments necessary to revise the flow pattern. We have to decide how much trade to reallocate in these adjustments. This can be accomplished easily. Because we cannot have negative flows, the highest flow change possible is equal to the flow in the minus cell that had the lowest original flow. If we take this amount away from this cell, it becomes a 0 (though not a negative) cell. Because our pivot cell is to receive flows, it will become a nonzero cell, and the number of nonzero cells thus remains constant at $n + m - 1$. We have two minus cells, and of these, cell 3,B has the lower original flow, of only 10. This is the magnitude of change (δ) for the revision of flows. All plus cells gain 10 units of flow, and all minus cells lose 10 units. The solution is a new, improved flow pattern that satisfies our constraints (Table 7.10). We can show that it is an improvement over the previous solution by computing ϕ:

$$\phi = (30 \cdot 16) + (20 \cdot 9) + (10 \cdot 10) + (20 \cdot 4) + (20 \cdot 2) = 880.$$

The question is now whether this new solution is optimum. We can answer as before by computing shadow prices and opportunity costs (Table 7.11). When we compare the new opportunity costs with the original transport costs (Table 7.7), we find that, in every case, $\bar{c}_{ij} < c_{ij}$. Thus, in this very simple example, we have found the optimum solution after one iteration from the initial feasible solution. With larger problems, the improvement procedure we have outlined must be taken through several iterations before we reach the optimum. The optimum flow pattern for our simple example is shown in Figure 7.10c.

Let's turn from the simple hypothetical case to the real world. The utility of the transportation problem for analyzing flow patterns in geography should be clear. Most goods in economic geography are not ubiquitous in supply or demand. On the one hand, we have actual commodity flows while, on the other hand, a linear programming analysis can provide the optimum flow pattern. Comparison of the two patterns may have theoretical or practical implications. First, we are led to ask why actual flows differ from the optimum pattern; the suggested answer is that there are factors other than transport costs to be considered in determining the patterns. Second, knowledge of optimum pattern max help minimize transport costs form and the basis for improving transport policy.

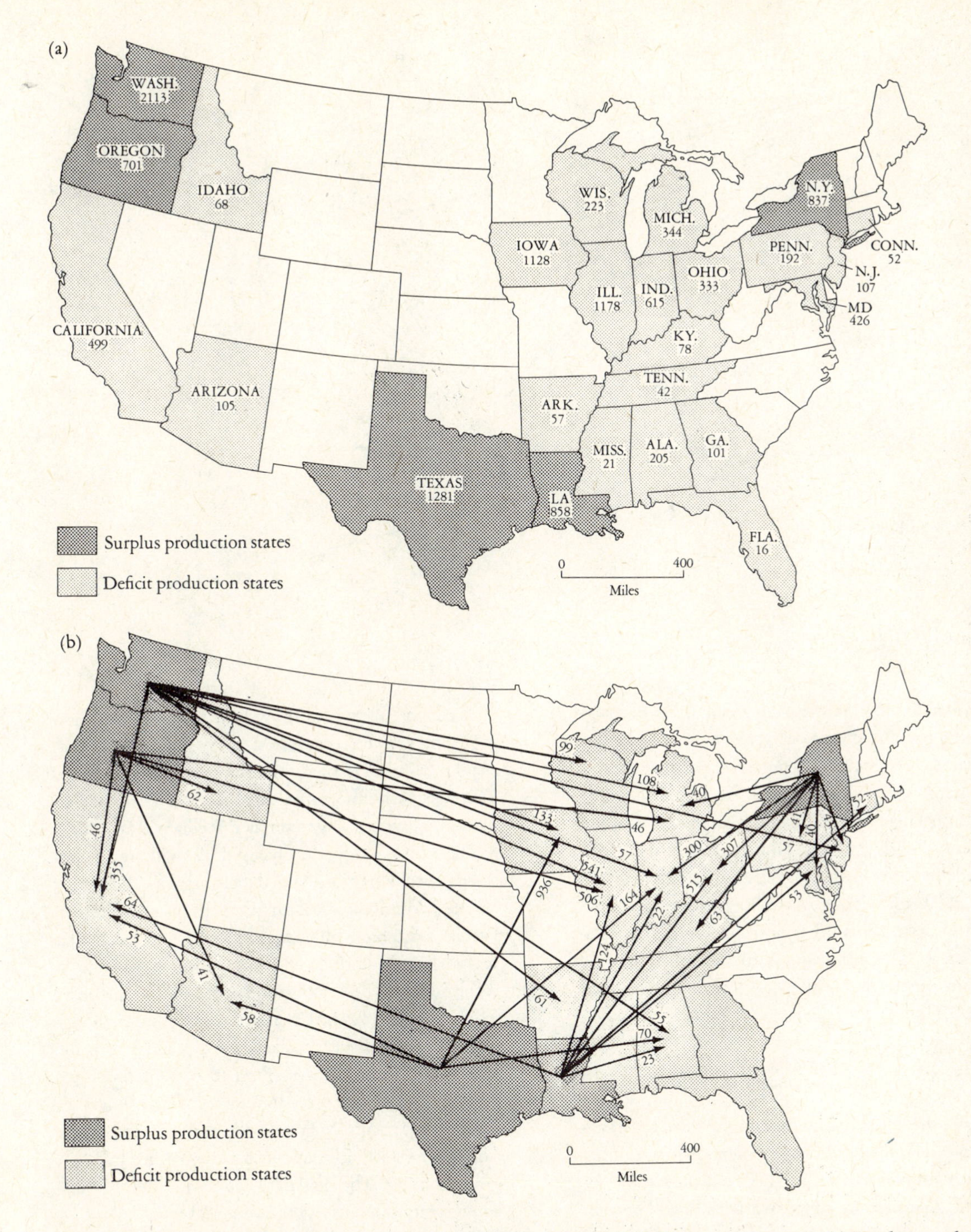

Figure 7.11 Interstate flows of aluminum bars in the United States, 1960 (figures denote tons). (a) Surpluses and deficit states, (b) actual interstate flow pattern, and (c) optimum interstate flow pattern.
From Kevin R. Cox. The application of linear programming to geographic problems. *Tijdschrift voor Economische en Sociale Geografie*, vol. 56, 1965.

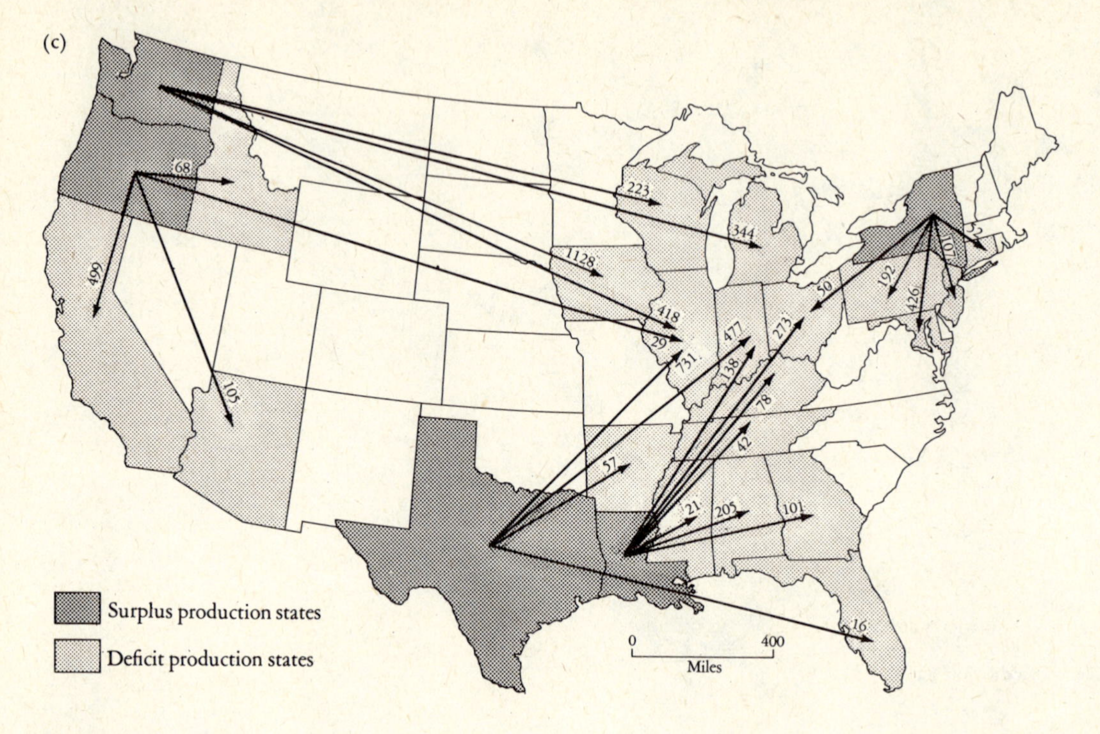

Figure 7.11 *(cont'd)*

Consider Kevin Cox's analysis of interstate flows of aluminum bars based on a one percent sample taken in the United States in 1960. The pattern of surplus and deficit states is shown in Figure 7.11a. The situation is interesting. There are three distinct surplus areas in the Northeast, Southwest, and Northwest and there is a major deficit area in the Midwest, in addition to the South and West. Figure 7.11b shows the complicated pattern of actual flows that distributes surplus to the deficit areas. This constrasts with the much simpler optimum pattern that minimizes total distance moved (Figure 7.11c).

Facility Deficits and Consumer Surpluses

We have introduced linear programming and the transportation problem but have not mentioned location-allocation problems. The link between these topics is, of course, that they are concerned with flows between geographical locations. The similarity extends to the search for spatial efficiency in a pattern of flows. Our purpose now is to show how location-allocation problems can be transferred to a framework that enables us to solve them as a transportation problem. This idea that this can be done is often attributed to pioneer quantitative geographer William Garrison.

The transportation problem deals with flows between surplus regions and deficit regions. We must cast the location-allocation problem in these terms. Facilities (hospitals, libraries, schools) become deficit regions that require flows from consumers located in surplus regions (the population areal units). The flows from surplus to deficit regions are consumer movements from residences to facilities. Application of the transportation problem method to data of this form produces an optimum assignment of consumers to facilities by minimizing total travel costs or distance. The result is a set of functional regions defined by linear programming. The general location-allocation problem is set up as a linear program in Table 7.12.

Let's illustrate with an example, using the work of Peter Gould and Thomas Leinbach (1966) in planning

Table 7.12 The Location-Allocation Problem in a Linear Program

Areal units of consumers	*Facility*					*Consumers*
	1	2	3	. . .	m	
1	c_{11}	c_{12}	c_{13}		c_{1m}	s_1
2	c_{21}					s_2
3						
.	.	.	.			.
.	.	.	.			.
.	.	.	.			.
n					c_{nm}	s_n
Capacity	d_1	d_2			d_m	

c_{ij} = cost from areal unit i to facility j.

hospital facilities in western Guatemala. Their purpose was to decide where to locate three new hospitals. Because the transport network was not well developed in this area, it was necessary to limit the number of possible locations to the five largest towns. The problem is to find the set of three possible locations of the ten combinations that minimizes total distance traveled by the population to the hospitals. This is to transport surplus population to deficit hospitals by minimizing total distance traveled. Notice how similar this problem is to Godlund's study of Swedish hospitals. Instead of the manual map overlay, we show a linear programming solution.

Each of the three hospital sites for any one proposed solution was defined as a deficit region. Each of these was allocated one third of the population of western Guatemala. Before it is assumed that this involves planning quite unbelievably huge hospital complexes, we must point out that this allocation of deficits is simply in order to comply with the constraint equations of the linear program. Realistic values of hospital sizes in terms of hospital beds may be obtained after the analysis by dividing the three deficit totals by an appropriate ratio between beds and population. Population surpluses were defined for eighteen areal units in western Guatemala, so that $\Sigma s_i = \Sigma d_j$. Distances were computed between population area centers and hospital sites along the road network, weighting unpaved roads double. Linear programming was then applied to each of the ten combinations of three sites in turn, producing the optimum allocation of population to hospitals for each set of three sites. The next step was to select the set of three sites that has the lowest total travel distance among the ten solutions. This was designated the optimum set of three sites for regional hospitals. The resulting sites and flows are shown in Figure 7.12a.

The pattern of flows in Figure 7.12a may seem at first somewhat puzzling. For an optimum allocation of consumers to facilities, there do seem to be some odd flows. The proposed hospital at Coatepeque, in particular, has a very wide catchment area, with potential patients passing by the two other hospitals in apparent enthusiasm to reach Coatepeque. In fact, these cross-flows are the consequence of the constraints we placed on the solution. We have allocated exactly the same deficits to each hospital, which means we postulate hospitals of equal size. Because Coatepeque's natural catchment area has a smaller population, it has to poach patients from other areas to fill beds. The solution can be improved by modifying the hospital sizes—making Coatepeque's deficit smaller and increasing the two other deficits. Experiments in this manner result in the much more logical pattern of flows in Figure 7.12b, in which the ratio between the three hospital sizes is 0.40/0.36/0.25 instead of the original 0.33/0.33/0.33.

With a little thought and minor modifications, many location-allocation problems are clearly amenable to solution by linear programming. For some quite fortuitous reason, all our examples of location-

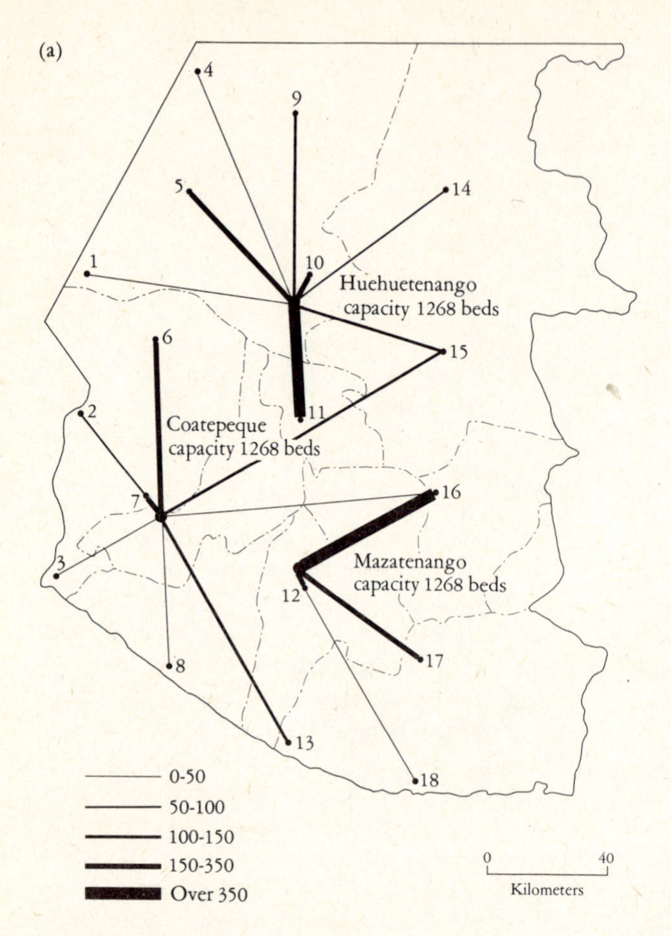

Figure 7.12 Additional hospital provision in western Guatemala: (a) equal-size constraint, and (b) final solution with varying sizes.
From Peter Gould and Thomas Leinbach. An approach to the geographic assignment of hospital services. *Tijdschrift voor Economische en Sociale Geografie*, vol. 57, 1966.

allocation problems have been medical. This reflects perhaps recent growth in medical geography. Other types of facility are equally amenable to the sorts of analyses we have described. School districting, in particular, poses many interesting location-allocation problems, some of which have been attacked as transportation problems.

School Districting Applications

The simplest application of linear programming to school districting is to allocate pupils to schools in the most efficient way by minimizing their travel. This was done by Maurice Yeates early in the 1960s in what is probably the earliest application of this model to location-allocation problems in geography. The problem was to allocate pupils in Grant County, Wisconsin, to the thirteen high schools in a spatially efficient manner. Spatial efficiency was defined in terms of the objective function of the transportation problem. Each school was defined as a deficit region to which pupils, aggregated into grid square surplus regions, were to be allocated. $\Sigma s_i = \Sigma d_j$. Transport costs were represented by simple straight-line distances. Given this

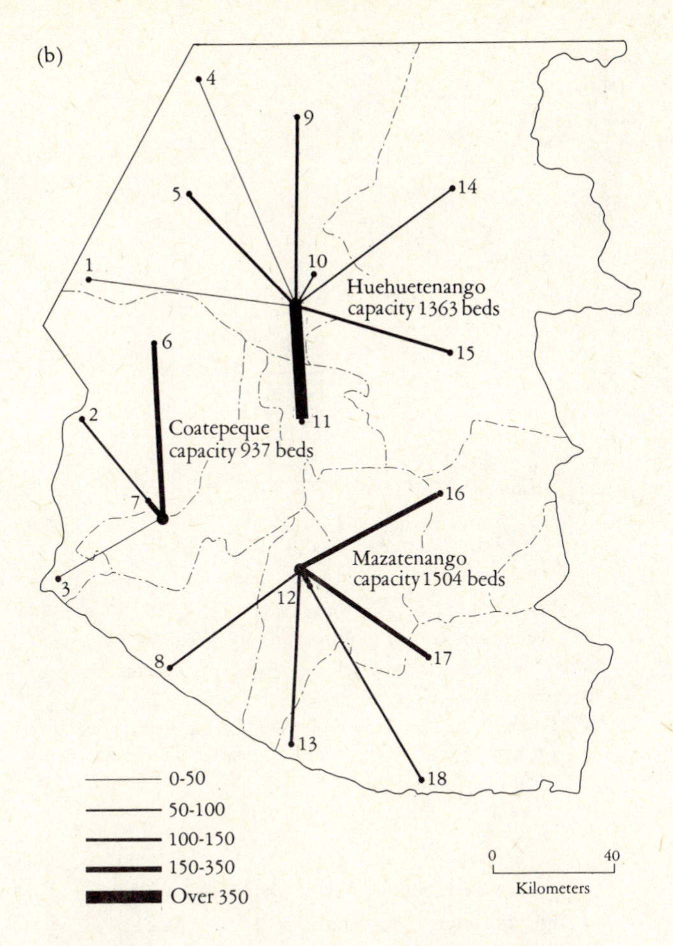

Figure 7.12 *(cont'd)*

data, the transportation problem solution is produced, as illustrated in Figure 7.13a. This set of functional regions can be compared with the actual school districts for 1961 in Figure 7.13b.

Comparison of the actual school districts and the optimum solution is of interest because the 1961 pattern is the result of many recent changes in districting resulting from the building of eight new schools between 1951 and 1961. Differences between the two maps are marked (Figure 7.13c) to such an extent that we can suggest that the actual districting pattern is quite spatially inefficient. In fact, 18 percent of the grid cells are allocated differently from the optimum solution, giving some idea of the degree of spatial inefficiency. In practical terms, this means that quite literally each day thousands of students are traveling longer distances to school than necessary, which, in addition, is costing their parents extra taxes for higher busing bills.

Before we consider more subtle applications of linear programming to school districting, we can consider briefly differences between the solution that minimizes total distances and the solution that would be produced simply by drawing Thiessen polygons around the thirteen high schools. Remember that Thiessen polygons produce distance-minimizing solutions. The difference between the two approaches is that the simple proximal solution of Thiessen polygons takes no account of the different sizes of the thirteen high schools. Almost inevitably, the numbers of pupils in each school district will be different from the number of places available in the respective high schools.

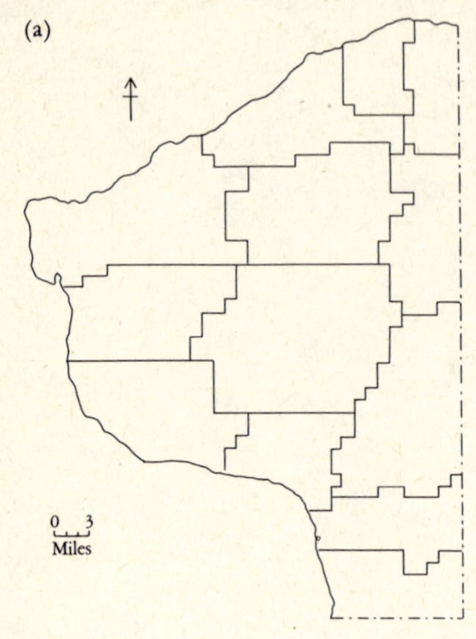

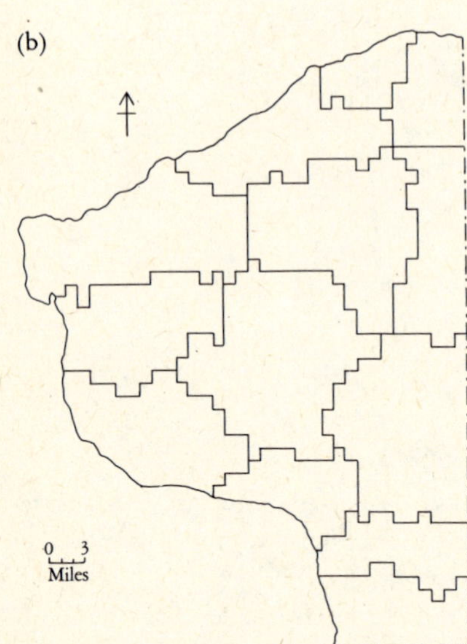

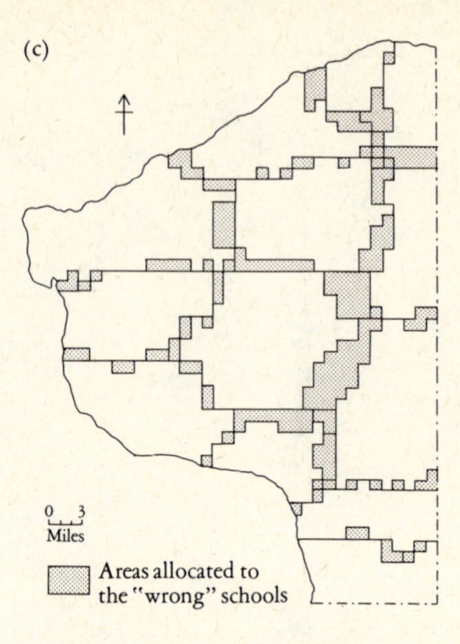

Figure 7.13 School districting for Grant County, Wisconsin: (a) linear programming solution, (b) actual school districts, and (c) the spatial inefficiency of the actual districts.

From M. Yeates. Hinterland delimitation. Reproduced by permission from the *Professional Geographer* of the Association of American Geographers, vol. 15, 1963.

The constraints in a transportation problem solution ensure that the number of pupils in a district is the same as the size of the school for the district. Hence, this second, more sophisticated, procedure is a much more useful approach in all situations similar to school districting in which sizes of facilities are given.

School administrators have allocation problems that are much more troublesome, of course, than locating students to the nearest schools. Donald Maxfield has treated such problems in a particularly well-designed use of linear programming. If a simple proximal policy is pursued, it will lead inevitably to overcrowding in some schools and underutilization of other school facilities. In Figure 7.14a, generalized school catchment areas are defined around thirteen primary schools in Georgia in the Athens School District. Each student is allocated to the nearest school in terms of road distance. The use of road distances means that this is not exactly a Thiessen polygon solution, but it has basically the same properties. Most of the central schools with compact catchment areas are allocated more students than their capacity; the more suburban schools have fewer than their capacities. This can be overcome by producing a transportation problem solution constraining the deficits of the schools to their actual capacities, as we have seen. A school's allocation problems do not necessarily end at this point. Many central areas have a predominance of black

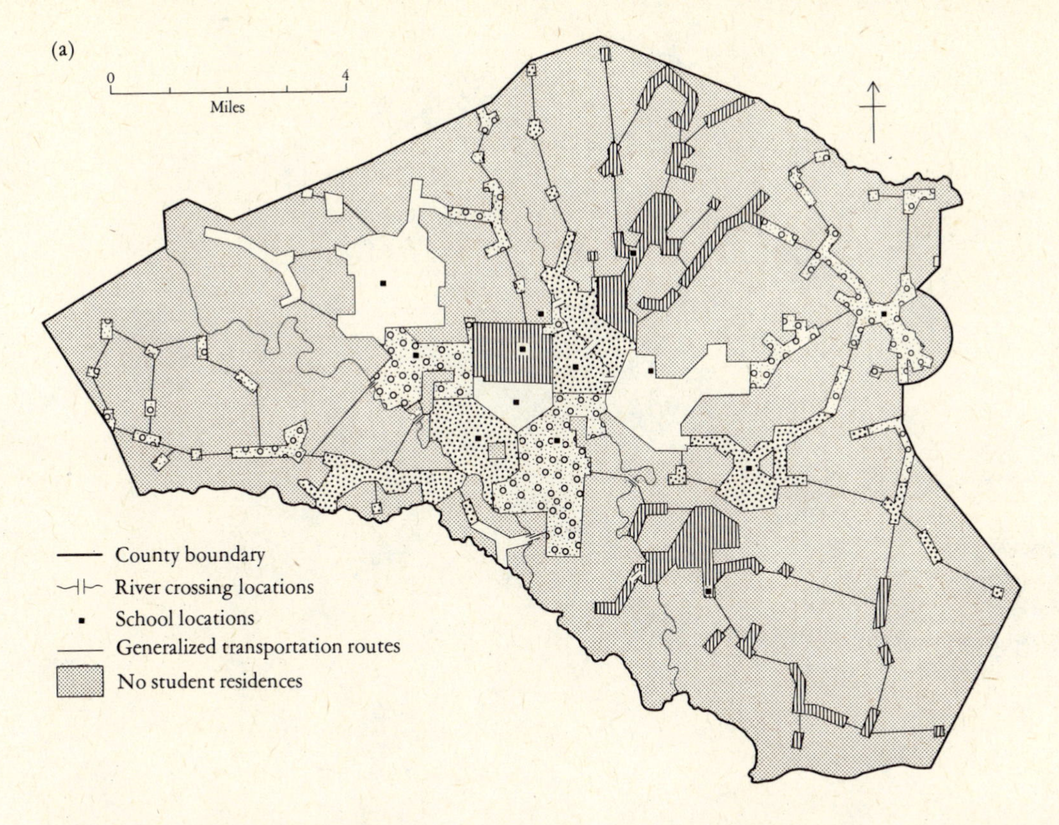

Figure 7.14 School districting patterns for Athens, Georgia, (a) Allocation of students to the nearest school with no capacity constraints.

From D.W. Maxfield. Spatial planning of school districts. Reproduced by permission from the *Annals* of the Association of American Geographers, vol. 62, 1972, Figs. 1, 7, and 8.

students whereas suburban areas have a predominance of white students. Simple least-cost solutions, even with capacity constraints, produce a high degree of racial segregation because of the traditional residential patterns of whites and blacks. We conclude this section with a practical example of the use of the transportation problem to design spatially efficient school catchment areas while simultaneously preventing overcrowding and providing racial balance in each school.

The overcrowding problem is tackled by allocating the same level of capacity to each of the 13 schools. The entire school district has a 92 percent student capacity and this level is assigned to every school. The surplus students equal the deficit school capacities and produce the same level of crowding in every school. Racial mixture is achieved by carrying out two separate analyses, one for allocating whites and another for allocating blacks. Because the ratio of white students to black students in the district as a whole is 66/34, this ratio is used for every school to give the fairest racial balance.

Ninety-two percent of the total capacity of every school is apportioned between whites and blacks in a 66/34 ratio. This produces two separate sets of school capacities as deficit regions. The locations of white and black students within grid squares are used as two different sets of surplus regions. Costs are defined as road distances. Linear programming solutions are computed separately for white and black students, giving two catchment areas for each school, as shown

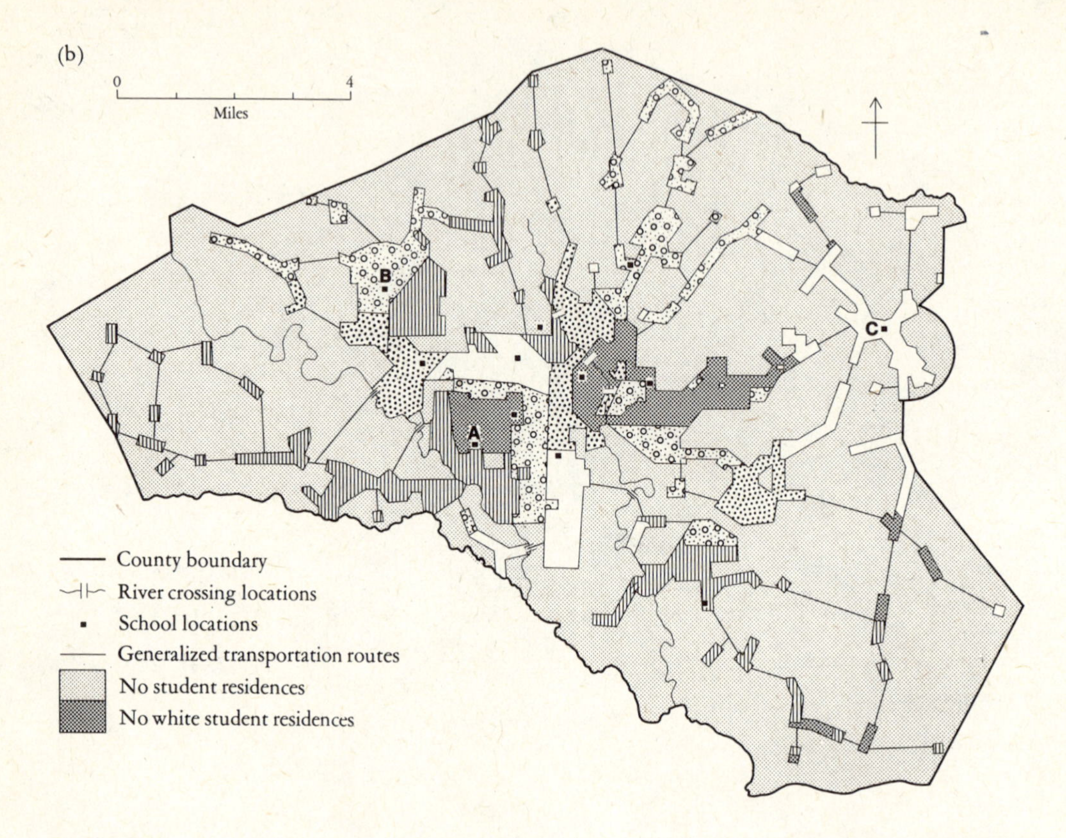

Figure 7.14 (b) Allocation of white students to the nearest school with capacity and integration constraints.

in Figures 7.14b and 7.14c. These solutions are indeed complex, so much so that only three of the thirteen catchment areas are shown on each map. This complexity is clearly the result of achieving the exact racial balance of the total district in every school despite acute residential segregation. In the solution for the white students, it was found that two of the central schools are located in areas with no white students. The resulting strung-out catchment area of one of these schools (A) is shown on Figure 7.14b. The suburban school's district (B) on this map is much more compact, although it loses some of its local white students, compared to Figure 7.14a, because they have to make way for black students. Finally, school C in Figure 7.14b shows a catchment area similar to the one in Figure 7.14a. This school is in a small community outside Athens and has a racial mixture similar to that of the district average. This last remark is confirmed in Figure 7.14c showing this same school (C) has very similar catchment areas for blacks and whites. This is in complete contrast to school B, which has to pick up its black students from scattered locations long distances from the school. The central area school (D) requires only a very small compact catchment area, smaller than its unconstrained area in Figure 7.14a, in order to serve its 34 percent quota of black students.

These maps are interesting and instructive. Remember that, despite their spatial complexity, each is a transportation problem solution and both are spatially the most efficient patterns of school districts, given the constraints and conditions we have imposed. The solutions are instructive in that they show us the degree of spatial distortion of catchment areas necessary to achieve goals of racial balance. There is clearly no need to assume that this very specific solution involving exactly the same capacities and racial mixtures in every

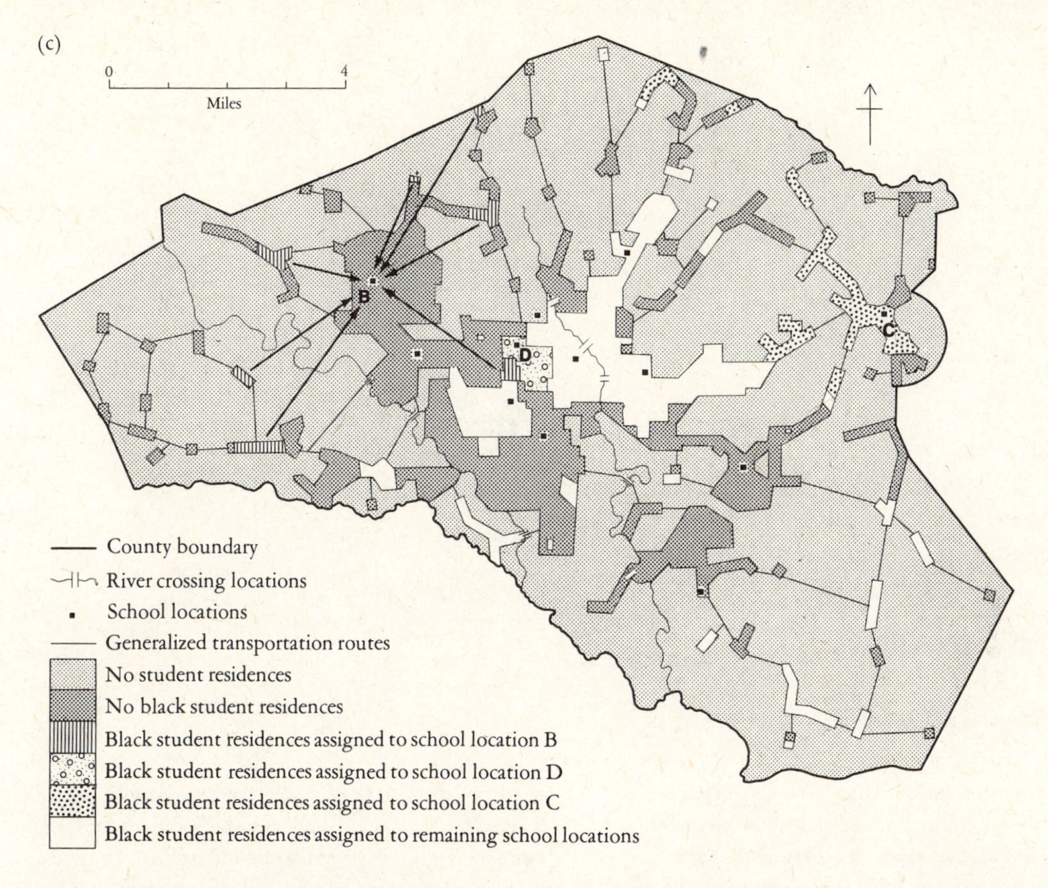

Figure 7.14 (c) Allocation of black students to the nearest school with capacity and integration constraints.

school should be followed to the letter. This is one type of solution about which there has been much controversy. Modifications to avoid the most extreme distortions do not necessarily violate current thinking on racial balance in schools. In linear programming, we have a tool that can provide alternative solutions for different sets of policy assumptions and goals. The linear programming approach is clearly a very useful spatial analysis technique for planning the allocation of children in actual school districts.

Heuristic Algorithms

All our examples of location-allocation problems have been relatively simple. This is because either facility sites have been given, such as in the Anglesey and school district examples, or we have been required to assess only a few alternative sets of locations. We have argued that hospitals in Sweden and western Guatemala are best located in large towns, and we restricted our search for optimum locations to these places. The result is that only twenty and ten sets of alternative sites had to be considered. In each case, the optimum set of locations for the facilities could be found simply by trying each of the alternatives in turn and selecting the one that produces the lowest total travel time. *Total enumeration* of alternatives is possible only in problems in which alternative sites for facilities are restricted. If we free our location-allocation problem from this restriction, total enumeration can become computationally infeasible. What we want is a straightforward mathematical procedure that will

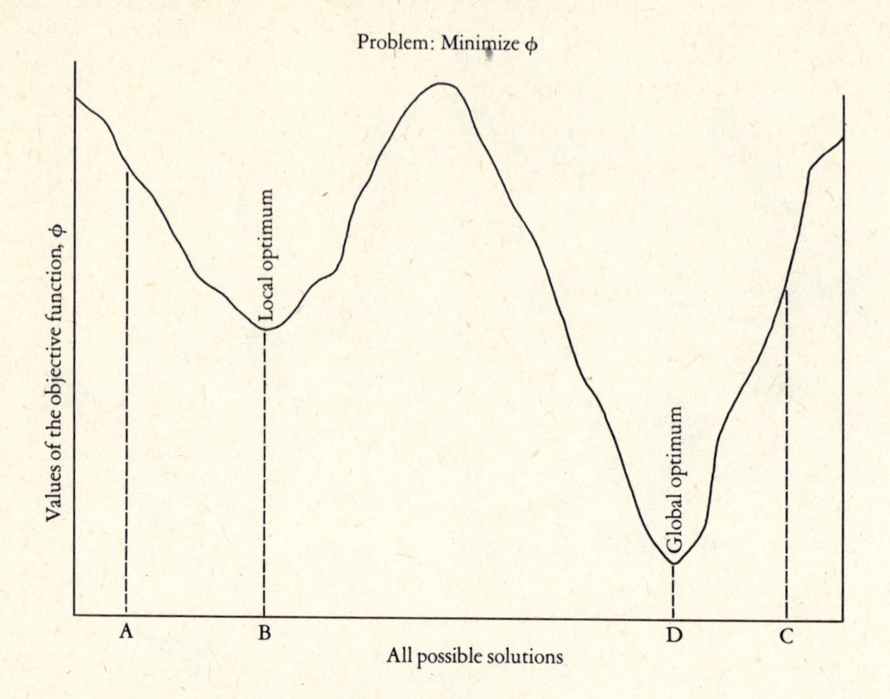

Figure 7.15 Local and global optima.

bypass the need for total enumeration and present us directly with optimum locations. Such a procedure unfortunately does not exist. We have come to what Peter Gould has termed "the geographer's unsolved problem." We ought not to despair at this point or throw in the towel. Exact mathematical solutions cannot be found for multiple location problems, but solutions can be arrived at by less elegant means—*heuristic algorithms.*

A heuristic algorithm is simply a set of rules defining a procedure for producing a solution to a given problem. The solution is not necessarily the optimum solution. An algorithm consists normally of an iterative process that tends to converge on the optimum solution. Because the procedure is not based on deductive mathematical foundations, optimum solutions cannot be guaranteed. It is in this sense that they differ from linear programming solutions, which can be proved, mathematically, always to produce the optimum solution.

It is instructive to consider the notion of local and global optima. A global optimum is the "true" optimum solution that we were concerned with above. This solution is the goal of heuristic methods also. But consider Figure 7.15. All solutions to a problem are arrayed along the x axis and their values on the objective function, ϕ, are on the y axis. If our purpose is to minimize this objective function, we can easily identify the global optimum on this diagram. We can use this simple schematic representation of problem solutions to show how and why many heuristic solutions do not always find the global optimum. Assume that our algorithm moves in short jumps always in the direction of a lower value for ϕ. If we start at point A, our algorithm will iterate in jumps down to B, and there it will become stuck. The solution presented will be the local optimum at B, not the global optimum solution. Notice, however, that if the iteration starts at C, then the global solution at D is produced from the heuristic algorithm. In real problems, the search space is obviously much more complicated than Figure 7.15 indicates, but the example shows nonetheless how converging heuristic algorithms may or may not reach the global optimum solution. "Good" heuristic pro-

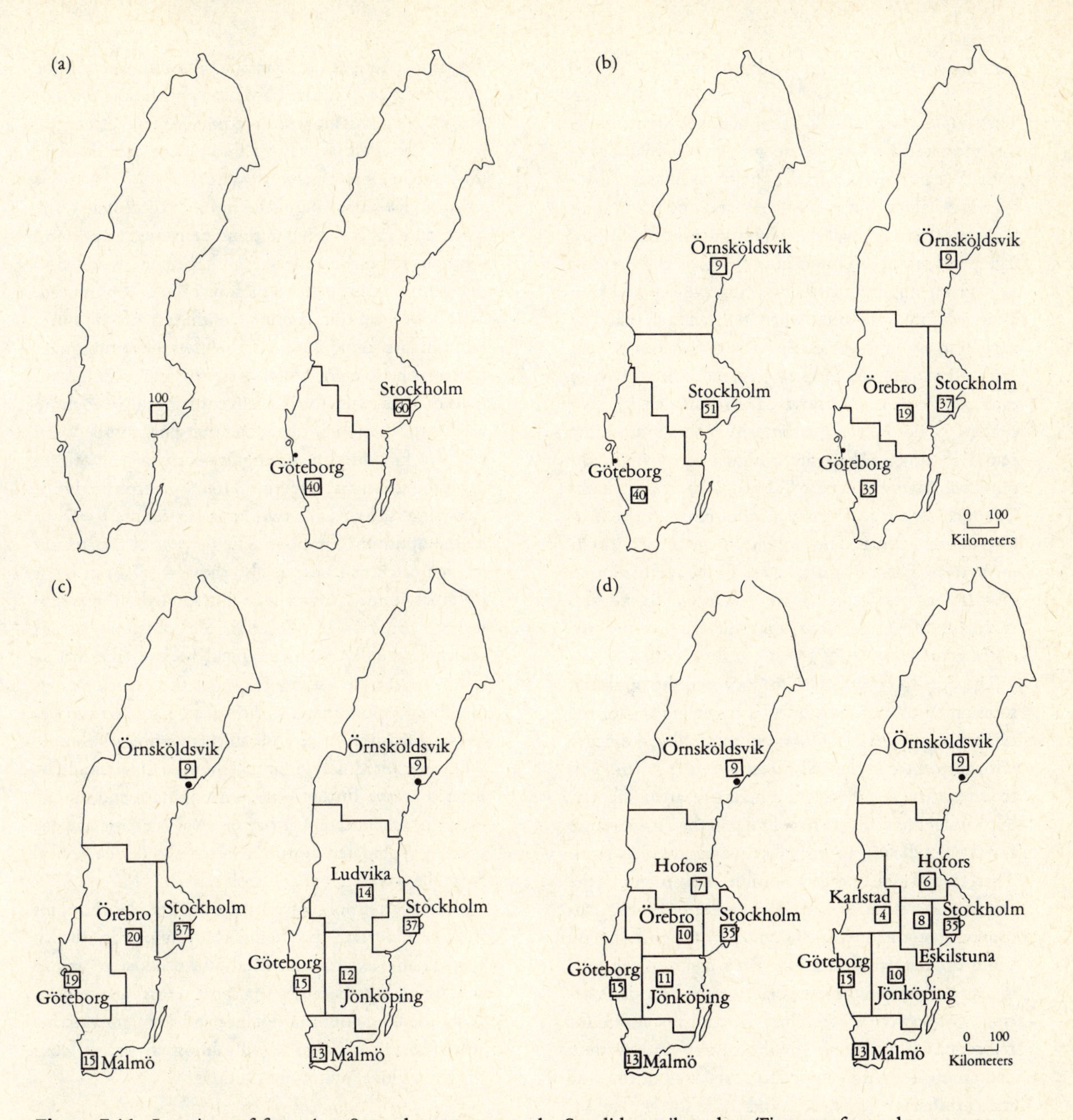

Figure 7.16 Locations of from 1 to 8 warehouses to serve the Swedish retail market. (Figures refer to the percentages of the market served by a center.)
From G. Tornquist *et al.* Multiple location analysis. *Lund Studies in Geography*, Series C, No. 12, 1971.

cedures clearly produce the global optimum in a large proportion of runs and, in any case, many local optimum solutions are very efficient solutions not radically different from the global optimum. Thus, with heuristic algorithms we can solve complicated problems such as for multiple locations.

The Contiguous Searching Algorithm

The simplest location-allocation heuristic algorithm is one produced by Swedish geographer Gunnar Tornquist. We must not let its simplicity deceive us, however. This algorithm is a very general one for finding locations for facilities at different numbers of locations. The problem that the algorithm solves is the location of a given number of facilities to serve a population arranged into grid cells when the facilities may be located in *any* of the grid cells. The population of each grid cell and the distances or transport costs between each pair of cells are known. Let's illustrate for two centers. Two cells are arbitrarily chosen as initial center locations. These are assumed to be optimum until we have evidence to disprove the assumption. The next step is to allocate all the remaining cells to their nearest center. This provides us with two catchment areas. The total distance for all the population to visit the nearest center is now calculated, and this represents the measure of spatial efficiency for the two initial centers.

The question is whether this is indeed the optimum solution. We can begin to test it by trying one of the centers in each of its four contiguous cells. We begin with the center in the cell directly to the north. This requires that we revise our catchment areas slightly, but having done this we may find that the total distance for the population to visit the two centers has risen. Therefore we rejecte this possible movement of the center and return to our initial solution. This procedure is repeated with the center in the cell directly south of its original location. Once again, if the total distance measure is greater than for the original location, we reject the move. This procedure is repeated for relocation of the center in the eastern and western contiguous cells. If in any of these relocations the total travel distance measure is found to be lower than for the original solution, we have found a better cell in which to locate our center. We turn to the other center and repeat the exercise, trying the center in the contiguous north, south, east, and west cells in turn. This whole procedure of trying each center in turn in contiguous cells is repeated until no further improvements on their locations can be made. We define this pair of locations as our heuristic optimum solution for two centers serving this particular population. (Notice that if we use this alogrithm to find *one* center, then we have an alternative way of finding the areal median discussed in Chapter One.)

Tornquist's next step is to add a third center to the space and repeat the iterations to provide optimum solutions for three centers. Another center is added so that an optimum pattern of four centers is produced. This sequence can continue until sets of optimum locations are found. Figure 7.16 shows the optimum set of from one to eight centers to serve the Swedish retail market. Retail sales for 182 cells covering all Sweden are used as the consumer totals, and transport cost between cells are known. The contiguity-searching algorithm gives the solutions in Figure 7.16. Not surprisingly for two centers, Sweden's two largest cities are found to be the optimum locations. The percent of the market in each catchment area is also shown—60/40 in favor of Stockholm. This can be interpreted in practical terms as the two best locations for warehouses from which to supply Sweden's retail market, the Stockholm warehouse having 1.5 times the capacity of its Goteborg counterpart. If, however, a supplier wishes to cut down transport costs further, more warehouses will be required. For eight warehouses, these should be located at five further centers in southern and middle Sweden, where the majority of retail sales are, and the sparsely populated northern areas should be served from the one center near Ornskoldsvik.

We have a simple converging heuristic algorithm for multiple location problems. It cannot be shown mathematically that it will always produce optimum solutions, but because transport costs are usually monotonic functions of distance and the algorithm is a spatial search process, it will almost inevitably converge on global optimum solutions.

The Center-Region Iterative Algorithm

One limitation of the contiguity-searching algorithm is that it produces unconstrained capacity solutions. Like elementary Thiessen polygon methods, the capacities of the facilities are not known until we produce the final solution. In this case, the limitation is not normally as severe as for the Thiessen polygon method, because we are finding new locations for facilities, and fixed plant capacities therefore do not exist. If we are planning a set of new warehouses, we can build them to the particular size required to serve the catchment area, but there are examples of multiple location problems in which some capacity constraints are necessary. In planning new administrative areas, we usually define threshold populations for minimum and maximum sizes. Administrative districts below the threshold are thought not to be able properly and efficiently to serve their areas with services; districts above the threshold are often thought of as too remote from the people they serve. The result is that administrative districts throughout the world are areally small in densely populated regions and much larger in sparsely populated parts of a country, producing a tendency toward equal population sizes. The center-region iterative algorithm has been developed for designing administrative districts and includes a capacity constraint.

From our previous discussions, we know that we can define single centers and functional regions efficiently. Given a specified region, we can derive iteratively the point of minimum aggregate travel (the areal median), as described in Chapter One, for example; this is by definition spatially the most efficient center for a region. From our previous discussion, we know too that given a set of centers we can allocate population units to the centers in the spatially most efficient manner by using the transportation problem solution. The center-region iterative procedure is a very neat combination of these two techniques and provides a multiple location solution. It is the result of a rare combination between mathematician and geographer, M.F. Goodchild and Bryan Massam.

As its name suggests the procedure iterates between defining centers for regions and then regions about centers, back to centers, and so on. It begins by specifying a set of n centers. These are interpreted as deficit regions; population areal units are surplus areas; and linear programming is used to define optimum catchment areas about the centers. If we want all catchment areas to have the same population, we set the deficit of each of the n centers to equal n/total population. We reconsider the centers. Points of minimum aggregate travel are found for each region. Because these areal medians are spatially the most efficient centers for each region, they define a new set of centers. We use these as deficit regions and run the transportation problem again, to produce new catchment areas. Points of minimum aggregate travel are produced for the new catchment areas, and the iterations from center to region and back again continue until the resulting small changes in center locations are considered unimportant. It is hoped that the algorithm has converged at this stage on what is the global optimum solution for n centers.

Goodchild and Massam applied this algorithm to the problem of designing the administrative map of southern Ontario. Many government services in southern Ontario are administered from eight centers, so eight was chosen as the number of centers in this application. The eight largest towns in southern Ontario were chosen initially as equal centers, and catchment areas were assigned to them from the 504 townships in the area, using linear programming. Figure 7.17a shows the original centers, optimum catchment areas, and new centers, the areal medians, of the catchment areas. Almost invariably, the centers are shifted toward the population concentration around Toronto. The final solution (Figure 7.17b) shows a pattern of equally populated administrative districts around eight proposed administrative centers. Notice that, even though the transportation problem always gives an optimum solution (we assume the areal medians are optimum), this combination of both optimizing approaches does not necessarily produce a global optimum for the multiple location problem with constrained capacities. It may well be that, if we start at a different initial set of eight locations, a different final solution will occur. We could, in fact, run this technique from different starting points to find what differences, if any, emerge. Nonetheless, there is no doubt that, with either a local or a global optimum, our heuristic device has given us a highly efficient administrative map in spatial terms.

Because the algorithm uses the transportation problem, it is just as flexible as this mathematical procedure in varying capacities. In hospital location for

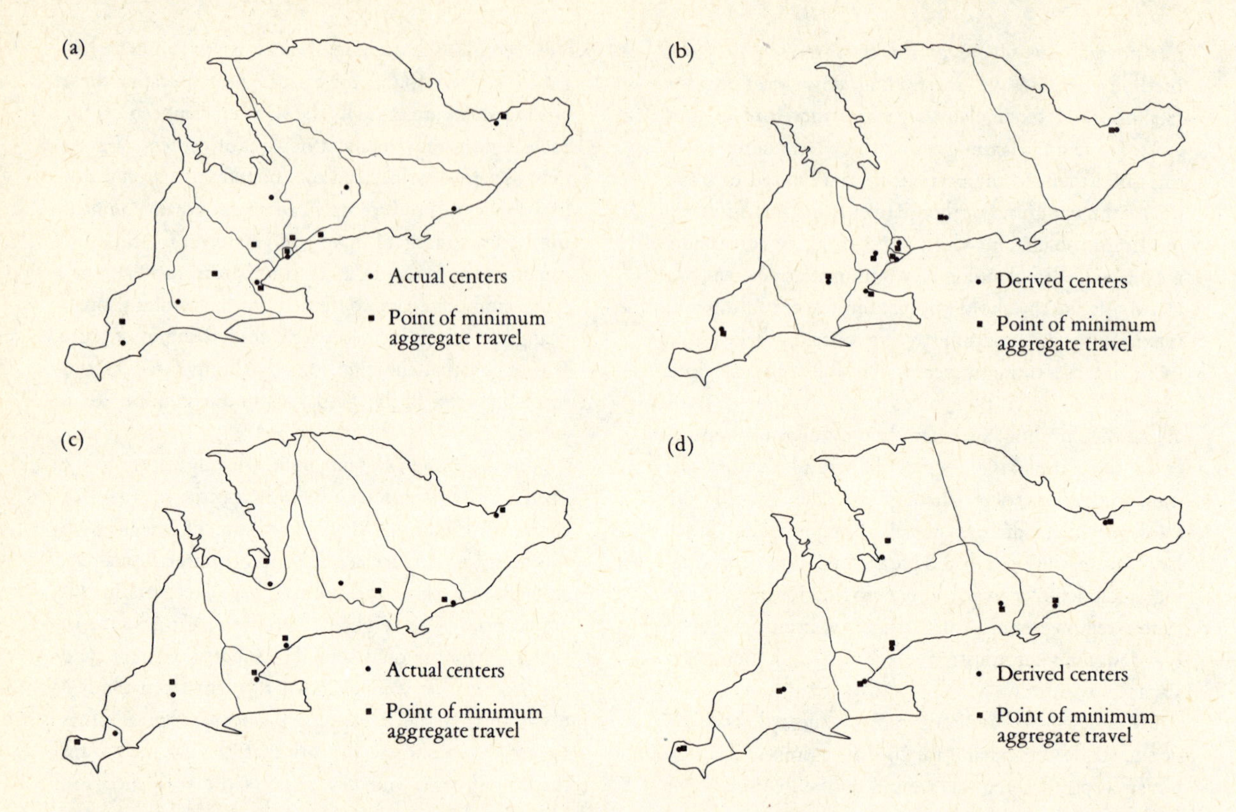

Figure 7.17 Least-cost administrative districts for southern Ontario. a. Initial least-cost solution: equal population constraints. b. Least-cost solution after final iteration: equal population constraints. c. Initial leastcost solution: scaled population constraint. d. Least-cost solution after final iteration: scaled population constraints.
From M.F. Goodchild and B.H. Massam. Some least cost models of spatial administration systems in southern Ontario. *Geografiska Annaler*, Series B, 1969.

western Guatemala, different catchment areas were produced by changing the hospital capacities. This sort of procedure might seem desirable for administrative districts in southern Ontario. In our equal capacity solution (Figure 7.17b), we can see that the Toronto Metropolitan District is divided into two parts. This division might be considered somewhat arbitrary, especially because the trend in administrative districts is to treat city regions as wholes. We can overcome this problem by allowing the capacity of the administrative center at Toronto to increase until it can cope with the whole city region. In fact, when the pattern of current administrative areas is looked at, it is found that Toronto administers on average an area comprising 50 percent of southern Ontario's population. Furthermore, the areas administered from London comprise, on the average, 15 percent of the total population of Ontario. These figures contrast with the 12.5 percent allocated to all eight centers in the initial problem. In fact, the average percentages for the first eight centers were found in practice to be 50, 15, 12, 12, 5, 2, 2, 2. These proportions were transformed into deficits and used in a second, more realistic run of the center-region algorithm. The first and last solutions are shown in Figures 7.17c and 7.17d, in which we find the whole central part of the area administered from Toronto.

How good are these solutions? In such research we

Table 7.13 Evaluation of Administrative District Patterns for Southern Ontario

Administrative district pattern	*Population served by nearest center (%)*	*Mean distance of population to center (mi)*
Equal population constraints		
Initial least-cost solution	59.8	39.76
Solution after first iteration	77.6	22.10
Final iteration solution	81.6	20.75
Scaled population constraints		
Initial least-cost solution	78.1	22.66
Solution after first iteration	85.4	20.64
Final iteration solution	87.1	19.82
Proximal solution about the initial eight centers	100.0	19.22
Actual administrative patterns		
Ontario Hydro Commission	54.6	52.51
Department of Education	49.2	39.68

Source: M.F. Goodchild and B.H. Massam. Some least cost models of spatial administration systems in southern Ontario. *Geografiska Annaler* 50:91, 1969.

must clearly go beyond the proposal of new arrangements and evaluate what we have produced against what is already being used. This evaluation is carried out in Table 7.13. Two criteria are used: the percentage of people served by the nearest center on the administrative map and the mean distance of the population from its administrative center. The first six assessments on Table 7.13 show the improving least-cost solutions for our two proposals, with equal cost constraints and scaled cost constraints, respectively. Because the latter solutions are sensitive to population distribution, we should not be surprised to find that the scaled population patterns do somewhat better on both criteria. Notice also how well the completely unconstrained proximal solution does on these criteria. Finally, we can compare the results against two existing administrative patterns for the Ontario Hydro Commission and the Education Department. For both criteria, the proposed solutions are far superior to both of the actual patterns. This table shows that quite literally the spatial inefficiency of the two administrative patterns means that people have to travel on average twice as far as they need to.

The center-region algorithm is clearly a useful and highly flexible tool. The pattern of capacity constraints that is used may be decided in terms of factors outside the model, for example. With the unconstrained contiguity-searching algorithm, geographers have succeeded in developing a complementary pair of heuristic procedures for tackling complicated multiple location problems. This practical achievement of modern computer-aided geographical research is an appropriate point on which to conclude this book.

Further Reading

The two major themes covered in this chapter are not normally considered together, and our readings consist of two basically separate lists. There is one exception. Because gravity models and allocation problems relate to flows, they have recently been brought together in transportation studies including "transport geography," for which there is an elementary textbook that presents both sets of ideas clearly and with substantive examples and discussion. Thus, for students who wish to ensure that their readings take them back into geography, Taaffe and Gauthier (1973) is recommended. Hay (1973) also presents these techniques in a substantive discussion of transport geography, and Chisholm and O'Sullivan (1973) present rare empirical comparisons in a case study of British road traffic flows.

Reilly's original work is generally available (1931), and his approach is widely reported. Strohkarck and Phelps (1948) give a very clear and concise description of Reilly's hinterland delimitation approach. Illeris (1967) developed a more sophisticated hinterland delimitation approach on similar lines and tested it empirically for Danish towns. More generally, Reilly's work has been incorporated into marketing geography and its role is developed and illustrated in introductory textbooks by Berry (1967) and Peter Scott (1969). These books also deal with the consumer probability approach, but for more complete information reference should be made to the original papers by Huff (1963) and Lakshmanan and Hansen (1965). Full details of the comparative study of these two approaches for the Haydock proposal can be found in the reports of the Department of Town and Country Planning, University of Manchester (1964 and 1966). This example and other similar studies are reviewed in Cole (1966), and further refinements are proposed by Lewis and Traill (1968). A critique of this literature has been presented recently by Bucklin (1971).

When we turn to gravity models in general, the literature we can cite expands enormously. Recognition of the necessity to consider constraints in these models can be traced to a basic transportation model by Wilson (1967). Many students find this and other papers by Wilson difficult to follow, but most of the basic ideas are presented in simple fashion by Masser (1972) which can be used to supplement our chapter and relate it to other modeling in planning. The use of the intervening opportunities model is introduced in one of the land use transportation studies (Chicago Area Transportation Study, 1959). This model is also presented separately by Schneider (1959) and Clark and Peters (1965).

Major supplementary readings on these models can take several routes. Parameter estimation is considered in some detail by Black and Larson (1972) and Hyman (1971). The major advances, however, have been in terms of applying entropy maximizing procedures. Much of this work has come from Wilson, and his early studies are brought together in Wilson (1970). A very clearly written introduction to Wilson's work is by Gould (1972), and applications of these approaches can be found in Batty (1970a and 1970b) and Wilson *et al.* (1969). The macro-micro-model debate was aired in a conference in London in 1971, and the discusion is presented briefly by Chisholm (1972); it gives some idea of the flavor of the controversy.

The literature on location-allocation problems is equally voluminous and also partly outside geography. We concentrate on geographical applications, and further inroads into this literature can be made with these references. The simple proximal solution is described by Haggett (1965) and was used earlier to define U.S. metropolitan regions by Bogue (1949). The map overlay approach was devised by Godlund (1961); and we presented his Swedish hospital application along with the Zambia flying doctor proposals of Jackman (1972). These simple location-allocation methods have not been widely applied and have been superseded by other, more sophisticated approaches, but they remain a good starting point for learning purposes.

We indicated that linear programming is a widely used problem-solving procedure beyond the particularly spatial transportation problem. Two textbooks can be recommended to introduce the full range of this approach, one by the technique's originator Dantzig (1963) and the other an economics introduction by Dorfman *et al.* (1958). Most students will want to begin further reading on this topic in geography, and elementary introductions for geographers are presented by Cox (1965) and Scott (1971a). Scott complements and extends our discussion. A clear and concise general introduction to linear programming is presented in Open University (1972). The casting of location-allocation problems as linear programs is illustrated by Yeates (1963) and by Gould and Leinbach (1966) and Maxfield (1972). All three applications are discussed in this chapter. Other applications of the transportation problem in geography can be found in Morrill and Garrison (1960) and Cassetti (1966).

The heuristic algorithms we presented have been described in Abler, Adams and Gould (1971). Tornquist's contiguous-searching algorithm was first presented in Swedish early in the 1960s and was not widely known until made more public in Abler, Adams, and Gould. Tornquist's work is available in English in Tornquist *et al.* (1971), and this volume includes tests of the method in various contexts outside Sweden. Robertson (1974) presents an application to the 1971 Scottish population. The center-region iterative algorithm is by Goodchild and Massam (1969), with an elementary introduction by Massam (1972). A further application of this approach is presented by Hirst (1973) in relation to Uganda administrative centers. This method was first proposed as essentially a warehouse location problem in operations research and was applied to political districting by Weaver and Hess (1963). The algorithm was used to produce compact, approximately equal, population electoral districts. This approach is illustrated and discussed in a British context by Mills (1967).

Supplementary reading on these topics in the geographical literature can be carried out by using Scott (1970 and 1971b) and Rushton *et al.* (1973). Scott's work is an overview of much of the operations research literature as it pertains to geography and planning applications. Rushton *et al.* (1973) is a set of papers from a conference on location-allocation problems in Iowa and includes discussion of a wide range of algorithms along with computer program listings.

Work Table 7.1 The Production-Attraction Constrained Gravity Model

Purpose

To calibrate A_i and B_j scaling factors in the equation

$$T_{ij} = A_i B_j \frac{O_i D_j}{d_{ij}^b}$$

so that $\Sigma\Sigma T_{ij} = \Sigma O_i = \Sigma D_j$

Data

Two zones E and W:

$O_e = 200 \quad O_w = 800$

$D_e = 900 \quad D_w = 100$

also: $d_{ew} = 1$ and $d_{ee} = d_{ww} = 0.5$; $b = 2$

Iterations

1. Set $A_e = 1$ and $A_w = 1$, which produce

2. $B_e = \dfrac{1}{1(200/0.5^2) + 1(800/1^2)} = \dfrac{1}{1{,}600} = 0.000625$

and

$B_w = \dfrac{1}{1(200/1^2) + 1 \cdot (800/0.5^2)} = \dfrac{1}{3{,}400} = 0.000294$

which produce

3. $A_e = \dfrac{1}{2.2794} = 0.438712$

and

$A_w = \dfrac{1}{0.680100} = 1.47037$

which produce

4. $B_e = 0.000655$ and $B_w = 0.000209$, which produce

5. $A_e = 0.420036$ and $A_w = 1.485663$, which produce

6. $B_e = 0.000656$ and $B_w = 0.000207$, which produce

7. $A_e = 0.419762$ and $A_w = 1.485443$, which produce

8. $B_e = 0.000656$ (as in step 6) and $B_w = 0.000207$ (as in step 6)

9. $A_e = 0.419762$ (as in step 7) and $A_w = 1.485443$ (as in step 7)

At an accuracy of 6 decimal places, the iteration converges at steps 8 and 9 with $A_e = 0.419762$, $A_w = 1.485443$, $B_e = 0.000656$, and $B_w = 0.000207$.

Check

$$T_{ew} = A_e B_w \frac{O_e D_w}{d_{ew}^2}$$

$$= \frac{0.419762 \cdot 0.000207 \cdot 200 \cdot 100}{1^2}$$

$$= 1.73782$$

Similarly,

$$T_{ee} = 198.06208$$

$$T_{we} = 701.60472$$

$$T_{ww} = 98.39384$$

Thus,

$$T_{we} + T_{ww} = 800 = O_w$$

$$T_{ee} + T_{we} = 900 = D_e$$

$$T_{ew} + T_{ww} = 100 = D_w$$

$$T_{ew} + T_{ee} = 200 = O_e$$

which satisfy production and attraction constraints.

Exercise

This work table is meant to be purely illustrative. Any actual application of these techniques would involve use of computer programs and would be beyond the scope of exercises in this book.

Table A.1 Random Numbers

Each one of the following digits is an independent sample from a population in which digits 0 to 9 are equally likely; that is, each has a probability of occurring of 0.1.

67 28	96 25	68 36	24 72	03 85	49 24	05 69	64 86	08 19	91 21
85 86	94 78	32 59	51 82	86 43	73 84	45 60	89 57	06 87	08 15
40 10	60 09	05 88	78 44	63 13	58 25	37 11	18 47	75 62	52 21
94 55	89 48	90 80	77 80	26 89	87 44	23 74	66 20	20 19	26 52
11 63	77 77	23 20	33 62	62 19	29 03	94 15	56 37	14 09	47 16
64 00	26 04	54 55	38 57	94 62	68 40	26 04	24 25	03 61	01 20
50 94	13 23	78 41	60 58	10 60	88 46	30 21	45 98	70 96	36 89
66 98	37 96	44 13	45 05	34 59	75 85	48 97	27 19	17 85	48 51
66 91	42 83	60 77	90 91	60 90	79 62	57 66	72 28	08 70	96 03
33 58	12 18	02 07	19 40	21 29	39 45	90 42	58 84	85 43	95 67
52 49	40 16	72 40	73 05	50 90	02 04	98 24	05 30	27 25	20 88
74 98	93 99	78 30	79 47	96 92	45 58	40 37	89 76	84 41	74 68
50 26	54 30	01 88	69 57	54 45	69 88	23 21	05 69	93 44	05 32
49 46	61 89	33 79	96 84	28 34	19 35	28 73	39 59	56 34	97 07
19 65	13 44	78 39	73 88	62 03	36 00	25 96	86 76	67 90	21 68
64 17	47 67	87 59	81 40	72 61	14 00	28 28	55 86	23 38	16 15
18 43	97 37	68 97	56 56	57 95	01 88	11 89	48 07	42 60	11 92
65 58	60 87	51 09	96 61	15 53	66 81	66 88	44 75	37 01	28 88
79 90	31 00	91 14	85 65	31 75	43 15	45 93	64 78	34 53	88 02
07 23	00 15	59 05	16 09	94 42	20 40	63 76	65 67	34 11	94 10
90 08	14 24	01 51	95 46	30 32	33 19	00 14	19 28	40 51	92 69
53 82	62 02	21 82	34 13	41 03	12 85	65 30	00 97	56 30	15 48
98 17	26 15	04 50	76 25	20 33	54 84	39 31	23 33	59 64	96 27
08 91	12 44	82 40	30 62	45 50	64 54	65 17	89 25	59 44	99 95
37 21	46 77	84 87	67 39	85 54	97 37	33 41	11 74	90 50	29 62

Source: Reproduced from M. G. Kendall and B. Babington-Smith. *Tables of Random Sampling Numbers.* Tracts for Computers, No. 24, Department of Statistics, University College, London, 1949.

Table A.2 The Normal Distribution

Ordinates of the Normal Distribution

Heights (y) of the normal distribution are given for values of z where z has a mean value of 0 and a standard deviation of 1.

z	y	z	y
0.0	0.39894	2.5	0.01753
0.1	0.39695	2.6	0.01358
0.2	0.39104	2.7	0.01042
0.3	0.38139	2.8	0.00792
0.4	0.36827	2.9	0.00595
0.5	0.35207	3.0	0.00443
0.6	0.33322	3.1	0.00327
0.7	0.31225	3.2	0.00238
0.8	0.28969	3.3	0.00172
0.9	0.26609	3.4	0.00123
1.0	0.24197	3.5	0.00087
1.1	0.21785	3.6	0.00061
1.2	0.19419	3.7	0.00042
1.3	0.17137	3.8	0.00029
1.4	0.14973	3.9	0.00020
1.5	0.12952	4.0	0.00013
1.6	0.11092	4.1	0.00009
1.7	0.09405	4.2	0.00006
1.8	0.07895	4.3	0.00004
1.9	0.06562	4.4	0.00002
2.0	0.05399	4.5	0.00002
2.1	0.04398	4.6	0.00001
2.2	0.03547	4.7	0.00001
2.3	0.02833	4.8	0.00000
2.4	0.02239		

Probabilities for z Tests

Probabilities are given for values as extreme as observed values of z in the normal distribution.

z	.00	.01	.02	.03	.04	.05	.06	.07	.08	.09
.0	.5000	.4960	.4920	.4880	.4840	.4801	.4761	.4721	.4681	.4641
.1	.4602	.4562	.4522	.4483	.4443	.4404	.4364	.4325	.4286	.4247
.2	.4207	.4168	.4129	.4090	.4052	.4013	.3974	.3936	.3897	.3859
.3	.3821	.3783	.3745	.3707	.3669	.3632	.3594	.3557	.3520	.3483
.4	.3446	.3409	.3372	.3336	.3300	.3264	.3228	.3192	.3156	.3121
.5	.3035	.3050	.3015	.2981	.2946	.2912	.2877	.2843	.2810	.2776
.6	.2743	.2709	.2676	.2643	.2611	.2578	.2546	.2514	.2483	.2451
.7	.2420	.2389	.2358	.2327	.2296	.2266	.2236	.2206	.2177	.2148
.8	.2119	.2090	.2061	.2033	.2005	.1977	.1949	.1922	.1894	.1867
.9	.1841	.1814	.1788	.1762	.1736	.1711	.1685	.1660	.1635	.1611
1.0	.1587	.1562	.1539	.1515	.1492	.1469	.1446	.1423	.1401	.1379
1.1	.1357	.1335	.1314	.1292	.1271	.1251	.1230	.1210	.1190	.1170
1.2	.1151	.1131	.1112	.1093	.1075	.1056	.1038	.1020	.1003	.0985
1.3	.0968	.0951	.0934	.0918	.0901	.0885	.0869	.0853	.0838	.0823
1.4	.0808	.0793	.0778	.0764	.0749	.0735	.0721	.0708	.0694	.0681
1.5	.0668	.0655	.0643	.0630	.0618	.0606	.0594	.0582	.0571	.0559
1.6	.0548	.0537	.0526	.0516	.0505	.0495	.0485	.0475	.0465	.0455
1.7	.0446	.0436	.0427	.0418	.0409	.0401	.0392	.0384	.0375	.0367
1.8	.0359	.0351	.0344	.0336	.0329	.0322	.0314	.0307	.0301	.0294
1.9	.0287	.0281	.0274	.0268	.0262	.0256	.0250	.0244	.0239	.0233
2.0	.0228	.0222	.0217	.0212	.0207	.0202	.0197	.0192	.0188	.0183
2.1	.0179	.0174	.0170	.0166	.0162	.0158	.0154	.0150	.0146	.0143
2.2	.0139	.0136	.0132	.0129	.0125	.0122	.0119	.0116	.0113	.0110
2.3	.0107	.0104	.0102	.0099	.0096	.0094	.0091	.0089	.0087	.0084
2.4	.0082	.0080	.0078	.0075	.0073	.0071	.0069	.0068	.0066	.0064
2.5	.0062	.0060	.0059	.0057	.0055	.0054	.0052	.0051	.0049	.0048
2.6	.0047	.0045	.0044	.0043	.0041	.0040	.0039	.0038	.0037	.0036
2.7	.0035	.0034	.0033	.0032	.0031	.0030	.0029	.0028	.0027	.0026
2.8	.0026	.0025	.0024	.0023	.0023	.0022	.0021	.0021	.0020	.0019
2.9	.0019	.0018	.0018	.0017	.0016	.0016	.0015	.0015	.0014	.0014
3.0	.0013	.0013	.0013	.0012	.0012	.0011	.0011	.0011	.0010	.0010
3.1	.0010	.0009	.0009	.0009	.0008	.0008	.0008	.0008	.0007	.0007
3.2	.0007									
3.3	.0005									
3.4	.0003									
3.5	.00023									
3.6	.00016									
3.7	.00011									
3.8	.00007									
3.9	.00005									
4.0	.00003									

Source: From *Tables of the Individual and Cumulative Terms of Poisson Distribution* by General Electric Company, © 1962. Reprinted by permission of Van Nostrand Reinhold Company.

Table A.3 Critical Values of D in the Kolmogorov-Smirnov One-Sample Test

For the Completely Specified Case

Computed values of D must exceed the following values in order to reject H_0 (The population values against which the sample is compared should not be based on properties of the sample.)

Sample size (N)	$\alpha =$ .20	.15	.10	.05	.01
1	.900	.925	.950	.975	.995
2	.684	.726	.776	.842	.929
3	.565	.597	.642	.708	.828
4	.494	.525	.564	.624	.733
5	.446	.474	.510	.565	.669
6	.410	.436	.470	.521	.618
7	.381	.405	.438	.486	.577
8	.358	.381	.411	.457	.543
9	.339	.360	.388	.432	.514
10	.322	.342	.368	.410	.490
11	.307	.326	.352	.391	.468
12	.295	.313	.338	.375	.450
13	.284	.302	.325	.361	.433
14	.274	.292	.314	.349	.418
15	.266	.283	.304	.338	.404
16	.258	.274	.295	.328	.392
17	.250	.266	.286	.318	.381
18	.244	.259	.278	.309	.371
19	.237	.252	.272	.301	.363
20	.231	.246	.264	.294	.356
25	.21	.22	.24	.27	.32
30	.19	.20	.22	.24	.29
35	.18	.19	.21	.23	.27
Over 35	$\frac{1.07}{\sqrt{N}}$	$\frac{1.14}{\sqrt{N}}$	$\frac{1.22}{\sqrt{N}}$	$\frac{1.36}{\sqrt{N}}$	$\frac{1.63}{\sqrt{N}}$

Source: Derived from F. J. Massey, Jr. The Kolmogorov-Smirnov test for goodness of fit. *Journal of the American Statistical Association* 46:70, 1951.

For Testing Against a Normal Distribution with $\bar{x}$ and σ Estimated from the Sample

Computed values of D must exceed the following values in order to reject H_0.

Sample size (N)	$\alpha =$ .20	.15	.10	.05	.01
4	.300	.319	.352	.381	.417
5	.285	.299	.315	.337	.405
6	.265	.277	.294	.319	.364
7	.247	.258	.276	.300	.348
8	.233	.244	.261	.285	.331
9	.223	.233	.249	.271	.311
10	.215	.224	.239	.258	.294
11	.206	.217	.230	.249	.284
12	.199	.212	.223	.242	.275
13	.190	.202	.214	.234	.268
14	.183	.194	.207	.227	.261
15	.177	.187	.201	.220	.257
16	.173	.182	.195	.213	.250
17	.169	.177	.189	.206	.245
18	.166	.173	.184	.200	.239
19	.163	.169	.179	.195	.235
20	.160	.166	.174	.190	.231
25	.149	.153	.165	.180	.203
30	.131	.136	.144	.161	.187
Over 30	$\frac{.736}{\sqrt{N}}$	$\frac{.768}{\sqrt{N}}$	$\frac{.805}{\sqrt{N}}$	$\frac{.886}{\sqrt{N}}$	$\frac{1.031}{\sqrt{N}}$

Source: Reproduced from H. W. Lilliefors. On the Kolmogorov-Smirnov test for normality with mean and variance unknown. *Journal of the American Statistical Association* 62: 400, 1967.

Table A.4 Critical Values of χ^2

Computed values of χ^2 must exceed the following values in order to reject H_0.

	$\alpha =$													
df	*.99*	*.98*	*.95*	*.90*	*.80*	*.70*	*.50*	*.30*	*.20*	*.10*	*.05*	*.02*	*.01*	*.001*
1	.00016	.00063	.0039	.016	.064	.15	.46	1.07	1.64	2.71	3.84	5.41	6.64	10.83
2	.02	.04	.10	.21	.45	.71	1.39	2.41	3.22	4.60	5.99	7.82	9.21	13.82
3	.12	.18	.35	.58	1.00	1.42	2.37	3.66	4.64	6.25	7.82	9.84	11.34	16.27
4	.30	.43	.71	1.06	1.65	2.20	3.36	4.88	5.99	7.78	9.49	11.67	13.28	18.46
5	.55	.75	1.14	1.61	2.34	3.00	4.35	6.06	7.29	9.24	11.07	13.39	15.09	20.52
6	.87	1.13	1.64	2.20	3.07	3.83	5.35	7.23	8.56	10.64	12.59	15.03	16.81	22.46
7	1.24	1.56	2.17	2.83	3.82	4.67	6.35	8.38	9.80	12.02	14.07	16.62	18.48	24.32
8	1.65	2.03	2.73	3.49	4.59	5.53	7.34	9.52	11.03	13.36	15.51	18.17	20.09	26.12
9	2.09	2.53	3.32	4.17	5.38	6.39	8.34	10.66	12.24	14.68	16.92	19.68	21.67	27.88
10	2.56	3.06	3.94	4.86	6.18	7.27	9.34	11.78	13.44	15.99	18.31	21.16	23.21	29.59
11	3.05	3.61	4.58	5.58	6.99	8.15	10.34	12.90	14.63	17.28	19.68	22.62	24.72	31.26
12	3.57	4.18	5.23	6.30	7.81	9.03	11.34	14.01	15.81	18.55	21.03	24.05	26.22	32.91
13	4.11	4.76	5.89	7.04	8.63	9.93	12.34	15.12	16.98	19.81	22.36	25.47	27.69	34.53
14	4.66	5.37	6.57	7.79	9.47	10.82	13.34	16.22	18.15	21.06	23.68	26.87	29.14	36.12
15	5.23	5.98	7.26	8.55	10.31	11.72	14.34	17.32	19.31	22.31	25.00	28.26	30.58	37.70
16	5.81	6.61	7.96	9.31	11.15	12.62	15.34	18.42	20.46	23.54	26.30	29.63	32.00	39.29
17	6.41	7.26	8.67	10.08	12.00	13.53	16.34	19.51	21.62	24.77	27.59	31.00	33.41	40.75
18	7.02	7.91	9.39	10.86	12.86	14.44	17.34	20.60	22.76	25.99	28.87	32.35	34.80	42.31
19	7.63	8.57	10.12	11.65	13.72	15.35	18.34	21.69	23.90	27.20	30.14	33.69	36.19	43.82
20	8.26	9.24	10.85	12.44	14.58	16.27	19.34	22.78	25.04	28.41	31.41	35.02	37.57	45.32
21	8.90	9.92	11.59	13.24	15.44	17.18	20.34	23.86	26.17	29.62	32.67	36.34	38.93	46.80
22	9.54	10.60	12.34	14.04	16.31	18.10	21.24	24.94	27.30	30.81	33.92	37.66	40.29	48.27
23	10.20	11.29	13.09	14.85	17.19	19.02	22.34	26.02	28.43	32.01	35.17	38.97	41.64	49.73
24	10.86	11.99	13.85	15.66	18.06	19.94	23.34	27.10	29.55	33.20	36.42	40.27	42.98	51.18
25	11.52	12.70	14.61	16.47	18.94	20.87	24.34	28.17	30.68	34.38	37.65	41.57	44.31	52.62
26	12.20	13.41	15.38	17.29	19.82	21.79	25.34	29.25	31.80	35.56	38.88	42.86	45.64	54.05
27	12.88	14.12	16.15	18.11	20.70	22.72	26.34	30.32	32.91	36.74	40.11	44.14	46.96	55.48
28	13.56	14.85	16.93	18.94	21.59	23.65	27.34	31.39	34.03	37.92	41.34	45.42	48.28	56.89
29	14.26	15.57	17.71	19.77	22.48	24.58	28.34	32.46	35.14	39.09	42.56	46.69	49.59	58.30
30	14.95	16.31	18.49	20.60	23.36	25.51	29.34	33.53	36.25	40.26	43.77	47.96	50.89	59.70

Source: Derived from R. A. Fisher and F. Yates. *Statistical Tables for Biological, Agricultural and Medical Research.* By permission of Longman Group Limited, London.

Table A.5 Critical Values of D in the Kolmogorov-Smirnov Two-Sample Case

For Small Samples

Where the two sample sizes are *equal and less than 40*, the computed value of D must exceed the following values in order to reject H_0.

	One-tailed test		*Two-tailed test*	
N	$\alpha = 0.05$	$\alpha = 0.01$	$\alpha = 0.05$	$\alpha = 0.01$
3	1.0000	—	—	—
4	1.0000	—	1.0000	—
5	0.8000	1.0000	1.0000	1.0000
6	0.8333	1.0000	0.8333	1.0000
7	0.7143	0.8571	0.8571	0.8571
8	0.6250	0.7500	0.7500	0.8750
9	0.6667	0.7778	0.6667	0.7778
10	0.6000	0.7000	0.7000	0.8000
11	0.5455	0.7273	0.6364	0.7273
12	0.5000	0.6667	0.5833	0.6667
13	0.5385	0.6154	0.5385	0.6923
14	0.5000	0.5714	0.5714	0.6429
15	0.4667	0.6000	0.5333	0.6000
16	0.4375	0.5625	0.5000	0.6250
17	0.4706	0.5294	0.4706	0.5882
18	0.4444	0.5556	0.5000	0.5556
19	0.4211	0.5263	0.4737	0.5263
20	0.4000	0.5000	0.4500	0.5500
21	0.3810	0.4762	0.4286	0.5238
22	0.4091	0.5000	0.4091	0.5000
23	0.3913	0.4783	0.4348	0.4783
24	0.3750	0.4583	0.4167	0.5000
25	0.3600	0.4400	0.4000	0.4800
26	0.3462	0.4231	0.3846	0.4615
27	0.3333	0.4444	0.3704	0.4444
28	0.3571	0.4286	0.3929	0.4643
29	0.3448	0.4138	0.3793	0.4483
30	0.3333	0.4000	0.3667	0.4333
35	0.3143	0.3714	0.3429	—
40	0.2750	0.3500	0.3250	—

Source: Derived from L. A. Goodman (1954). Kolmogorov-Smirnov tests for psychological research. *Psychological Bulletin* 51:167, and F. J. Massey, Jr. (1951). The distribution of the maximum deviation between two sample cumulative step functions. *Annals of Mathematical Statistics* 22:126–127.

For Large Samples

Where sample sizes are larger than 40 (it is not necessary for sample sizes to be equal), the computed value of D must exceed critical values calculated as follows.

$\alpha =$	$D =$
.10	$1.22\sqrt{\frac{n_1 + n_2}{n_1 n_2}}$
.05	$1.36\sqrt{\frac{n_1 + n_2}{n_1 n_2}}$
.025	$1.48\sqrt{\frac{n_1 + n_2}{n_1 n_2}}$
.01	$1.63\sqrt{\frac{n_1 + n_2}{n_1 n_2}}$
.005	$1.73\sqrt{\frac{n_1 + n_2}{n_1 n_2}}$
.001	$1.95\sqrt{\frac{n_1 + n_2}{n_1 n_2}}$

Source: Derived from N. Smirnov (1948). Tables for estimating the goodness of fit of empirical distributions. *Annals of Mathematical Statistics* 19:280–281.

Table A.6 Critical Values of U in the Mann-Whitney Test

Where the sample sizes n_1 and n_2 ($n_1 \leq n_2$) are 20 or less, the computed value of U must be *less than* the following values in order to reject H_0. Unbracketed values are for one-tailed tests at $\alpha = 0.05$ or two-tailed tests at $\alpha = 0.10$, and bracketed values are for two-tailed tests at $\alpha = 0.05$ or one-tailed tests at $\alpha = 0.025$.

Where n_2 is larger than 20 a z test can be performed by computing

$$z_U = \frac{U - \frac{n_1 n_2}{2}}{\frac{n_1 n_2 (n_1 + n_2 + 1)}{12}}$$

and by using Table A.2.

n_1	$n_2 =$																	
	3	4	5	6	7	8	9	10	11	12	13	14	15	16	17	18	19	20
1	*	*	*	*	*	*	*	*	*	*	*	*	*	*	*	*	*	*
	(*)	(*)	(*)	(*)	(*)	(*)	(*)	(*)	(*)	(*)	(*)	(*)	(*)	(*)	(*)	(*)	(*)	(*)
2	*	*	0	0	0	1	1	1	1	2	2	2	3	3	3	4	4	4
	(*)	(*)	(*)	(*)	(*)	(0)	(0)	(0)	(0)	(1)	(1)	(1)	(1)	(1)	(2)	(2)	(2)	(2)
3	0	0	1	2	2	3	3	4	5	5	6	7	7	8	9	9	10	11
	(*)	(*)	(0)	(1)	(1)	(2)	(2)	(3)	(3)	(4)	(4)	(5)	(5)	(6)	(6)	(7)	(7)	(8)
4		1	2	3	4	5	6	7	8	9	10	11	12	14	15	16	17	18
		(*)	(0)	(2)	(3)	(4)	(4)	(5)	(6)	(7)	(8)	(9)	(10)	(11)	(12)	(12)	(13)	(13)
5			4	5	6	8	9	11	12	13	15	16	18	19	20	22	23	25
			(1)	(3)	(5)	(6)	(7)	(8)	(9)	(11)	(12)	(13)	(14)	(15)	(17)	(18)	(19)	(20)
6				7	8	10	12	14	16	17	19	21	23	25	26	28	30	32
				(5)	(6)	(8)	(10)	(11)	(13)	(14)	(16)	(17)	(19)	(21)	(22)	(24)	(25)	(27)
7					11	13	15	17	19	21	24	26	28	30	33	33	37	39
					(8)	(10)	(12)	(14)	(16)	(18)	(20)	(22)	(24)	(26)	(28)	(30)	(32)	(34)
8						15	18	20	23	26	28	31	33	36	39	41	44	47
						(13)	(15)	(17)	(19)	(22)	(24)	(26)	(29)	(31)	(34)	(36)	(38)	(41)
9							21	24	27	30	33	36	39	42	45	48	51	54
							(17)	(20)	(23)	(26)	(28)	(31)	(34)	(37)	(39)	(42)	(45)	(48)
10								27	31	34	37	41	44	48	51	55	58	62
								(23)	(26)	(29)	(33)	(36)	(39)	(42)	(45)	(48)	(52)	(55)
11									34	38	42	46	50	54	57	61	65	69
									(30)	(33)	(37)	(40)	(44)	(47)	(51)	(55)	(58)	(62)
12										42	47	51	55	60	64	68	72	77
										(37)	(41)	(45)	(49)	(53)	(57)	(61)	(65)	(69)
13											51	56	61	65	70	75	80	84
											(45)	(50)	(54)	(59)	(63)	(67)	(72)	(76)
14												61	66	71	77	82	87	92
												(55)	(59)	(64)	(67)	(74)	(78)	(83)
15													72	77	83	88	94	100
													(64)	(70)	(75)	(80)	(85)	(90)
16														83	89	95	101	107
														(75)	(81)	(86)	(92)	(98)
17															96	102	109	115
															(87)	(93)	(99)	(105)
18																109	116	123
																(99)	(106)	(112)
19																	123	130
																	(113)	(119)
20																		138
																		(127)

Source: Derived from H. B. Mann and D. R. Whitney. On a test of whether one of two random variables is stocastically larger than the other. *Annals of Mathematical Statistics* 18:52–54, 1947.

Table A.7 Critical Values of H in the Kruskal-Wallis Test

Where there are three samples with sizes of up to five items each, the computed value of H must equal or exceed the following values to reject H_0. When there are more than five items per sample, H is distributed as χ^2 with *df* one less than the number of samples, and Table A.4 should be used.

Sample sizes			*Significance levels*	
n_1	n_2	n_3	$\alpha = 0.05$	$\alpha = 0.01$
3	2	2	4.7143	—
3	3	1	5.1429	—
3	3	2	5.3611	—
3	3	3	5.6000	7.2000
4	2	2	5.3333	—
4	3	1	5.2083	—
4	3	2	5.4444	6.4444
4	3	3	5.7273	6.7455
4	4	1	4.9667	6.6667
4	4	2	5.4545	7.0364
4	4	3	5.5985	7.1439
4	4	4	5.6923	7.6538
5	2	1	5.0000	—
5	2	2	5.1600	6.5333
5	3	1	4.9600	—
5	3	2	5.2509	6.8218
5	3	3	5.6485	7.0788
5	4	1	4.9855	6.9545
5	4	2	5.2727	7.1182
5	4	3	5.6308	7.4449
5	4	4	5.6176	7.7604
5	5	1	5.1273	7.3091
5	5	2	5.3385	7.2692
5	5	3	5.7055	7.5429
5	5	4	5.6429	7.7914
5	5	5	5.7800	7.9800

Source: Derived from W. H. Kruskal and W. A. Wallis. Use of ranks in one-criterion variance analysis. *Journal of the American Statistical Association* 47:614–617, and "Errata," *Journal of the American Statistical Association* 48:910, 1952.

Table A.8 Critical Values of t in Student's t Tests

The computed value of t should exceed the following values in order to reject H_0.

	Level of significance for one-tailed test					
	.10	*.05*	*.025*	*.01*	*.005*	*.0005*
	Level of significance for two-tailed test					
df	*.20*	*.10*	*.05*	*.02*	*.01*	*.001*
1	3.078	6.314	12.706	31.821	63.657	636.619
2	1.886	2.920	4.303	6.965	9.925	31.598
3	1.638	2.353	3.182	4.541	5.841	12.941
4	1.533	2.132	2.776	3.747	4.604	8.610
5	1.476	2.015	2.571	3.365	4.032	6.859
6	1.440	1.943	2.447	3.143	3.707	5.959
7	1.415	1.895	2.365	2.998	3.499	5.405
8	1.397	1.860	2.306	2.896	3.355	5.041
9	1.383	1.833	2.262	2.821	3.250	4.781
10	1.372	1.812	2.228	2.764	3.169	4.587
11	1.363	1.796	2.201	2.718	3.106	4.437
12	1.356	1.782	2.179	2.681	3.055	4.318
13	1.350	1.771	2.160	2.650	3.012	4.221
14	1.345	1.761	2.145	2.624	2.977	4.140
15	1.341	1.753	2.131	2.602	2.947	4.073
16	1.337	1.746	2.120	2.583	2.921	4.015
17	1.333	1.740	2.110	2.567	2.898	3.965
18	1.330	1.734	2.101	2.552	2.878	3.922
19	1.328	1.729	2.093	2.539	2.861	3.883
20	1.325	1.725	2.086	2.528	2.845	3.850
21	1.323	1.721	2.080	2.518	2.831	3.819
22	1.321	1.717	2.074	2.508	2.819	3.792
23	1.319	1.714	2.069	2.500	2.807	3.767
24	1.318	1.711	2.064	2.492	2.797	3.745
25	1.316	1.708	2.060	2.485	2.787	3.725
26	1.315	1.706	2.056	2.479	2.779	3.707
27	1.314	1.703	2.052	2.473	2.771	3.690
28	1.313	1.701	2.048	2.467	2.763	3.674
29	1.311	1.699	2.045	2.462	2.756	3.659
30	1.310	1.697	2.042	2.457	2.750	3.646
40	1.303	1.684	2.021	2.423	2.704	3.551
60	1.296	1.671	2.000	2.390	2.660	3.460
120	1.289	1.658	1.980	2.358	2.617	3.373
∞	1.282	1.645	1.960	2.326	2.576	3.291

Source: Reproduced from S. Siegel *Nonparametric Statistics for the Behavioral Sciences.* New York: John Wiley, 1956. Derived from R. A. Fisher and F. Yates. *Statistical Tables for Biological, Agricultural and Medical Research.* By permission of Longman Group Limited, London.

Table A.9 The Poisson Distribution

The following distributions are selected for largely illustrative purposes to indicate variations in Poisson distributions with changes in the expectation, λ. The three distributions for each selected value of λ are defined as follows. $p(x)$: individual Poisson probabilities for values of x; $c_0(x)$: cumulated Poisson probabilities for values of x from $x = 0$; $c_\infty(x)$: cumulated Poisson probabilities for values of x from $x = \infty$. More detailed tables for large numbers of λ values can be found in the source given below.

x	$p(x)$	$c_0(x)$	$c_\infty(x)$
	$\lambda = 0.001$		
0	.99900050	.99900050	1.00000000
1	.00099900	.99999950	.00099950
2	.00000050	.99999999	.00000050
	$\lambda = 0.01$		
0	.99004983	.99004983	1.00000000
1	.00990050	.99995033	.00995017
2	.00004950	.99999983	.00004967
3	.00000016	.99999999	.00000017
	$\lambda = 0.1$		
0	.90483742	.90483742	1.00000000
1	.09048374	.99532116	.09516258
2	.00452419	.99984535	.00467884
3	.00015081	.99999615	.00015465
4	.00000377	.99999992	.00000385
5	.00000008	.99999999	.00000008
	$\lambda = 0.2$		
0	.81873076	.81873076	1.00000000
1	.16374615	.98247690	.18126925
2	.01637461	.99885152	.01752310
3	.00109164	.99994316	.00114848
4	.00005458	.99999774	.00005684
5	.00000218	.99999992	.00000226
6	.00000007	.99999999	.00000007
	$\lambda = 0.3$		
0	.74081822	.74081822	1.00000000
1	.22224547	.96306369	.25918178
2	.03333682	.99640051	.03693631
3	.00333368	.99973419	.00359949
4	.00025003	.99998421	.00026581
5	.00001500	.99999922	.00001578
6	.00000075	.99999996	.00000078
7	.00000003	.99999999	.00000003
	$\lambda = 0.4$		
0	.67032005	.67032005	1.00000000
1	.26812802	.93844806	.32967995
2	.05362560	.99207366	.06155194
3	.00715008	.99922375	.00792633
4	.00071501	.99993876	.00077625
5	.00005720	.99999595	.00006124
6	.00000381	.99999977	.00000404
7	.00000022	.99999999	.00000023

x	$p(x)$	$c_0(x)$	$c_\infty(x)$
	$\lambda = 0.5$		
0	.60653067	.60653067	1.00000000
1	.30326533	.90979599	.39346934
2	.07581633	.98561233	.09020401
3	.01263606	.99824838	.01438768
4	.00157951	.99982788	.00175162
5	.00015795	.99998583	.00017212
6	.00001316	.99999899	.00001416
7	.00000094	.99999993	.00000100
8	.00000006	.99999999	.00000006
	$\lambda = 0.6$		
0	.54881164	.54881164	1.00000000
1	.32928698	.87809862	.45118836
2	.09878609	.97688472	.12190138
3	.01975722	.99664193	.02311528
4	.00296358	.99960551	.00335806
5	.00035563	.99996115	.00039448
6	.00003556	.99999671	.00003885
7	.00000305	.99999975	.00000329
8	.00000023	.99999999	.00000024
	$\lambda = 0.7$		
0	.49658531	.49658531	1.00000000
1	.34760971	.84419502	.50341470
2	.12166340	.96585841	.15580498
3	.02838813	.99424654	.03414158
4	.00496792	.99921446	.00575345
5	.00069551	.99990997	.00078553
6	.00008114	.99999112	.00009002
7	.00000811	.99999923	.00000888
8	.00000071	.99999994	.00000076
9	.00000006	.99999999	.00000005
	$\lambda = 0.8$		
0	.44932897	.44932897	1.00000000
1	.35946317	.80879214	.55067103
2	.14378527	.95257740	.19120786
3	.03834274	.99092014	.04742259
4	.00766855	.99858869	.00907985
5	.00122697	.99981566	.00141131
6	.00016360	.99997925	.00018434
7	.00001870	.99999795	.00002074
8	.00000187	.99999982	.00000025
9	.00000017	.99999999	.00000018

x	$p(x)$	$c_0(x)$	$c_\infty(x)$
	$\lambda = 0.9$		
0	.40656966	.40656966	1.00000000
1	.36591269	.77248235	.59343034
2	.16466071	.93714307	.22751765
3	.04939821	.98654128	.06285693
4	.01111460	.99765588	.01345872
5	.00200063	.99965651	.00234412
6	.00030009	.99995660	.00034349
7	.00003858	.99999518	.00004340
8	.00000434	.99999952	.00000482
9	.00000043	.99999996	.00000048
10	.00000004	.99999999	.00000004
	$\lambda = 1$		
0	.36797944	.36787944	1.00000000
1	.36787944	.73575888	.63212056
2	.18393972	.91969860	.26424112
3	.06131324	.98101184	.08030140
4	.01532831	.99634015	.01898816
5	.00306566	.99940581	.00365985
6	.00051094	.99991675	.00059418
7	.00007299	.99998975	.00008324
8	.00000912	.99999887	.00001025
9	.00000101	.99999989	.00000113
10	.00000010	.99999999	.00000011
	$\lambda = 2$		
0	.13533528	.13533528	1.00000000
1	.27067056	.40600585	.86466472
2	.27067056	.67667641	.59399416
3	.18044704	.85712346	.32332359
4	.09022352	.94734698	.14287654
5	.03608941	.98343639	.05265302
6	.01202980	.99546619	.01656361
7	.00343709	.99890327	.00453381
8	.00085927	.99976255	.00109672
9	.00019095	.99995350	.00023745
10	.00003819	.99999169	.00004650
11	.00000694	.99999863	.00000831
12	.00000116	.99999979	.00000136
13	.00000018	.99999997	.00000021
14	.00000003	.99999999	.00000003
	$\lambda = 3$		
0	.04978707	.04978707	1.00000000
1	.14936120	.19914827	.95021293
2	.22404181	.42319008	.80085173
3	.22404181	.64723188	.57680992
4	.16803136	.81526324	.35276811
5	.10081881	.91608205	.18473676
6	.05040941	.96649146	.08391794
7	.02160403	.98809549	.03350854
8	.00810151	.99619701	.01190450
9	.00270050	.99889751	.00380299
10	.00081015	.99970766	.00110249
11	.00022095	.99992861	.00029234
12	.00005524	.99998385	.00007139
13	.00001275	.99999660	.00001615
14	.00000273	.99999933	.00000340
15	.00000055	.99999987	.00000067
16	.00000010	.99999998	.00000012
17	.00000002	.99999999	.00000002
	$\lambda = 4$		
0	.01831564	.01831654	1.00000000
1	.07326256	.09157819	.98168436
2	.14652511	.23810330	.90842181
3	.19536681	.43347012	.76189670
4	.19536681	.62883693	.56652988
5	.15629345	.78513038	.37116307
6	.10419563	.88932602	.21486961
7	.05954036	.94886638	.11067398
8	.02977018	.97863656	.05113362
9	.01323119	.99186775	.02136343
10	.00529248	.99716023	.00813224
11	.00192454	.99908477	.00283977
12	.00064151	.99972628	.00091523
13	.00019739	.99992367	.00027372
14	.00005640	.99998006	.00007633
15	.00001504	.99999510	.00001993
16	.00000376	.99999886	.00000489
17	.00000088	.99999975	.00000113
18	.00000020	.99999995	.00000025
19	.00000004	.99999999	.00000005
	$\lambda = 5$		
0	.00673795	.00673795	1.00000000
1	.03368973	.04042768	.99326206
2	.08422434	.12465202	.95957232
3	.14037389	.26502591	.87534799
4	.17546737	.44049328	.73497409
5	.17546737	.61596065	.55950672
6	.14622281	.76218346	.38403935
7	.10444486	.86662832	.23781654
8	.06527804	.93190636	.13337168
9	.03626558	.96817194	.06809364
10	.01813279	.98630473	.03182806

x	$p(x)$	$c_0(x)$	$c_\infty(x)$
	$\lambda = 5$		
11	.00824218	.99454691	.01369527
12	.00343424	.99798115	.00545309
13	.00132086	.99930201	.00201885
14	.00047174	.99977374	.00069799
15	.00015725	.99993099	.00022625
16	.00004914	.99998013	.00006901
17	.00001445	.99999458	.00001987
18	.00000401	.99999859	.00000542
19	.00000106	.99999965	.00000140
20	.00000026	.99999992	.00000035
21	.00000006	.99999998	.00000008
22	.00000001	.99999999	.00000002
	$\lambda = 6$		
0	.00247875	.00247875	1.00000000
1	.01487251	.01735127	.99752125
2	.04461754	.06196880	.98264874
3	.08923508	.15120388	.93803120
4	.13385262	.28505650	.84879612
5	.16062314	.45567964	.71494351
6	.16062314	.60630278	.55432037
7	.13767698	.74397976	.39369722
8	.10325773	.84723749	.25602024
9	.06883849	.91607598	.15276251
10	.04130309	.95737907	.08392402
11	.02252896	.97990803	.04262092
12	.01126448	.99117251	.02009196
13	.00519899	.99637150	.00882748
14	.00222814	.99859964	.00362849
15	.00089126	.99949089	.00140035
16	.00033422	.99982512	.00050910
17	.00011796	.99994308	.00017488
18	.00003932	.99998240	.00005692
19	.00001242	.99999481	.00001760
20	.00000373	.99999854	.00000518
21	.00000106	.99999961	.00000146
22	.00000029	.99999990	.00000039
23	.00000008	.99999997	.00000010
24	.00000002	.99999999	.00000002
	$\lambda = 7$		
0	.00091188	.00091188	1.00000000
1	.00638317	.00729506	.99908812
2	.02234111	.02963616	.99270495
3	.05212925	.08176542	.97036384
4	.09122619	.17299161	.91823459
5	.12771667	.30070828	.82700840
6	.14900278	.44971105	.69929173
7	.14900278	.59871383	.55028895
8	.13037743	.72909126	.40128617
9	.10140467	.83049593	.27090874
10	.07098327	.90147920	.16950406
11	.04517117	.94665037	.09852079
12	.02634985	.97300022	.05334962
13	.01418838	.98718860	.02699977
14	.00709419	.99428280	.01281139
15	.00331062	.99759342	.00571720
16	.00144840	.99904181	.00240658
17	.00059640	.99963821	.00095818
18	.00023193	.99987014	.00036178
19	.00008545	.99995559	.00012985
20	.00002991	.99998550	.00004440
21	.00000997	.99999547	.00001450
22	.00000317	.99999864	.00000453
23	.00000097	.99999961	.00000135
24	.00000028	.99999989	.00000039
25	.00000008	.99999997	.00000011
26	.00000002	.99999999	.00000003
	$\lambda = 8$		
0	.00033546	.00033546	1.00000000
1	.00268370	.00301916	.99966454
2	.01073480	.01375397	.99698084
3	.02862614	.04238011	.98624603
4	.05725229	.09963240	.95761989
5	.09160366	.19123606	.90036760
6	.12213822	.31337428	.80876394
7	.13958653	.45296081	.68662573
8	.13958653	.59254733	.54703920
9	.12407692	.71662425	.40745266
10	.09926153	.81588579	.28337574
11	.07219021	.88807599	.18411421
12	.04812680	.93620280	.11192400
13	.02961649	.96581929	.06379720
14	.01692371	.98274300	.03418070
15	.00902598	.99176899	.01725699
16	.00451299	.99628197	.00823101
17	.00212376	.99840573	.00371802
18	.00094389	.99934963	.00159426
19	.00039743	.99974706	.00065037
20	.00015897	.99990603	.00025294
21	.00006056	.99996659	.00009397
22	.00002202	.99998861	.00003341
23	.00000766	.99999627	.00001139
24	.00000255	.99999882	.00000373
25	.00000082	.99999964	.00000117
26	.00000025	.99999990	.00000036
27	.00000007	.99999997	.00000010
28	.00000002	.99999999	.00000003

x	$p(x)$	$c_0(x)$	$c_\infty(x)$
	$\lambda = 9$		
0	.00012341	.00012341	1.00000000
1	.00111069	.00123410	.99987660
2	.00499810	.00623220	.99876591
3	.01499429	.02122649	.99376781
4	.03373716	.05496364	.97877352
5	.06072688	.11569052	.94503636
6	.09109032	.20678084	.88430949
7	.11711612	.32389696	.79321916
8	.13175564	.45565260	.67610304
9	.13175564	.58740824	.54434740
10	.11858008	.70598832	.41259176
11	.09702006	.80300838	.29401168
12	.07276505	.87577342	.19699162
13	.05037580	.92614923	.12422657
14	.03238444	.95853367	.07385077
15	.01943067	.97796433	.04146633
16	.01092975	.98889409	.02203566
17	.00578634	.99468043	.01110591
18	.00289317	.99757359	.00531957
19	.00137045	.99894404	.00242640
20	.00061670	.99956074	.00105595
21	.00026430	.99982505	.00043925
22	.00010812	.99993317	.00017495
23	.00004231	.99997548	.00006683
24	.00001587	.99999134	.00002452
25	.00000571	.99999706	.00000865
26	.00000198	.99999903	.00000294
27	.00000066	.99999969	.00000096
28	.00000021	.99999990	.00000031
29	.00000007	.99999997	.00000009
30	.00000002	.99999999	.00000003

x	$p(x)$	$c_0(x)$	$c_\infty(x)$
	$\lambda = 10$		
0	.00004540	.00004540	1.00000000
1	.00045400	.00049940	.99995460
2	.00227000	.00276940	.99950060
3	.00756665	.01033605	.99723060
4	.01891664	.02925269	.98966395
5	.03783327	.06708596	.97074731
6	.06305546	.13014142	.93291404
7	.09007923	.22022065	.86985859
8	.11259903	.33281968	.77977936
9	.12511003	.45792971	.66718032
10	.12511003	.58303975	.54207029
11	.11373640	.69677614	.41696025
12	.09478033	.79155647	.30322386
13	.07290795	.86446442	.20844352
14	.05207710	.91654152	.13553558
15	.03471807	.95125959	.08345847
16	.02169879	.97295839	.04874040
17	.01276400	.98572239	.02704161
18	.00709111	.99281349	.01427761
19	.00373216	.99654566	.00718650
20	.00186608	.99841174	.00345434
21	.00088861	.99930035	.00158826
22	.00040391	.99970426	.00069965
23	.00017561	.99987987	.00029574
24	.00007317	.99995305	.00012012
25	.00002927	.99998231	.00004695
26	.00001126	.99999357	.00001768
27	.00000417	.99999774	.00000642
28	.00000149	.99999923	.00000225
29	.00000051	.99999975	.00000076
30	.00000017	.99999992	.00000025
31	.00000006	.99999997	.00000008
32	.00000002	.99999999	.00000002

Source: From *Tables of the Individual and Cumulative Terms of Poisson Distribution* by General Electric Company, © 1962. Reprinted by permission of Van Nostrand Reinhold Company.

Table A.10 Critical Values of r_S (The Spearman Rank Correlation Coefficient)

The computed value of r_s should exceed the following values in order to reject H_0.

	Significance level (one-tailed test)	
N	.05	.01
4	1.000	
5	.900	1.000
6	.829	.943
7	.714	.893
8	.643	.833
9	.600	.783
10	.564	.746
12	.506	.712
14	.456	.645
16	.425	.601
18	.399	.564
20	.377	.534
22	.359	.508
24	.343	.485
26	.329	.465
28	.317	.448
30	.306	.432

Source: Derived from E. G. Olds (1938). Distributions of sums of squares of rank differences for small numbers of individuals. *Annals of Mathematical Statistics* 9:133–148, and E. G. Olds (1949). The 5% significance levels for sums of squares of rank differences and a correction. *Annals of Mathematical Statistics* 20:117–118.

Table A.11 Critical Values of F in Snedecor's Variance Ratio Test

The computed value of F should exceed the following values in order to reject H_0. (df_1 refers to the degrees of freedom of the sample with the *larger* variance, and df_2 to the degrees of freedom of the sample with the *smaller* variance.)

$\alpha = 0.05;\ df_1 =$

df_2	1	2	3	4	5	6	7	8	10	12	24	∞
1	161.4	199.5	215.7	224.6	230.2	234.0	236.8	238.9	241.9	243.9	249.0	254.3
2	18.5	19.0	19.2	19.2	19.3	19.3	19.4	19.4	19.4	19.4	19.5	19.5
3	10.13	9.55	9.28	9.12	9.01	8.94	8.89	8.85	8.79	8.74	8.64	8.53
4	7.71	6.94	6.59	6.39	6.26	6.16	6.09	6.04	5.96	5.91	5.77	5.63
5	6.61	5.79	5.41	5.19	5.05	4.95	4.88	4.82	4.74	4.68	4.53	4.36
6	5.99	5.14	4.76	4.53	4.39	4.28	4.21	4.15	4.06	4.00	3.84	3.67
7	5.59	4.74	4.35	4.12	3.97	3.87	3.79	3.73	3.64	3.57	3.41	3.23
8	5.32	4.46	4.07	3.84	3.69	3.58	3.50	3.44	3.35	3.28	3.12	2.93
9	5.12	4.26	3.86	3.63	3.48	3.37	3.29	3.23	3.14	3.07	2.90	2.71
10	4.96	4.10	3.71	3.48	3.33	3.22	3.14	3.07	2.98	2.91	2.74	2.54
11	4.84	3.98	3.59	3.36	3.20	3.09	3.01	2.95	2.85	2.79	2.61	2.40
12	4.75	3.89	3.49	3.26	3.11	3.00	2.91	2.85	2.75	2.69	2.51	2.30
13	4.67	3.81	3.41	3.18	3.03	2.92	2.83	2.77	2.67	2.60	2.42	2.21
14	4.60	3.74	3.34	3.11	2.96	2.85	2.76	2.70	2.60	2.53	2.35	2.13
15	4.54	3.68	3.29	3.06	2.90	2.79	2.71	2.64	2.54	2.48	2.29	2.07
16	4.49	3.63	3.24	3.01	2.85	2.74	2.66	2.59	2.49	2.42	2.24	2.01
17	4.45	3.59	3.20	2.96	2.81	2.70	2.61	2.55	2.45	2.38	2.19	1.96
18	4.41	3.55	3.16	2.93	2.77	2.66	2.58	2.51	2.41	2.34	2.15	1.92
19	4.38	3.52	3.13	2.90	2.74	2.63	2.54	2.48	2.38	2.31	2.11	1.88
20	4.35	3.49	3.10	2.87	2.71	2.60	2.51	2.45	2.35	2.28	2.08	1.84
21	4.32	3.47	3.07	2.84	2.68	2.57	2.49	2.42	2.32	2.25	2.05	1.81
22	4.30	3.44	3.05	2.82	2.66	2.55	2.46	2.40	2.30	2.23	2.03	1.78
23	4.28	3.42	3.03	2.80	2.64	2.53	2.44	2.37	2.27	2.20	2.00	1.76
24	4.26	3.40	3.01	2.78	2.62	2.51	2.42	2.36	2.25	2.18	1.98	1.73
25	4.24	3.39	2.99	2.76	2.60	2.49	2.40	2.34	2.24	2.16	1.96	1.71
26	4.23	3.37	2.98	2.74	2.59	2.47	2.39	2.32	2.22	2.15	1.95	1.69
27	4.21	3.35	2.96	2.73	2.57	2.46	2.37	2.31	2.20	2.13	1.93	1.67
28	4.20	3.34	2.95	2.71	2.56	2.45	2.36	2.29	2.19	2.12	1.91	1.65
29	4.18	3.33	2.93	2.70	2.55	2.43	2.35	2.28	2.18	2.10	1.90	1.64
30	4.17	3.32	2.92	2.69	2.53	2.42	2.33	2.27	2.16	2.09	1.89	1.62
32	4.15	3.29	2.90	2.67	2.51	2.40	2.31	2.24	2.14	2.07	1.86	1.59
34	4.13	3.28	2.88	2.65	2.49	2.38	2.29	2.23	2.12	2.05	1.84	1.57
36	4.11	3.26	2.87	2.63	2.48	2.36	2.28	2.21	2.11	2.03	1.82	1.55
38	4.10	3.24	2.85	2.62	2.46	2.35	2.26	2.19	2.09	2.02	1.81	1.53
40	4.08	3.23	2.84	2.61	2.45	2.34	2.25	2.18	2.08	2.00	1.79	1.51
60	4.00	3.15	2.76	2.53	2.37	2.25	2.17	2.10	1.99	1.92	1.70	1.39
120	3.92	3.07	2.68	2.45	2.29	2.18	2.09	2.02	1.91	1.83	1.61	1.25
∞	3.84	3.00	2.60	2.37	2.21	2.10	2.01	1.94	1.83	1.75	1.52	1.00

$\alpha = 0.01$; $df_1 =$

df_2	1	2	3	4	5	6	7	8	10	12	24	∞
1	4052	5000	5403	5625	5764	5859	5928	5981	6056	6106	6235	6366
2	98.5	99.0	99.2	99.2	99.3	99.3	99.4	99.4	99.4	99.4	99.5	99.5
3	34.1	30.8	29.5	28.7	28.2	27.9	27.7	27.5	27.2	27.1	26.6	26.1
4	21.2	18.0	16.7	16.0	15.5	15.2	15.0	14.8	14.5	14.4	13.9	13.5
5	16.26	13.27	12.06	11.39	10.97	10.67	10.46	10.29	10.05	9.89	9.47	9.02
6	13.74	10.92	9.78	9.15	8.75	8.47	8.26	8.10	7.87	7.72	7.31	6.88
7	12.25	9.55	8.45	7.85	7.46	7.19	6.99	6.84	6.62	6.47	6.07	5.65
8	11.26	8.65	7.59	7.01	6.63	6.37	6.18	6.03	5.81	5.67	5.28	4.86
9	10.56	8.02	6.99	6.42	6.06	5.80	5.61	5.47	5.26	5.11	4.73	4.31
10	10.04	7.56	6.55	5.99	5.64	5.39	5.20	5.06	4.85	4.71	4.33	3.91
11	9.65	7.21	6.22	5.67	5.32	5.07	4.89	4.74	4.54	4.40	4.02	3.60
12	9.33	6.93	5.95	5.41	5.06	4.82	4.64	4.50	4.30	4.16	3.78	3.36
13	9.07	6.70	5.74	5.21	4.86	4.62	4.44	4.30	4.10	3.96	3.59	3.17
14	8.86	6.51	5.56	5.04	4.70	4.46	4.28	4.14	3.94	3.80	3.43	3.00
15	8.68	6.36	5.42	4.89	4.56	4.32	4.14	4.00	3.80	3.67	3.29	2.87
16	8.53	6.23	5.29	4.77	4.44	4.20	4.03	3.89	3.69	3.55	3.18	2.75
17	8.40	6.11	5.18	4.67	4.34	4.10	3.93	3.79	3.59	3.46	3.08	2.65
18	8.29	6.01	5.09	4.58	4.25	4.01	3.84	3.71	3.51	3.37	3.00	2.57
19	8.18	5.93	5.01	4.50	4.17	3.94	3.77	3.63	3.43	3.30	2.92	2.49
20	8.10	5.85	4.94	4.43	4.10	3.87	3.70	3.56	3.37	3.23	2.86	2.42
21	8.02	5.78	4.87	4.37	4.04	3.81	3.64	3.51	3.31	3.17	2.80	2.36
22	7.95	5.72	4.82	4.31	3.99	3.76	3.59	3.45	3.26	3.12	2.75	2.31
23	7.88	5.66	4.76	4.26	3.94	3.71	3.54	3.41	3.21	3.07	2.70	2.26
24	7.82	5.61	4.72	4.22	3.90	3.67	3.50	3.36	3.17	3.03	2.66	2.21
25	7.77	5.57	4.68	4.18	3.86	3.63	3.46	3.32	3.13	2.99	2.62	2.17
26	7.72	5.53	4.64	4.14	3.82	3.59	3.42	3.29	3.09	2.96	2.58	2.13
27	7.68	5.49	4.60	4.11	3.78	3.56	3.39	3.26	3.06	2.93	2.55	2.10
28	7.64	5.45	4.57	4.07	3.75	3.53	3.36	3.23	3.03	2.90	2.52	2.06
29	7.60	5.42	4.54	4.04	3.73	3.50	3.33	3.20	3.00	2.87	2.49	2.03
30	7.56	5.39	4.51	4.02	3.70	3.47	3.30	3.17	2.98	2.84	2.47	2.01
32	7.50	5.34	4.46	3.97	3.65	3.43	3.26	3.13	2.93	2.80	2.42	1.96
34	7.45	5.29	4.42	3.93	3.61	3.39	3.22	3.09	2.90	2.76	2.38	1.91
36	7.40	5.25	4.38	3.89	3.58	3.35	3.18	3.05	2.86	2.72	2.35	1.87
38	7.35	5.21	4.34	3.86	3.54	3.32	3.15	3.02	2.83	2.69	2.32	1.84
40	7.31	5.18	4.31	3.83	3.51	3.29	3.12	2.99	2.80	2.66	2.29	1.80
60	7.08	4.98	4.13	3.65	3.34	3.12	2.95	2.82	2.63	2.50	2.12	1.60
120	6.85	4.79	3.95	3.48	3.17	2.96	2.79	2.66	2.47	2.34	1.95	1.38
∞	6.63	4.61	3.78	3.32	3.02	2.80	2.64	2.51	2.32	2.18	1.79	1.00

Source: Reproduced from D. V. Lindley and J. C. P. Miller. *Cambridge Elementary Statistical Tables*. London: Cambridge University Press, Table 7, pp. 8 and 10, 1971. (Table 7 is based on Biometrika tables and acknowledgment is made to the Biometrika Trustees for permission to use them.)

APPENDIX B. ANSWERS TO WORK TABLES

Work Table 1.1

1. a. $Var(x) = 218.73$; $\sigma = 14.79$.

Clearly these two new extreme observations have increased the *variation* in the data and hence both statistics have been increased.

b. $Var(x) = 114.18$; $\sigma = 10.69$.

In this case the new observations are *central* and so the level of variation in the data, as measured by the two statistics, has declined.

2. a. $Var(x) = 181.82$; $\sigma = 13.48$.

The *aggregation* effect has produced lower values here than for 1a, but these new measures of variation are still much larger than those computed for the original data.

b. $Var(x) = 109.09$; $\sigma = 10.44$.

Once again there is an aggregation effect (compared with 1b) and together with the influence of the two new central observations, the overall lowest measures of variation are produced in this case.

Work Table 1.2

1. The mean center for ungrouped data is marked (1) on Figure A.1a. The mean center for grouped data is marked (1) on Figure A.1b.
2. The new mean center is located at $\bar{x} = 2.5$, $\bar{y} = 2.75$ and is marked (2) on Figure A.1a. The new point draws the mean center toward itself.
3. The new mean center is located at $\bar{x} = 2.25$, $\bar{y} = 3.5$ and is marked (3) on Figure A.1b. The decline in area B's frequency means that its dominance of the mean center's location is lessened and it is pulled slightly towards the locationally extreme area C.

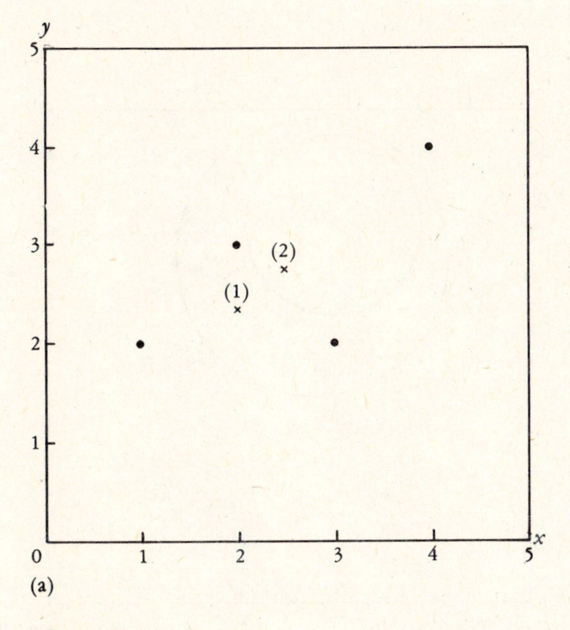

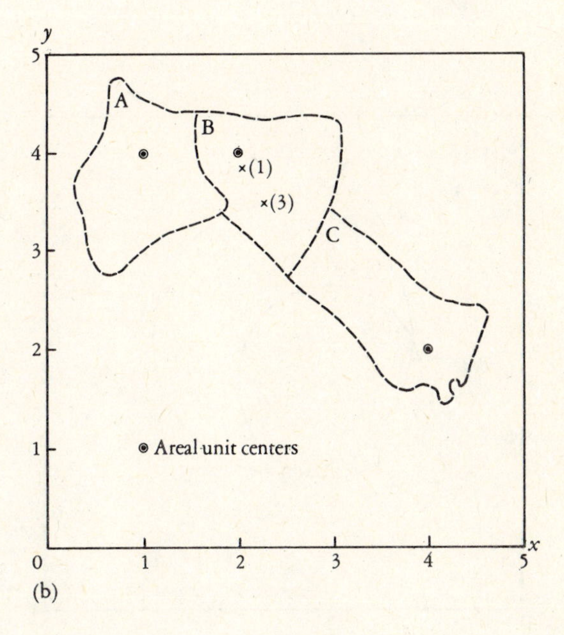

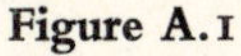

Figure A.1

Work Table 1.3

1. The standard distance circle for ungrouped data is marked (1) on Figure A.2a. The standard distance circle for grouped data is marked (1) on Figure A.2b.
2. The new standard distance is 1.37 and is marked (2) on Figure A.2a. The additional point has clearly made the distribution more spread out.
3. The new standard distance is 1.39 so that it is much less concentrated than before the reduction of area B's frequency. The new standard distance is marked (3) on Figure A.2b.

Work Table 2.1

With just one placement of the circle lattices at each of the four scales you may not get a consistently increasing length as ϵ becomes smaller. This is probably because of the known imprecision of this measurement technique. In this case the exercise should be repeated several times to confirm an increase in length. If this feature still does not occur, then you should check that you have carried out the exercise correctly.

Work Table 2.2

1. The graphical solution to the problem is shown in Figure A.3a.
2. The new resultant has the bearing 158° 24′, illustrated in Figure A.3b. The effect of these two bearings is to pull the resultant into a more southerly direction.

Work Table 2.3

1. The surface of potential is shown on Figure A.4a.
2. The revised potentials are as follows:

$v_A = 218.536$; $v_B = 128.536$;
$v_C = 122.071$; $v_D = 95.711$

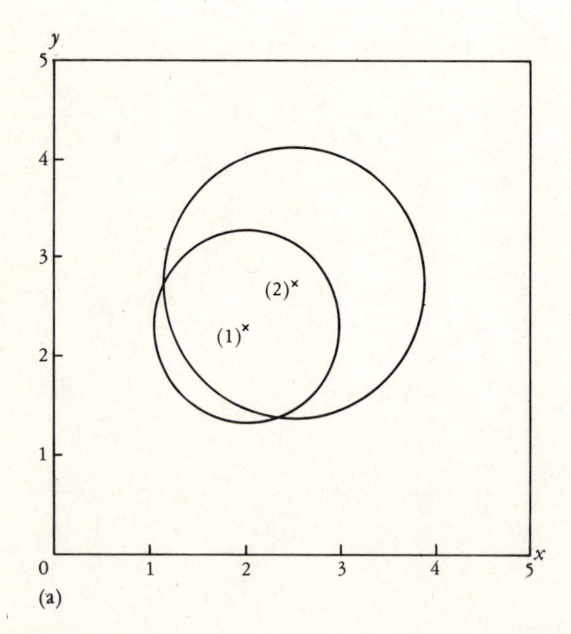

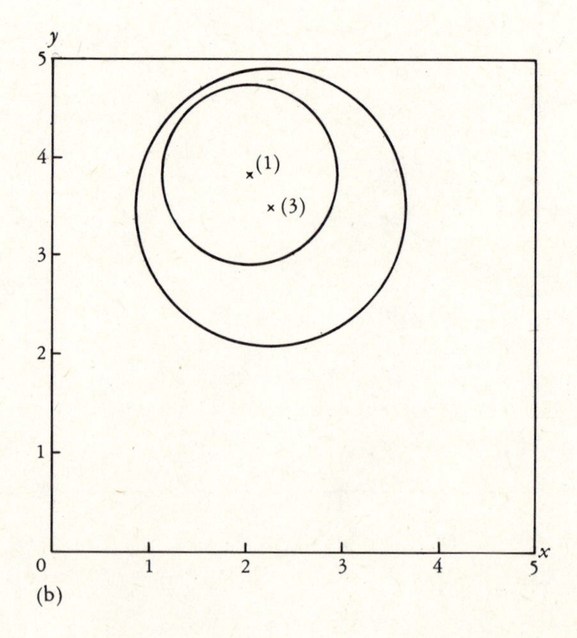

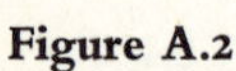
Figure A.2

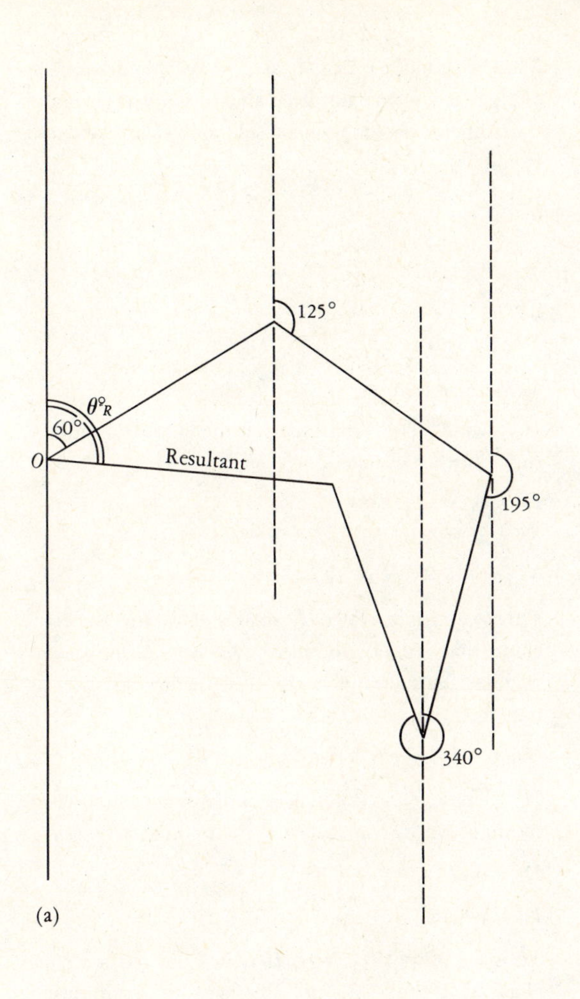

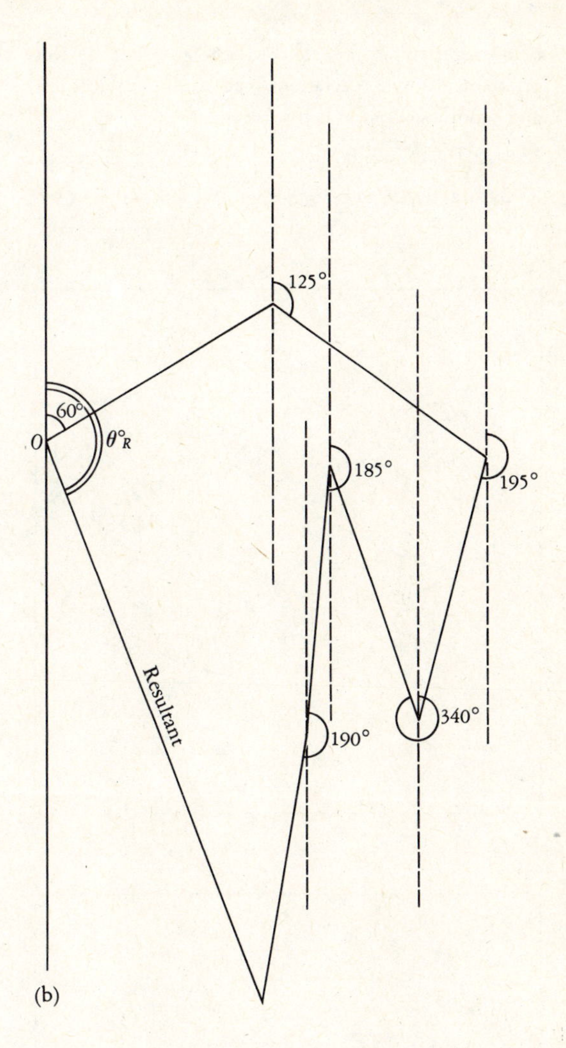

Figure A.3

The new surface of potential is shown on Figure A.4b and we can see that the additional population causes the *whole* surface to rise although this effect is naturally most intensified in area A.

Work Table 3.1

H_0 There are no differences between hillside and upland in terms of land use types.

$\chi^2 = 9.2$; $df = 3$

χ^2 $(\alpha = 0.05, df = 3) = 7.82$

Thus we can reject H_0 at $\alpha = 0.05$, and we conclude that there are differences between hillside and upland in terms of land use types.

H_0 There are no differences between hillside and valley in terms of land use types. If you have used all four classes to compute χ^2 you will have utilized an expectation table with three out of eight expectations below 5. Thus you should have modified your data before computing χ^2. The easiest way out of this problem of low expectations is to combine the forest and moorland classes. If this is done:

$\chi^2 = 2.7$; $df = 2$

$\chi^2\ (\alpha = 0.05, df = 2) = 5.99$

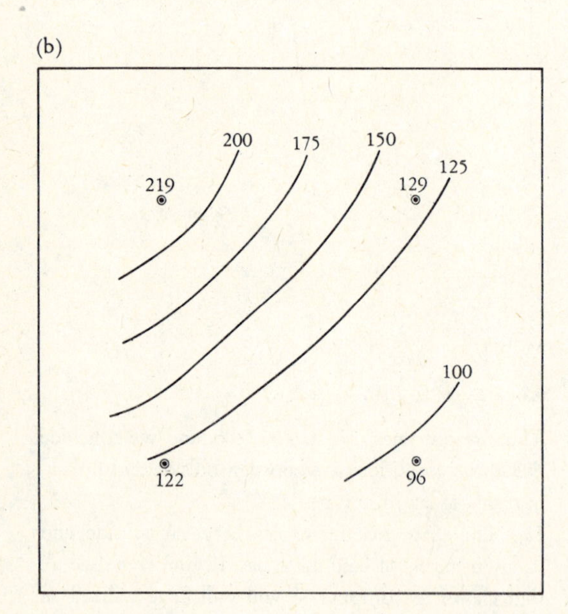

Figure A.4

Thus we cannot reject H_0 at $\alpha = 0.05$, and we conclude that we have not been able to show any significant differences between hillside and valley land use types.

Work Table 3.2

One-tailed test

H_0 Middle class commuters at the later date did not travel further distances.

$D = 0.17$

$\chi^2 = 3.47$; $df = 2$

$\chi^2\ (\alpha = 0.05, df = 2) = 5.99$

Thus we cannot reject H_0 at $\alpha = 0.05$, and we conclude that we do not have evidence to show that middle class commuters travelled further distances at the later date.

Two-tailed test

H_0 There are no differences in the distances travelled by middle class commuters at the two dates.

$D = 0.17$

$D(\alpha = 0.05, n_1 = n_2 = 60) = 0.25$

Thus we cannot reject H_0 at $\alpha = 0.05$, and we conclude that there are no differences in commuting distances between the two periods.

Work Table 3.3

One-tailed test

H_0 There is no southern bias in the preference rankings.

$U = 13$

$U(\alpha = 0.05, n_1 = n_2 = 8) = 15$

Thus we can reject H_0 at $\alpha = 0.05$, and we conclude that there is a southern bias.

Two-tailed test

H_0 There are no differences in rankings between northern and southern cities.

$U = 13$

$U(\alpha = 0.05, n_1 = n_2 = 8) = 13$

Thus we can just reject H_0 at $\alpha = 0.05$, and we conclude that there are differences in the two sets of rankings.

Both of these results suggest that the very small sample sizes used in the work table produced type II errors.

Work Table 3.4

H_0 There are no differences between regions in terms of preference rankings.

$H = 4.75$; n_j's > 5; $df = 2$

$H(\alpha = 0.05, df = 2, n_j\text{'s} > 5) = 5.99$

Thus we cannot reject H_0 at $\alpha = 0.05$, and we conclude that we do not have evidence for regional differences; so our earlier analysis does not seem to have involved a type II error.

Work Table 3.5

With $\bar{x} = 66.55$ and $\sigma = 17.11$, the standardized frequencies for the central city sample are

$(< -2\sigma)$	$(-2\sigma$ *to* $-\sigma)$	$(-\sigma$ *to* $\bar{x})$
0	3	7
$(\bar{x}$ *to* $+\sigma)$	$(+\sigma$ *to* $+2\sigma)$	$(> +2\sigma)$
6	4	0

The resulting value of $D = 0.0415$ which is identical to that derived using the suburban sample. Since $D(\alpha = 0.20, \text{N} = 20) = 0.16$, this D value is extremely low and we conclude that we are in no position to reject a null hypothesis of no differences between our sample scores and a normal distribution.

Work Table 3.6

H_0 There is no difference between the sample distances for $\epsilon = 6.4$ km and the population represented by the Portuguese estimate.

$\bar{x} = 1032$ km; $\hat{\sigma} = 64.4$; $\text{M} = 1214$ km

$t = 9.4$; $df = 10$

$t(\alpha = 0.05, df = 10) = 2.23$

Thus we can easily reject H_0 at $\alpha = 0.05$, and so it would seem that the Portuguese estimate represents a different population from that from which our sample was drawn.

Work Table 3.7

H_0

1. The number of WW joins is not different from random expectations.
2. The number of WB joins is not different from random expectations.
3. The number of BB joins is not different from random expectations.

$E(\text{WW}) = 2.0207$; $\text{Var(WW)} = 4.0800$

$E(\text{WB}) = 5.3878$; $\text{Var(WB)} = 3.0124$

$E(\text{BB}) = 3.5915$; $\text{Var(BB)} = 6.7389$

$z_{\text{WW}} = -0.0102$; $p(z = 0.01) = 0.4960$

$z_{\text{WB}} = -0.7998$; $p(z = 0.80) = 0.2119$

$z_{\text{BB}} = +0.5426$; $p(z = 0.54) = 0.2946$

We cannot reject any of our three H_0's at $\alpha = 0.05$ Thus the additional area does not alter our conclusion that there is no spatial autocorrelation in the map pattern.

Work Table 3.8

H_0 There is no spatial autocorrelation between contiguity relations in the new map pattern:

$c = 0.4678$; $z_c = 1.0786$; $p(z = 1.08) = 0.1401$

We still cannot reject H_0 at $\alpha = 0.05$ and we conclude that there is no evidence of spatial autocorrelation in the map pattern.

Work Table 3.9

H_0 There is no spatial autocorrelation in the map pattern:

$r_a = 0.5965$; $I = 0.8947$; $E(I) = -0.3$;
$E(I^2) = 0.7890$; $\text{Var}(I) = 0.699$; $z_I = 1.4290$;
$p(z = 1.43) = 0.0764$

We still cannot reject H_0 at $\alpha = 0.05$, although in this case we do approach this arbitrary threshold—a map pattern as autocorrelated as ours will only occur 7 times in every 100 random generations.

Work Table 4.1

In this case the following parameters are found:

$\bar{x} = 1$; $\text{Var}(x) = 0.2222$; $\hat{p} = 0.8819$;
$\hat{\gamma} = 0.1181$; $\hat{q} = 0.1181$

x	$p(x)$	$p(x)\Sigma f$	f
0	0.1049	0.9441	1
1	0.7961	7.1649	7
2	0.0931	0.8379	1
3	0.0054	0.0486	0

The third and fourth columns represent the expected frequencies with this model and the observed frequencies, respectively, and a very close fit can be observed. Once again the frequencies are too small for a χ^2 test, so we employ a Kolmogorov-Smirnov test. In this case $D = 0.0121$ whereas $D(\alpha = 0.2, n = 9) = 0.339$, indicating a very high degree of similarity between the expected and observed frequencies.

Work Table 4.2

For this new map pattern $\bar{x} = 1$ and $\text{Var}(x) = 8.0000$, so the parameters of the model are as follows:

$\hat{p} = 0.1250$; $\hat{k} = 0.1429$; $\hat{q} = 0.8750$; $c = 0.7429$

x	$p(x)$ ($p = 0.1250$; $k = 0.1429$)	*Expected* $p(x)\Sigma f$	*Observed* (f)
0	0.7429	6.6861	8
1	0.0929	0.8361	0
2	0.0465	0.4185	0
3	0.0290	0.2610	0
4	0.0200	0.1800	0
5	0.0145	0.1305	0
6	0.0109	0.0981	0
7	0.0083	0.0747	0
8	0.0065	0.0585	0
9	0.0052	0.0468	1
10	0.0041	0.0369	0

From the final two columns we can see that the degree of fit is not very impressive. For our goodness of fit test the frequencies are too small to employ χ^2, so we use the Kolmogorov-Smirnov test. In this case $D = 0.1460$ compared with $D(\alpha = 0.02, n = 9) = 0.339$. Thus the observed frequencies and the calibrated model are not so different in terms of this test.

Work Table 4.3

In this case the following average nearest neighbor in sectors were found:

$\bar{r}_{(1)}$	$\bar{r}_{(2)}$	$\bar{r}_{(3)}$	$\bar{r}_{(4)}$	$\bar{r}_{a4}$
18.6	19.6	20.8	24.8	20.95

where $\bar{r}_{a4}$ is the average for all four sectors. Notice that the average distance for the first three sectors is the same as for the first three sectors in the 6-sector analysis. This merely means that both analyses are picking up the same neighbors in these lower ranked sectors.

With $k = 4$ and $n = 5$, $A = 1444$ as before, $\bar{r}_{e4} = 16.8919$ so that $R_4 = 1.2402$.

This suggests a tendency towards regularity since it is greater than the Poisson expectation of 1. However, interpretation of this single statistic from the analysis is difficult since the expectation from a perfectly regular lattice of points is given by $2.1491/\sqrt{4} =$

1.0746. Thus our pattern exceeds both random and regular expectations! In this situation we produce the whole table of average nearest neighbor predictions for sectors. Poisson expectations are from Table 4.13; regular expectations are found from $E_h = (1.075 \cdot \sqrt{A/n}) = 18.2687$ so that we have

	Sectors 1	2	3	4
Observed	18.6	19.6	20.8	24.8
Random	8.4459	13.6740	18.9814	26.4679
Regular	18.2687	18.2687	18.2687	18.2687

Thus, our observed values lie between the random and regular expected distances. Finally, it must be remembered that this example is illustrative and that its simplicity has produced boundary problems that make full interpretation very difficult.

Work Table 5.1

1. $r_s = 0.6$; the addition of area E has caused the level of correlation to decline
2. $r = 0.1449$; the addition of area E has caused the level of correlation to decline to quite a large degree. Notice that the decline in r is much greater than for r_s. This reflects the extra information used in computing r by taking into account the relatively large interval between the x value of 19 for the new area E and all the other values of x.

Work Table 5.2

1. When $x = 10$, $y = 4.7886$. The points of the y intercept ($y = 0.5576$, $x = 0$), the two means ($\bar{y} = 2.25$, $\bar{x} = 4$), and the computed prediction of ($y = 4.7886$, $x = 10$) are shown in Figure A.5a and are marked (1), (2), and (3) respectively on the regression line.
2. With the addition of area E, $a = 1.9282$ and $b = 0.0388$. The regression line, $y = 1.9284 + 0.0388x$ is drawn on Figure A.5b. Area E causes the regression line to be much flatter, which is numerically represented by the smaller b value.

Work Table 5.3

The recalibrated multiple regression equation is

$x_1 = 12.4999 - 0.9651x_2 - 1.1279x_3$

Notice that the new observation has not changed either of the regression coefficients. All that has happened is that the regression "plane" has shifted upward, as reflected by the slightly higher base constant.

Work Table 5.4

$R_{1.23} = 0.7090$

$r_{12.3} = 0.4607$

$r_{13.2} = 0.6279$

from $r_{12} = 0.4231$ and $r_{13} = -0.6072$. The addition of area F has reduced the overall level of correlation while maintaining the relative order of importance of x_2 and x_3 as determinants of x_1.

Work Table 5.5

H_0

1. r is not significantly different from zero.
2. r_s is not significantly different from zero.
3. b is not significantly different from zero.

Statistics

1. $t = 1.811$; $t(\alpha = 0.05, df = 10) = 2.228$
2. $r_s = 0.498$; $r_s(\alpha = 0.05, n = 12) = 0.506$
3. $t = 1.653$; $t(\alpha = 0.05, df = 10) = 2.228$

We cannot reject H_0 in all three tests so that we conclude that this sample does not indicate population coefficients that differ from zero.

Work Table 6.1

H_0 There are no differences between the three sectors in terms of economic status.

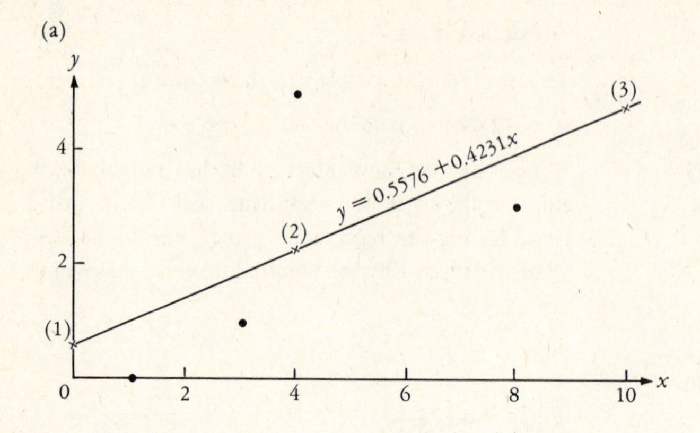

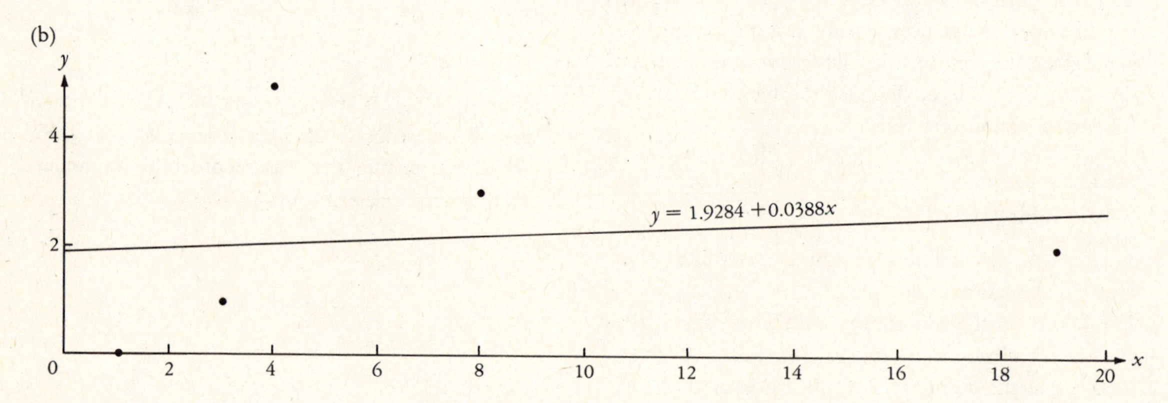

Figure A.5

Analysis of variance table

Source of variation	SS	*df*	*Variance*
Total	21.3600	17	1.2565
Between	10.6431	2	5.3216
Within	10.7169	15	0.7145

$F = 7.4480$

$F(df_1 = 2, df_2 = 15, \alpha = 0.05) = 3.68$

Thus we can reject H_0 and we conclude that economic status is sectorally arranged in this particular city.

Work Table 6.2

H_0

1. There are no differences between sectors in terms of economic status.
2. There are no differences between zones in terms of economic status.
3. There is no significant pattern of combinations of zones and sectors in terms of economic status.

Analysis of variance table

Source of variation	SS	*df*	*Variance*
Total	18.56	15	1.2373
Between cell	18.40	7	2.6286
Between sector	16.96	3	5.6533
Between zone	1.44	1	1.4400
Interaction	0.00	3	0.0000
Unexplained	0.16	8	0.0200

1. $F = 282.6650;$
   $$F(df_1 = 3, df_2 = 8, \alpha = 0.05) = 4.07$$
2. $F = 72.0000;$
   $$F(df_1 = 1, df_2 = 8, \alpha = 0.05) = 5.32$$
3. $F = 0.0000;$
   $$F(df_1 = 3, df_2 = 8, \alpha = 0.05) = 4.07$$

Clearly in the first two cases we can reject H_0 and we conclude that there is sectoral, zonal, but no interaction patterns of economic status in this particular city. The relative sizes of the F ratios suggest that the sector pattern is dominant although the zonal pattern is highly significant.

BIBLIOGRAPHY

Abler, R., J. Adams, and P. Gould. *Spatial Organization.* Englewood Cliffs, N.J.: Prentice-Hall, 1971.

Ackoff, R.L., S.K. Gupta, and J.S. Minas. *Scientific Method: Optimizing Applied Research Decisions.* New York: Wiley, 1962.

Adler, I. *The New Mathematics.* New York: John Day, 1959.

Alexander, J.W., and J.B. Lindberg. Measurement of manufacturing: Coefficients of correlation. *Journal of Regional Science* 3:71–81, 1961.

Alker, H.R. A typology of ecological fallacies. In M. Doggan and S. Rokkan, eds. *Quantitative Ecological Analysis in the Social Sciences.* Cambridge, Mass.: M.I.T. Press, 1969.

Angel, S., and G.M. Hymen. Transformations and geographic theory. *Geographical Analysis* 4:350–367, 1972.

Anuchin, V.A. Mathematization and the geographic method. *Soviet Geography* 9(2):71–81, 1970.

Arrow, K.J. Mathematical models in the social sciences. In D. Lerner and H.D. Lasswell, eds. *The Policy Sciences.* Stanford, Cal.: Stanford University Press, 1951.

Bachi, R. Standard distance measures and related methods for spatial analysis. *Papers and Proceedings, Regional Science Association* 10:83–132, 1963.

Batty, M. Models and projections of the space economy: A sub-regional study in Northwest England. *Town Planning Review* 41:121–148, 1970a.

Batty, M. An activity allocation model for the Nottinghamshire-Derbyshire subregion. *Regional Studies* 4: 307–332, 1970b.

Bell, W. Economic, family and ethnic status: An empirical test. *American Sociological Review* 20:45–52, 1955.

Berry, B.J.L. Comments on the use of chi-square. *Annals, Association of American Geographers* 49:89, 1959.

Berry, B.J.L. *Essays on Commodity Flows and the Spatial*

Structure of the Indian Economy. Research Paper 111, Department of Geography, University of Chicago, 1966.

Berry, B.J.L. *Geography of Market Centres and Retail Distribution*. Englewood Cliffs, N.J.: Prentice-Hall, 1967.

Berry, B.J.L. Introduction: The logic and limitations of comparative factorial ecology. *Economic Geography* 47:209–219, 1971a.

Berry, B.J.L. Dido data analysis: Gigo or pattern recognition. In *Perspectives in Geography* 1 (Models of Spatial Variation):105–131, 1971b.

Berry, B.J.L., and A. Baker. Geographic sampling. In B.J.L. Berry and D.F. Marble, eds. *Spatial Analysis: A Reader in Statistical Geography*. Englewood Cliffs, N.J.: Prentice-Hall, 1968.

Berry, B.J.L., H.G. Barnum, and R.J. Tennant. Retail location and consumer behavior. *Papers and Proceedings, Regional Science Association* 9:65–105, 1962.

Berry, B.J.L., and D.F. Marble, eds. *Spatial Analysis: A Reader in Statistical Geography*. Englewood Cliffs, N.J.: Prentice-Hall, 1968.

Black, W.R., and R. Larson. A comparative evaluation of alternative friction factors in the gravity model. *Professional Geographer* 24:335–337, 1972.

Blalock, H.M. *Social Statistics*. New York: McGraw-Hill, 1960.

Blalock, H.M. *Causal Inferences in Non-Experimental Research*. Chapel Hill: University of North Carolina Press, 1964.

Blaut, J.M. Micro-geographic sampling: A quantitative approach to regional agriculture geography. *Economic Geography* 36:254–259, 1959.

Blaut, J.M. Object and relationship. *Professional Geographer* 14:1–7, 1962.

Board, C. Maps as models. In R.J. Chorley and P. Haggett, eds. *Models in Geography*. London: Methuen, 1966.

Bogue, D.J. *The Structure of the Metropolitan Community*. Ann Arbor: University of Michigan School of Graduate Studies, 1949.

Boyce, R.R., and W.A.V. Clark. The concept of shape in geography. *Geographical Review* 54:561–572, 1964.

Bucklin, L.P. Retail gravity models and consumer choice: A theoretical and empirical critique. *Economic Geography* 47:489–497, 1971.

Bunge, W. *Theoretical Geography*, revised ed. Lund Studies in Geography, C1, 1966.

Burton, I. The quantitative revolution and theoretical geography. *Canadian Geographer* 7:151–162, 1968. Reprinted in B.J.L. Berry and D.F. Marble, eds. *Spatial Analysis: A Reader in Statistical Geography*. Englewood, Cliffs, N.J.: Prentice-Hall, 1968.

Caprio, R.J. Centrography and geostatistics. *Professional Geographer* 22:15–19, 1970.

Carey, G.W. The regional interpretation of Manhattan population and housing patterns through factor analysis. *Geographical Review* 56:551–569, 1966.

Cassetti, E. Optimum location of steelmills serving the Quebec and Southern Ontario steel market. *Canadian Geographer* 10:27–39, 1966.

Cattell, R.B. Factor analysis: An introduction to essentials, I. *Biometrics* 21:190–215, 1965a.

Cattell, R.B. Factor analysis: An introduction to essentials, II. *Biometrics* 21:405–435, 1965b.

Chapman, G.P. The application of information theory to the analysis of population distributions in space. *Economic Geography* 46:317–331, 1970.

Chicago Area Transportation Study. *Final Report*. Chicago: Western Engraving and Embossing Company, 1959.

Child, D. *The Essentials of Factor Analysis*. London: Holt, Rinehart and Winston, 1970.

Chisholm, M. The geography of commuting. *Annals, Association of American Geographers* 50:187–188 and 491–492, 1960.

Chisholm, M. *Rural Settlement and Land Use*. London: Hutchinson, 1962.

Chisholm, M. Macro- and micro-approaches to urban systems research. *Geographical Journal* 138:60–63, 1972.

Chisholm, M., and P. O'Sullivan. *Freight Flows and the Spatial Structure of the British Economy*. Cambridge: Cambridge University Press, 1973.

Chorley, R.J., ed. *Spatial Analysis in Geomorphology*. London: Methuen, 1972.

Chorley, R.J., and P. Haggett. Trend-surface mapping in geographical research. *Transactions, Institute of British Geographers* 37:47–67, 1965.

Chorley, R.J., and P. Haggett, eds. *Models in Geography*. London: Methuen, 1966.

Chorley, R.J., and P. Haggett. *Network Analysis in Geography*. London: Arnold, 1969.

Clark, C., and G.H. Peters. The "intervening opportunities" method of traffic analysis. *Traffic Quarterly* 19:101–119, 1965.

Clark, C., F. Wilson, and J. Bradley. Industrial location and economic potential. *Regional Studies* 3:197–212, 1969.

Clark, D. The formal and functional structure of Wales. *Annals, Association of American Geographers* 63:71–84, 1973a.

Clark, D. Normality, transformation and the principal components solution: An empirical note. *Area* 5: 110–113, 1973b.

Clark, D., W.K.D. Davies, and R.J. Johnston. The application of factor analysis in human geography. *The Statistician* 23:259–281, 1974.

Clark, P.J., and F.C. Evans. Distance to nearest neighbour as a measure of spatial relationships in populations. *Ecology* 35:445–453, 1954.

Clark, W.A.V. The use of residuals from regression in geographical research. *New Zealand Geographer* 23:64–67, 1967.

Clarke, C.G. Residential segregation and intermarriage in San Fernando, Trinidad. *Geographical Review* 61: 198–218, 1971.

Cliff, A.D., P. Haggett, J.K. Ord, K. Bassett, and R. Davies. *Elements of Spatial Structure.* Cambridge: Cambridge University Press, 1975.

Cliff, A.D., R.L. Martin, and J.K. Ord. A test for spatial autocorrelation in choropleth maps based upon a modified χ^2 statistic. *Transactions, Institute of British Geographers* 65:109–129, 1975.

Cliff, A.D., and J.K. Ord. The problems of spatial autocorrelation. In A.J. Scott, ed. *Studies in Regional Science: London Papers in Regional Science.* London: Pion, 1969.

Cliff, A.D., and J.K. Ord. Spatial autocorrelation: a review of existing and new measures with applications. *Economic Geography* 46:269–292, 1970.

Cliff, A.D., and J.K. Ord. *Spatial Autocorrelation.* London: Pion, 1973.

Cliff, A.D., and J.K. Ord. The comparison of means when samples consist of spatially autocorrelated observations. *Environment and Planning* 7A:725–734, 1975a.

Cliff, A.D., and J.K. Ord. The choice of a test for spatial autocorrelation. In J.C. Davis and M.J. McCullagh, eds. *Display and Analysis of Spatial Data.* London: Wiley, 1975b.

Cochran, W.G. *Sampling Techniques.* New York: Wiley, 1953.

Cole, H.R. Shopping assessments at Haydock and elsewhere: A review. *Urban Studies* 3:147–156, 1966.

Cole, J.P., and C.A.M. King. *Quantitative Geography.* London: Wiley, 1968.

Coleman, J.S. *Introduction to Mathematical Sociology.* New York: Glencoe, 1964.

Comrey, A.L. *A First Course in Factor Analysis.* London: Academic Press, 1973.

Conkling, E.C. South Wales: A case study in industrial diversification. *Economic Geography* 39:258–272, 1963.

Coppock, J.T. The parish as a geographical-statistical unit. *Tidjschrift voor Economische en Sociale Geografie* 51:317–326, 1960.

Coppock, J.T., and J. Johnson. Measurement in human geography. *Economic Geography* 38:130–137, 1962.

Courant, R., and H. Robbins. *What is Mathematics?* London: Oxford University Press, 1941.

Cox, K.R. The application of linear programming to geographical problems. *Tidjschrift voor Economische en Sociale Geografie* 56:228–236, 1965.

Cox, K.R. Suburbia and voting behavior in a London metropolitan area. *Annals, Association of American Geographers* 58:111–127, 1968.

Curry, L. A note on spatial association. *Professional Geographer* 18:97–99, 1966.

Curry, L. Quantitative geography, 1967. *Canadian Geographer* 11:265–279, 1967.

Dacey, M.F. Analysis of central place and point patterns by a nearest neighbour method. In K. Norborg, ed. *Proceedings of the I.G.U. Symposium on Urban Geography.* Lund Studies in Geography, B24, 1962.

Dacey, M.F. Order neighbour statistics for a class of random patterns in multidimensional space. *Annals, Association of American Geographers* 53:505–515, 1963.

Dacey, M.F. Modified Poisson probability law for point pattern more regular than random. *Annals, Association of American Geographers* 54:559–565, 1964a. Reprinted in B.J.L. Berry and D.F. Marble, eds. *Spatial Analysis: A Reader in Statistical Geography.* Englewood Cliffs, N.J.: Prentice-Hall, 1968.

Dacey, M.F. Two dimensional random point patterns: A review and an interpretation. *Papers and Proceedings, Regional Science Association* 13:41–58, 1964b.

Dacey, M.F. A compound probability law for a pattern more dispersed than random with areal inhomogeneity. *Economic Geography* 42:172–179, 1966a.

Dacey, M.F. A county seat model for the areal pattern of an urban system. *Geographical Review* 56:527–542, 1966b.

Dacey, M.F. Description of line pattern. In W.L. Garrison and D.F. Marble, eds. *Quantitative Geography, Part I.* Northwestern University, Studies in Geography, no. 13, 1967.

Dacey, M.F. An empirical study of the areal distribution

of houses in Puerto Rico. *Transactions, Institute of British Geographers* 45:51–69, 1968a.

Dacey, M.F. A review on measures of contiguity for two and *k*-colour maps. In B.J.L. Berry and D.F. Marble, eds. *Spatial Analysis: A Reader in Statistical Geography*. Englewood Cliffs, N.J.: Prentice-Hall, 1968b.

Dacey, M.F. Similarities in the areal distribution of houses in Japan and Puerto Rico. *Area* 1:35–37, 1969.

Dantzig, G.B. *Linear Programming and Extensions*. Princeton: Princeton University Press, 1963.

Davies, W.K.D. Varimax and the destruction of generality. *Area* 3:112–118, 1971a.

Davies, W.K.D. Varimax and generality. *Area* 3:254–259, 1971b.

Davies, W.K.D. Varimax and generality: A second reply. *Area* 4:209–210, 1972.

Davies, W.K.D., and G.J. Lewis. The urban dimensions of Leicester. *Institute of British Geographers, Special Publication* 5:71–85, 1973.

Department of Town and Country Planning, University of Manchester. *Regional Shopping Centres in North West England*, 1964.

Department of Town and Country Planning, University of Manchester. *Regional Shopping Centres in North West England. Part Two: A Retail Shopping Model*, 1966.

Dixon, W.J., and F.J. Massey, Jr. *Introduction to Statistical Analysis*. New York: McGraw-Hill, 1969.

Dogan, M., and S. Rokkan, eds. *Quantitative Ecological Analysis in the Social Sciences*. Cambridge: M.I.T. Press, 1969.

Dorfman, R., P.A. Samuelson, and R.M. Solow. *Linear Programming and Economic Analysis*. New York: McGraw-Hill, 1958.

Draper, N.R., and H. Smith. *Applied Regression Analysis*. New York: Wiley, 1966.

Driscol, E.M., and D. Hopley. Coastal development in a part of tropical Queensland, Australia. *Journal of Tropical Geography* 26:17–28, 1968.

Duncan, O.D., R.P. Cuzzort, and B. Duncan. *Statistical Geography: Problems in Analyzing Areal Data*. New York: Free Press, 1961.

Duncan, O.D., and B. Duncan. A methodological analysis of segregation indices. *American Sociological Review* 20:210–217, 1955.

Duncan, O.D., and S. Lieberson. Ethnic segregation and assimilation. *American Journal of Sociology* 64:364–374, 1959.

Ehrenberg, A.S.C. Some questions about factor analysis. *The Statistician* 12:191–208, 1962.

Ellis, B. *Basic Concepts of Measurement*. Cambridge: University Press, 1966.

Etter, A.G. Mathematics, ecology, and a piece of land. *Landscape* 12 (3):28–31, 1963. Reprinted in C.L. Salter, ed. *The Cultural Landscape*. Belmont, California: Wadsworth.

Fararo, T.J. The nature of mathematical sociology: A non-technical essay. *Social Research* 36:75–92, 1969.

Fay, W.T., and R.C. Klove. The 1970 census. *Professional Geographer* 22:284–289, 1970.

Feller, W. *An Introduction to Probability Theory and Its Applications*. New York: Wiley, 1957.

French, H.M. Quantitative methods and non-parametric statistics. In H.M. French and J.B. Racine, eds. *Quantitative and Qualitative Geography*. Ottawa: University of Ottawa Press, 1971.

Garner, B.J. The internal structure of retail nucleations. *Northwestern University Studies in Geography*, No. 12, 1966.

Garrison, W.L., and D.F. Marble. Factor analytic study of the connectivity of a transportation network. *Papers and Proceedings, Regional Science Association* 12: 231–238, 1964.

Geary, R.C. The contiguity ratio and statistical mapping. *The Incorporated Statistician* 5:115–141, 1954. Reprinted in B.J.L. Berry and D.F. Marble, eds. *Spatial Analysis: A Reader in Statistical Geography*. Englewood Cliffs, N.J.: Prentice-Hall, 1968.

General Electric Company. *Tables of the Individual and Cumulative Terms of the Poisson Distribution*. Princeton: Princeton University Press, 1962.

Getis, A. The determination of the location of retail activities with the use of a map transformation. *Economic Geography* 39:14–22, 1963.

Getis, A. Temporal land use pattern analysis with the use of nearest neighbour and quadrat methods. *Annals, Association of American Geographers* 54:391–399, 1964.

Giggs, J.A. The distribution of schizophrenics in Nottingham. *Transactions, Institute of British Geographers* 59:77–98, 1973.

Gittus, E. An experiment in the definition of urban sub-areas. *Transactions of the Bartlett Society* 2:109–135, 1964.

Goddard, J.B. Functional regions within the city: A study by factor analysis of taxi flows in central London. *Transactions, Institute of British Geographers* 49: 161–182, 1970.

Godlund, S. Population, regional hospitals, transport facilities and regions: Planning the location of regional hospitals in Sweden. *Lund Studies in Geography* B21, 1961.

Golledge, R.G., and D.H. Amadeo. Some introductory notes on regional division and set theory. *Professional Geographers* 18:14–19, 1966.

Golledge, R.G., and G. Rushton. *Multidimensional Scaling: Review and Geographical Applications.* Technical Paper No. 10. A.A.G. Commission on College Geography, 1972.

Goodchild, M.F., and B. Massam. Some least-cost models of spatial administrative systems in southern Ontario. *Geografiska Annaler* 52B:86–94, 1969.

Goodman, L.A. Some alternatives to ecological correlation. *American Journal of Sociology* 64:610–625, 1959. Reprinted in B.J.L. Berry and D.F. Marble, eds. *Spatial Analysis: A Reader in Statistical Geography.* Englewood Cliffs, N.J.: Prentice-Hall, 1968.

Gould, P.R. On the geographical interpretation of eigenvalues. *Transactions, Institute of British Geographers* 42:53–92, 1967.

Gould, P.R. Methodological developments since the fifties. In C. Board et al., eds. *Progress in Geography,* vol. 1, pp. 3–49. London: Arnold, 1969.

Gould, P.R. Is *Statistix Inferens* the geographical name for a wild goose? *Economic Geography* 46:439–448, 1970a.

Gould, P.R. Computers and spatial analysis: Extensions of geographic research. *Geoforum* 1:53–69, 1970b.

Gould, P.R. Pedagogic review. *Annals, Association of American Geographers* 62:689–700, 1972.

Gould, P.R., and T.R. Leinbach. An approach to the geographic assignment of hospital services. *Tidjschrift voor Economische en Sociale Geographie* 57:203–206, 1966.

Greer-Wootten, B. *A Bibliography of Statistical Applications in Geography.* Technical Paper No. 9. A.A.G. Commission on College Geography, 1972.

Greig-Smith, P. The use of random and contiguous quadrats in the study of the structure of plant communities. *Annals of Botany* (*London*) N.S. 16:293–316, 1952.

Greig-Smith, P. *Quantitative Plant Ecology.* London: Butterworths, 1964.

Gregory, S. *Statistical Methods and the Geographer.* London: Longmans, 1968.

Gregory, S. The quantitative approach in geography. In H.M. French and J.B. Racine, eds. *Quantitative and Qualitative Geography.* Ottawa: University of Ottawa Press, 1971.

Griffiths, J.C. Current trends in geomathematics. *Earth Science Review* 6:121–140, 1970.

Gudgin, G., and J.B. Thornes. Probability in geographic research: Applications and problems. *Statistician* 123: 157–178, 1974.

Guest, P.G. *Numerical Methods of Curve Fitting.* Cambridge: Cambridge University Press, 1961.

Gurevich, B.L., and Y.G. Saushkin. The mathematical method in geography. *Soviet Geography* 7 (4):3–34, 1966.

Hagerstrand, T. The computer and the geographer. *Transactions, Institute of British Geographers* 42:1–19, 1967.

Haggett, P. *Locational Analysis in Human Geography.* London: Arnold, 1965.

Haggett, P. On geographical research in a computer environment. *Geographical Journal* 135:497–507, 1969.

Hagood, M.J., N. Danilevsky, and C.O. Beum. An examination of the use of factor analysis in the problem of sub-regional delineation. *Rural Sociology* 6:216–233, 1941.

Haight, F.A. *Handbook of the Poisson Distribution.* New York: Wiley, 1967.

Harris, C.D. The market as a factor in the localization of industry in the United States. *Annals, Association of American Geographers* 44:315–348, 1954.

Hamill, L. A note on tree diagrams, set theory and symbolic logic. *Professional Geographer* 18:224–226, 1966.

Hammond, R., and P.S. McCullagh. *Quantitative Techniques in Geography.* Oxford: Clarendon Press, 1974.

Harman, H.H. *Modern Factor Analysis.* Chicago: University of Chicago Press, 1967.

Hart, J.F., and N.E. Salisbury. Population change in Middle Western villages: a statistical approach. *Annals, Association of American Geographers* 55:140–160, 1965.

Hartman, G.W., and J.C. Hook. Substandard housing in United States: A quantitative analysis. *Economic Geography* 32:95–114, 1956.

Harvey, D.W. Geographic processes and the analysis of point patterns. Testing models of diffusion by quadrat sampling. *Transactions, Institute of British Geographers* 40:81–95, 1966.

Harvey, D.W. Some methodological problems in the use of Neyman Type A and Negative Binomial probability distributions for the analysis of spatial point patterns. *Transactions, Institute of British Geographers* 44:85–95, 1967.

Harvey, D.W. Pattern, process and the scale problem in geographical research. *Transactions, Institute of British Geographers* 45:71–78, 1968.

Harvey, D.W. *Explanation in Geography*. London: Arnold, 1969.

Haworth, J., and P. Vincent. Calculation of predicted limits in linear regression. *Area* 6(2):113–116, 1974.

Hay, A. *Transport for the Space Economy: A Geographical Study*. London: Macmillan, 1973.

Hempel, C.G. *Fundamentals of Concept Formation in Empirical Science*. Chicago: University of Chicago Press, 1952.

Henshall, J.D., and L.J. King. Some structural characteristics of peasant agriculture in Barbados. *Economic Geography* 42:74–84, 1966.

Hepple, L.W. The impact of stocastic process theory upon spatial analysis in human geography. In C. Board, *et al.* (eds.), *Progress in Geography*, vol. 6. London: Edward Arnold, 1974.

Hidore, J.J. The relationship between cash grain farming and land-forms. *Economic Geography* 39:84–89, 1963.

Hirst, M.A. The changing pattern of district administrative centres in Uganda since 1900. *Geographical Analysis* 3:90–98, 1971a.

Hirst, M.A. Tanzania's population distribution, 1948–67: Some macroscopic comments. *Area* 3:2–7, 1971b.

Hirst, M.A. Administrative reorganization in Uganda: Toward a more efficient solution. *Area* 5:177–181, 1973.

Hodge, G. The use and mis-use of measurement scales in city planning. *Journal of the American Institute of Planners* 29:112–121, 1963.

Hogben, L. *Statistical Theory*. London: Norton, 1957.

Holmes, J. Problems in location sampling. *Annals, Association of American Geographers* 57:757–780, 1967.

Holmes, J. The theory of plane sampling and its application in geographic research. *Economic Geography* 46:379–392, 1970.

Horst, P. *Matrix Algebra for Social Scientists*. New York: Holt, Rinehart and Winston, 1963.

Houston, C. Market potential and potential transport cost: An evaluation of the concepts and their surface patterns in the U.S.S.R. *Canadian Geographer* 13:216–236, 1969.

Hudson, J.C. A location theory of rural settlement. *Annals, Association of American Geographers* 59:365–381, 1969a.

Hudson, J.C. Pattern recognition in empirical map analysis. *Journal of Regional Science* 9:189–199, 1969b.

Huff, D.L. A probabilistic analysis of shopping centre trade areas. *Land Economics* 39:81–90, 1963.

Hughes, J.G., and G.W. Carey. Factorial ecology: Oblique and orthogonal solutions. *Environment and Planning* 4:147–162, 1972.

Hyman, G.M. The calibration of trip distribution models. *Environment and Planning* 1:105–112, 1971.

Illeris, S. Hierarchies of functional regions: Theoretical models and empirical evidence for Denmark. Reprinted in B.J.L. Berry and F.E. Horton, eds. *Geographic Perspectives on Urban Systems*, pp. 200–207. Englewood Cliffs, N.J.: Prentice-Hall, 1967.

Isard, W. *Methods of Regional Analysis*. New York: Wiley, 1960.

Jackman, M.E. Flying doctor services in Zambia. In N.D. McGlashan, ed. *Medical Geography*. London: Methuen, 1972.

Jacobs, H.R. *Mathematics: A Human Endeavour*. San Francisco: Freeman, 1970.

Janson, C-G. Some problems in ecological factor analysis. In M. Dogan and S. Rokkan, eds. *Quantitative Ecological Analysis in the Social Sciences*, pp. 301–341. Cambridge: M.I.T. Press, 1969.

Jay, L.S. Data collection systems for metropolitan planning. *Papers and Proceedings, Regional Science Association* 16:77–92, 1966.

Johnson, I. Agricultural diversification in Georgia, 1939–1959. *Southeastern Geographer* 7:34–49, 1967.

Johnston, R.J. Multi-variate regions: A further approach. *Professional Geographer* 17 (5):9–12, 1965.

Johnston, R.J. Components analysis in geographical research. *Area* 2:68–71, 1970a.

Johnston, R.J. On spatial patterns in the residential structure of cities. *Canadian Geographer* 14:361–367, 1970b.

Joshi, T.R. Toward computing factor scores. In W.P. Adams and F.M. Helleiner, eds. *International Geography*. Montreal: I.G.U., 1972.

Kansky, K.L. *Structure of Transport Networks: Relationships between Network Geometry and Regional Characteristics*. University of Chicago Department of Geography, Research Papers, no. 84, 1963.

Kemeny, J.G. Mathematics without numbers. In D. Lerner, ed. *Quantity and Quality*. New York: Free Press, 1961.

Kendall, M.G. The geographical distribution of crop productivity in England. *Journal of the Royal Statistical Society* 102:21–48, 1939. Reprinted in B.J.L. Berry and D.F. Marble, eds. *Spatial Analysis: A Reader in*

Statistical Geography. Englewood Cliffs, N.J.: Prentice-Hall, 1968.

Kershaw, K.A. *Quantitative and Dynamic Ecology*. London: Arnold, 1964.

King, C.A.M. Map interpretation of the Nottingham area by means of factor analysis. *East Midland Geographer* 4:400–413, 1969.

King, L.J. A quantitative expression of the pattern of urban settlements in selected areas in the U.S. *Tidjschrift voor Economische en Sociale Geografie* 53:1–7, 1962. Reprinted in B.J.L. Berry and D.F. Marble, eds. *Spatial Analysis: A Reader in Statistical Geography*. Englewood Cliffs, N.J.: Prentice-Hall, 1968.

King, L.J. *Statistical Analysis in Geography*. Englewood Cliffs, N.J.: Prentice-Hall, 1969a.

King, L.J. The analysis of spatial form and its relation to geographic research. *Annals, Association of American Geographers* 59:573–595, 1969b.

Kitagawa, T. *Tables of the Poisson Distribution*. Tokyo: Baifukan, 1952.

Knos, D.S. *Distribution of land values in Topeka, Kansas*. Lawrence: University of Kansas, 1962. Reprinted, in part, in B.J.L. Berry and D.F. Marble, eds. *Spatial Analysis: A Reader in Statistical Geography*. Englewood Cliffs, N.J.: Prentice-Hall, 1968.

Korner, S. *The Philosophy of Mathematics*. London: Hutchinson, 1960.

Krumbein, W.C. Measurement and error in regional stratigraphic analysis. *Journal of Sedimentary Petrology* 28:175–185, 1958.

Krumbein, W.C., and F.A. Graybill. *An Introduction to Statistical Models in Geology*. New York: McGraw-Hill, 1965.

Kuhn, H.W., and R.E. Kuenne. An efficient algorithm for the numerical solution of the generalized Weber problem in spatial economics. *Journal of Regional Science* 4 (2):21–33, 1962.

Lakshmanan, T.R., and W.G. Hansen. A retail market potential model. *Journal of the American Institute of Planners* 31:134–143, 1965.

Langley, R. *Practical Statistics*. London: Pan, 1968.

Lavalle, P., H. McConnell, and R.G. Brown. Certain aspects of the expansion of quantitative methodology in American geography. *Annals, Association of American Geographers* 57:423–436, 1967.

Lawley, D.N., and A.E. Maxwell. *Factor Analysis as a Statistical Method*. London: Butterworths, 1963.

Lee, C. Data banks for planning. *Planning Outlook* 10:24–35, 1971.

Lee, D.R., and G.T. Sallee. A method of measuring shape. *Geographical Review* 60:555–563, 1970.

Lewis, J. Parry, and A.L. Trail. The assessment of shopping potential. *Town Planning Review* 38:317–326, 1968.

Lewis, P.F. Impact of negro migration on the electoral geography of Flint, Michigan, 1932–62: A cartographic analysis. *Annals, Association of American Geographers* 55:1–25, 1965.

Lieberson, S. *Ethnic Patterns in American Cities*. New York: Glencoe, 1963.

Lipschutz, S. *Theory and Problems of Probability*. New York: McGraw-Hill (Schaum's Outline Series), 1968.

Lukerman, F. The "calcul des probabilities" and the Ecole Française. *Canadian Geographer* 9:128–137, 1965.

Mackay, J.R. Chi-square as a tool for regional studies. *Annals, Association of American Geographers* 48:164–166, 1958.

Mackay, J.R. Comments on the use of chi-square. *Annals, Association of American Geographers* 49:89, 1959.

Mardia, K.V. *Statistics of Directional Data*. London: Academic, 1972.

Martin, R.L. On spatial dependence, bias, and the use of first spatial differences in regression analysis. *Area* 6:185–194, 1974.

Massam, B.H. The spatial structure of administrative systems. *Resource Paper* No. 12, A.A.G., Commission on College Geography, 1972.

Masser, I. *Analytic Models for Urban and Regional Planning*. New York: Halsted Press, 1972.

Mather, P.M. Varimax and generality. *Area* 3:252–254, 1971.

Mather, P.M. Varimax and generality. *Area* 4:27–30, 1972.

Maxfield, D.W. Spatial planning of school districts. *Annals, Association of American Geographers* 62:582–590, 1972.

McCarty, H.H. An approach to a theory of economic geography. *Economic Geography* 30:95–101, 1954.

McCarty, H.H. Toward a more general economic geography. *Economic Geography* 35:283–289, 1959.

Medvedkov, Y.V. The concept of entropy in settlement pattern analysis. *Papers and Proceedings, Regional Science Association* 18:165–168, 1967a.

Medvedkov, Y.V. The regular component in settlement patterns as shown on a map. *Soviet Geography* 8:150–168, 1967b.

Medvedkov, Y.V. Applications of mathematics to population geography. In G.J. Demko, H.M. Rose, and

G.A. Schnell, eds. *Population Geography: A Reader,* 1967c.

Medvedkov, Y.V. Entropy: An assessment of potentialities in geography. *Economic Geography* 46:308–316, 1970.

Meyer, D.R. Geographical population data: Statistical descriptions not statistical inference. *Professional Geography* 24:26–27, 1972.

Mills, G. The determination of local government electoral boundaries. *Operations Research Quarterly* 18:243–257, 1967.

Minnick, R.F. A method for the measurement of areal correspondence. *Papers of the Michigan Academy of Science, Arts and Letters* 49:333–344, 1964.

Mitchell, B. A comparison of chi-square and Kolmogorov-Smirnov tests. *Area* 3:237–241, 1971.

Moran, P.A. The interpretation of statistical maps. *Journal of the Royal Statistical Society* B10: 243–251, 1948.

Morrill, R.L., and W.L. Garrison. Projections of interregional patterns of trade in wheat and flour. *Economic Geography* 36:116–126, 1960.

Morgenstern, O. *On the Accuracy of Economic Observations.* Princeton: Princeton University Press, 1965.

Morrison, D.E., and R.E. Henkel. *The Significance Test Controversy: A Reader.* London: Butterworths, 1970.

Murdie, R.A. *Factorial Ecology of Metropolitan Toronto, 1951–1961.* Research Paper No. 116. Chicago: University of Chicago, Department of Geography, 1969.

National Research Council. *The Mathematical Sciences: A Report.* Publication 1681. Washington, D.C.: National Academy of Sciences, 1968.

Neft, D.S. Macrogeography and the realms of influence in Asia. *Journal of Conflict Resolution* 5:254–273, 1961.

Neft, D. *Statistical Analysis for Areal Distributions.* Regional Science Research Institute, Monograph Series No. 2, Philadelphia, 1966.

Norcliffe, G.B. On the use and limitations of trend surface models. *Canadian Geographer* 13:338–348, 1969.

Nystuen, J.D. Identification of some fundamental spatial concepts. *Papers and Proceedings of the Michigan Academy of Science, Arts and Letters* 48:373–384, 1963. Reprinted in B.J.L. Berry and D.F. Marble, eds. *Spatial Analysis: A Reader in Statistical Geography.* Englewood Cliffs, N.J.: Prentice-Hall, 1968.

Nystuen, J.D. Boundary shapes and boundary problems. *Peace Research Society Papers* 8:107–128, 1967.

The Open University. *Linear Programming.* Bletchley, Bucks: The Open University Press, 1972.

Palm, R., and D. Caruso. Labelling in factorial ecology. *Annals, Association of American Geographers* 62:122–133, 1972.

Papageorgiou, G.T. Description of a basis necessary to the analysis of spatial systems. *Geographical Analysis* 1:213–215, 1969.

Pielou, E.C. *An Introduction to Mathematical Ecology.* New York: Wiley, 1969.

Pincus, H.J. Some vector and arithmetic operations on two dimensional orientation variates with application to geological data. *Journal of Geology* 64:533–557, 1956.

Pitts, F.R. A graph theoretic approach to historical geography. *Professional Geographer* 17 (5):15–20, 1965.

Poole, M.A., and P.N. O'Farrell. The assumptions of the linear regression model. *Transactions, Institute of British Geographers* 52:145–158, 1971.

Poulsen, T.M. Centrography in Russian geography. *Annals, Association of American Geographers* 49:326–327, 1959.

Ray, D.M. The spatial structure of economic and cultural differences: A factorial ecology of Canada. *Papers and Proceedings, Regional Science Association* 23:7–12, 1969.

Rees, P.H. Factorial ecology: An extended definition, survey and critique of the field. *Economic Geography* 47:220–233, 1971.

Reilly, W. *The Law of Retail Gravitation.* New York: Pilsbury, 1931.

Roberts, M.C., and K.W. Rumage. The spatial variations in urban left-wing voting in England and Wales in 1951. *Annals, Association of American Geographers* 55:161–178, 1965.

Robertson, I.M.L. Scottish population distribution: implications for locational decisions. *Transactions, Institute of British Geographers* 63:111–124, 1974.

Robertson, I.M.L. The census and research: Ideals and realities. *Transactions, Institute of British Geographers* 48:173–187, 1969.

Robinson, A.H. The necessity of weighting values in correlation of areal data. *Annals, Association of American Geographers* 47:379–391, 1956.

Robinson, W.S. Ecological correlations and the behavior of individuals. *American Sociological Review* 15: 351–357, 1950.

Robinson, A.H. Mapping the correspondence of isarithmic maps. *Annals, Association of American Geographers* 52:414–425, 1962. Reprinted in B.J.L. Berry and D.F. Marble, eds. *Spatial Analysis: A Reader in Sta-*

tistical Geography. Englewood Cliffs, N.J.: Prentice-Hall, 1968.

Robinson, A.H., and R.A. Bryson. A method for describing quantitatively the correspondence of geographical distributions. *Annals, Association of American Geographers* 47:379–391, 1957.

Robinson, A.H., J.B. Lindberg, and L.W. Brinkman. A correlation and regression analysis applied to rural farm densities in the Great Plains. *Annals, Association of American Geographers* 51:211–221, 1961. Reprinted in B.J.L. Berry and D.F. Marble, eds. *Spatial Analysis: A Reader in Statistical Geography*. Englewood Cliffs, N.J.: Prentice-Hall, 1968.

Roscoe, J.T. *Fundamental Research Statistics for the Behavioural Sciences*. New York: Holt, Rinehart and Winston, 1969.

Rose, J.K. Corn yields and climate in the Corn Belt. *Geographical Review* 26:88–102, 1936.

Rosing, K.E., and P.A. Wood. *Character of a Conurbation: A Computer Atlas of Birmingham and the Black Country*. London: University of London Press, 1971.

Rummel, R.J. Understanding factor analysis. *Journal of Conflict Resolution* 11:444–480, 1969.

Rummel, R.J. *Applied Factor Analysis*. Evanston, Ill.: Northwestern University Press, 1970.

Rushton, G. Map transformations of point patterns: Central place patterns in areas of variable population density. *Papers and Proceedings, Regional Science Association* 28:111–129, 1972.

Rushton, G., R.G. Golledge, and W.A.V. Clark. Formulation and test of normative model for the spatial allocation of grocery expenditures by a dispersed population. *Annals, Association of American Geographers* 57:389, 400, 1967.

Rushton, G., M.F. Goodchild, and L.M. Ostresh, Jr., eds. *Computer Programs for Location-Allocation Problems*. Monograph No. 6, Department of Geography, University of Iowa, 1973.

Russett, B.M. *International Regions and the International System*. Chicago: Rand McNally, 1967.

Saushkin, Y.G. Results and prospects of the use of mathematical methods in economic geography. *Soviet Geography* 12 (7):416–427, 1971.

Sawyer, W.W. *A Path to Modern Mathematics*. Harmondsworth: Penguin, 1966.

Schaaf, W.L. *Basic Concepts of Elementary Mathematics*, 3d ed. New York: Wiley, 1969.

Schmid, C.F., and K. Tagashira. Ecological and demographic indices: A methodological analysis. *Demography* 1:194–211, 1965.

Schneider, M. Gravity models and trip distribution theory. *Papers and Proceedings, Regional Science Association* 5:51–56, 1959.

Scott, A.J. Location-allocation systems: A review. *Geographical Analysis* 2:95–119.

Scott, A.J. *An introduction to Spatial Allocation Analysis*. Resource Paper No. 9. A.A.G., Commission of College Geography, 1971a.

Scott, A.J. *Combinatorial Programming, Spatial Analysis, and Planning*. London: Metheun, 1971b.

Scott, P. Areal variations in the class structure of the central place hierarchy. *Australian Geographical Studies* 2:73–86, 1964.

Scott, P. *Geography and Retailing*. London: Hutchinson, 1969.

Selltiz, C., M. Jahoda, M. Deutch, and S.W. Cook. *Research Methods in Social Relations*. New York: Holt, Rinehart and Winston, 1959.

Seymour, D.R. The polygon of forces and the Weber problem. *Journal of Regional Science* 8 (2):243–246, 1968.

Shachar, A. Some applications of geo-statistical methods in urban research. *Papers and Proceedings, Regional Science Association* 18:197–202, 1967.

Shimwell, D.W. *Description and Classification of Vegetation*. London: Sidgwick and Jackson, 1971.

Siegel, S. *Nonparametric Statistics for the Behavioural Sciences*. New York: McGraw-Hill, 1956.

Sinnhuber, K.A. The representation of disputed political boundaries in general atlases. *Cartographic Journal* 1:20–28, 1964.

Slonim, M.J. *Guide to Sampling*. London: Pan, 1960. Initially published as *Sampling in a Nutshell*.

Smirnov, L.Y. The role and significance of objective (mathematical) methods in geographic research. *Soviet Geography* 9 (1):55–67, 1968.

Smith, C.T. Problems of regional development in Peru. *Geography* 53:260–281, 1968.

Smith, D.M. Identifying the "Grey" Areas — a multivariate approach. *Regional Studies* 2: 183–193, 1968.

Smith, R.D.P. The changing urban hierarchy. *Regional Studies* 2:1–19, 1968.

Snedecor, G.W. and W.G. Cochran. *Statistical Methods*. Ames: Iowa State University Press, 1967.

Snedecor, G.W., and W.G. Cochran. *Statistical Methods*. Ames: Iowa State University Press, 1966.

Spate, O.H.K. Quality and quantity in geography.

Annals, Association of American Geographers 50:377–394, 1960.

Spence, N.A. A multifactor uniform regionalization of British counties on the basis of employment data for 1961. *Regional Studies* 2:87–104, 1968.

Spence, N.A., and P.J. Taylor. Quantitative methods in regional taxonomy. In C. Board *et al.*, eds. *Progress in Geography*, vol. 2, pp. 1–63. London: Arnold, 1970.

Steger, J.A. *Readings in Statistics for the Behavioural Scientist*. New York: Holt, Rinehart and Winston, 1971.

Stevens, S.S. On the theory of scales of measurement. *Science* 103:677–680, 1946.

Stewart, J.Q., and W. Warntz. Macrogeography and social science. *Geographical Review* 48:167–184, 1958.

Strohkarck, F., and K. Phelps. The mechanics of constructing a market area map. *Journal of Marketing* 13:493–496, 1948.

Sviatlovsky, E.E., and W.C. Eells. The centrographical method and regional analysis. *Geographical Review* 27:240–254, 1937.

Sweetser, F.L. Factor structure as ecological structure in Helsinki and Boston. *Acta Sociologica* 8:202–225.

Taaffe, E.J., and H.L. Gauthier. *Geography of Transportation*. Englewood Cliffs: Prentice-Hall, 1973.

Taaffe, E.J., R.L. Morrill, and P.R. Gould. Transport expansion in underdeveloped countries: A comparative analysis. *Geographical Review* 53:503–529, 1963.

Taeuber, K.E., and A.F. Taeuber. *Negroes in Cities*. Chicago: Aldine, 1965.

Taylor, P.J. Causal models in geographic research. *Annals, Association of American Geographers* 59:402–404, 1969.

Taylor, P.J. Distances within shapes: An introduction to a family of finite frequency distributions. *Geografiska Annaler* B53:40–54, 1971.

Tegsjo, B., and S. Oberg. Concept of potential applied to price formation. *Geografiska Annaler* 47B:51–58, 1965.

Thomas, E.N. Areal associations between population growth and selected factors in the Chicago urbanized area. *Economic Geography* 36:158–170, 1960.

Thomas, E.N. Maps of residuals from regression. In B.J.L. Berry and D.F. Marble, eds. *Spatial Analysis: A Reader in Statistical Geography*. Englewood Cliffs, N.J.: Prentice-Hall, 1968.

Thomas, E.N. and D.L. Anderson. Additional comments on weighting values in correlation analysis of areal data. *Annals, Association of American Geographers* 55:492–505, 1965.

Thompson, J.H., S.C. Sufrin, P.R. Gould, and M.A. Buck. Toward a geography of economic health: The case of New York State. *Annals, Association of American Geographers* 52:1–20, 1962.

Till, R. The use of linear regression in geomorphology. *Area* 5:303–307, 1973.

Timms, D.W.G. *The Urban Mosaic: Towards a Theory of Residential Differentiation*. Cambridge, England: University Press, 1969.

Tinkler, K.J. The physical interpretation of eigen functions of dichotomous matrices. *Transactions, Institute of British Geographers* 55:17–46, 1972.

Tobler, W. Geographic area and map projections. *Geographical Review* 53:59–78, 1968. Reprinted in B.J.L. Berry and D.F. Marble, eds. *Spatial Analysis: A Reader in Statistical Geography*. Englewood Cliffs, N.J.: Prentice-Hall, 1968.

Tornquist, G., S. Nordbeck, B. Rystedt, and P. Gould. Multiple location analysis. *Lund Studies in Geography*, C12, 1971.

Udo, R.K. Population and politics in Nigeria. In J.C. Caldwell and C. Okonjo, eds. *The Population of Tropical Africa*. London: Longmans, 1968.

Unwin, D.J., and L.W. Hepple. The statistical analysis of spatial series. *The Statistician* 23:211–228, 1974.

Warntz, W. A methodological consideration of some geographic aspects of the Newfoundland Referendum on Confederation with Canada. *Canadian Geographer* 1 (6):39–49, 1955.

Warntz, W. Measuring spatial association with special consideration of the case of market orientation of production. *Journal of the American Statistical Association* 51:597–604, 1956.

Warntz, W. The geography of prices and spatial interaction. *Papers, Regional Science Association* 3:118–129, 1957.

Warntz, W. *Toward a Geography of Price*. Philadelphia: University of Pennsylvania Press, 1959.

Warntz, W. Macroscopic analysis and some patterns of the geographical distribution of population in the United States, 1790–1950. In W.L. Garrison and D.F. Marble, eds. *Quantitative Geography, Part I*. Northwestern University, Studies in Geography, no. 13, Chicago, 1967.

Warntz, W., and D. Neft. Contributions to a statistical methodology for areal distributions. *Journal of Regional Science* 2:47–66, 1960.

Weaver, J.B., and S.W. Hess. A procedure for nonpartisan districting. *Yale Law Journal* 73:288–308, 1963.

Wilkinson, H.R. *Maps and Politics: A Review of the Ethnographic Cartography of Macedonia*. Liverpool: University of Liverpool Press, 1951.

Williams, K. Do you seriously want to be a factor analyst? *Area* 3:228–229, 1971.

Williamson, E., and M.H. Bretherton. *Tables of the Negative Binomial Distribution*. New York: Wiley, 1964.

Williamson, J.G. Regional inequality and the process of national development: A description of the patterns. *Economic Development and Cultural Change* 13:3–45, 1965. Reprinted in L. Needham, ed. *Regional Analysis*. London: Penguin.

Willis, K.G. Regression models of migration. *Geografiska Annaler* 57B:42–54, 1975.

Wilson, A.G. A statistical theory of spatial distribution models. *Transportation Research* 1:253–269, 1967.

Wilson, A.G. *Entropy in Urban and Regional Modelling*. London: Pion, 1970.

Wilson, A.G., A.F. Hawkins, G.J. Hill, and D.J. Wagon. Calibration and testing of the SELNEC Transport Model. *Regional Studies* 3:337–350, 1969.

Wilson, E.B. *An Introduction to Scientific Research*. New York: McGraw-Hill, 1951.

Wood, W.F. Use of stratified random samples in a land use survey. *Annals, Association of American Geographers* 45:350–367, 1955.

Wright, J.K. Some measures of distributions. *Annals, Association of American Geographers* 27:177–211, 1937.

Wrigley, N. The use of percentages in geographical research. *Area* 5:183–186, 1973.

Wrigley, N. Analyzing multiple alternative dependent variables. *Geographical Analysis* 7:187–196, 1975.

Yates, F. *Sampling for Censuses and Surveys*. London: Griffin, 1953.

Yeates, M. Hinterland delimitation: A distance minimizing approach. *Professional Geographer* 15 (6):7–10, 1963.

Yuill, R.S. The standard deviational ellipse: An updated tool for spatial description. *Geografiska Annaler* 53B: 28–29, 1971.

Zetterberg, H.L. *On Theory and Verification in Sociology*, revised ed. New York: Bedminster Press, 1963.

Zobler, L. Statistical testing of regional boundaries. *Annals, Association of American Geographers* 47:83–95, 1957.

Zobler, L. Decision making in regional construction. *Annals, Association of American Geographers* 48:140–148, 1958.

Zobler, L. The distinction between absolute and relative frequencies in using chi-square for regional analysis. *Annals, Association of American Geographers* 49:456–457, 1959.

LENGTH AND AREA CONVERSIONS

Metric

1 kilometer	= 0.6213 mile = 3,281 feet = 1,000 meters
1 centimeter	= 0.3937 inch
1 meter	= 39.37 inches = 3.2808 feet = 1.9036 yards
1 hectare	= 0.01 square kilometer = 10,000 square meters = 2.4711 acres
1 square kilometer	= 247.11 acres = 0.386 square miles

English

1 inch	= 2.54 centimeters
1 foot	= 0.3048 meter = 30.48 centimeters
1 yard	= 0.9144 meter = 91.44 centimeters
1 yard	= 3 feet = 36 inches
1 mile	= 1.609 kilometers = 1,609 meters
1 mile	= 1,760 yards = 5,280 feet
1 fathom	= 6.08 feet
1 square mile	= 2.59 square kilometers = 259 hectares = 640 acres
1 acre	= 43,560 square feet

INDEX

CDEFGHIJ-VB-8210